Thomas G. Mezger

The
Rheology Handbook

For users of
rotational and oscillatory rheometers

3rd revised edition

Thomas G. Mezger: The Rheology Handbook
© Copyright 2011 by Vincentz Network, Hanover, Germany
ISBN 978-3-86630-864-0

Cover: Wacker Chemie, Burghausen, Germany

Mezger, Thomas G.
The Rheology Handbook, 3rd revised edition
Hanover: Vincentz Network, 2011
European Coatings Tech Files
ISBN 3-86630-864-7
ISBN 978-3-86630-864-0

Please ask for our book catalogue
Vincentz Network, Plathnerstr. 4c, 30175 Hanover, Germany
T (202) 684-6630, F (202) 380-9129
E-mail: books@american-coatings.com, www.american-coatings.com

Layout: Vincentz Network, Hanover, Germany

ISBN 3-86630-864-7
ISBN 978-3-86630-864-0

Foreword

Why was this book written?

People working in industry are often confronted with the effects of rheology, the science of deformation and flow behavior. When looking for appropriate literature, they find either short brochures which give only a few details and contain little useful information, or highly specialized books overcharged of physical formulas and mathematical theories. There is a lack of literature between these two extremes which reduces the discussion of theoretical principles to the necessary topics, providing useful instructions for experiments on material characterization. This book is intended to fill that gap.

The practical use of rheology is presented in the following areas: quality control (QC), production and application, chemical and mechanical engineering, industrial research and development, and materials science. Emphasis is placed on current test methods related to daily working practice. After reading this book, the reader should be able to perform useful tests with rotational and oscillatory rheometers, and to interpret the achieved results correctly.

How did this book come into existence?

The first computer-controlled rheometers came into use in industrial laboratories in the mid 1980s. Ever since then, test methods as well as control and analysis options have improved with breathtaking speed. In order to organize and clarify the growing mountain of information, company Anton Paar Germany – and previously Physica Messtechnik – has offered basic seminars on rheology already since 1988, focused on branch-specific industrial application. During the "European Coatings Show" in Nuremberg in April 1999, the organizer and publishing director Dr. Lothar Vincentz suggested expanding these seminar notes into a comprehensive book about applied rheology.

What is the target audience for this book? For which industrial branches will it be most interesting?

"The Rheology Handbook" is written for everyone approaching rheology without any prior knowledge, but is also useful to people wishing to update their expertise with information about recent developments. The reader can use the book as a course book and read from beginning to end or as a reference book for selected chapters. The numerous cross-references make connections clear and the detailed index helps when searching. If required, the book can be used as the first step on the ladder towards theory-orientated rheology books at university level. In order to break up the text, there are as well many figures and tables, illustrative examples and small practical experiments, as well as several exercises for calculations. The following list reflects how the contents of the book are of interest to rheology users in many industrial branches.

- **Polymers:** Solutions, melts, solids; film emulsions, cellulose solutions, latex emulsions, solid films, sheetings, laminates; natural resins, epoxies, casting resins; silicones, caoutchouc, gums, soft and hard rubbers; thermoplastics, elastomers, thermosets, blends, foamed materials; unlinked and cross-linked polymers containing or without fillers or fibers; polymeric compounds and composites; solid bars of glass-fibre, carbon-fibre and synthetic-fibre reinforced polymers (GFRP, CFRP, SFRP); polymerization, cross-linking, curing, vulcanization, melting and hardening processes

Thomas G. Mezger: The Rheology Handbook
© Copyright 2011 by Vincentz Network, Hanover, Germany
ISBN 978-3-86630-864-0

- **Adhesives and sealants:** Glues, single and multi-component adhesives, pressure sensitive adhesives (PSA), UV curing adhesives, hotmelts, plastisol pastes (e.g. for automotive underseals and seam sealings), construction adhesives, putties; uncured and cured adhesives; curing process; tack, stringiness
- **Coatings, paints, lacquers:** Spray, brush, dip coatings; solvent-borne, water-based coatings; metallic effect, textured, low solids, high solids, photo-resists, UV (ultra violet) radiation curing, powder coatings; glazes and stains for wood; coil coatings; solid coating films
- **Printing inks and varnishes:** Gravure, letterpress, flexographic, planographic, offset, screen printing inks, UV (ultra violet) radiation curing inks; ink-jet printer inks; writing inks for pens; millbase premix, color pastes, "thixo-pastes"; liquid and pasty pigment dispersions; printing process; misting; tack
- **Paper coatings:** Primers and topcoats; immobilization process
- **Foodstuffs:** Water, vegetable oils, aroma solvents, fruit juices, baby food, liquid nutrition, liqueurs, syrups, purees, thickeners an stabilizing agents, gels, pudding, jellies, ketchup, mayonnaise, mustard, dairy products (such as yogurt, cream cheese, cheese spread, soft and hard cheese, curds, butter), emulsions, chocolate (melt), soft sweets, ice cream, chewing gum, dough, whisked egg, cappuccino foam, sausage meat, sauces containing meat chunks, jam containing fruit pieces, animal feed; bio-technological fluids; gel formation of hydrocolloids (e.g. of corn starch and gelatin); interfacial rheology (e.g. for emulsions); food tribology (e.g. for creaminess); tack
- **Cosmetics, pharmaceuticals, medicaments, bio-tech products, personal care, health and beauty care products:** Cough mixtures, perfume oils, wetting agents, nose sprays, X-ray film developing baths, blood (hemo-rheology), blood-plasma substitutes, emulsions (e.g. skin care, hair-dye), lotions, nail polish, roll-on fluids (deodorants), saliva, mucus, hydrogels, shampoo, shower gels, dispersions containing viscoelastic surfactants, skin creams, peeling creams, hair gels, styling waxes, shaving creams, tooth-gels, toothpastes, makeup dispersions, lipsticks, mascara, ointments, vaseline, biological cells, tissue engineered medical products (TEMPs), natural and synthetic membranes, silicone pads and cushions, dental molding materials, tooth filling, sponges, contact lenses, medical adhesives (for skin plasters or diapers), denture fixative creams, hair, bone cement, implants, organic-inorganic compounds (hybrids); interfacial rheology (e.g. for emulsions)
- **Agrochemicals:** Plant or crop protection agents, solutions and dispersions of insecticides and pesticides, herbicides and fungicides
- **Detergents, home care products:** Household cleaning agents, liquid soap, disinfectants, surfactant solutions, washing-up liquids, dish washing agents, laundry, fabric conditioners, washing powder concentrate, fat removers; interfacial rheology
- **Surface technology:** Polishing and abrasive suspensions; cooling emulsions
- **Electrical engineering, electronics industry:** Thick film pastes, conductive, resistance, insulating, glass paste, soft solder and screen printing pastes; SMD adhesives (for surface mounted devices), insulating and protective coatings, de-greasing agents, battery fluids and pastes, coatings for electrodes
- **Petrochemicals:** Crude oils, petroleum, solvents, fuels, mineral oils, light and heavy oils, lubricating greases, paraffines, waxes, petrolatum, vaseline, natural and polymer-modified bitumen, asphalt binders, distillation residues, and from coal and wood: tar and pitch; interfacial rheology (e.g. for emulsions)
- **Ceramics and glass:** Casting slips, kaolin and porcelain suspensions, glass powder and enamel pastes, glazes, plastically deformable ceramic pastes, glass melts, aero-gels, xero-gels, sol/gel materials, composites, organo-silanes (hybrids), basalt melts
- **Construction materials:** Self-leveling cast floors, plasters, mortar, cement suspensions, tile adhesives, dispersion paints, sealants, floor sheeting, natural and polymer-modified bitumen, asphalt binders for pavements
- **Metals:** Melts of magnesium, aluminum, steel, alloys; moulding process in a semi-solid state ("thixo-forming", "thixo-casting", "thixo-forging"), ceramic fibre reinforced light-weight metals

- **Waste industry:** Waste water, sewage sludges, animal excrements (e.g. of fishes, poultry, cats, dogs, pigs), residues from refuse incineration plants
- **Geology, soil mechanics, mining industry:** Soil sludges, muds; river and lake sediment masses; soil deformation due to mining operations, earthwork, canal and drain constructions; drilling fluids (e.g. containing "flow improvers")
- **Disaster control:** Behavior of burning materials, soil deformation due to floods and earthquakes
- **Materials for special functions** (e.g. as "smart fluids"): Magneto-rheological fluids (MRF), electro-rheological fluids (ERF), di-electric (DE) materials, self-repairing coatings, materials showing self-organizing superstructures (e.g. surfactants), dilatant fabrics (shock-absorbing, "shot-proof"), liquid crystals (LC), ionic fluids, micro-capsule paraffin wax (e.g. as "phase-change material" PCM)

It is pleasing that the first two editions of "The Rheology Handbook", published in 2002 and 2006, sold out so unexpectedly quickly. It was positive to hear that the book met with approval, not only from laboratory technicians and practically oriented engineers, but also from teachers and professors of schools and colleges of applied sciences. Even at universities, "The Rheology Handbook" is meanwhile taken as an introductory teaching material for explaining the basics of rheology in lectures and practical courses, and as a consequence, many students worldwide are using it when writing their final paper or thesis.

Also for the third edition, further present-day examples have been added resulting as well from contacts to industrial users as well as from corporation with several working groups, e.g. for developing modern standardizing measuring methods for diverse industrial branches. Here in a nutshell, the following additional chapters of this new edition are: Types of flow in the Two-Plates-Model (Chapter 2.4), the effects of rheological additives in aqueous dispersions (Chapter 3.3.7), SAOS and LAOS tests and Lissajous diagrams (Chapter 8.3.6), nano-structures and complex rheological behavior such as shear-banding explained by using surfactant systems as an example (Chapter 9), and special measuring devices for rheo-optical systems and extensional tests (Chapter 10.8). Also the references and standards have been updated (Chapter 14). This textbook is also available in German language, and also here, three editions were published up to now, in 2000, 2006 and in July of 2010 (title: "Das Rheologie Handbuch").

I hope that "The Rheology Handbook" will prove itself a useful source of information for characterizing the above mentioned products in an application-oriented way, assuring their quality and helping to improve them wherever possible.

Acknowledgements
I would like to say a big "Thank You" to all those people without whose competent ideas and suggestions for improvement it would not have been possible to produce such a comprehensive and understandable book. The following people have been of special help to me: Heike Audehm, Monika Bernzen, Stefan Büchner, Marcel de Pender, Gerd Dornhöfer, Andreas Eich, Elke Fischle, Ingrid Funk, Patrick Heyer, Siegfried Huck, Jörg Läuger, Thomas Litters, Sabine Neuber, Matthias Prenzel, Hubert Reitberger, Michael Ringhofer, Oliver Sack, Michael Schäffler, Carmen Schönhaar, Werner Stehr, Heiko Stettin, Jürgen Utz, Detlef van Peij, Simone Will and Klaus Wollny. Sarah Knights translated the original German text into English. Besides the support of my colleagues worldwide, I would also like to mention the managers of Anton Paar GmbH in Graz, Austria, and Anton Paar Germany GmbH in Ostfildern (near Stuttgart).

Stuttgart, December 2010

Thomas G. Mezger

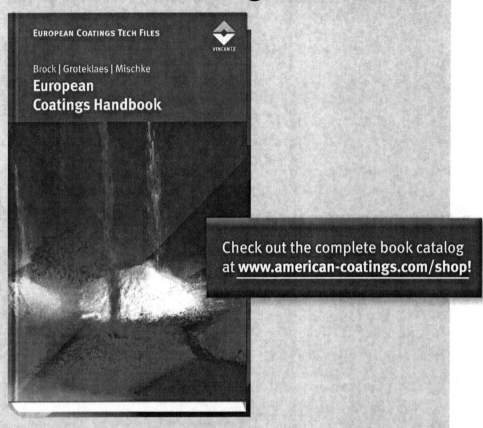

Contents

Thomas G. Mezger: The Rheology Handbook
© Copyright 2011 by Vincentz Network, Hanover, Germany
ISBN 978-3-86630-864-0

1   Introduction

1.1   Rheology, rheometry and viscoelasticity

a) Rheology

Rheology is the science of deformation and flow. It is a branch of physics and physical chemistry since the most important variables come from the field of mechanics: forces, deflections and velocities. The term "rheology" originates from the Greek: "rhein" meaning "to flow". Thus, rheology is literally "flow science". However, rheological experiments do not merely reveal information about **flow behavior of liquids** but also about **deformation behavior of solids**. The connection here is that a large deformation produced by shear forces causes many materials to flow.

All kinds of shear behavior, which can be described rheologically in a scientific way, can be viewed as being in between two extremes: flow of ideally viscous liquids on the one hand and deformation of ideally elastic solids on the other. Two illustrative examples for the extremes of ideal behavior are low-viscosity mineral oils and rigid steel balls. Flow **behavior of ideally viscous fluids** is explained in Chapter 2. **Ideally elastic deformation** behavior is described in Chapter 4.

Behavior of all real materials is based on the combination of both a viscous and an elastic portion and therefore, it is called viscoelastic. Wallpaper paste is a viscoelastic liquid, for example, and a gum eraser is a viscoelastic solid. Information on **viscoelastic behavior** can be found in Chapter 5. **Complex and extraordinary rheological behavior** is presented in Chapter 9 using the example of surfactant systems.

Table 1.1 shows the most important terms, all of which will be covered in this book. This chart can also be found at the beginning of Chapters 2 to 8, with those terms given in bold print being discussed in the chapter in hand.

Table 1.1: Overview on different kinds of rheological behavior

Liquids		Solids	
(ideal-) viscous flow behavior Newton's law	viscoelastic flow behavior Maxwell's law	viscoelastic deformation behavior Kelvin/Voigt's law	(ideal-) elastic deformation behavior Hooke's law
flow/viscosity curves	creep tests, relaxation tests, oscillatory tests		

Rheology was first seen as a science in its own right not before the beginning of the 20th century. However, scientists and practical users have long before been interested in the behavior of liquids and solids, although some of their methods have not always been very scientific. A list of important facts of the historical development in rheology is given in Chapter 13. Of special interest are here the various attempts to classify all kinds of different rheological behavior, such as the classification of Markus Reiner in 1931 and 1960, and of George W. Scott Blair in 1942. The aim of the rheologists' is to measure deformation and flow behavior of a great variety of matters, to present the obtained results clearly and to explain it.

Thomas G. Mezger: The Rheology Handbook
© Copyright 2011 by Vincentz Network, Hanover, Germany
ISBN 978-3-86630-864-0

b) Rheometry

Rheometry is the measuring technology used to determine rheological data. The emphasis here is on **measuring systems, instruments,** and **test** and **analysis methods**. Both, liquids and solids can be investigated using rotational and oscillatory rheometers. **Rotational tests** which are performed to characterize viscous behavior are presented in Chapter 3. In order to evaluate viscoelastic behavior, **creep tests** (Chapter 6), **relaxation tests** (Chapter 7) and **oscillatory tests** (Chapter 8) are performed. Chapter 10 contains information on **measuring systems** and **special measuring devices**, and Chapter 11 gives an overview on diverse **instruments** used.

Analog programmers and on-line recorders for plotting flow curves have been on the market since around 1970. Around 1980, digitally controlled instruments appeared which made it possible to store measuring data and to use a variety of analysis methods, including also complex ones. Developments in measuring technology are constantly pushing back the limits. At the same time, thanks to standardized measuring systems and procedures, measuring results can be compared world-wide today. Meanwhile, several rheometer manufacturers can offer test conditions to customers in many industrial branches which come very close to simulate even complex process conditions in practice.

A short **guideline for rheological measurements** is presented in Chapter 12 in order to facilitate the daily laboratory work for practical users.

Chapter 14 (Appendix) shows all the used **signs, symbols and abbreviations** with their units. The **Greek alphabet** and a **conversion table for units** (SI and cgs system) can also be found there.

References are listed in Chapter 15. The publications and books can be identified by the number in brackets (e.g. as [123]). Here, also more than 400 **standards** are listed (**ISO, ASTM, EN and DIN**).

c) Information for "Mr. and Ms. Cleverly"

Throughout this textbook, the reader will find sections for "Mr. and Ms. Cleverly" which are marked with a symbol showing glasses: ᓂ

These sections are written for those readers who wish to go deeper into the theoretical side and who are not afraid of a little extra mathematics and fundamentals in physics. However, these "Cleverly" explanations are not required to understand the information given in the normal text of later chapters, since this textbook is also written for beginners in the field of rheology. Therefore, for those readers who are above all interested in the practical side of rheology, the "Cleverly" sections can simply be ignored.

1.2 Deformation and flow behavior

We are confronted with rheological phenomena every single day. Some experiments are listed below to demonstrate this point. The examples given will be discussed in detail in the chapters mentioned in brackets.

Experiment 1.1: Behavior of mineral oil, plasticine, and steel

Completely different types of behavior can be seen when the following three subjects hit the floor (see Figure 1.1):

a) The **mineral oil** is flowing and spreading until it shows a very thin layer finally (**ideally viscous flow behavior**: see Chapter 2.3.1)

b) The **plasticine** will be deformed when it hits the floor, and afterwards, it remains deformed permanently (inhomogeneous **plastic behavior** outside the linear viscoelastic deformation range: see Chapter 3.3.4.3d)

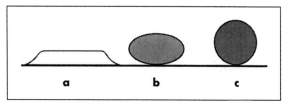

Figure 1.1: Deformation behavior after hitting the floor: a) mineral oil, b) plasticine, c) steel ball

c) The **steel ball** bounces back, and exhibits afterwards no deformation at all (**ideally elastic behavior**: see Chapter 4.3.1)

Experiment 1.2: Playing with "bouncing putty" (some call it "Silly Putty")
The **silicone polymer** (unlinked PDMS) displays different **rheological behaviors depending on the period of time under stress** (viscoelastic behavior of polymers: see Chapter 8.4, frequency sweep):
a) **When stressed briefly and quickly**, the putty behaves like a rigid and elastic **solid**: If you mold a piece of it to the shape of a ball and throw it on the floor, it is bouncing back.
b) **When stressed slowly at a constantly low force** over a longer period of time, the putty shows the behavior of a flexible and yielding, highly viscous **liquid**: If it is in the state of rest, thus, if you leave it untouched for a certain period of time, it is spreading very slowly under its own weight due to gravity to show an even layer with a homogeneous thickness finally.

Experiment 1.3: Do the rods remain in the position standing up straight?
Three wooden rods are put into three glasses containing different materials and left for gravity to do its work.
a) In the glass of **water**, the rod changes its position immediately and falls to the side of the glass (**ideally viscous flow behavior**: see Chapter 2.3.1).
Additional observation: All the air bubbles which were brought into the water when immersing the rod are rising quickly within seconds.
b) In the glass containing a **silicone polymer** (unlinked PDMS), the rod moves very, very slowly, reaching the side of the glass after around 10 minutes (polymers showing **zero-shear viscosity**: see Chapters 3.3.2.1a, 6.3.4.1 and 8.4.2.1a).
Additional observation concerning the air bubbles which were brought into the polymer sample by the rod: Large bubbles are rising within a few minutes, but the smaller ones seem to remain suspended without visible motion. However, after several hours even the smallest bubble has reached the surface. Therefore, indeed long-term but complete de-aeration of the silicone occurs finally.
c) In the glass containing a **hand cream**, the rod still remains standing straight in the initial position even after some hours (**yield point**: see Chapters 3.3.4 and 8.3.4.1; and **structure strength at rest of dispersions**: see Chapter 8.4.4a).
Additional observation concerning the air bubbles: All bubbles, independent of their size, remain suspended, and therefore here, no de-aeration takes place at all.

Summary
Rheological behavior depends on many influences. Above all, the following test conditions are important:
- Type of loading (preset of deformation, velocity or force; or shear strain, shear rate or shear stress, respectively)
- Degree of loading (low-shear or high-shear conditions)
- Duration of loading (the periods of time under load and at rest)
- Temperature (see Chapters 3.5 and 8.6)

Further important parameters are, for example:
- Concentration (of solid particles in a suspension: see Chapter 3.3.3; of polymer molecules in a solution: see Chapter 3.3.2.1a; of surfactants in a dispersion: see Chapter 9). Using an "Immobilization Cell," the amount of liquid can be reduced under controlled conditions (e.g. when testing dispersions such as paper coatings: see Chapter 10.8.1.3).
- Ambient pressure (see Chapter 3.6)
- pH value (e.g. of surfactant systems: see Chapter 9).
- Strength of a magnetic or an electric field when investigating magneto-rheological fluids MRF or electro-rheological fluids ERF, respectively (see Chapters 10.8.1.1 and 2).
- UV radiation curing (e.g. of resins, adhesives and inks: see Chapter 10.8.1.4).

2 Flow behavior and viscosity

In this chapter are explained the following terms given in bold:

Liquids		Solids	
(ideal-) viscous flow behavior Newton's law	viscoelastic flow behavior Maxwell's law	viscoelastic deformation behavior Kelvin/Voigt's law	(ideal-) elastic deformation behavior Hooke's law
flow/viscosity curves	creep tests, relaxation tests, oscillatory tests		

2.1 Introduction

Before 1980 in industrial practice, rheological experiments on pure liquids and dispersions were carried out almost exclusively in the form of rotational tests which enabled the characterization of flow behavior at medium and high flow velocities. Meanwhile since measurement technology has developed, many users have expanded their investigations on deformation and flow behavior performing measurements which cover also the low-shear range.

2.2 Definition of terms

The **Two-Plates-Model** is used to define fundamental rheological parameters (see Figure 2.1). The upper plate with the (shear) area A is set in motion by the (shear) force F and the resulting velocity v is measured. The lower plate is stationary (v = 0). Between the plates there is the distance h, and the sample is sheared in this shear gap. It is assumed that the following **shear conditions** are occurring:

1) The sample shows **adhesion** to both plates **without any wall-slip effects**.
2) There are **laminar flow conditions**, i.e. flow can be imagined in the form of layers. Therefore, there is **no turbulent flow**, i.e. no vortices are appearing.

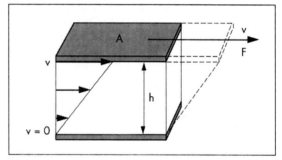

Figure 2.1: The Two-Plates-Model for shear tests to illustrate the velocity distribution of a flowing fluid in the shear gap

Accurate calculation of the rheological parameters is only possible if both conditions are met.

Experiment 2.1: The stack of beer mats
Each one of the individual beer mats represents an individual flowing layer. The beer mats are showing a laminar

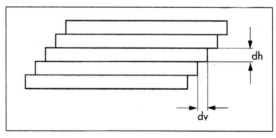

Figure 2.2: Laminar flow in the form of planar fluid layers

Thomas G. Mezger: The Rheology Handbook
© Copyright 2011 by Vincentz Network, Hanover, Germany
ISBN 978-3-86630-864-0

shape, and therefore, they are able to move in the form of layers along one another (see Figure 2.2). Of course, this process takes place without vortices, thus without showing any turbulent behavior.

The real geometric conditions in rheometer measuring systems are not as simple as in the Two-Plates-Model. However, if a shear gap is narrow enough, the necessary requirements are largely met and the definitions of the following rheological parameters can be used.

2.2.1 Shear stress

Definition of the shear stress:

Equation 2.1: $\tau = F/A$

τ (pronounced: "tou"); with the shear force F [N] and the shear area A [m^2], see Figure 2.1. The following holds: $1N = 1kg \cdot m/s^2$

The unit of the shear stress is [Pa], ("pascal").
Blaise Pascal (1623 to 1662 [278]) was a mathematician, physicist, and philosopher.
For conversions: $1Pa = 1N/m^2 = 1kg/m \cdot s^2$
A previously used unit was [dyne/cm^2]; with: $1dyne/cm^2 = 0.1Pa$

Note: [Pa] is also the unit of pressure
$100Pa = 1hPa$ (= 1mbar); or $100,000Pa = 10^5Pa = 0.1MPa$ (= 1bar)
Example: In a weather forecast, the air pressure is given as 1070hPa (hectopascal; = 107kPa).

Some authors take the symbol σ for the shear stress (pronounced: "sigma") [18, 215]. However, this symbol is usually used for the tensile stress (see Chapters 4.2.2 and 10.8.4.1). To avoid confusion and in agreement with the majority of current specialized literature and standards, here, the symbol τ will be used to represent the shear stress (see e.g. ASTM D4092 and DIN 1342-1).

2.2.2 Shear rate

Definition of the shear rate:

Equation 2.2: $\dot{\gamma} = v/h$

$\dot{\gamma}$ (pronounced: "gamma-dot"); with the velocity v [m/s] and the distance h [m] between the plates, see Figure 2.1.
The unit of the shear rate is [1/s] or [s^{-1}], called "reciprocal seconds".

Sometimes, the following terms are used as synonyms: **shear gradient, velocity gradient, strain rate,** and **rate of deformation**.

Previously, the symbol D was often taken instead of $\dot{\gamma}$. Nowadays, almost all current standards are recommending the use of $\dot{\gamma}$ (see e.g. ASTM D4092). Table 2.1 presents typical shear rate values occurring in industrial practice.

$\mathcal{GP}$ For "Mr. and Ms. Cleverly"

a) Definition of the shear rate using differential variables

Equation 2.3: $\dot{\gamma} = dv/dh$

with the "infinitely" (differentially) small velocity difference dv between two neighboring flowing layers, and the "infinitely" (differentially) small thickness dh of a single flowing layer (see Figure 2.2).

Table 2.1: Typical shear rates of technical processes

Process	Shear rates $\dot{\gamma}$ (s^{-1})	Practical examples
Physical aging, long-term creep within days and up to several years	10^{-8} ... 10^{-5}	Polymers, asphalt
Cold flow	10^{-8} ... 0.01	Rubber mixtures, elastomers
Sedimentation of particles	≤ 0.001 0.01	Emulsion paints, ceramic suspensions, fruit juices
Surface leveling of coatings	0.01 ... 0.1	Coatings, paints, printing inks
Sagging of coatings, dripping, flow under gravity	0.01 ... 1	Emulsion paints, plasters, chocolate coatings (couvertures)
Self-leveling at low-shear conditions in the range of the zero-shear viscosity	≤ 0.1	Silicone polymers (PDMS)
Dip coating	1 ... 100	Dip coatings, candy masses
Applicator roller, at the coating head	1 ... 100	Paper coatings
Thermoforming	1 ... 100	Polymers
Mixing, kneading	1 ... 100	Rubber mixtures, elastomers
Chewing, swallowing	10 ... 100	Jelly babies, yogurt, cheese
Spreading	10 ... 1,000	Butter, toothpastes
Extrusion	10 ... 1,000	Polymer melts, dough, ceramic pastes, tooth paste
Pipe flow, capillary flow	10 ... 10^4	Crude oils, paints, juices, blood
Mixing, stirring	10 ... 10^4	Emulsions, plastisols, polymer blends
Injection moulding	100 ... 10^4	Polymer melts, ceramic suspensions
Coating, painting, brushing, rolling, blade coating (manually)	100 ... 10^4	Brush coatings, emulsion paints, wall paper paste, plasters
Spraying	1,000 ... 10^4	Spray coatings, fuels, nose spray aerosols, adhesives
Impact	1,000 ... 10^5	Solid polymers
Milling pigments in fluid bases	1,000 ... 10^5	Pigment pastes for paints and printing inks
Rubbing	1,000 ... 10^5	Skin creams, lotions, ointments
Spinning process	1,000 ... 10^5	Polymer melts, polymer fibers
Blade coating (by machine), high-speed coating	1,000 ... 10^7	Paper coatings, adhesive dispersions
Lubrication of engine parts	1,000 ... 10^7	Mineral oils, lubricating greases

There is a **linear velocity distribution** between the plates, since the velocity v decreases linearly in the shear gap. Thus, for laminar and ideally viscous flow, the velocity difference between all neighboring layers are showing the same value: dv = const. All the layers are assumed to have the same thickness: dh = const. Therefore, the shear rate is showing a constant value everywhere between the plates of the Two-Plates-Model since

$\dot{\gamma}$ = dv/dh = const/const = const (see Figure 2.3).

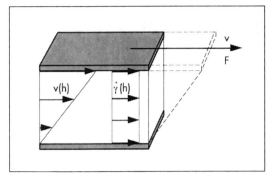

Figure 2.3: Velocity distribution and shear rate in the shear gap of the Two-Plates-Model

Both $\dot{\gamma}$ and v provide information about the velocity of a flowing fluid. The advantage of selecting the shear rate is that it shows a constant value throughout the whole shear gap. Therefore, the shear rate is independent of the position of any flowing layer in the shear gap. Of course, this applies only if the shear conditions are met as mentioned in the beginning of Chapter 2.2. However, this does not apply to the velocity v which decreases from the maximum value v_{max} on the upper, movable plate to the minimum value $v_{min} = 0$ on the lower, immovable plate. Therefore when testing pure liquids, sometimes as a synonym for shear rate the term **velocity gradient** is used (e.g. in ASTM D4092).

b) Calculation of shear rates occurring in technical processes
The shear rate values which are given below are calculated using the mentioned formulas and should only be seen as rough estimations. The main aim of these calculations is to get merely an idea of the dimension of the relevant shear rate range.

1) Coating processes: painting, brushing, rolling or blade-coating
$\dot{\gamma} = v/h$, with the coating velocity v [m/s] and the wet layer thickness h [m]

Examples:
1a) Painting with a brush:
With v = 0.1m/s and h = 100µm = 0.1mm = 10^{-4}m; result: $\dot{\gamma}$ = 1000s^{-1}

1b) Buttering bread:
With v = 0.1m/s and h = 1mm = 10^{-3}m; result: $\dot{\gamma}$ = 100s^{-1}

1c) Applying emulsion paint with a roller
With v = 0.2m/s (or 5s per m), and h = 100µm = 0.1mm = 10^{-4}m; result: $\dot{\gamma}$ = 2000s^{-1}

1d) Blade-coating of adhesive emulsions (e.g. for pressure-sensitive adhesives PSA):
with the application rate AR (i.e. mass per coating area) m/A [g/m²]; for the coating volume V [m³] applies, with the mass m [kg] and the density ρ [1g/cm³ = 1000kg/m³]: V = m/ρ
Calulation: h = V/A = (m/ρ)/A = (m/A)/ρ = AR/ρ; with AR = 1g/m² = 10^{-3}kg/m² holds: h =10^{-6}m = 1µm; and then: $\dot{\gamma}$ = v/h. See Table 2.2 for shear rates occurring in various kinds of blade-coating processes [14, 76].

Table 2.2: Shear rates of various kinds of blade-coating processes for adhesive emulsions

Coating process	Application rate AR [g/m²]	Coating velocity v [m/min]	Coating velocity v [m/s]	Layer thickness h [µm]	Approx. shear rate range $\dot{\gamma}$ [s⁻¹]
Metering blade	2 to 50	up to 250	up to 4.2	2 to 50	80,000 to 2 mio.
Roller blade	15 to 100	up to 100	up to 1.7	15 to 100	10,000 to 100,000
Lip-type blade	20 to 100	20 to 50	0.33 to 0.83	20 to 100	3,000 to 50,000
Present maximum	2 to 100	700	12	2 to 100	120,000 to 6 mio.
Future plans		up to 1,500	up to 25		250,000 to 12.5 mio.

2) Flow in pipelines, tubes and capillaries

Assumptions: horizontal pipe, steady-state and laminar flow conditions (for information on laminar and turbulent flow see Chapter 3.3.3), ideally viscous flow, incompressible liquid. According to the **Hagen/Poiseuille relation**, the following holds for the maximum shear stress τ_w and the maximum shear rate $\dot\gamma_w$ in a pipeline (index "w" for "at the wall"):

Equation 2.4 $\tau_w = (R \cdot \Delta p)/(2 \cdot L)$

Equation 2.5 $\dot\gamma_w = (4 \cdot \dot V)/(\pi \cdot R^3)$

with the pipe radius R [m]; the pressure difference Δp [Pa] between inlet and outlet of the pipe or along the length L [m] of the measuring section, respectively (Δp must be compensated by the pump pressure); and the volume flow rate $\dot V$ [m³/s]. This relation was named in honor to *Gotthilf Heinrich Ludwig Hagen* (1797 to 1848) [172] and *Jean Louis Marie Poiseuille* (1799 to 1869) [291].

Examples:

2a) Pipeline transport of automotive coatings [105], [350]

For a closed circular pipeline with the diameter DN 26 (approx. R = 13mm = $1.3 \cdot 10^{-2}$m), and the volume flow rate $\dot V$ = 1.5 to 12L/min = $2.51 \cdot 10^{-5}$ to $2.00 \cdot 10^{-4}$m³/s; results: $\dot\gamma_w$ = 14.6 to 116s⁻¹ = approx. 15 to 120s⁻¹. For a pipeline branch with DN 8 (approx. R = 4mm = $4 \cdot 10^{-3}$m), and $\dot V$ = 0.03 to 1L/min = $5.06 \cdot 10^{-7}$ to $1.67 \cdot 10^{-5}$m³/s; results: $\dot\gamma_w$ = 10.1 to 332s⁻¹ = approx. 10 to 350s⁻¹

2b) Drinking water supply, transport in pipelines [61]

For a pipeline with the diameter DN 1300 (approx. R = 650mm = 0.65m), and a volume flow rate of max. $\dot V$ = 3300L/s = 3.30m³/s; and for a second pipeline with DN 1600 (approx. R = 800mm = 0.80m) with max. $\dot V$ = 4700L/s = 4.70m³/s; results: max. $\dot\gamma_w$ = 15.3 and 11.7s⁻¹, respectively.

2c) Filling bottles using a filling machine (e.g. drinks in food industry):

Filling volume per bottle: V = 1L = 0.001m³; filling time per bottle: t = 5s, then: $\dot V$ = V/t = $2 \cdot 10^{-4}$m³/s; diameter of the circular geometry of the injection nozzle: d = 2R = 10mm; result: $\dot\gamma_w$ = 2037s⁻¹ = approx. 2000s⁻¹

2d) Squeezing an ointment out of a tube (e.g. pharmaceuticals):

Pressed out volume: V = 1cm³ = 10^{-6}m³; time to squeeze out: t = 1s; then: $\dot V$ = V/t = 10^{-6}m³/s; diameter of the tube nozzle: d = 2R = 6mm; result: $\dot\gamma_w$ = 47.2s⁻¹ = approx. 50s⁻¹

2e) Filling ointment into tubes using a filling machine (e.g. medicine):

Filling volume per tube: V = 100ml = 10^{-4}m³; filling time per tube (at 80 work-cycles per minute, where 50% is filling time): t = (60s/2)/80 = 0.375s; then: $\dot V$ = V/t = $2.67 \cdot 10^{-4}$m³/s, using an injection nozzle with an annular geometry and a cross sectional area of A = $24 \cdot 10^{-6}$m², which for a rough estimation, corresponds to a circular area showing R = $2.76 \cdot 10^{-3}$m (since A = $\pi \cdot R^2$); result: $\dot\gamma_w$ = 16,200s⁻¹

3) Sedimentation of particles in suspensions

Assumptions: fluid in a state-at-rest; the particles are almost suspended and therefore they are sinking very, very slowly in a steady-state process (laminar flow, at a Reynolds number Re ≤ 1; more about Re numbers: see Chapter 10.2.2.4b); spherical particles; the values of the weight force F_G [N] and the flow resistance force F_R [N] of a particle are approximately equal in size.

According to **Stokes' law** (*Georges Gabriel Stokes*, 1819 to 1903 [348]):

Equation 2.6: $F_G = \Delta m \cdot g = F_R = 3 \cdot \pi \cdot d_p \cdot \eta \cdot v$

with the mass difference Δm [kg] between a particle and the surrounding fluid, the gravitation constant g = 9.81m/s², the mean particle diameter d_p [m], the shear viscosity of the dispersion fluid η [Pas], and the particles' settling velocity v [m/s].

The following applies: $\Delta m = V_p \cdot \Delta\rho$, with the volume V_p [m³] of a particle, and the density difference $\Delta\rho$ [kg/m³] = $(\rho_p - \rho_{fl})$ between the particles and the dispersion fluid; particle density ρ_p [kg/m³] and fluid density ρ_{fl} [kg/m³].

The following applies for spheres: $V_p = (\pi \cdot d_p^3)/6$; and therefore, for the settling velocity

Equation 2.7: $\qquad v = \dfrac{\Delta m \cdot g}{3 \cdot \pi \cdot d_p \cdot \eta} = \dfrac{V_p \cdot \Delta\rho \cdot g}{3 \cdot \pi \cdot d_p \cdot \eta} = \dfrac{\pi \cdot d_p^3 \cdot \Delta\rho \cdot g}{6 \cdot 3 \cdot \pi \cdot d_p \cdot \eta} = \dfrac{d_p^2 \cdot g}{18 \cdot \eta} \cdot (\rho_p - \rho_{fl})$

Assumption for the shear rate: $\dot\gamma = v/h$
with the thickness h of the boundary layer on a particle surface, which is sheared when in motion against the surrounding liquid (the shear rate occurs on both sides of the particle). This equation is valid only if there are neither interactions between the particles, nor between the particles and the surrounding dispersion fluid.
Assuming simply, that $h = 0.1 \cdot d$, then: $\dot\gamma = (10 \cdot v)/d$

Examples: Sedimentation of sand particles in water
3a) With $d_p = 10\mu m = 10^{-5}m$, $\eta = 1mPas = 10^{-3}Pas$, and $\rho_p = 1.5g/cm^3 = 1500kg/m^3$ (e.g. quartz silica sand), and $\rho_{fl} = 1g/cm^3 = 1000kg/m^3$ (pure water); results: v = approx. $2.7 \cdot 10^{-5}m/s$
Such a particle is sinking a maximum path of approx. 10cm in 1h (or approx. 2.3m per day). With $h = 1\mu m$ results: $\dot\gamma = v/h$ = approx. $27s^{-1}$
3b) With $d_p = 1\mu m = 10^{-6}m$, and $\eta = 100mPas$ (e.g. water containing a thickener, measured at $\dot\gamma = 0.01s^{-1}$), and with the same values for ρ_p and ρ_{fl} as above in Example (3a), results: v = approx. $2.7 \cdot 10^{-9}m/s$ (or v = 0.23mm per day). With $h = 0.1\mu m$ results: $\dot\gamma = 0.03s^{-1}$ approximately.

Note 1: Calculation of a too high settling velocity if interactions are ignored
Stokes' sedimentation formula only considers a single particle sinking, undisturbed on a straight path. Therefore, relatively high shear rate values are calculated. These values do not mirror the real behavior of most dispersions, since usually interactions are occurring. The layer thickness h is hardly determinable. We know from colloid science: It depends on the strength of the ionic charge on the particle surface, and on the ionic concentration of the dispersion fluid (interaction potential). Due to ionic adsorption, a diffuse double layer of ions can be found on the particle surface. For this reason, in reality the result is usually a considerably lower settling velocity. Therefore, and since the shear rate within the sheared layer is not constant: **It is difficult to estimate the corresponding shear rate values occurring with sedimentation processes.**

Note 2: Particle size of colloid dispersions
Colloid particles are showing diameters in the range of 10^{-9} to $10^{-7}m$ (or 1 to 100nm) [397], or from 10^{-9} to $10^{-6}m$ (or 1nm to 1μm) [128]. Due to Brownian motion, the particles usually are remaining in a suspended state and do not tend to sedimentation. Above all, the limiting value of the settling particle size depends on the density difference of particles and dispersing fluid.

∞   End of the Cleverly section

2.2.3 *Viscosity*

For all flowing fluids, the molecules are showing relative motion between one another, and this process is always combined with internal frictional forces. Therefore for all fluids in motion, a certain flow resistance occurs which may be determined in terms of the viscosity. All materials which clearly show flow behavior are referred to as **fluids** (thus: **liquids and gases**).

a) Shear viscosity

For ideally viscous fluids measured at a constant temperature, the value of the ratio of shear stress τ and corresponding shear rate $\dot{\gamma}$ is a material constant. Definition of the shear viscosity:

Equation 2.8: $\eta = \tau / \dot{\gamma}$

η (eta, pronounced: "etah" or "atah"), **the unit of shear viscosity is [Pas], ("pascal-seconds")**.

The following holds: $1Pas = 1N \cdot s/m^2 = 1kg/s \cdot m$

For low-viscosity liquids, the following unit is usually used:
1mPas ("milli-pascal-seconds") = 10^{-3}Pas
Sometimes, for highly viscous samples the following units are used:
1kPas ("kilo-pascal-seconds") = 1000Pas = 10^3Pas, or even
1MPas ("mega-pascal-seconds") = 1,000kPas = 1,000,000Pas = 10^6Pas
A previously used unit was [P], ("poise"; at best pronounced in French); and: 1P = 100cP; however, this is not an SI unit [59]. This unit was named in honor to the doctor and physicist *Jean L.M. Poiseuille* (1799 to 1869) [291].
The following holds: 1cP ("centi-poise") = 1mPas, and 1P = 0.1Pas = 100mPas

Sometimes, the term **"dynamic viscosity"** is used for η (as in DIN 1342-1). However, some people use the same term to describe either the complex viscosity determined by oscillatory tests, or to mean just the real part of the complex viscosity (the two terms are explained in Chapter 8.2.4b). To avoid confusion and in agreement with the majority of current international authors, here, the term "viscosity" or "shear viscosity" will be used for η. Table 2.3 lists viscosity values of various materials.

The inverse value of viscosity is referred to as **fluidity** φ (phi, pronounced: "fee" or "fi") [303]. However today, this parameter is rarely used. The following holds:

Equation 2.9: $\varphi = 1/\eta$ with the unit $[1/Pas] = [Pas^{-1}]$

&ℐ   For "Mr. and Ms. Cleverly"

Note 1: Usually, samples are viscoelastic when showing high viscosity values.
Many rheological investigations showed that at values of $\eta > 10kPas$, the elastic portion should no longer be ignored. These kinds of samples should no longer be considered simply viscous only, but visco-elastic (see also Chapter 5).

Note 2: Shear viscosity η and extensional viscosity η_E
For ideally viscous fluids under uniaxial tension the following applies for the values of the extensional viscosity (in Pas) and shear viscosity η (also in Pas): $\eta_E(\dot{\varepsilon}) = 3 \cdot \eta(\dot{\gamma})$, if the values of the extensional strain rate $\dot{\varepsilon}$ $[s^{-1}]$ and shear rate $\dot{\gamma}$ $[s^{-1}]$ are equal in size (see also Chapter 10.8.4.1: Trouton relation).

&ℐ   End of the Cleverly section

b) Kinematic viscosity

Definition of the kinematic viscosity:

Equation 2.10: $v = \eta / \rho$

v (ny, pronounced: "nu" or "new"), with the density ρ $[kg/m^3]$, (rho, pronounced: "ro").
For the unit of density holds: $1g/cm^3 = 1000kg/m^3$
The unit of kinematic viscosity is $[mm^2/s]$; and: $1mm^2/s = 10^{-6}m^2/s$

Table 2.3: Viscosity values, at T = +20°C when without further specification; own data and from [18, 215, 276]

Material	Viscosity η [mPas]
Gases/air	0.01 to 0.02 / 0.018
Pentane/acetone/gasoline, petrol (octane)/ethanol	0.230 / 0.316 / 0.538 / 1.20
Water at 0 / +10 / +20 / +30 / +40 / +50 / +60 / + 70 / +80 / +90 / +100°C	1.79 / 1.31 / 1.00 / 0.798 / 0.653 / 0.547 / 0.467 / 0.404 / 0.354 / 0.315 / 0.282
Mercury	1.55
Blood plasma at +20 / +37°C	1.7 / 1.2
Wine, fruit juices (undiluted)	2 to 5
Milk, coffee cream	2 to 10
Blood (from a healthy body) at +20 / +37°C	5 to 120 / 4 to 15 (at $\dot{\gamma}$ = 0.01 to 1000s^{-1})
Light oils	10
Glycol	20
Sulphuric acid	25
Sugar solutions (60%)	57
Motor oils SAE 10W-30, at +23 / +50 / +100°C	50 to 1000 175 / 52 / 20
Olive oils	Approx. 100
Gear oils	300 to 800
Glycerine	1480
Honey, concentrated syrups	Approx. 10Pas
Polymer melts (at processing conditions, e.g. between T = + 150 and 300°C, and at $\dot{\gamma}$ = 10 bis 1000s^{-1})	10 to 10,000Pas
Polymer melts: zero-shear viscosity at $\dot{\gamma} \leq 0.1s^{-1}$ and at T = +150 to 300°C	1kPas to 1MPas
Silicone (PDMS, unlinked, zero-shear viscosity)	10 to 100kPas
Hotmelts (maximum processing viscosity for melt extruders)	100kPas
Bitumen (example): at T = +80 / +60 / +40 / +20°C and at T = 0°C	200Pas / 1kPas / 20kPas / 0.5MPas and 1MPas, i.e., then almost like a viscoelastic solid

A previously used unit was [St] ("stokes"); with: 1St = 100cSt. This unit was named in honor to the mathematician and physicist *George G. Stokes* (1819 to 1903) [348]. The following holds: 1cSt ("centi-stokes") = 1mm^2/s.

Example: Conversion of the values of kinematic viscosity and shear viscosity
Preset: A liquid shows ν = 60mm^2/s = 60 · 10^{-6}m^2/s, and ρ = 1.1g/cm^3 = 1100kg/m^3
Calculation: $\eta = \nu \cdot \rho$ = 60 · 10^{-6} · 1100 (m^2/s) · (kg/m^3) = 66 · 10^{-3}kg/(s · m) = 66mPas

Usually, kinematic viscosity values are measured by use of flow cups, capillary viscometers, falling-ball viscometers or Stabinger viscometers (see Chapters 11.2.11L and 11.3 to 11.5).

2.3   Shear load-dependent flow behavior

Experiment 2.2: The double-tube test, or the contest of the two fluids (see Figure 2.4)
In the beginning, fluid F1 is flowing faster than fluid F2. With decreasing fluid level, F1 shows reduced flow velocity. F2, however, continues to flow with a hardly visible change in velocity. Therefore finally, F2 empties its tube before F1 does. F1, a wallpaper paste, is an aqueous methyl-cellulose solution, and F2 is a mineral oil. Flow behavior of polymer solutions such as the wallpaper paste is explained in Chapter 3.3.2.1: shear-thinning flow behavior.

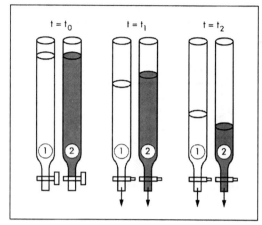

Figure 2.4: Double-tube test

2.3.1   Ideally viscous flow behavior according to Newton

a) Newton's law
Formally, ideally viscous (or Newtonian) flow behavior is presented by Newton's law:

Equation 2.11 $\tau = \eta \cdot \dot{\gamma}$

Isaac Newton (1643 to 1727) is often called the "father of classic physics". He proposed, in a quite inaccurate form (in 1687, in his textbook "principia" [266]): The resistance to flow ("defectus lubricitatus") of liquids is proportional to the flow velocity. Based on later research on fluid dynamics by *Daniel Bernoulli* (in 1738, "Hydrodynamica" [34]), *Leonhard Euler* (in 1739/1773, "Scientia Navalis", and "construction des vaisseaux" [126]), *Johann Bernoulli* (in 1740, "Hydraulica" [35]), and *Claude Louis Marie Henri Navier* (in 1823 [263]), finally *Georges Gabriel Stokes* (in 1845 [348]) stated the accurate and modern form of Newton's hypothesis. Therefore sometimes, the viscosity law is also termed the "Newton/Stokes law" [237].

Examples of ideally viscous materials:
Low-molecular liquids (and this means here: with a molar mass below 10,000g/mol) such as water, solvents, mineral oils (without polymer additives), silicone oils, viscosity standard oils (of course!), pure and clean bitumen (without associative superstructures and at sufficiently high temperatures), blood plasma

Flow behavior is illustrated graphically by flow curves (previously sometimes also called "rheograms") and viscosity curves. **Flow curves** are showing the interdependence of shear stress τ and shear rate $\dot{\gamma}$. Usually, $\dot{\gamma}$ is presented on the x-axis (abscissa), and τ on the y-axis (ordinate). However, τ might also be displayed on the x-axis and $\dot{\gamma}$ on the y-axis, but this is meanwhile rarely made in industrial laboratories.

Viscosity curves are derived from flow curves. Usually, η is presented on the y-axis and $\dot{\gamma}$ on

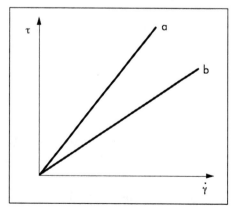

Figure 2.5: Flow curves of two ideally viscous fluids

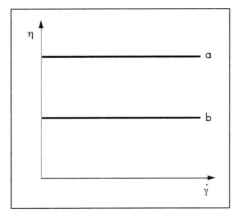

Figure 2.6: Viscosity curves of two ideally viscous fluids

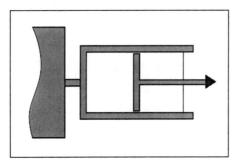

Figure 2.7: The dashpot model to illustrate ideally viscous behavior

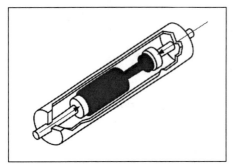

Figure 2.8: A shock absorber which can be loaded from both sides [121]

the x-axis. Alternatively, the function $\eta(\tau)$ can be shown with η on the y-axis and τ on the x-axis, however, this is less frequently carried out in industrial labs.

Generally, the slope value of each point (x; y) of a curve can be calculated as: y/x. This counts for each point of a flow curve with the pair of values $(\dot\gamma; \tau)$. The result of this calculation again corresponds to the viscosity value, this is because: $\eta = \tau/\dot\gamma$. Therefore, the $\eta(\dot\gamma)$-curve can be calculated point by point from the $\tau(\dot\gamma)$-curve. Correspondingly, a steeper slope of the flow curve results in a higher level of the viscosity curve (see Figures 2.5 and 2.6). Usually today, this calculation is performed by a software program.

The values of shear viscosity of ideally viscous or Newtonian fluids are independent of the degree and duration of the shear load applied.

Viscosity values of ideally viscous liquids are often measured using flow cups, capillary viscometers, falling-ball viscometers or Stabinger viscometers (see Chapters 11.2.11L, 11.3 to 11.5). However, when using these simple devices, the results do not accurately mirror the more complex behavior of non-Newtonian liquids (see for example Chapter 11.3.1.2c: change of shear rates in capillaries).

$\mathcal{GG}$ For "Mr. and Ms. Cleverly"

b) The dashpot model

The dashpot model is used to illustrate the behavior of ideally viscous or Newtonian liquids (see Figure 2.7). Mechanically similar examples are gas or liquid shock absorbers (see Figure 2.8).

Ideally viscous flow behavior, explained by the behavior of a dashpot
1) When loading
Under a constant force, the piston is moving continuously as long as the force is applied, pressing the dashpot fluid (e.g. an oil) through the narrow annular gap between the piston and the cylinder wall of the dashpot. When applying forces of differing strength to the dashpot, it can be observed in all cases: The resulting velocity of the piston is proportional the driving force. The proportionality factor corresponds to the internal friction of the dashpot, and therefore, to the fluid's flow resistance or viscosity, respectively.

2) When removing the load
As soon as the force is removed, the piston immediately stops to move and remains in the position reached.

Summary: Behavior of the dashpot model

Under a constant load, the dashpot fluid is flowing with a constant velocity or deformation rate. After removing the load, the deformation applied to the fluid remains to the full extent. In other words: **After a load cycle, an ideally viscous fluid completely remains in the deformed state**. These kinds of fluids show absolutely no sign of elasticity.

Comparison: Dashpot fluid and Newton's law

For a dashpot, the force/velocity law holds (flow resistance according to *Newton*):

$F = C_N \cdot v \ (= C_N \cdot \dot{s})$

with the force F [N], the dashpot constant C_N [Ns/m = kg/s], the index "N" is due to Newton; the piston velocity v [m/s], and the time derivative of the piston's deflection path $\dot{s}$ [m/s]. Here: F corresponds to the shear stress τ, C_N corresponds to the viscosity η, v or $\dot{s}$ correspond to the shear rate $\dot{\gamma}$, and s corresponds to the deformation γ.

Note: Flow behavior, viscous heating, and lost deformation energy

Deformation energy acting on a fluid leads to relative motion between the molecules. As a consequence, **in flowing fluids** frictional forces are occurring **between the molecules**, causing **frictional heating**, also called **viscous heating**. For fluids showing ideally viscous flow behavior, **the applied deformation energy is completely used up** and can be imagined as deformation work. A part of this thermal energy may heat up the fluid itself and another part may be released as heat to the surrounding environment. During a flow process, the applied deformation energy is consumed completely by the fluid, and therefore, it is no longer available for the fluid afterwards, i.e., **it is lost**.

When the load is removed, the state of deformation which was reached finally by the fluid is remaining to the full extent. Not even a partial elastic re-formation effect can be observed. Therefore here, **an irreversible process** has taken place since the shape of the sample remains permanently changed finally, after the load is released from the fluid.

If fluids are showing ideally viscous flow behavior, there are absolutely no or at least no significant interactions between their mostly small molecules or particles. Examples are pure solvents, oils and water; and there might be also some polymer solutions and dispersions, however, only if they show a really very low concentration. Since these kinds of fluids do not show any visco-elastic gel-like structure, they may tend to separation, and therefore, effects like sedimentation or flotation may occur in mixtures of fluids and in dispersions.

    ⌒   End of the Cleverly section

2.4   Types of flow illustrated by the Two-Plates Model

Figure 2.9 illustrates seven different types of laminar flow which may occur in a shear gap: (1) state- at- rest; (2) **homogeneous laminar flow, showing a constant shear rate** (see also Chapter 2.2); (3) **wall-slip**, the sample displays very pronounced cohesion while slipping along the walls without adhesion; (4) **"plastic behavior"**, only a part of the sample is sheared homogeneously (see also Chapter 3.3.4.3d); (5) **transient behavior, showing a start-up effect** as time-dependent transition until steady-state viscosity is reached (occurring above all at low shear rates; see also Chapter 3.3.1b); (6) **shear-banding,** exhibiting here pronounced cohesion of the medium band (see also Chapter 9.2.2); (7) **shear-banding, showing here three different flow velocities** or viscosities, respectively (see also here Chapter 9.2.2)

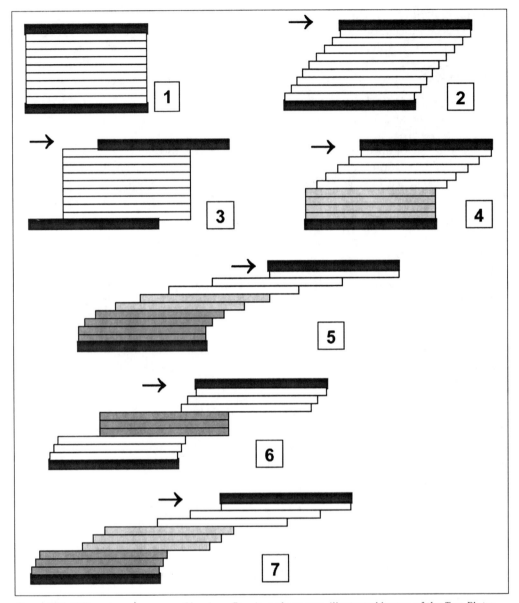

Figure 2.9: Different appearances of laminar flow in a shear gap, illustrated by use of the Two-Plates Model

3   Rotational tests

In this chapter are explained the following terms given in bold:

Liquids		Solids	
(ideal-) viscous	viscoelastic	viscoelastic	(ideal-) elastic
flow behavior	flow behavior	deformation behavior	deformation behavior
Newton's law	Maxwell's law	Kelvin/Voigt's law	Hooke's law
flow/viscosity curves	creep tests, relaxation tests, oscillatory tests		

3.1   Introduction

In Chapter 2 using Newton's law, the rheological background of fluids showing ideally viscous flow behavior was explained. Chapter 3 concentrates on rheometry: The performance of rotational tests to investigate the mostly more complex, non-Newtonian flow behavior of liquids, solutions, melts and dispersions (suspensions, emulsions, foams) used in daily practice in industry will be described here in detail.

With most of the rheometers used in industrial laboratories, the bob is the rotating part of the measuring system (Searle method, see Chapter 10.2.1.2a). But there are also types of instruments where the cup (Couette method) or the lower plate, respectively, is set in rotational motion (see also Chapters 10.2.12b and 11.6.1 with Figure 11.6).

3.2   Basic principles

3.2.1   Test modes controlled shear rate (CSR) and controlled shear stress (CSS), raw data and rheological parameters

a) Tests with controlled shear rate (CSR tests)
When performing CSR tests, the rotational speed or shear rate, respectively, is preset and controlled by the rheometer (see Table 3.1). This test method is called a "controlled shear rate test", or briefly, "CSR test" or "CR test".

The test method with controlled shear rate is usually selected if the liquid to be investigated shows self-leveling behavior (i.e. no yield point), and if viscosity should be measured at a desired flow velocity or shear rate, respectively. This is the case, if certain process conditions have to be simulated, for example, occurring with pipe flow, or when painting and spraying. Shear rates which are occurring in industrial practice are listed in Table 2.1 (see Chapter 2.2.2).

Table: 3.1: Raw data and rheological parameters of rotational tests with controlled shear rate (CSR)

Rotation CSR	Test preset	Results
Raw data	Rotational speed n [min^{-1}]	Torque M [mNm]
Rheological parameters	Shear rate $\dot{\gamma}$ [s^{-1}]	Shear stress τ [Pa]
Viscosity calculation		$\eta = \tau / \dot{\gamma}$ [Pas]

Thomas G. Mezger: The Rheology Handbook
© Copyright 2011 by Vincentz Network, Hanover, Germany
ISBN 978-3-86630-864-0

Table: 3.2: Raw data and rheological parameters of rotational tests with controlled shear stress (CSS)

Rotation CSS	Test preset	Results
Raw data	Torque M [mNm]	Rotational speed n [min⁻¹]
Rheological parameters	Shear stress τ [Pa]	Shear rate γ̇ [s⁻¹]
Viscosity calculation		η = τ / γ̇ [Pas]

ISO 3219 standard recommends to measure and to compare viscosity values preferably at defined shear rate values. For this purpose, the following two alternative series are specified. Dividing or multiplying these values by 100 provides further γ̇-values.

1) $1.00/2.50/6.30/16.0/40.0/100/250 s^{-1}$. This geometric series shows a multiplier of 2.5.
2) $1.00/2.50/5.00/10.0/25.0/50.0/100 s^{-1}$

b) Tests with controlled shear stress (CSS tests)
When performing CSS tests, the torque or shear stress, respectively, is preset and controlled by the rheometer (see Table 3.2). This test method is called a "controlled shear stress test", or briefly, "CSS test" or "CS test".

This is the "classic" method to determine yield points of dispersions, pastes or gels (see also Chapter 3.3.4.1b). In nature, almost all flow processes are shear stress controlled, since any motion – creep or flow – is mostly a reaction to an acting force.

Examples from nature
Rivers, avalanches, glaciers, landslides, ocean waves, the motion of clouds or of leaves on a tree, earthquakes, blood circulation. The acting forces here are gravitational force, the forces of the wind and of the continental drift or the pumping power of the heart.

3.3   *Flow curves and viscosity functions*

Flow curves are usually measured at a constant measuring temperature, i.e. at isothermal conditions. In principle, flow behavior is combined always with flow resistance, and therefore with an internal friction process occurring between the molecules and particles. In order to perform accurate tests in spite of the resulting viscous heating of the sample, the use of a temperature control device is required, for example in the form of a water bath or a Peltier element (see also Chapter 11.6.6: Temperature control systems).

3.3.1   *Description of the test*

Preset
1) With controlled shear rate (CSR): Profile γ̇(t) in the form of a step-like function, see Figure 3.1; or as a shear rate ramp, see Figure 3.2.
2) With controlled shear stress (CSS): Profile τ(t), similar to Figures 3.1 and 3.2. The shear stress ramp test is the "classic method" to determine the yield point of a sample (see Chapter 3.3.4).

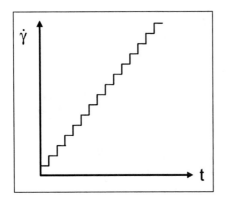

Figure 3.1: Preset profile: Time-dependent shear rate ramp in the form of a step-like function

Measuring result: Flow curve τ(γ̇) or γ̇(τ), respectively, see Figure 3.3

Usually, flow curves are plotted showing $\dot\gamma$ on the x-axis and τ on the y-axis; and rarely reversed. This also applies to curves which are obtained when controlling the shear stress.

Further results: Viscosity function $\eta(\dot\gamma)$, see Figure 3.4; or $\eta(\tau)$, respectively.

Usually, viscosity curves are presented showing $\dot\gamma$ on the x-axis and η on the y-axis. This also applies to curves which are obtained when controlling the stress. Therefore in industrial labs, τ is rarely presented on the x-axis.

a) Extended test programs (including intervals at rest, for temperature equilibration, and pre-shear)

Sometimes in industry, test programs are used showing besides shear rate ramps also other intervals without ramps, for example:

1) **Rest intervals**, presetting constantly $\dot\gamma = 0$ either after a pre-shear interval or at the start of the test to enable **relaxation of the sample** after gap setting of the measuring system which may cause a high internal stress particularly when testing highly viscous and viscoelastic samples. Simultaneously, this period is suited to enable **temperature equalibration**.

2) **Pre-shear intervals**, presetting a constant low shear rate to distribute the sample in the shear gap homogeneously, and to equalize or even to reduce possibly still existing pre-stresses deriving from the preparation of the sample.

Example 1: Testing resins

Test program consisting of three intervals, preset:
1st interval: pre-shear phase (for t = 3min): at $\dot\gamma = 5s^{-1}$ = const
2nd interval: rest phase (for t = 1min): at $\dot\gamma = 0$
3rd interval: upward shear rate ramp
(in t = 2min): $\dot\gamma = 0$ to $\dot\gamma_{max}$
at $\dot\gamma_{max} = 100s^{-1}$ (or, for a rigid consistency:
at $\dot\gamma_{max} = 20s^{-1}$ only)
Analysis: viscosity value at $\dot\gamma = 50s^{-1}$ (or at $10s^{-1}$, respectively)

Example 2: Testing chocolate melts (at the test temperature T = +40°C)

Test program consisting of four intervals, preset:
1st interval: pre-shear phase (for t = 500s):
at $\dot\gamma = 5s^{-1}$ = const
2nd interval: upward shear rate ramp
(in t = 180s): $\dot\gamma = 2$ to $50s^{-1}$
3rd interval: high-shear phase (for t = 60s): at $\dot\gamma = 50s^{-1}$ = const
4th interval: downward shear rate ramp (in t = 180s): $\dot\gamma = 50$ to $2s^{-1}$

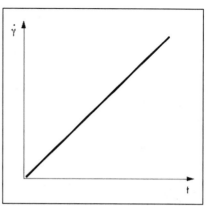

Figure 3.2: Preset profile: Time-dependent shear rate ramp

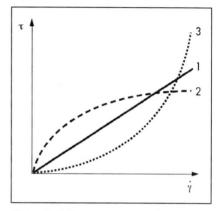

Figure 3.3: Flow curves, overview:
(1) ideally viscous/(2) shear-thinning/
(3) shear-thickening behavior

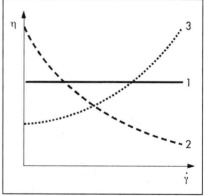

Figure 3.4: Viscosity functions, overview:
(1) ideally viscous/(2) shear-thinning/
(3) shear-thickening behavior

Analysis: According to the ICA method (International Confectionery Association), the following two values are determined of the downward curve:
1) Viscosity value at $\dot{\gamma} = 40s^{-1}$ as the so-called "apparent viscosity η_{40}"
2) Shear stress value at $\dot{\gamma} = 5s^{-1}$ as the so-called "yield stress YS_5" [197]

b) Time-dependent effects, steady-state viscosity and transient viscosity (at low shear rates)

When measuring at shear rates of $\dot{\gamma} < 1s^{-1}$, it is important to ensure that the measuring point duration is long enough. This is especially true when testing highly viscous and viscoelastic samples at these low-shear conditions. Otherwise **start-up effects or time-dependent transition effects** are obtained, i.e., values of the transient viscosity function will be determined instead of the desired constant value of steady-state viscosity at each measuring point. **Steady-state viscosity is only dependent on the shear rate (or the shear stress) applied**, resulting point by point in the viscosity function $\eta(\dot{\gamma})$ or $\eta(\tau)$, respectively. The values of **transient viscosity**, however, **are dependent on both, shear rate (or shear stress) and passing time**. Therefore, they are presented in the form of $\eta^+(\dot{\gamma}, t)$ or $\eta^+(\tau, t)$, respectively.

When performing tests at $\dot{\gamma} > 1s^{-1}$, only samples with pronounced viscoelastic properties are still influenced by transient effects. Therefore here, for liquids showing low or medium viscosity values, a duration of $t = 5s$ is sufficient for each single measuring point in almost all cases. However, transient effects should always be taken into account for polymers when measuring at shear rates of $\dot{\gamma} < 1s^{-1}$ (i.e. in the low-shear range). **As a rule of thumb: The measuring point duration should be selected to be at least as long as the value of the reciprocal shear rate $(1/\dot{\gamma})$.**

Illustration, using the Two-Plates-Model (see also Figure 2.9, no.5)

When setting the upper plate in motion at a constant speed, **during a certain start-up time** not all flowing layers of the sample are already shifted to the same extent along the neighboring layers. Initially, the resulting shear rate is not constant in the entire shear gap then since at first, only those layers are shifted which are close to the moving upper shear area. It takes a certain time until all the other layers are also set in motion, right down to the stationary bottom plate. Of course, this process will take a considerably longer time when presetting a considerably lower velocity to the upper plate.

The full shear force representing the flow resistance of the whole and homogeneously sheared sample is measured not before the shear rate is reaching a constant value throughout the entire shear gap as illustrated in Figures 2.3 and 2.9 (no. 2). Only in this case, steady-state viscosity $\eta = \eta(\dot{\gamma})$ will be obtained. Until reaching this time point, the lower flow resistance of the – up to then only partially flowing – liquid will be detected. Up to then, measured are still the clearly lower values of the still time-dependent, transient viscosity $\eta^+ = \eta(\dot{\gamma}, t)$. Considering the entire shear gap in this period of time, the shear gradient still will be not constant and therefore, **the shear process will be still inhomogeneously**.

Example 1: Useful measuring point duration to avoid transient effects

According to the rule of thumb (see above): When presetting $\dot{\gamma} = 0.1s^{-1}$, the measuring point duration should be set to $t \geq 10s$; and when $\dot{\gamma} = 0.01s^{-1}$, then $t \geq 100s$. These values should only be taken as a rough guide for a first try: Sometimes a longer duration is necessary and in some few cases a shorter duration is already sufficient.

Example 2: Suggestions for test profile presets, with $\dot{\gamma}(t)$ as a ramp function
1) Linear preset

Preset: $\dot{\gamma} = 0$ to $1000s^{-1}$ with 20 measuring points in a period of $t = 120s$ for the total test interval. The shear rates are increased in steps showing always the same distance between their values when presenting the curves on a linear scale (they are therefore equidistant; see Figure 3.1). In this case, each one of **the measuring points is generated by the rheometer in the same period of time**. In this example, the test program results in a duration of t = 6s for each single measuring point.

2) Logarithmic preset

Preset: $\dot{\gamma}$ = 0.01 to 1000s^{-1} with 25 measuring points. The shear rates are increased in logarithmic steps, now showing the same distance in between neighboring points, but of course, only when using a logarithmic scale. Here, a **variable logarithmic measuring point duration** should be preset, for example, beginning with t = 100s and ending with t = 5s. Therefore, when using this method, **longer measuring point durations are occurring at lower shear rates, and shorter durations at the higher ones.** As a consequence, for each single measuring point in the low-shear range the transient effects of the sample will be either reduced to a minimum or they have already completely decayed at the end of the prolonged duration of each measuring point.

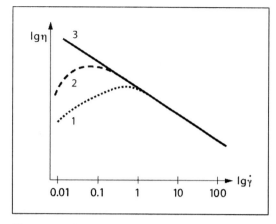

Figure 3.5: Viscosity functions of a dispersion, occurrence of time-dependent effects in the low-shear range such as the "transient viscosity peak", when presetting too short measuring point duration: (1) clearly too short; (2) better, but still too short; (3) sufficiently long

    For "Mr. and Ms. Cleverly"

Example 3: Occurrence of a "transient viscosity peak" when presetting a too short measuring point duration in the low-shear range (e.g. when testing dispersions and gels)
Figure 3.5 presents three viscosity functions of the same dispersion, all curves are measured in the range between $\dot{\gamma}$ = 0.01 and 100s^{-1}.
Preset for the duration of each individual measuring point:
Test 1: t = 10s = const (i.e., with the same time for each one of the measuring points),
Test 2: t = 60 to 5s (i.e., with variable time, decreasing towards higher shear rates),
Test 3: t = 120 to 5s (similar to Test 2, but beginning with a longer time)

Results: Dispersions and gels when showing stability at rest, thus, a gel-like viscoelastic structure, usually display a constantly falling η-curve from low to high shear rates, at least in the low-shear range (i.e. at $\dot{\gamma}$ < 1s^{-1}). This behavior is indicated by the curve of Test 3 (see Figure 3.5). In contrast, the η-curves of Tests 1 and 2 are showing "transient viscosity peaks", since here, still time-dependent effects are measured. The shorter the selected measuring point duration in the low-shear range, the lower are the η-values determined in this range, and the higher are the shear rate values at the occurrence of the peaks.

Summary: When testing dispersions and gels

The longer the measuring point duration in the low-shear range, the greater is the chance to avoid transient effects.

Note: Unlinked polymers in the form of solutions and melts usually show a constant η-value in the low-shear range, which is the **plateau value of the zero-shear viscosity** η_0 (see Chapter 3.3.2.1, Figure 3.10). However, also here a "transient viscosity peak" might appear when presetting too short measuring point durations in this shear rate range.

    End of the Cleverly section

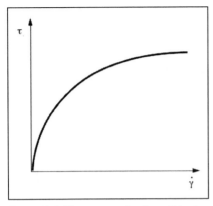

Figure 3.6: Flow curve of a shear-thinning liquid

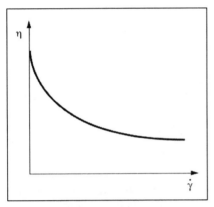

Figure 3.7: Viscosity curve of a shear-thinning liquid

3.3.2 Shear-thinning flow behavior

Viscosity of a shear-thinning material depends on the degree of the shear load (shear rate or shear stress, respectively). The flow curve shows a decreasing curve slope (see Figure 3.6), i.e., viscosity decreases with increasing load (see Figure 3.7).

Examples of shear-thinning materials
Polymer solutions (e.g. methylcellulose), unfilled polymer melts, most coatings, glues, shampoos

The terms **shear-thinning** and **pseudoplastic** are identical in their meaning (see also Chapter 13.3: 1925, with the concepts of *Eugen C. Bingham* [42], *Wolfgang Ostwald jun.* [273] – the latter used the German term "strukturviskos" – and others on this subject [378]). However, "pseudoplastic" contains the word "plastic", a behavior which cannot be exactly determined in a scientific sense since it is the result of inhomogeneous deformation and flow behavior (see also Chapter 3.3.4.3d and Figure 2.9: no. 4). This is the reason why the use of the term "pseudoplastic" is diminishing more and more in current literature.

Apparent shear viscosity
If the ratio of shear stress to shear rate varies with the shear load, the corresponding values are often called the "apparent shear viscosity" at the corresponding shear rate, to illustrate that these kinds of values are different from constant viscosity values of ideally viscous fluids (according to ASTM D4092 and DIN 1342-1). **Each one of these viscosity values obtained represents a single point of the viscosity function only.** Therefore, these viscosity values can only evaluated in the appropriate form if information is also given about the shear conditions. Examples for accurate specifications are as follows:
$\eta(\dot\gamma = 100s^{-1}) = 345mPas$, or $\eta(\tau = 500Pa) = 12.5Pas$

Note 1: Shear-thinning, time-dependent and independent of time
Sometimes, the term "shear-thinning" is used to describe time-dependent flow behavior at a constant shear load (see Figure 3.39: no. 2, Figure 3.40: left-hand interval, and Figure 3.43: medium interval). There is a difference between time-dependent shear-thinning behavior (see Chapter 3.4) and shear-thinning behavior which is independent of time (as explained in this section). If no other information is given, the term should be understood in common usage as the latter one.

Note 2: Very simple evaluation methods
1) Speed-dependent "viscosity ratio"(VR) and "shear-thinning index"
Some users still perform the following simple test and analysis method which consists of two intervals. In the first part a constantly low rotational speed n_1 [min^{-1}] is preset, and in the second part a constantly high speed n_2 (usually with $n_2 = 10 \cdot n_1$, for example, at $n_1 = 3min^{-1}$ and $n_2 = 30min^{-1}$). Afterwards the "viscosity ratio" (VR) is calculated as follows [55]:
$VR = \eta_1(n_1)/\eta_2(n_2)$.

Sometimes this ratio is called the "shear-thinning index" [91]. For ideally viscous (Newtonian) flow behavior VR =1, for shear-thinning VR > 1, and for shear-thickening VR < 1.

Example: with η_1 = 250mPas at n_1 = 3min[-1], and η_2 = 100mPas at n_2 = 30min[-1], results:
VR = 250/100 = 2.5

In order to avoid confusion, this ratio should better be called "speed-dependent (or shear rate-dependent) viscosity ratio". Sometimes in out-of-date literature, this speed-dependent ratio is named "thixotropy index", TI. But this term is misleading, since VR quantifies non-Newtonian behavior independent of time, but not thixotropic behavior which is a time-dependent effect. For TI, see also Note 3 in Chapter 3.4.2.2a, and Chapter 3.4.2.2c.

2) "Pseudoplastic index (PPI)"

Some users still use the following simple test and analysis method consisting of two intervals, presetting in the first part a constantly high rotational speed n_H [min[-1]] for a period of t_{10} = 10min, and in the second part a constantly low speed n_L for another 10min = t_{20} (e.g. when testing ceramic suspensions, with n_L = n_H/10, for example, at n_H = 100min[-1] and n_L = 10min[-1]). Afterwards the "pseudoplastic index" (PPI) is calculated as follows [99]:

PPI = [lg $\eta_L(n_L, t_{20})$ - lg $\eta_H(n_H, t_{10})$]/(lg n_L - lg n_H)

For ideally viscous (Newtonian) flow behavior PPI = 0, for shear-thinning (pseudoplastic) PPI < 0, and for shear-thickening (dilatant) PPI > 0.

Example: with η_H = 0.3Pas at n_H = 100min[-1], and η_L = 1.2Pas at n_L = 10min[-1], then:
PPI = (lg 1.2 - lg 0.3)/(lg 10 - lg 100) = [0.0792 - (-0.523)]/(1 - 2) = 0.602/(-1) ≈ -0.6

Please be aware that η-values are relative viscosity values if the test is performed using a spindle (which is a relative measuring system, see also Chapter 10.6.2). Here, instead of the shear stress often is taken the dial reading DR which is the relative torque value M_{rel} in %. Then, the viscosity value is calculated simply as η = DR/n (with the rotational speed n in min[-1]). Usually here, all units are ignored.

Thus, here: PPI = [lg (DR_L/n_L) - lg (DR_H/n_H)]/(lg n_L - lg n_H)

Example: with n_H = 100 and n_L = 10, and with DR_H = 50 and DR_L = 40, results:
PPI = [lg (40/10) - lg (50/100)]/(lg 10 - lg 100) = (lg 4 - lg 0.5)/(1 - 2)
PPI = [0.602 - (-0.301)]/(-1) ≈ -0.9

Comment: Both determinations, as well VR as well PPI are not scientific methods.

3.3.2.1 *Structures of polymers showing shear-thinning behavior*

The entanglement model

Example: A chain-like macromolecule of a linear polyethylene (PE) with a molar mass of M = approx. 100,000g/mol shows a length L of approx. $1\mu m$ = 10^{-6}m = 1000nm and a diameter of approx. d = 0.5nm (macromolecule: Greek "makros" means large) [119]. Therefore, the ratio L/d = 2000:1. Using an illustrative dimensional comparison, this corresponds to a piece of spaghetti which is 1mm thick – and 2000mm = 2m long! So it is easy to imagine that in a polmyer melt or solution these relatively long molecules would entangle loosely with others many times. As a second comparison: A hair with the diameter of d = $50\mu m$ showing the length L = 10cm. For an ultra-high molecular weight PE (UHMW) with M = 3 to 6 mio g/mol, then L/d = 50,000:1 to 100,000:1 approximately. Here, using the illustrative comparison, the piece of spaghetti would be approx. 50 to 100m long, and the hair with 2.5 to 5m would be permanently out of control!

At rest, each individual macromolecule can be found in the state of the lowest level of energy consumption: Therefore, without any external load it will show the shape of a three-dimensional coil (see Figure 3.8). Each coil shows an approximately spherical shape and each one is entangled many times with neighboring macromolecules.

During the shear process, the molecules are more or less oriented in shear direction, and their orientation is also influenced by the direction of the shear gradient. When in motion,

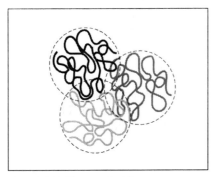

Figure 3.8: Macromolecules at rest, showing coiled and entangled chains

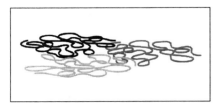

Figure 3.9: Macromolecules under high shear load, showing oriented and partially disentangled chains

the molecules disentangle to a certain extent which reduces their flow resistance. For low concentrated polymer solutions, the chains may even become completely disentangled finally if they are oriented to a high degree. Then, the individual molecules are no longer in the same close contact as before, therefore moving nearly independently of each other (see Figures 3.9 and 3.34: no. 2).

Using this concept, **the result of the double-tube test** can be explained now as follows (see Chapter 2.3, Experiment 2.2 and Figure 2.4): Fluid F1 shows shear-thinning and fluid F2 ideally viscous flow behavior. At the beginning, there is a certain load on the molecules at the bottom of each tube due to the weight of the column of liquid which is compressing vertically onto them due to the hydrostatic pressure. Regarding the polymer molecules of F1, shortly after the beginning of the Experiment they are moving faster because they are stretched into flow direction now. As a consequence, they are disentangled to a high degree then. Hence, they are able to glide off each other more easily also during the passage through the valve.

Along with the falling liquid levels, also the shear load or shear stress, respectively, is decreasing continuously. Therefore, the macromolecules are recoiling more and more, and as a consequence, the viscosity of F1 is increasing now. However F2, the mineral oil, still shows constant viscosity, independent of the continuously changing shear load, since for this ideally viscous liquid with its very short molecules counts, that there is no significant shear load-dependent change in the flow resistance.

Figure 3.10 presents the viscosity function of a polymer displaying three intervals on a double logarithmic scale. These three distinct ranges of the viscosity curve only occur for unlinked and unfilled polymers showing loosely entangled macromolecules. However, this does not apply for polymer solutions with a concentration which is too low to form entanglements,

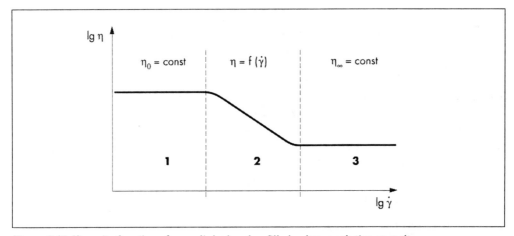

Figure 3.10: Viscosity function of an unlinked and unfilled polymer solution or melt

and also not for gels and paste-like dispersions exhibiting a network of chemical bonds or physical-chemical interactive forces between the molecules or particles. The three intervals are in detail:

(1) The first Newtonian range with the **plateau value of the zero-shear viscosity** η_0
(2) The shear-thinning range with the shear rate-dependent viscosity function $\eta = f(\dot{\gamma})$
(3) The second Newtonian range with the **plateau value of the infinite-shear viscosity** η_∞

In order to explain this, you can imagine a volume element containing many entangled polymer molecules. Of course, in each sample there are millions and billions of them.

a) Shear range 1 at low-shear conditions: the "low-shear range"

For many polymers, the upper limit of this range occurs around the shear rate $\dot{\gamma} = 1s^{-1}$. For some polymers, however, this limit can be found already at $\dot{\gamma} = 0.01s^{-1}$ or even lower.

Superposition of two processes

When shearing, a certain number of **macromolecules** are oriented into shear direction. For some of them, this results in partial **disentanglements**. As a consequence, viscosity decreases in these parts of the volume element. Simultaneously however, some other macromolecules which were already oriented and disentangled in the previous time interval are **recoiling and re-entangling** now again. This is a consequence of their visco-**elastic** behavior which the polymer molecules still are able to show under the occurring low-shear conditions. As a result, viscosity increases again in these parts of the volume element.

Summary of this superposition

In the observed period of time, the sum of the partial orientations and re-coilings of the macromolecules, and as a consequence, the sum total of all disentanglements and re-entanglements, **results in no significant change of the flow resistance related to the behavior of the whole volume element**. Therefore here, the sum of the viscosity decrease and increase results in a constant value. Thus finally, the total η-value in this shear rate range is still measured as a constant value, which is referred to as the zero-shear viscosity η_0. For unfilled and unlinked molecules, **zero-shear viscosity η_0 is occurring as a constant limiting value of the viscosity function towards sufficiently low shear rates** which are "close to zero-shear rate".

&⌒   For "Mr. and Ms. Cleverly"

1) Zero-shear viscosity in mathematical notation

Equation 3.1 $\eta_0 = \lim_{\dot{\gamma} \to 0} \eta(\dot{\gamma})$

Zero-shear viscosity is the limiting value of the shear rate-dependent viscosity function at an "infinitely low" shear rate (see also Chapter 6.3.4.1a: η_0 via creep tests, and Chapter 8.4.2.1a: η_0 via oscillation, frequency sweeps).

2) Dependence of η_0 on the polymer concentration c

For a polymer solution in the low-shear range, the η_0 value is attained only under the condition that the polymer concentration c [g/l] is high enough. Only in this case the molecule chains are able to get in contact with one another to form entanglements. For polymers having a constant molar mass and using the value of the critical concentration c_{crit}, the following applies (see Figure 3.11):

For $c < c_{crit}$: $\eta/c = c_1 = const$

Low-concentrated polymer solutions having no effective entanglements between the individual molecule chains display ideally viscous flow behavior, and this counts also for the low-shear range. In this case, the viscosity value is directly proportional to the concentration (with the material-specific factor c_1).

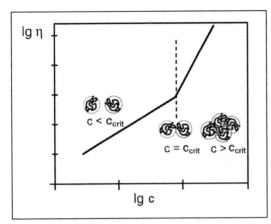

Figure 3.11: Polymer solutions: Dependence of low-shear viscosity on the polymer concentration (with the critical concentration c_{crit})

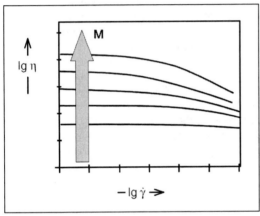

Figure 3.12: Polymer solutions and melts: Dependence of zero-shear viscosity on the average molar mass

For c > c_{crit}

Concentrated solutions of polymers and polymer melts showing effective entanglements between the individual molecule chains indicate a zero-shear viscosity value in the low-shear range. With increasing concentration, there is a stronger increase of the η_0-value as can be seen by the higher slope of the $\eta(c)$-function in this concentration range (see Figure 3.11).

3) **Dependence of η_0 on the average molar mass M** [18, 276]

For M < M_{crit}

Equation 3.2 $\eta / M = c_2 = $ const

with the material-specific factor c_2, the molar mass M [g/mol], and the critical molar mass M_{crit} for the formation of effective entanglements between the macromolecules.

Polymers with smaller molecules, therefore showing no effective entanglements between the individual molecule chains, display ideally viscous flow behavior. This counts also for the low-shear range), as illustrated by the bottom curve of Figure 3.12. In this case, viscosity is directly proportional to the molar mass.

For M > M_{crit}

Equation 3.3 $\eta_0 = c_2 \cdot M^{3.4}$

This relation is often associated with the name of *T.G. Fox* [37]. Polymers with larger molecules, therefore showing effective entanglements between the individual molecule chains, display a zero-shear viscosity in the low-shear range. The higher the molar mass, the higher is the plateau value of η_0. Proportionality of η_0 and M usually shows the exponent 3.4 (to 3.5). This value is approximately the same for all unlinked polymers (although it is possible to find also values between 3.2 and 3.9). Using the above relation, **it is possible to calculate the average molar mass from the η_0-value**, if the factor c_2 is known. Information about this factor is documented for almost all polymers.

Summary: The higher the average molar mass M, the higher is the η_0-value (see Figure 3.12).

Note: Different M_{crit} values for diverse polymers

M_{crit} is considered in physics a limiting value between materials showing a low molar mass and polymers. In order to make a rough estimate, often is taken here M_{crit} = 10,000g/mol. However, this value depends on the kind of polymer; see the following M_{crit}-values (in g/mol) [237]: polyethylene PE (4000); polybutadiene PB or butyl (5600); polyisobutylene PIB (17,000); polymethylmethacrylate PMMA (27,500); polystyrene PS (35,000)

☞   End of the Cleverly section

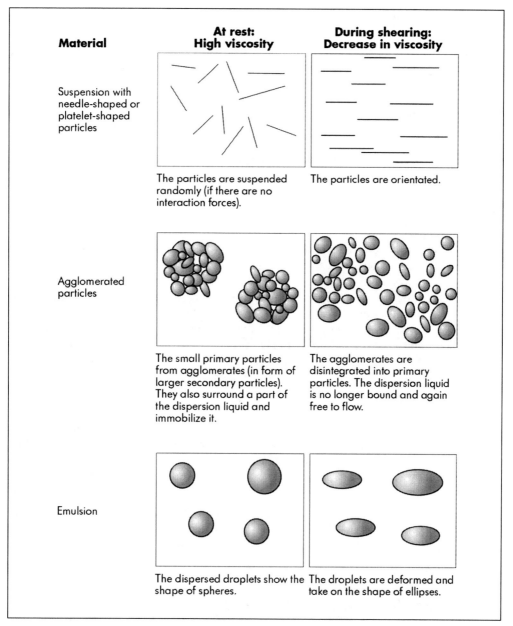

Figure 3.13: Structural changes of dispersions showing shear-thinning behavior

b) Shear range 2 at medium shear rates: the "flow range"

At increased shear rates, the number of disentanglements is more and more exceeding the number of the re-entanglements. As a consequence, the polymer shows shear-thinning behavior, and therefore, the curve of the viscosity function $\eta(\dot{\gamma})$ is decreasing now continuously.

c) Shear range 3 at high-shear conditions: the "high-shear range"

For polymer solutions the high-shear range may begin at around $\dot{\gamma} = 1000s^{-1}$, however, for some of them $\dot{\gamma} = 10,000s^{-1}$ has to be exceeded. Finally, all macromolecules are almost fully oriented and disentangled. Flow resistance is reduced to a minimum value now and cannot be decreased any further, corresponding to the friction between the individual disentangled molecules gliding off each other. Viscosity is measured then as a constant value which is referred to as the infinite-shear viscosity η_∞.

Infinite-shear viscosity is occurring as a constant limiting value of the viscosity function towards sufficiently high shear rates which are "close to an infinitely high shear rate".

  For "Mr. and Ms. Cleverly"

Infinite-shear viscosity in mathematical notation

Equation 3.4 $\eta_\infty = \lim_{\dot{\gamma} \to \infty} \eta(\dot{\gamma})$

Infinite-shear viscosity is the limiting value of the shear rate-dependent viscosity function at an "infinitely high" shear rate.

  End of the Cleverly section

Note: Degradation of polymer molecules
At these extreme shear conditions, there is always the risk of degradation of macromolecules. The polymer chains may be torn and devided into pieces, a process which might destroy their original chemical structure. However, this should be avoided when performing rheological tests. Using the previously highly sheared sample again, it is easy to check if degradation has really taken place by carrying out a second measurement at the same low-shear conditions of shear range 1. If the η_0-value remains still unchanged, no effective degradation has taken place. If the occurring η_0-value is significantly lower then, polymer molecules have been destroyed under the previously applied high shear load, thus, the value of the average molar mass is significantly lower now.

3.3.2.2 Structures of dispersions showing shear-thinning behavior

Besides polymers, also other materials may show shear-thinning behavior. For dispersions, a shear process may cause orientation of particles into flow direction. Shearing may also cause disintegration of agglomerates or a change in the shape of particles (see Figure 3.13). Usually then, the effectivity of interactive forces between the particles is more and more reduced, which results in a decreasing flow resistance.

Note: Observation and visualization of flowing emulsions using a rheo-microscope
Deformation and flow behavior of droplets in flowing emulsions at defined shear conditions can be observed when using a rheo-microscope or other rheo-optical devices (like SALS; see Chapter 10.8.2.2). Corresponding photographic images are shown e.g. in [45, 138, 390].

3.3.3 Shear-thickening flow behavior

Experiment 3.1: Shear-thickening of a plastisol dispersion (see Figure 3.14)

1) Placed in a beaker, a wooden rod inclines very slowly under its own weight through the dispersion.
2) Pulling the rod very quickly, causes the sample to solidify immediately.
3) Therefore then, the rod, the plastisol mass, and even the beaker are lifted all together now.

Viscosity of a shear-thickening material is dependent on the degree of the shear load. The flow curve shows an increasing curve slope (see Figure 3.15), i.e., viscosity increases with increasing load (see Figure 3.16).

Examples of shear-thickening materials
Dispersions showing a high concentration of solid matter or gel-like particles; ceramic suspensions, starch dispersions, paper coatings, plastisol pastes (containing not enough plasticizer),

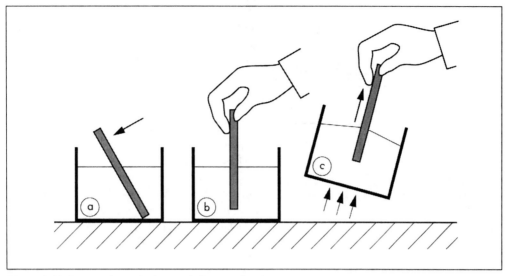

Figure 3.14: Testing shear-thickening behavior of a plastisol dispersion

natural rubber (NR), highly filled elastomers (such as polybutadiene or butyl rubber BR), dental composites, shock-resistant "smart fluids"

The terms **shear-thickening** and **dilatant** are identical in their meaning; sometimes the terms shear-hardening, shear-stiffening or solidifying can be heard. The note on the term **"apparent viscosity"** of Chapter 3.3.2 also applies here.

Problems with flow processes should always be taken into account when working with shear-thickening materials. Flow should be observed carefully for the occurrence of **wall-slip effects and separation of the material**, e.g. on surfaces of measuring systems, along pipeline walls or between individual layers of the sample. This can be investigated by repeating the test several times under identical measuring conditions, comparing the results with regards to reproducibility.

For dispersions, shear-thickening flow behavior should be taken into account
- at a high particle concentration
- at a high shear rate

Figure 3.17 presents the **dependence of viscosity on the particle concentration** (here with the **volume fraction** Φ). For $\Phi = 0$ (i.e. pure fluid without any particle), the liquid shows ideally viscous flow behavior. For $0.2 \leq \Phi \leq 0.4$, the suspension displays shear-thinning, particularly in the low-shear range. For $\Phi > 0.4$, on the one hand there is shear-thinning in the low-shear range,

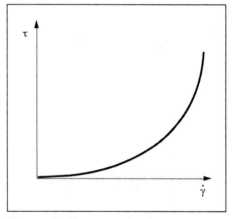

Figure 3.15: Flow curve of a shear-thickening liquid

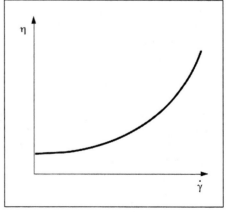

Figure 3.16: Viscosity curve of a shear-thickening liquid

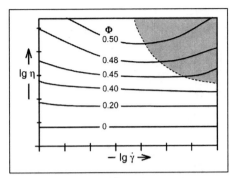

Figure 3.17: Viscosity functions of dispersions: Dependence on the particle concentration (with the volume concentration Φ)

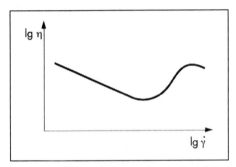

Figure 3.18: Viscosity curve of a shear-thickening material showing a "dilatancy peak" at a high shear rate

but on the other hand occurs shear-thickening in the medium and high-shear range. At higher concentrations, the range of shear-thickening behavior is beginning at lower shear rate values already.

Shear-thickening materials are much less common in industrial practice compared to shear-thinning materials. Nevertheless, **shear-thickening behavior** is desirable for special applications and is therefore encouraged in these cases (example: dental masses). Usually, however, this behavior is undesirable and should never be ignored since it **may lead to enormous technical problems and in some cases even to destruction of equipment, e.g. pumps or stirrers**.

Note 1: Shear-thickening, time-dependent and independent of time

Sometimes, the term "shear-thickening" is used to describe time-dependent flow behavior at a constant shear load (see Figure 3.39: no. 3; and Figure 3.41: left-hand interval). There is a difference between time-dependent shear-thickening behavior (see Chapter 3.4) and shear-thickening behavior which is independent of time (as explained in this section; see also Note 1 in Chapter 3.3.2). If no other information is given, the term should be understood in common usage as the latter one.

Note 2: Dilatancy peak

Sometimes with highly concentrated dispersions, shear-thickening does not occur until higher shear rates are reached. If this behavior is presented in a diagram on a logarithmic scale, the viscosity curve often shows initially shear-thinning behavior up to medium shear rates before a "dilatancy peak" is occurring at higher shear rates finally (see Figure 3.18).

Example 1: Plastisols in automotive industry

A PVC plastisol – as a paste-like micro-suspension – showed shear-thinning behavior in the range of $\dot{\gamma} = 1$ to $100s^{-1}$, and then a **dilatancy peak** at $\dot{\gamma} = 500s^{-1}$. This may cause problems when it is sprayed as an automotive underbody coating or seam sealing (e.g. by blocking the flatstream spray nozzle).

Example 2: Paper coatings

Several highly concentrated preparations of paper coatings (suspensions showing a pigment content of about 70 weight-%) displayed – after a certain shear-thinning range – shear-thickening behavior beginning at around $\dot{\gamma} = 1000s^{-1}$, often followed by a **dilatancy peak**. The peak was found to be shifted towards higher shear rate values for coatings showing a lower pigment concentration. Shear-thickening may cause problems during a coating process leading to **coating streaks** including the danger of tearing the paper.

Example 3: Hair shampoos containing surfactant superstructures

In the range of $\dot{\gamma} = 1$ to $15s^{-1}$ a shampoo displayed shear-thinning behavior and at $\dot{\gamma} = 30s^{-1}$ occurred a **dilatancy peak**. The shampoo manufacturer wanted the shampoo to show a certain superstructure in this shear rate range, which corresponds to the process when flowing out of the shampoo bottle. Consumers then subconsciously believe that they have bought a liquid showing "body". In this case, the goal was reached via superstructures, built up by viscoelastic surfactants (VES; see also Chapter 9.1.1d: worm-like micelles).

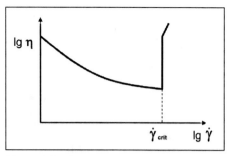

Figure 3.19: Viscosity curve of a highly filled suspension showing immediate shear thickening when reaching the critical shear rate $\dot{\gamma}_{crit}$ ("dilatant switch")

Note 3: Temperature-dependent shear thickening of elastomers

Different temperatures can lead to fundamentally different kinds of flow behaviors of filled and unfilled elastomers and rubber mixtures.

Example 4: Shear-induced dilatant behavior of elastomers

A filled elastomer showed shear-thinning behavior at T = +80°C across the whole shear rate range of $\dot{\gamma}$ = 0 to 100s^{-1}. At T = +40°C it first indicated shear-thinning, followed by shear-thickening when exceeding $\dot{\gamma}$ = 50s^{-1}. At T = +23°C finally, already at around $\dot{\gamma}$ = 10s^{-1} pronounced shear-thickening was occurring. Possible reason: **At low temperatures, shear-induced crystallization** can be expected for this material. A result of this is **shear thickening** and **hardening**,[398]. See also Chapter 9.2.2: Shear-induced effects.

Note 4: Composite materials as a "dilatant switch"

Shear-thickening fluids (STF) were developed with the aim of immediate thickening as soon as a defined limiting value of loading is exceeded.

Example 5: "Nano-fluids" for shock-resistant or bullet-proof materials

These aqueous dispersions are mixtures of PEG (polyethylene glycol) and colloidal silica particles (60 weight-%, with d = 400nm, monodispersely distributed). After initial shear-thinning behavior showing η = 100Pas at $\dot{\gamma}$ = 0.001s^{-1} and η = 2Pas at $\dot{\gamma}$ = 40s^{-1}, the viscosity value at the **"critical shear rate value"** of $\dot{\gamma}_{crit}$ = 50s^{-1} immediately steps upwards to η > 500Pas; see Figure 3.19. At this **limiting value of the loading velocity**, the silica particles are agglomerating, abruptly forming a rigid "hydro-cluster" due to interparticle interactions. **Applications: Shock-proof, stab-proof and bullet-proof protective clothing** as a combination of this "nano-dispersion" with synthetic technical textile fabrics; reinforced technical polymers for special functions (e.g. as "nano-composite STF-Kevlar")[226].

For "Mr. and Ms. Cleverly"

Increased flow resistance due to flow instabilities and turbulences

Increased flow resistance can also occur due to hydrodynamic flow instabilities which may lead to **secondary flow effects** and even to **turbulent flow behavior** showing vortices at high shear rates. In this case, flow and viscosity curves will display as well higher shear stress and viscosity values as well as higher curve slope values compared to curves measured at regular (i.e. laminar) flow conditions, therefore giving at the first glance an impression of shear-thickening behavior.

When performing tests on liquids using concentric cylinder measuring systems with a rotating inner cylinder (Searle method, see Chapter 10.2.1.2a) there is a critical upper limit between laminar and turbulent flow conditions in the circular gap. Exeeding this limit, secondary flow effects may occur for the reason of centrifugal or inertial forces due to the mass of the fluid. The critical limiting value can be calculated in the form of a **Taylor number (Ta)**. The range of turbulent flow is also reached when the critical **Reynolds number (Re)** is exceeded. Re numbers represent the ratio between the forces of inertia and flow resistance. (More about Ta and Re number: see Chapters 10.2.2.4 and 11.3.1.3.)

Example 6: Turbulent flow of water
Water was measured at different temperatures using a double-gap measuring system. The limiting value of the shear rate range of ideally viscous flow behavior was found at

$\dot{\gamma} = 1300s^{-1}$ at T = +10°C showing η = 1.3mPas
$\dot{\gamma} = 1000s^{-1}$ at T = +20°C showing η = 1.0mPas
$\dot{\gamma} = 800s^{-1}$ at T = +30°C showing η = 0.80mPas

In each viscosity curve at the mentioned upper limit of the shear rate a clear bend was observed, followed by a distinctive increase in the slope of the viscosity curve, indicating the begin of the turbulent flow range.

Note 5: Observation and visualization of turbulent behavior
Using a special measuring device, flow behavior of dispersions at defined shear conditions can be observed simultaneously as well in the form of a measured viscosity function as well as visually, for example, in order to observe the onset of vortex formation. This process can be recorded via digital photography or video, measuring point by measuring point. (See also Chapter 10.8.2.7: Rheo-optics, **velocity profile of a flow field**).

     End of the Cleverly section

Note 6: Daniel wet point (WP), flow point (FP), and dilatancy index [86, 281]
The Daniel WP and FP technique used for millbase premix pigment pastes (pigment powder and vehicle), dispersions, paints and other coatings with a high pigment concentration, is **a simple hand-mixing method** for characterizing two consistency stages in the take-up of vehicle (mixture of solvent and binder) by a bed of pigment particles. WP is defined as the stage in the titration of a specified amount of a pigment mass (e.g. 20g) with vehicle, where just sufficient vehicle as incorporated by vigorous kneading with a glass rod or a spatula is present to form a soft, coherent paste-like mass showing a putty-like consistency. FP is determined by noting what further vehicle is required to produce a mixture that just drops, flows or falls off under its own weight from a horizontally held spatula. Between WP and FP, the mass hangs on a spatula with no sign of flow. The unit of WP and FP is volume of vehicle per mass (weight) of pigment [cm^3/g].

"Daniel dilatancy index" (DDI) is defined as DDI (in %) = [(FP – WP)/WP] · 100%. This is the proportion of the additional vehicle required to reach the FP from the WP. A DDI of 5 to 15% is considered strongly dilatant, does not disperse well although fluid, showing no tack; a DDI = 15 to 30% is considered moderately to weakly dilatant, an excellent dispersion, showing some tack; and a DDI > 30% is considered substantially non-dilatant, a dispersion obtained but with difficulty, showing tacky behavior.

Comment: These three test methods WP, FP and DDI are not scientific since this is a very simple and manually performed technique, and the result depends on the subjective evaluation of the testing person. Even for a given pigment mixture as well WP as well FP obtained may vary significantly if the same pigment paste is used.

3.3.3.1 Structures of polymers showing shear-thickening behavior

Shear-thickening flow behavior may occur when shearing highly concentrated, chemically unlinked polymer solutions and melts due to mechanical **entanglements between the molecule chains**, particularly if they are branched and therefore often relatively rigid. The higher the shear load (shear rate or shear stress, respectively), the more the molecule chains may prevent relative motion between neighbored molecules.

3.3.3.2 Structures of dispersions showing shear-thickening behavior

Usually with highly filled suspensions during a process at increasing shear rates, the particles may more and more come into contact to one another, and particularly softer and gel-like particles may become more or less compressed. In this case, flow resistance will be increased. Here, the particle shape plays a crucial role. Due to the shear gradient which occurs in each flowing liquid, **the particles are rotating as they move** into shear direction [128, 276]. Even **rod-like particles and fibers** are showing now and then rotational motion (photographic images e.g. in [138]).

Illustration, using the Two-Plates-Model (see Figure 2.1)
Rotation of a particle occurs clockwise when using a Two-Plates-Model with a stationary lower plate and the upper plate moving to the right. **Cube-shaped particles** are requiring of course more space when rotating compared to the state-at-rest. As a consequence, between the particles there is less free volume left for the dispersion liquid. On the other hand, **spherical particles** require the same amount of volume when rotating or when at rest; these kinds of dispersions are less likely to show shear-thickening. A material's ability to flow can be improved by increasing the amount of free volume available between the particles. This can be achieved by changing the shape of the particles, – and of course also by adding more dispersion liquid.

Note 1: Droplet subdivision when testing emulsions
When shearing emulsions, with increasing shear rates sometimes sloping up of the viscosity function can be observed. This may be assumed to be an indication of shear-thickening behavior. However, this effect is often occurring due to a reduction of the average droplet size, caused by **droplet subdivision** during a continued dispersing process due to the shear forces. Here, corresponding **increase of the volume-specific surface** (which is the ratio of droplet surface and droplet volume) and, as a consequence, the resulting increase in the interactions between the now smaller droplets may lead to higher values of the flow resistance (more on emulsions: see also Chapter 9.1.2 and [217, 220]).

Example: "Creaming" of pharmaceutical or cosmetic products
The "creaming effect" is a result of this continued dispersion process. When spreading and rubbing corresponding emulsions such as **creams, lotions and ointments** on the skin, a "whitening effect" may occur which is often leading to **tacky** and even **stringy behavior**, therefore causing of course an **unpleasant skin sensation**.

Note 2: Observation and visualization of flowing emulsions using a rheo-microscope
Using special measuring devices, flow behavior of emulsions at defined shear conditions can be observed simultaneously as well in the form of the measured viscosity function as well as visually, for example, in order to observe the onset of breaking up the droplets. This process can be recorded via digital photography or video, measuring point by measuring point. (See also Chapters 10.8.2.2 and 10.8.2.4: Rheo-optics, microscopy and SALS).

3.3.4 Yield point

Experiment 3.2: Squeezing toothpaste out of the tube (see Figure 3.20)
A certain amount of force must be applied before the toothpaste starts to flow.

A sample with a yield point begins to flow not before the external forces F_{ext} acting on the material are larger than the internal structural forces F_{int}. Below the yield point, the material shows elastic behavior, i.e. it behaves like a rigid solid, exhibiting under load only a very small degree of deformation, which however recovers completely after removing the load. The

Figure 3.20: Toothpaste – our daily struggle with the yield point

following applies: If $F_{ext} < F_{int}$, the material is deformed to such a small degree only that it is hardly perceptible to the human eye. The sample does not begin to flow before $F_{ext} > F_{int}$. The yield point is also referred to as **yield stress** or **yield value**.

Experiment 3.3: The rods in hand cream and silicone
Small rods are put into the following two materials, in order to observe their motion.
a) Hand cream: The rod remains standing straight in the cream, thus here, a yield point exists.
b) Silicone polymer (unlinked PDMS): The rod moves very slowly to the side, i.e., although the silicone displays a very high flow resistance it shows no yield point. It exhibits behavior of a highly viscous, viscoelastic liquid indicating a high zero-shear viscosity in the low-shear range (see also Chapter 3.3.2.1a).

Examples of materials which may show a yield point
Gels, dispersions with a high concentration of solid particles such as plastisol pastes, conductor pastes (electrotechnics), toothpastes, sealants, putties, emulsion paints, printing pastes, ceramic masses, lipsticks, creams, ketchups, mayonnaises, chocolate melts, margarines, yogurts; semi-solid materials, concentrated surfactant systems

Yield points have great importance for practical users, and therefore, various methods for the acquisition of appropriate measurement values have been developed over the years – with quite a lot of creativity, obtaining more or less useful results (see also Chapters 3.3.6.4a, 11.2.3d, 11.2.4a/c, 11.2.6d/e, 11.2.7c, 11.2.8b2, 11.2.9, 11.2.11a/e/i). Sometimes, there is distinguished between tests to determine "apparent" or "really existing" yield points (for more information on this discussion, see also [18]). Materials with a yield point often show "plastic behavior", they tend to flow inhomogeneously, and then, wall-slip effects should be taken into account (see also Chapter 3.3.4.3d).

3.3.4.1 Yield point determination using the flow curve diagram

a) With controlled shear rate (CSR): Yield point calculation via curve fitting models
Here, rotational speeds (or shear rates, resp.) are preset in the form of steps or as a ramp (see Figures 3.1 and 3.2). However, using this kind of testing, a yield point cannot be determined directly. Therefore, it is calculated by use of a fitting function which is adapted to the available measuring points of the flow curve. Curve fitting is carried out using one of the various model functions, e.g. according to Bingham, Casson or Herschel/Bulkley (see Chapter 3.3.6.4). For all these approximation models, the yield point value τ_y is determined by extrapolation of the flow curve towards the shear rate value $\dot{\gamma} = 0$, or at the intersection point of the fitting function and the τ-axis, respectively (as described in the meanwhile withdrawn DIN 53214). The different model functions usually produce different yield point values because each model uses a different basis of calculation. Today, **this method should only be used for simple QC tests** but

no longer for modern research and development work since a yield point value obtained in this way is not measured but merely calculated by a more or less exactly fitting approximation.

b) With controlled shear stress (CSS): Yield point as the stress value at the onset of flow

This is the "classic" method for the determination of a yield point: When increasing the shear stress with time in the form of steps or as a ramp (similar to Figures 3.1 and 3.2), the shear stress value is taken as the yield point, at which the measuring device is still detecting no sign of motion. This is the last measuring point at which the rotational speed n (or shear rate $\dot\gamma$, resp.) is still displayed as n = 0 (or as $\dot\gamma$ = 0, respectively). The yield point value occurs as an intersection on the τ-axis when plotted on a linear scale (see Figure 3.21). If presented on a logarithmic scale, the yield point is the τ-value at the lowest measured shear rate, e.g. at $\dot\gamma$ = 1 or 0.1 or 0.01s^{-1} (see Figure 3.22).

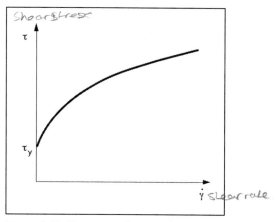

Figure 3.21: Flow curve showing a yield point (on a linear scale)

Summary

Using the flow curve analysis methods mentioned above, the resulting yield point is dependent on the speed resolution of the viscometer or rheometer used. An instrument which can detect lower rotational speeds (e.g. n_{min} = 10^{-4}min^{-1}) will display a lower yield point value compared to a device which cannot detect such low minimum speeds (e.g. displaying n_{min} = 0.5min^{-1} only). Of course, the latter device cannot detect any motion below its measuring limits, therefore still evaluating any speed in this range as n = 0. As a result, **a lower value of the yield point will be obtained by the more sensitive instrument**. This can be illustrated clearly when presenting flow curves on a logarithmic scale (see Figure 3.22): The lower the smallest shear rate which can be detected, the lower is the corresponding shear stress. Therefore, counts the following: **A yield point is not a material constant since this value is always dependent on the options of the measuring instrument used.**

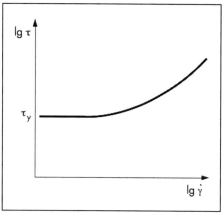

Figure 3.22: Flow curve showing a yield point (on a logarithmic scale)

For this reason, the two methods (a and b) mentioned before should only be taken for simple quality assurance tests, thus, just for a rough estimation of yield point values. For users in R & D, however, modern methods are recommended as explained in the next Chapter or, even better, in Chapter 8.3.4: Determination of as well the **yield point** as well as the **flow point** (oscillatory tests, amplitude sweeps).

3.3.4.2 Yield point determination using the shear stress/deformation diagram

a) Yield point at the limit of the linear-elastic range, using a single fitting line

Preset is a controlled shear stress function (similar to Figure 3.1 or 3.2). The steps or the ramp, respectively, should begin at shear stress values at least one decade below the assumed yield point and end at least one decade above that point [101]. The yield point is the shear stress value at which the linear-elastic deformation range is exceeded. The measuring points are usu-

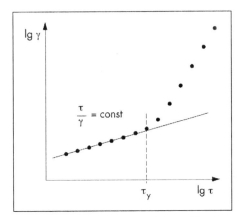

Figure 3.23: Determination of the yield point τ_y at the limit of the linear-elastic deformation range, using a single straight fitting line in the logarithmic tau-gamma diagram

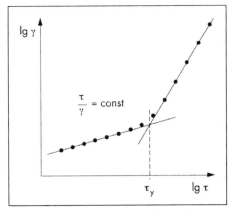

Figure 3.24: Determination of the yield point τ_y using the "tangent crossover point method" in the logarithmic tau-gamma diagram

ally presented on a logarithmic scale (see Figure 3.23), with the shear stress τ [Pa] on the x-axis and the shear deformation γ [%] on the y-axis, as a logarithmic stress/deformation or "tau/gamma diagram" (see also Chapter 4.2.1: Definition of the shear deformation γ).

The rising measuring curve shows a constant slope in the range of low values of τ and γ. For the analysis, a straight line is fitted in this curve interval based on the following consideration: In this first interval, the sample shows linear-elastic deformation behavior. Sometimes, this line is also called a "tangent", and correspondingly, the procedure is termed the "tangent method". When properly speaking however, tangents are usually adapted to curves and not to points of straight lines. Hooke's elasticity law applies here since the values of τ and γ are increasing proportionally. In this deformation range, these kinds of samples behave like homogeneously deformable, gel-like or soft solids. (Detailed information on elastic behavior and Hooke's law can be found in Chapter 4.)

Summary
The yield point is the one shear stress value at which the range of the reversible elastic deformation behavior ends and the range of the irreversible deformation behavior begins. Then, viscoelastic or viscous flow is occurring.

A sample's structure is already deformed below the yield point, however, this deformation is only very small. The structure would completely reform after removing the load as long as the yield point has not been exceeded. In analysis, usually by use of a software program, the τ-value is specified as the one point just before the measuring curve deviates significantly from the straight fitting line (see Figure 3.23). Before starting the analysis, the user has to define the bandwidth of the tolerated deviation (e.g. as 5 or 10%).

When using this method, the very small deflection of the measuring bob in the range below the yield point can only be detected by very sensitive **instruments showing a high resolution** for the values of torque and deflection angle (or rotational speed, respectively). Therefore it is senseful to use here an **air-bearing rheometer** (see also Chapter 11.6.5b). This method should be selected for scientific experiments to be on the safe side to get the yield value still in the **reversible elastic deformation range**. Some examples of tests results obtained with ketchups and coatings are shown in [251].

b) Yield point by the "tangent crossover method", using two fitting lines
Also here, the measuring points are presented in a logarithmic stress/deformation or "tau/gamma diagram" on a logarithmic scale (see Figure 3.24). The first line is fitted in the curve interval of the linear-elastic range at low deformation values as explained above. A second fitting line is adapted in the flow range, i.e., in the curve interval showing the high values of

τ and γ. This straight line in the flow range indicates clearly a higher slope value compared to the first one. The shear stress value at the crossover point of the two lines is taken as the yield point τ_y. Sometimes, instead of a second (linear) straight line a non-linear fitting function is chosen to be fitted to the measuring points in the flow range, for example, in the form of a curved polynomial function.

Please note: Using this method, the yield point value determined may occur already in the range of viscoelastic or viscous flow behavior (yield zone, see the following Chapter 3.3.4.3a). A certain part of the sample's deformation would remain permanently if the up to then increasing shear stress were removed within that range, and the internal structure at rest would have been changed already irreversibly. If this should be avoided in order to be for sure on the safe side, then it is better to use the method with a single fitting line only, as explained above.

c) Yield point by the method of maximum deviation from the fitting line
Here, a straight line is fitted through the **entire** range of measuring points of the logarithmic stress/deformation diagram or "tau/gamma diagram". Then the location is determined at which the distance between the measuring points and this straight line reaches a maximum value. The corresponding shear stress value is then taken as the yield point τ_y. An advantage of this method is that also the whole deviation function can be presented versus the shear stress τ showing the distinctness of the bend of the logarithmic tau-gamma function. Another advantage is that several bends may be determined if occurring.

Note: Yield point and flow point
A further method for determining yield points and **flow points**, respectively, is explained in Chapter 8.3.4 (oscillatory tests, amplitude sweeps). An overview on even further methods which might be used for yield point determinations is given in [101].

3.3.4.3 Further information on yield points

a) Yield zone
The transition between the elastic deformation range and the flow range is often not occurring as a clear bend but merely as a gradual change in the slope of the $\gamma(\tau)$-curve. In this case it is better to speak of a yield zone or a yield/flow transition range and not of a single yield point value; for more information see also Chapter 8.3.4.3.

b) Time-dependence of the yield point
Yield point values depend on the duration of the test. With each new measuring point at the beginning of each new step on the stress ramp, the structure of the sample is stretched at first under the applied constant shear load. As a consequence, a constant, steady-state value of the shear deformation or shear rate is resulting – but only after a certain delay. To avoid this time-dependent **start-up effect** which is also called **transient effect**, the user should wait sufficiently long at each measuring point (see also Chapter 3.3.1b and Figure 2.9: no. 5). For samples showing clearly time-dependent behavior, differing measuring times for an otherwise identical preset test profile may result in different yield point values. A yield value also depends on the **sample preparation** before the test (e.g. concerning shear load, time effects, temperature).

Summary
A yield point is not a material constant. Since it is time-dependent, it depends on the conditions during the preparation of the sample as well as on the test conditions.

Note: Structural strength at rest and frequency sweeps
In order to investigate long-term storage stability of dispersions, frequency sweeps (oscillatory tests) should be preferred since these kinds of tests take best into account the influence of

time when determining "structural strength" or "consistency" at rest. Here, when presetting frequencies, in principle time-dependent results are obtained since frequencies are "inversed times" (see also Chapter 8.4.4).

&ↄ   For "Mr. and Ms. Cleverly"

c) Interaction forces and network of forces

Dispersions and gels are showing yield points due to **intermolecular forces (Van-der-Waals forces)**. This includes dipole-dipole interactions between particles, and between particles and the surrounding dispersion agent. There are different kinds of interaction forces: Electrostatic interactions between permanent dipoles (Keesom forces), interactions due to induction between permanent and induced dipoles (Debye forces), and dispersion forces between mutually induced dipoles (London forces) [220, 246]. They are all based on **physical-chemical bonds (secondary bonds)** between the molecules and have a considerably lower bonding energy, usually below 20kJ/mol, compared to the **chemical primary valency bonds** which are acting within the molecules. These primary bonds are covalent electron-pair bonds, ionic or metallic "electron gas" bonds which are usually showing energy values of 50 to 400, and maximum values of 1000kJ/mol [128, 397]. Bonds via intermolecular **hydrogen bridges** are an exceptional type of physical-chemical bonds. In large numbers, however, they can have a great effect on rheological behavior.

Interactions may build up a **three-dimensional network of forces**. In the low-deformation range, this network occurs as a stable and solid-like structure resulting in elastic behavior ("gel-like character", see also Chapter 8.3.2a).

d) Plastic behavior

DIN 1342-1 states the following: "For a plastic material, rheological behavior is characterized by a yield point." And: "Plasticity is the ability of a material to show remaining deformation (and flow) only if the yield point is exceeded. Below the yield point occurs no or only elastic deformation." Further: "A deformable material is called plastic if it behaves in the range of low shear stresses as a rigid, elastic or viscoelastic solid, in a higher shear stress range however, as a liquid. The shear stress value at which the transition takes place is called the yield point (or yield stress)." Sometimes further terms can be found in literature such as "plastic deformation", "plastic creep", or "plastic flow".

In 1916, *Eugen C. Bingham* (1878 to 1945) described the behavior of dispersions showing a yield point [42]. He called this behavior to be "plastic". **Ideally plastic behavior** was illustrated using the **Saint Venant model** (*A.J.B. de Saint Venant*, 1797 to 1886 [314]); see also DIN 1342-1. Drawings of this model are shown in the meanwhile withdrawn DIN 13342 of 1976, and in [276, 303]. This model consists of a **friction element**, which does not begin to move until the shear force overcomes the resistance caused by static friction. Then the structure of the stressed sample yields, showing an increasing deformation (creep or flow), but the motion is still slowed down by the sliding friction of the friction element. In the past, various rheologists designed a lot of different concepts to explain the behavior in the transition range between rest and flow:

1) Some rheologists assumed that a material remained completely rigid and undeformed under increasing shear load until the yield point was exceeded. This behavior was called **"plastic-rigid"** or **"inelastic"** [18]. Above the yield point the material showed "plastic flow" and finally, under higher shear load, "viscous flow". This behavior was described in 1919 using the Bingham model [42] (see also Chapters 3.3.6.4a and 13.3: 1916). Both steps of this behavior were termed **"viscoplastic"** (as in the redrawn DIN 13342) and [303], or as **"plastic-viscous"** [276]. After the load is removed, no reformation occurs at all.

2) Other rheologists were convinced that the sample under increasing shear load first showed reversible elastic behavior in a very limited deformation range until the yield

point was exceeded. Then an irreversible plastic deformation occurred. This behavior was described in 1924 by the Prandtl model [303], or in 1930 by the Prandtl/Reuss model (DIN 1342-3 and [276], see also Chapter 13.3). Both stages of that behavior were termed **"elasto-plastic"** (withdrawn DIN 13342), as **"elastic-plastic"** [276], or as **"plastoelastic"** [382]. The extent of reformation after removing the load represents the elastic portion. Using modern terms, these kinds of materials should be called viscoelastic liquids.

3) Another concept was that under a low shear load below the yield point the sample was deformed reversible-elastically (Hookean behavior). After exceeding the yield point, the material showed plastic behavior (slow flow, slowed down by the friction element according to Saint Venant), and finally, under increased shear load it showed viscous flow (Newtonian behavior, without any effect of the friction element). This behavior was described using an extended Bingham model [303], and all three stages were named **"elastico-plastico-viscous"** or **"elasto-visco-plastic"** or similar terms were used [155, 352]. The extent of reformation after removing the load corresponds to the elastic portion. Using modern terms, these kinds of materials should be called either viscoelastic liquids if there is only partial reformation even after a sufficiently long period of time, or viscoelastic solids if they are recovering completely.

For people working scientifically, the term "plastic behavior" used in rheology is a synonym for "inhomogeneous behavior". For practical users performing rheological tests, the following can be stated: **Plastic behavior is shown by materials which do not exhibit homogeneous shear behavior, related to the entire shear gap. These kinds of materials often display effects like wall-slip, sliding and plug flow, when forced to move through capillaries, tubes and pipes.** Similar effects may also occur in the gap of a rheometer measuring system. For these kinds of samples, rheological behavior is not constant throughout the whole volume of the test material: **a part of the sample is flowing, and another one does not** [233]. Inhomogeneous shear effects should always be expected when testing dispersions showing a high concentration of solid matter since here, **phase separation** of the sample may occur.

Example 1: Plastic behavior of metals for cold forging processes
Plastic deformation occurs in **cold forging processes of metals**, or with other crystal-forming materials, if **lattice dislocation** takes place between the atomic levels or crystals [160, 198]. Below the yield point, there is elastic and reversible deformation behavior. Above the yield point, however, behavior is irreversible and inelastic and then, deformation is no more homogeneously distributed throughout the entire shear gap [220].

Example 2: Plastic deformation of the landscape and soil flow (solifluction)
In the Ice Age (which, for example, came to an end in Northern Germany around 14,000 years ago), the shear force of glaciers moved loose layers of soil masses and pieces of solid rock, e.g. boulders, weighing several tons. The so-called "solifluction" of the partially thawed permafrost soil resulted in plastic deformation of the landscape.

Example 3: Plastic land subsidence caused by mining or dike construction
As a result of mining, land subsidence may occur when loosened sediment layers meet ground water. Here, and also in the construction of water protection dikes, the soil-mechanical, rheologically plastic behavior is a crucial point (creep, sliding, flow).

Example 4: Plastic flow of debris and mud avalanches in the mountains
Debris and mud avalanches are an inhomogeneous mixture of water, fine sediment (max. particle size up to d = 0.1mm), mud (clay, silt, sand, with 0.1mm < d < 20mm), and larger "particles" (granules, gravel, stones, boulders, with max. d = 20cm or even 1m, - which ideed was a little bit too large for a normal rheometer ...). They may show mean velocities of v = 1 to 30m/s, thus, more than 100km/h. Considering a longitudinal section, two zones of the velocity distribution occur: In the lower range **close to the bottom a shear zone showing clearly increasing**

velocity values from the bottom upwards (i.e. flowing layers showing a velocity gradient), **and above a zone close to the surface of almost constant velocity** (i.e. v = const, showing **"plug flow"**) [319]. For corresponding rheological tests of dipersions showing particel size up to 10mm, a ball measuring system can be used (see Chapter 10.6.5).

Illustration, using the Two-Plates-Model (see Figure 2.9: no. 4)
Here, not all flowing layers in the Two-Plates-Model are shifting along one another to the same extent. Therefore in the shear gap, the resulting deflection and velocity, or shear gradient, shear deformation and shear rate, respectively, are not constant. Only some layers are in motion and others remain still at rest. After removing the load, any re-formation of the layers depends on the elastic portion. Related to the entire gap this re-formation is also inhomogeneously. In comparison to this, Figures 2.1, 2.9 (no.2) and 4.1 exhibit the behavior desired by rheologists at homogeneous flow and deformation conditions.

Examples of plastic materials
Solids or dispersions with a high concentration of solids showing high interactive and cohesive forces; i.e. plasticine (see Experiment 1.1b of Chapter 1.2), wax, a piece of soap, sealants, sludges, loam, mocha coffee grounds, plasters, surfactant systems showing shear-banding (see Chapter 9.2.2).

When performing rheological experiments in a scientific sense, however, it is assumed that there are homogeneous shear conditions in the entire shear gap. Only then, a constant shear rate or shear deformation can be expected to occur during the whole shearing process and after removing the load, and viscous, viscoelastic or elastic behavior. And only under these preconditions, the behavior can be described formally and mathematically by the relations according to Newton, Hooke, Maxwell, Kelvin/Voigt and Burgers.

Summary
Plastic behavior cannot be described unequivocally using the usual scientific fundamentals of mathematics and physics since it is an **inhomogeneous behavior**. Therefore, it can only be presented in terms of relative values obtained from empirical tests.
Unfortunately, in many industrial laboratories the terms **plastic, ideally plastic, viscoplastic or elastoplastic** still are used to mean a lot of different things. It is useful to understand what these terms might mean, but their use should be avoided when performing and analyzing scientific rheological tests. Within a limited deformation range, in most cases, samples can be characterized as **viscoelastic** (DIN 13343). However, **if a material cannot be sheared homogeneously it is often necessary to use special relative measuring systems** (see Chapter 10.6). **In this case, it is better to work only with the measured raw data such as torque, rotational speed and deflection angle,** instead of any rheological parameter such as shear stress, shear rate, shear deformation, viscosity, and shear modulus.

Note by the way
Both, the plastic surgeon and the sculptor (working with wood, stone or plaster to create plastic sculptures), are performing usually inhomogeneous and irreversible deformation processes, related to the structure of the material as a whole. Hopefully, the patients, operated beauties and lovers of fine arts are pleased with the end-products of these "plastic deformations".

e) Practical example: Yield point and wet layer thickness of a coating
Using the yield point value, it is possible to make a simple, rough estimation of the wet layer thickness on a vertical wall (see Figure 3.25).

The following holds: $\tau = F/A$, with the area to be coated: $A = b \cdot c$,
and the weight force of the layer due to gravity: $F = F_G = m \cdot g = V \cdot \rho \cdot g$
with the mass m [kg] and the volume V [m^3] = $a \cdot b \cdot c$ of the volume element,
the gravitation constant g = 9.81m/s^2, and the density ρ [kg/m^3] of the coating,

it follows that: $F_G = a \cdot b \cdot c \cdot \rho \cdot g$

Result: $\tau = \tau_y = F_G/A = (a \cdot b \cdot c \cdot \rho \cdot g)/(b \cdot c) = a \cdot \rho \cdot g$

Summary

A coating layer with the layer thickness "a" remains on the wall only then, if the limiting value of the shear stress between the states of rest and flow is not exceeded. This τ-value is the yield point τ_y [Pa]. As the calculation shows, the yield point is independent of the layer width and length. Thus, the wet layer thickness of a coating which remains on a vertical wall can be calculated as:

Equation 3.5 $a = \tau_y/(\rho \cdot g)$

Examples

1) for τ_y = 200Pa and ρ = 2.0g/cm³ = 2000kg/m³, results: a = 10mm

2) for τ_y = 3Pa and ρ = 1.0g/cm³ = 1000kg/m³, results: a = 0.3mm

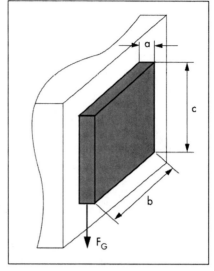

Figure 3.25: Volume element of a coating layer on a wall, with weight F_G, wet layer thickness a, layer width b and layer length c

Comment: The obtained values should only be seen as a rough estimation since other factors, such as **roughness of the wall, more or less (in-) homogeneous sagging behavior**, and **surface tension of the coating** additionally may have a crucial effect on the result. If the yield point value τ_y is determined according to Chapter 3.3.4.1, the restrictions should also be taken into consideration as discussed in that section. However, **the thickness of the wet layer also depends strongly on its time-dependent behavior during structural regeneration directly after the shear-intensive application ("thixotropic behavior")**. In order to obtain useful results for R & D it is recommended to evaluate **leveling and sagging behavior** as explained in Chapter 3.4.2.2 (and Table 3.3), or even better as shown in Chapter 8.5.2.2 (and Table 8.4).

      End of the Cleverly section

3.3.5 *Overview: Flow curves and viscosity functions*

In this section, an overview is presented by Figures 3.26 to 3.31, showing the above discussed flow and viscosity curves. The labels used are as follows: (1) ideally viscous (Newtonian), (2) shear-thinning, (3) shear-thickening, (4) without a yield point, (5) showing a yield point

a) Diagrams on a linear scale
See Figures 3.26 to 3.28.

b) Diagrams on a logarithmic scale
Logarithmic scaling is recommended if it is desired to present the shape of the curves also at very low values of τ and $\dot{\gamma}$, see Figures 3.29 to 3.31. In this case, the diagrams usually are displayed on a double-logarithmic scale.

c) Three-dimensional diagrams of flow curves and viscosity functions
Results of several tests measured at different, but constant temperatures (i.e. isothermal) can be presented in the form of a three-dimensional (3D) diagram, for example, with the shear rate $\dot{\gamma}$ on the x-axis, shear stress τ or viscosity η on the y-axis, and the temperature T on the z-axis.

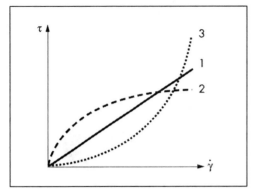

Figure 3.26: Comparison of flow curves

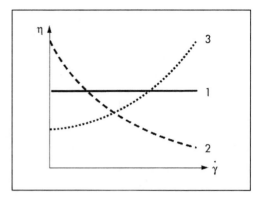

Figure 3.27: Comparison of viscosity functions

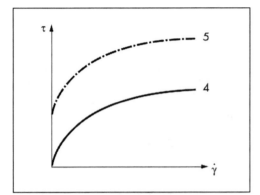

Figure 3.28: Comparison of flow curves with and without a yield point

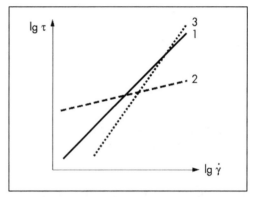

Figure 3.29: Comparison of flow curves: (1) showing the slope s = 1:1, (2) with s < 1, (3) with s > 1

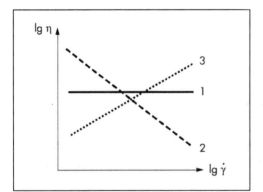

Figure 3.30: Comparison of viscosity functions: (1) showing the slope s = 0, (2) with s < 0, (3) with s > 0

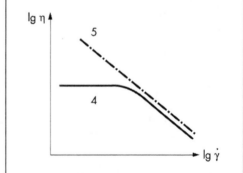

Figure 3.31: Comparison of viscosity functions; (4) showing a zero-shear viscosity plateau (i.e. there is no yield point), (5) without a zero-shear viscosity plateau (i.e. there is a yield point)

3.3.6 Fitting functions for flow and viscosity curves

After a test, measuring data mostly are available in the form of diagrams and tables. For each single measuring point there are usually values available for temperature, measuring time, shear rate, shear stress, and, calculated from these, viscosity. If measuring data should be compared, for example, when performing quality assurance tests, it is not useful – and in most cases it is also not possible – to compare all values of one test with those of another due to the mostly large number of individual measuring points.

Mathematical **model functions for curve fitting** are therefore used to characterize complete flow or viscosity curves resulting in **a small number of curve parameters** only. This simplifies to compare measuring curves, since there are only a few model parameters left for comparison then. Fitting is also called **approximation** and the corresponding functions are often referred to as **regression models**.

In the past, these fitting functions were used more frequently than today because only few users had access to rheometers which enabled the user to control or to detect such low deflection angles or rotational speeds as required to determine technically important parameters like yield points and zero-shear viscosities with sufficient accuracy. At that time, these kinds of rheometers were usually too expensive for industrial users. They therefore resorted to these model functions to characterize samples in terms of the above parameters. In this way, analysis was at least possible using approximately calculated data. Since around 1985, the increased use of computers in industrial laboratories has facilitated analysis of flow and viscosity curves, and above all, made it less time-consuming. Before, analysis of curves had to be performed manually with the aid of rulers, multicurves or so-called nomograms.

Fitting functions are still used today in many laboratories, especially for quality assurance, where financial support is at a minimum, and therefore small, inexpensive instruments are still in use. However, if data in the low-shear range below $\dot{\gamma} = 1s^{-1}$ are of real interest, it is better to measure in this range instead of calculating whatsoever values via fitting functions. Appropriate instruments are affordable meanwhile, even for small companies due to considerable improvements in the price/performance ratio.

Not each model function can be used for each kind of flow behavior. If the correlation value (e.g. in %) indicates insufficient agreement between measuring data and model function, it is useful to try another model function. It is also important to keep in mind that both, model-specific coefficients and exponents are purely mathematical variables and do not represent real measuring data in principle.

Since there are a lot of fitting functions – because it seems in the past "almost every rheologist" designed his own one – it is only possible to mention frequently used models here. In the following there are listed more than 20 models. **Often are useful those of Newton, Ostwald/de Waele, Carreau/(Yasuda) and Herschel/Bulkley.** Further information on model functions can be found e.g. in DIN 1342-3 and [18, 46, 155, 233, 276, 352].

3.3.6.1 Model function for ideally viscous flow behavior

Newton: $\tau = \eta \cdot \dot{\gamma}$ (see also Chapter 2.3.1a, with Figures 2.5 and 2.6)

3.3.6.2 Model functions for shear-thinning and shear-thickening flow behavior

Here are explained three model functions for flow curves without a yield point.

a) Ostwald/de Waele, or "power-law": $\tau = c \cdot \dot{\gamma}^p$
Flow curve model function according to W. Ostwald jun. (of 1925 [273]) and A. de Waele (of 1923 [96]) with "flow coefficient" c [Pas] and exponent p. Sometimes, c is referred to as "consistency", and p to as "flow index" or "power-law index". It counts: $p < 1$ for shear-thinning, $p > 1$ for shear-thickening, $p = 1$ for ideally viscous flow behavior.

A disadvantage of this model function is that for flow curves of most polymer solutions and melts it cannot be fitted as well in the low-shear range as well as in the high-shear range. These are the ranges of zero-shear viscosity and infinite-shear viscosity. Despite this, the model function is often used in the polymer industry to be fitted in the medium shear rate range (see Figures 3.6 and 3.7, 3.15 and 3.16, and the curve overview of Chapter 3.3.5).

    For "Mr. and Ms. Cleverly"

b) Steiger/Ory: $\dot{\gamma} = c_1 \cdot \tau + c_2 \cdot \tau^3$
Flow curve model function with the "(Steiger/Ory) coefficients" c_1 [1/Pas] and c_2 [1/Pa^3 · s], (of 1961 [346])

c) Eyring/Prandtl/Ree or Ree-Eyring: $\dot{\gamma} = c_1 \cdot \sinh(\tau/c_2)$
Flow curve model function with "(EPR) factor" c_1 [1/s] and "scaling factor" c_2 [Pa], (of 1936/1955 [130, 300])

3.3.6.3 Model functions for flow behavior with zero-shear and infinite-shear viscosity

Listed below are ten model functions showing the following parameters:
Zero-shear viscosity $\eta_0 = \lim_{\dot{\gamma} \to 0} \eta(\dot{\gamma})$ and **infinite-shear viscosity** $\eta_\infty = \lim_{\dot{\gamma} \to \infty} \eta(\dot{\gamma})$

These model functions have been designed for unlinked and unfilled polymers and are **not suitable for dispersions and gels** (see also Chapter 3.3.2.1).

a) Cross: $\dfrac{\eta(\dot{\gamma}) - \eta_\infty}{\eta_0 - \eta_\infty} = \dfrac{1}{1 + (c \cdot \dot{\gamma})^p}$ and simplified: $\dfrac{\eta(\dot{\gamma})}{\eta_0} = \dfrac{1}{1 + (c \cdot \dot{\gamma})^p}$

Viscosity curve model function with "Cross constant" c [s] and "Cross exponent" p (of 1965 [83]). For the simplified version "Cross 0", it is assumed that η_∞ is very low in contrast to η_0 and can therefore be ignored. This is usually the case for all concentrated polymer solutions and melts, since here, the viscosity values usually decrease at least by two decades.

b) Carreau: $\dfrac{\eta(\dot{\gamma}) - \eta_\infty}{\eta_0 - \eta_\infty} = \dfrac{1}{(1 + (c_1 \cdot \dot{\gamma})^2)^p}$ or as $\tau = \dfrac{\eta_0 \cdot \dot{\gamma}}{(1 + \frac{\dot{\gamma}}{\dot{\gamma}_c})^c}$

simplified: $\dfrac{\eta(\dot{\gamma})}{\eta_0} = \dfrac{1}{(1 + (c_1 \cdot \dot{\gamma})^2)^p}$

Viscosity curve model function with "Carreau constant" c_1 [s], "Carreau exponent" p, slope value c of the viscosity curve at high shear rates on a log/log scale, and the shear rate value $\dot{\gamma}_c$ at the bend between the plateau of η_0 and the falling η-curve in the range of shear-thinning behavior (of 1968/1972 [69]).

For the simplified version "Carreau 0", the same assumptions are made as for the "Cross 0" model above. The Cross and Carreau model functions are similar and they are often used by people working in R & D in the polymer industry. Modifications of the Carreau model:

1) **Carreau/Gahleitner:** $\dfrac{\eta(\dot{\gamma}) - \eta_\infty}{\eta_0 - \eta_\infty} = \dfrac{1}{(1 + (c_1 \cdot \dot{\gamma})^{p_1})^p}$

Viscosity curve model function with "Gahleitner exponent" p_1. For $p_1 = 2$, the model is identical to the Carreau model (in 1989 [147]).

2) **Carreau/Yasuda:** $\dfrac{\eta(\dot{\gamma}) - \eta_\infty}{\eta_0 - \eta_\infty} = \dfrac{1}{(1 + (\lambda \cdot \dot{\gamma})^{p_1})^{\frac{1-p}{p_1}}}$

Viscosity curve model function with "Yasuda exponent" p_1, relaxation time λ [s], and "power-law-index" p. It counts: $p < 1$ for shear-thinning, $p > 1$ for shear-thickening, $p = 1$ for ideally viscous flow behavior (of 1981 [401; 257]).

c) Krieger/Dougherty: $\dfrac{\eta(\dot\gamma) - \eta_\infty}{\eta_0 - \eta_\infty} = \dfrac{\tau_c}{\tau_c + \tau}$

Viscosity curve model function with the shear stress value τ_c at the viscosity value ($\eta_0/2$), assuming that the value of η_∞ is low in comparison to η_0 (of 1959 [213]).

d) Vinogradov/Malkin: $\dfrac{\eta(\dot\gamma) - \eta_\infty}{\eta_0 - \eta_\infty} = \dfrac{1}{1 + c_1 \cdot \dot\gamma^p + c_2 \cdot \dot\gamma^{2p}}$

Viscosity curve model function with coefficients c_1 [s] and c_1 [s], and exponent p (of 1980 [373]).

e) Ellis and Sisko:

Ellis: $\tau = \eta_0 \cdot \dot\gamma + c \cdot \dot\gamma^p$ and Sisko: $\tau = c \cdot \dot\gamma^p + \eta_\infty \cdot \dot\gamma$

Flow curve model functions with "consistency" c and "index" p. These model functions are especially designed to describe the behavior at low shear rates (according to Ellis, in 1927 [120]), and at high shear rates (acc. to Sisko, in 1958 [338]).

f) Exponential or e-function: $\eta(\dot\gamma) = \eta_0 \cdot \exp(-c \cdot \dot\gamma)$

Viscosity curve model function showing an exponential shape, with the "factor" c [s].
Note: $\exp(xyz)$ means e^{xyz}, using Euler's number e = 2.718...

g) Philipps/Deutsch: $\tau = c_1 \cdot \dfrac{(1 + c_2 \cdot \dot\gamma^2)}{(1 + c_3 \cdot \dot\gamma^2)} \cdot \dot\gamma$

Flow curve model function with "viscosity factor" c_1 [Pas], "numerator coefficient" c_2 [s^2] and "denominator coefficient" c_3 [s^2]. Assumptions: For very low shear rates the viscosity plateau value c_1 is reached, corresponding to η_0; and for very high shear rates the viscosity plateau value is ($c_1 \cdot c_2/c_3$) [Pas], corresponding to η_∞.

h) Reiner/Philippoff: $\dot\gamma = \dfrac{\tau}{\eta_0} \cdot \dfrac{1 + (\tau/c_2)^2}{1 + (\tau/c_1)^2}$

Flow curve model function with the coefficients c_1 [Pa] and c_2 [Pa], (of 1936 [288, 303]).

The following applies for $\dot\gamma \to \infty$:

$(\dot\gamma/\tau) = (1/\eta_\infty) = (\tau \cdot c_1)^2/[(c_2 \cdot \tau)^2 \cdot \eta_0] = (c_1/c_2)^2/\eta_0$

or $(\eta_0/\eta_\infty) = (c_1/c_2)^2$ or $\eta_\infty = (c_2/c_1)^2 \cdot \eta_0$

    End of the Cleverly section

3.3.6.4 Model functions for flow curves with a yield point

Detailed information on the yield point can be found in Chapter 3.3.4. Figure 3.21 presents a possible shape of flow curves explained below in sections b) to e).

a) Bingham: $\tau = \tau_B + \eta_B \cdot \dot\gamma$

Flow curve model function with the "Bingham yield point" τ_B which occurs at the intersection of "Bingham straight line" and τ-axis in a diagram on a linear scale (see Figure 3.32), and "Bingham viscosity" η_B (of 1916 [42]). Note: Despite of the denomination, this equation first was proposed by T. Schwedoff already in 1880 [237, 366]. Please note: η_B is not a viscosity value of an investigated sample, since it is not more than a calculated coefficient used for curve fitting (it would be better to speak of the "Bingham flow coefficient").

Before computers became widely used for analysis of flow curves, the Bingham model was often selected because analysis is very simple via the "Bingham straight line", merely requiring a ruler. However, the "Bingham yield point" describes the transition from the state of rest

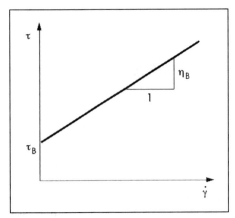

Figure 3.32: Flow curve fitting according to Bingham

to flow rather inaccurately. This model should therefore only be used for very simple QC tests (see also Chapter 3.3.4.3d: Plastic behavior, and Chapter 13.3: 1916).

Note: Simple evaluation methods ("according to Bingham")
Some users still perform the following simple test and analysis method consisting of two intervals. In the first part, a constantly high rotational speed n_H [min^{-1}] is preset for a period of t_{10} = 10min, and in the second part, a constantly low speed n_L for another 10min = t_{20} (e.g. for ceramic suspensions: with nL = n_H/10, for example, at n_H = 100min^{-1} and at n_L = 10min^{-1}). Please be aware that the values of τ and η are relative stress and viscosity values if the test is performed with a spindle which is a relative measuring system (see also Chapter 10.6.2). Here, instead of the shear stress often is used dial reading DR (which is the relative torque value M_{rel} in %), and the viscosity values are calculated then simply as η = DR/n (with the rotational speed n in min^{-1}). Usually here, all units are ignored [99].

1) "Yield stress" (YS) using a straight line through two measuring points
Preparation of a flow curve diagram with DR on the y-axis and the speed n on the x-axis (both on a linear scale). The two measuring points are plotted: (DR$_H$; n_H) and (DR$_L$; n_L) showing the values from the end of the two test intervals. A straight line is drawn through the two points, and then extrapolated towards n = 0. YS is the point at which this line crosses the y-axis (or DR-axis, respectively).
Example: with n_H = 100, n_L = 10, DR$_H$ = 50, DR$_L$ = 40, then: YS is read off as 38.

2) "Plasticity index" (PI)
Calculation: PI = YS/η_B
with the "Bingham plastic viscosity" η_B = (DR$_H$ - DR$_L$)/(n_H - n_L), or ΔDR/Δn
Example: with n_H, n_L, DR$_H$, DR$_L$ and YS as above,
then: η_B = (50 - 40)/(100 - 10) = (10/90) = 0.90, and PI = 38/0.9 = 42.2
Another, but also very simple evaluation method called "Bingham Build Up" is mentioned in Chapter 3.4.2.2c: structure recovery and thixoptropic behavior.

b) Casson: $\sqrt{\tau} + \sqrt{\tau_C} + \sqrt{\eta_C \cdot \dot{\gamma}}$

or in the following form: $\tau^{1/2} = \tau_C^{1/2} + (\eta_C \cdot \dot{\gamma})^{1/2}$

Flow curve model function with "Casson yield point" τ_C, i.e. where the fitting curve meets the τ-axis in a diagram on a linear scale, and "Casson viscosity" η_C (of 1959 [70]). For η_C the same applies as above for η_B, thus, η_C is not more than a coefficient to be used for curve fitting but it is not a real viscosity value.

This model function was originally designed for printing pastes. In 1973, the OICC (Office International du Cacao et du Chocolat) recommended the Casson model to evaluate chocolate melts at T = +40°C in the range of $\dot{\gamma}$ = 5 to 60s^{-1}; see also Chapter 3.3.1a [271].

Modifications of the Casson model:

1) "Generalized Casson model":
Here, instead of exponent $\frac{1}{2}$ (which corresponds to the square root), the exponent 1/p is used.

2) **Casson/Steiner:** $\quad \sqrt{\tau} = [2/(1 + a)]\ \sqrt{\tau_C} + \sqrt{\eta_C \cdot \dot{\gamma}}$

with form-factor $a = (R_i/R_e) = (1/\delta_{cc})$ of cylinder measuring systems, bob radius R_i (inner radius) and cup radius R_e (external radius), and ratio of radii δ_{cc} (of 1958 [347]). For ISO cylinder measuring systems: $\delta_{cc} = 1.0847$, or $a = (1/\delta_{cc}) = 0.9219$ (see Chapter 9.2.2.1 a)

Then: $\quad\quad\quad\quad\quad \sqrt{\tau} = 1.04 \cdot \sqrt{\tau_C} + \sqrt{\eta_C \cdot \dot{\gamma}}$

Summary: Using the Casson/Steiner model function, there is only a change in the "Casson yield point" when comparing to the original Casson model.

c) Herschel/Bulkley: $\quad \tau = \tau_{HB} + c \cdot \dot{\gamma}^p$

Flow curve model function with "yield point according to Herschel/Bulkley" τ_{HB}, "flow coefficient" c [Pas], (also called "Herschel/Bulkley viscosity" η_{HB}), and exponent p (also called "Herschel/Bulkley index"; of 1925 [182]). It counts: $p < 1$ for shear-thinning, $p > 1$ for shear-thickening, $p = 1$ for "Bingham behavior"

    ✂   For "Mr. and Ms. Cleverly"

d) Windhab: $\quad\quad\quad \tau = \tau_0 + (\tau_1 - \tau_0) \cdot [1 - \exp(-\dot{\gamma}/\dot{\gamma}^*)] + \eta_\infty \cdot \dot{\gamma}$

The predecessor organization of the ICA (International Confectionery Association), the IOCCC (International Office of Cocoa, Chocolate and Sugar Confectionery), recommended in 2001 to use this model to evaluate chocolate melts at T = +40°C in the range of $\dot{\gamma}$ = 2 to 50s⁻¹ (or in the "extended range" of $\dot{\gamma}$ = 1 to 100s⁻¹) [199, 390]; see also Chapter 3.3.1a.

Flow curve model function with yield point τ_0 [Pa]; shear stress τ_1 [Pa] which leads to the "maximum shear-induced structural change" (at the intersection of τ-axis and the straight line fitted to the flow curve in the high shear rate range); shear rate $\dot{\gamma}^*$ [s⁻¹] at the point $\tau^* = \tau_0 + (\tau_1 - \tau_0) \cdot (1 - 1/e)$, and slope value of the flow curve at high shear rates termed η_∞ [Pas], ("steady-state viscosity"; this is usually a constant value for most chocolate melts in the range of $\dot{\gamma}$ = 60 to 100s⁻¹); see Figure 3.33.

The following applies for $\dot{\gamma}$ = 0:
$\tau = \tau_0 + (\tau_1 - \tau_0) \cdot (1 - e^0) + 0 = \tau_0 + (\tau_1 - \tau_0) \cdot (1 - 1) = \tau_0 + 0 = \tau_0$
Result: "Structural strength at rest" is represented by the "yield point" in the range of $\tau \leq \tau_0$.

The following applies for $\dot{\gamma}$ = ∞:
$\tau = \tau_0 + (\tau_1 - \tau_0) \cdot (1 - e^{-\infty}) + \eta_\infty \cdot \dot{\gamma} = \tau_0 + (\tau_1 - \tau_0) \cdot (1 - 0) + \eta_\infty \cdot \dot{\gamma} = \tau_1 + \eta_\infty \cdot \dot{\gamma}$
Result: In the "high-shear" range, the slope of the flow curve is constant, which corresponds to "Bingham behavior".

The following applies for $\dot{\gamma} = \dot{\gamma}^*$:
$\tau = \tau_0 + (\tau_1 - \tau_0) \cdot (1 - 1/e) + 0 = \tau_0 + 0.632 \cdot (\tau_1 - \tau_0) = \tau^*$
Result: Between τ_0 and τ_1 is the "range of shear-induced structural change" (with structural decomposition at increasing shear load, i.e. in the "upwards ramp"; or with structural regeneration at decreasing load, i.e. in the "downwards ramp"). Above the yield point τ_0 the point τ^* (or $\dot{\gamma}^*$, resp.) is reached at the shear stress value τ = 63.2% $(\tau_1 - \tau_0)$.

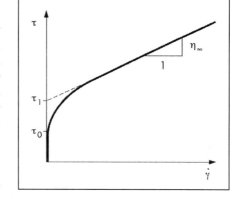

Figure 3.33: Flow curve fitting according to Windhab

e) Tscheuschner: $\tau = \tau_0 + c_1 \cdot \dot{\gamma}^p + c_2 \cdot \dot{\gamma}$

Flow curve model function with yield point τ_0 [Pa], coefficient c_1 [Pas] for the low and medium shear range to describe the shear-induced structure change, coefficient c_2 [Pas] as slope value of the flow curve at high shear rates, and exponent p. This model function was designed for chocolate melts at T = +40°C [366].

f) Polynomials

These model functions are purely mathematical descriptions of flow curve functions (e.g. according to **Williamson, Rabinowitsch, Weissenberg**). Polynomial models represent the most general approach to curve analysis and can be fitted to each type of curves. However, when using this method, a relatively large number of coefficients have to be determined, although this is no longer a problem with current software analysis programs.

Example: third order polynomial $\tau = c_1 + c_2 \cdot \dot{\gamma} + c_3 \cdot \dot{\gamma}^2 + c_4 \cdot \dot{\gamma}^3$

with the coefficients c_1 [Pa] as 0th order coefficient, representing the yield point; c_2 [Pa · s], as 1st order coefficient; c_3 [Pa · s²], as 2nd order coefficient; c4 [Pa · s³], as 3rd order coefficient. If a second order polynomial is used for analysis, then the last term of the function is ignored. Polynomials of higher orders (e.g. 5th) show correspondingly more terms.

There are also polynomials which are solved to the shear rate:

$\dot{\gamma} = c_1 \cdot \tau + c_2 \cdot \tau^2$

with the coefficients c_1 [1/Pas] and c_2 [1/Pa² · s]

A comparable model is the Steiger/Ory model (see Chapter 3.3.6.2b).

~~   End of the Cleverly section

3.3.7 *The effects of rheological additives in aqueous dispersions*

Legislative restrictions concerning industrial products increasingly are forcing the reduction of volatile organic compounds (VOC). As a result, solutions and dispersions containing organic solvents are more and more replaced by aqueous dispersions; examples are coatings of all kinds such as paints and adhesives. Therefore, adapted rheological additives have been developed which of course also influence flow behavior, sometimes even considerably [170, 176, 258, 339]. Typical examples of **nanostructures and microstructures of rheological additives** are shown in Figure 3.34 [313]; the black bar in the Figure indicates the dimension of 100nm to illustrate the order of magnitude when comparing the material systems.

a) Aqueous dispersions containing clay as the thickening agent

Inorganic primary particles of clay exhibit the shape of thin platelets. A "house of cards structure" with a network of secondary forces is built up between the surfaces and edges of the platelets when at rest; see Figure 3.34. This is due to the fact that the large flat surfaces of the platelets show negative electrical charge, whereas the narrow sides and edges are positively charged. Therefore, **at rest** a superstructure in the form of a **gel structure** is occurring if the additive is incorporated in an appropriate way. In this case, in the low-shear range there is a **yield point** or an "infinitely high" viscosity, respectively.

The network of forces breaks if the yield point is exceeded under sufficiently high shear force. With increasing shear rates the viscosity values are decreasing continuously since finally, the "house of cards structure" will be completely destroyed. Increasingly, the individual sheet-like and very flat particles of the additive are oriented into shear direction. At high shear rates, and at a typical concentration of around 0.1 to 0.3% as it is used in practice, the clay particles **in a flowing liquid** display **no longer a significant thickening effect**.

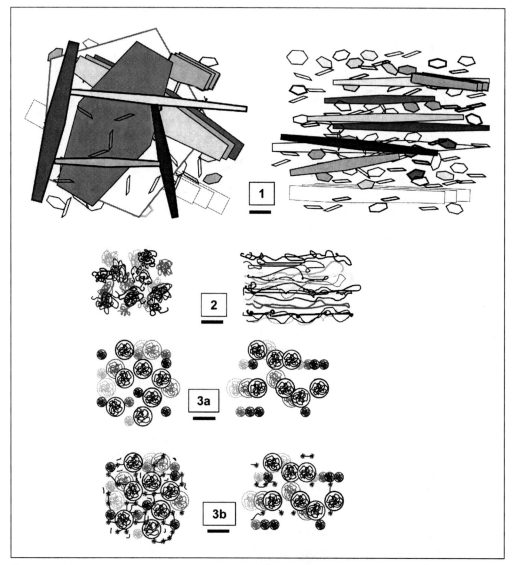

Figure 3.34: Nano- and microstructures in water-based coating systems, left: at rest, and right: in a sheared state: 1) clay (as an inorganic gellant), 2) polymer molecules in solution, 3a) dispersed polymer particles, without an additive, 3b) polymer dispersion with a polymeric associative thickener, here, also surfactant molecules are integrated in the bridge-like clusters. (The black bar indicates the dimension of 100nm.)

Dimensions: Primary particles of **bentonite** are around $800 \cdot 800 \cdot 1$ (in nm) and those of hectorite are around $800 \cdot 80 \cdot 1$; with an organic surface modification the distance between individual platelets is around 4nm (as primary particles).

Figure 3.35 presents a typical viscosity function of a pigmented water-based coating with clay as the thickener. On the one hand there is a very pronounced thickening effect in the low-shear range at $\dot{\gamma} < 1s^{-1}$ which for example may cause problems when starting to pump the dispersion from a state of rest. On the other hand, there is no more thickening effect at high shear rates, e.g. at $\dot{\gamma} > 1000s^{-1}$, and this may cause uncontrolled splashing and spattering of droplets when the coating is applied using a brush or a roller. Therefore, this coating shows **pronounced shear-thinning behavior** throughout the whole shear rate range.

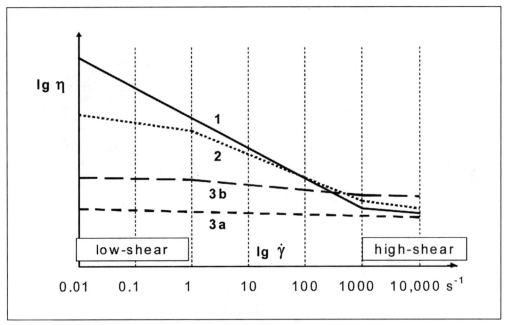

Figure 3.35: Principal shape of viscosity curves of pigmented water-based coatings containing different thickener additives: 1) clay, as an inorganic gellant, 2) dissolved polymer, 3a) polymer dispersion without an additive, 3b) polymer dispersion including an associative thickener

b) Aqueous dispersions containing dissolved polymer as the thickening agent

Dissolved polymer macromolecules exhibit the form of coils when at rest, see Figure 3.34. At a sufficiently high concentration there are entanglements between the long, thread-like molecules when **at rest**, and a loose **network of entanglements** is built up as a continuous phase (see also Chapter 3.3.2.1). Its strength is only based on mechanical forces, such as the relatively high frictional forces between the long polymer molecules. However, this network is not held together by chemical or physical forces. For hydrophilic polymers, e.g. hydrocolloids (such as corn starch, carrageen, agar or carob bean gum), cellulose derivatives or polyacrylic acid, a pronounced **hydration shell** may develop additionally like an envelope around the individual molecules. Hydrogen bridges can also build up a superstructure surrounding the molecules "like a fog". In this case, there is a network of interactions which is also strengthened by the physical-chemical secondary forces of the hydrogen bridges. Therefore, if the additive is incorporated in an appropriate way, a relatively **high** – but not in all cases "infinitely high" – **viscosity value** may result **in the state of rest**. Then towards the low-shear range, the shape of the viscosity curve is becoming flatter on a high level (see Figure 3.35), showing the tendency to end on the **plateau value of zero-shear viscosity**. Indeed, this plateau only appears for unlinked polymer solutions, thus without showing a gel-like structure (see Chapter 3.3.2.1a). With a high thickener concentration, however, a gel-like structure with a yield point may occur as a superstructure. In this case, the shape of the vicosity curve towards low shear rates is steadily increasing towards an "infinitely high" value.

Under a sufficiently high shear force, the network disintegrates and the molecules are disentangling more and more. Therefore, at increasing shear rates the viscosity values are decreasing considerably and the individual molecules of the additive are more and more oriented into shear direction. At a typical concentration of around 0.2% as used in practice, dissolved polymer molecules **in flowing liquids display no longer a significant thickening effect** at high shear rates.

Dimensions: The chains of individual **polymer molecules** exhibit diameters of around 0.5nm. At rest, typical diameters of individual polymer coils are between 5 and 100nm depending on the degree of dissolution, which of course again is influenced by temperature and pH value. Here as a measure, often is taken the **"hydrodynamic radius"** R_H which is derived from the "hydrodynamic volume" of a freely moveable molecule appearing in a spherical shape when rotating together with its hydration shell. In this case however, instead of a "radius" it would be better to talk of a diameter.

Figure 3.35 presents a typical viscosity function of a pigmented water-based coating containing a dissolved polymer as the thickener. On the one hand it shows a pronounced thickening effect in the range of $\dot\gamma < 1s^{-1}$, which may lead to the same problems as described above for the clay thickener. On the other hand, there is no more thickening effect at high shear rates, e.g. at $\dot\gamma > 1000s^{-1}$, with the same consequences as described above. The coating displays **pronounced shear-thinning behavior** throughout the whole shear rate range.

c) Aqueous polymer dispersion without a thickener

In polymer dispersions, the macromolecules are a component of discrete particles which consist of a shell of surface-active small molecules (surfactants), whose hydrophilic part points outwards into the water phase, whereas the hydrophobic part points inwards into the core of the particles containing the organic polymer molecules (**polymer dispersion particles**, "latex particles"; see also Chapter 9.1.3). Despite of their relatively high molar mass and also at a high polymer concentration, there is no possibility for the macromolecules to entangle like in polymer solutions, because they are enclosed in their spheres and therefore are shielded from their surrounding; see Figure 3.34. Without additional additive, no network can occur between the particles. If the polymer concentration remains within the corresponding limits, for example at a volume fraction of $\Phi < 50$ %, the individual particles do not prevent each other's motion by direct contact and corresponding friction. Therefore, the **viscosity values remain relatively low, as well at rest as well as also in the sheared state**. Since the individual dispersed polymer particles do not break even at higher shear rates, also here, the **viscosity values are still remaining almost on a constant level**. As a comparison: When flowing, dissolved polymer molecules are oriented and stretched out in shear direction showing continuously decreasing flow resistance as they glide increasingly easier along one another. Dimensions: Individual particles of polymer dispersion may exhibit diameters of 20nm to 20μm, and often are around 50 to 500nm.

Figure 3.35 presents a typical viscosity function of a pigmented water-based coating without a thickener (at $\Phi < 50$%). As well in the range of $\dot\gamma < 1s^{-1}$ as well as at higher shear rates, e.g. at $\dot\gamma > 1000s^{-1}$, viscosity remains usually relatively low. As this value hardly depends on the shear rate, these dispersions are showing **almost Newtonian behavior**. Usually under these conditions, polymer dispersions without thickener are not useful in practice since their viscosity is mostly too low. Neither their stability against sedimentation nor their matrix is strong enough to prevent spattering during application. They usually also display a highly pronounced tendency for sagging after application.

d) Aqueous polymer dispersion containing an associative thickener

Associative thickeners consist of molecules which are able to connect to one another. This occurs in aqueous systems **via hydrophobic molecular groups**, whose connections and **association** takes place via non-permanent secondary bonds. Connections of these **surface-active groups** of the thickener molecules can take place between one another or via other hydrophobic components such as particle surfaces (see also Chapter 9.1.3: Surfactant-like polymers). **At rest a non-permanent network** is built up which is only held together by the physical-chemical secondary forces of the hydrophobic interactions; see Figure 3.34. This Figure also displays small surfactant molecules which by self-organization can integrate themselves into these clusters. The **thickener molecules** are not remaining permanently in one place, they are continuously changing positions; they **show fluctuation**. The result is a weaker network, if compared as well to the network of entangled polymers as well as to the gel structure mentioned above; see Figure 3.35. **In the range of very low shear rates, viscosity can be adjusted to the required value** if the concentration of the dispersion components remains within the corresponding limits. Therefore, flow behavior can be controlled and optimized as desired for any requirement. A higher proportion of hydrophobic groups or longer hydrophobic molecule segments will lead to a higher structural strength at rest. In a sheared state, an increased number of hydrophobic molecule clusters and associative bridges are breaking and the thickener molecules are oriented more and more into shear direction. The viscosity values may still decrease but only to a limited extent since also here, the particles of the polymer dispersion are not destroyed, even at high shear rates. Due to their size they are showing also here considerably more friction between one another compared to dissolved polymer molecules, and therefore, correspondingly higher viscosity values.

Dimensions: Surface-active polymers exhibit a length of around 50 to 100nm when stretched. Comparison: Many surfactant molecules are merely 0.5 to 5nm long.

Figure 3.35 presents a typical viscosity function of a pigmented aqueous coating (at $\Phi < 50\%$) including an associative thickener (with a concentration of around 0.1 to 0.2% of the active components). In the range of $\dot{\gamma} < 1s^{-1}$ there is a limited, not excessive thickening effect. But also in the range of high shear rates (at $\dot{\gamma} > 1000s^{-1}$) the shear-thinning effect is limited. Thus, the viscosity values are changing only comparatively little also in the sheared state. As the viscosity here is not very dependent on the shear rate, therefore showing **only moderate shear-thinning flow behavior**, some users talk about "quasi-Newtonian behavior". Even when this term is not scientific, sometimes for practical users it is useful to a certain degree. Associative thickener molecules with a comparatively higher proportion of hydrophobic groups are causing a more pronounced shear-thinning effect, whereas associative thickener molecules with a comparatively higher hydrophilic proportion are leading to more ideally viscous (Newtonian) flow behavior.

The great advantage can be found in both: On the one hand, in the range of low shear rates the viscosity values are not too high but are usually sufficient as required for sedimentation stability. On the other hand, in the range of high shear rates the viscosity values are usually significantly higher compared to the other two thickener types mentioned above (clay and dissolved polymers). This may help, for example, to prevent spattering when performing fast application processes.

3.4   Time-dependent flow behavior and viscosity function

This section informs about time-dependent flow behavior. These tests are performed at constant test conditions for each test interval, since here, as well the degree of the shear load as well as the measuring temperature are kept at a constant value.

3.4.1 Description of the test

Preset
1) With controlled shear rate (CSR): constant shear rate $\dot{\gamma}$ = const. (see Figure 3.36)
2) With controlled shear stress (CSS): constant shear stress τ = const. (similar to Figure 3.36)

Measuring result
1) Time-dependent shear stress: $\tau(t)$, see Figure 3.37
2) Time-dependent shear rate: $\dot{\gamma}(t)$, see Figure 3.38

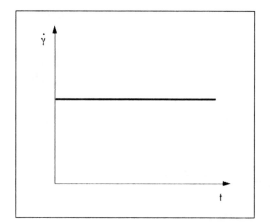

Figure 3.36: Preset of a constant shear rate

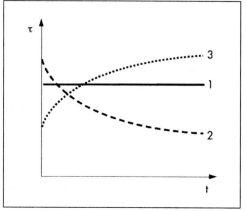

Figure 3.37: Time-dependent shear stress curves:
(1) with no change in viscosity with time (e.g. viscosity standard oils)
(2) decreasing viscosity with time (e.g. emulsion paints, ketchups, yogurts)
(3) increase in viscosity with time (e.g. hardening process of adhesives, gel formation)

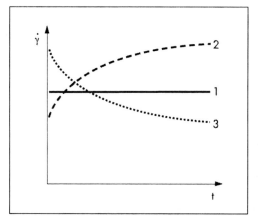

Figure 3.38: Time-dependent shear rate curves: for (1) to (3) see text of Figure 3.37

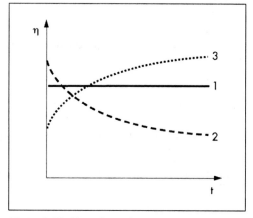

Figure 3.39: Time-dependent viscosity curves: for (1) to (3) see text of Figure 3.37

Further results: Time-dependent viscosity $\eta(t)$, see Figure 3.39
The shapes of the curves of $\tau(t)$ and $\eta(t)$ are similar because τ and η are proportional since $\tau = \eta \cdot \dot{\gamma}$, here with $\dot{\gamma}$ = const. However, the curves of $\dot{\gamma}(t)$ and $\eta(t)$ are inversely, because $\dot{\gamma}$ and η are inversely proportional since $\eta = \tau/\dot{\gamma}$ here with τ = const.

3.4.2 Time-dependent flow behavior of samples showing no hardening

When presetting low shear rates or shear stresses, the duration for each individual measuring point must be long enough. Otherwise particularly with highly viscous and visco-elastic samples, time-dependent **start-up effects or transient effects** are obtained. In this case, the measured curve approaches the final value of τ or η from below (see also Chapters 3.3.1b and 3.4.2.2a). **When applying high shear rates, viscous heating may occur to a** higher extend and then, the time-dependent viscosity values may decrease additionally due to thermal effects. In this case, with low-viscosity liquids also **turbulent flow effects** must be taken into account (see also Chapter 3.3.3).

Extended test profiles (including intervals at rest, for temperature equilibration, and pre-shearing)

Sometimes, test programs are used showing several intervals with preset of a constant shear rate in each; for example:

1) **Rest intervals,** presetting constantly $\dot{\gamma} = 0$, either after a pre-shear interval or at the start of the test to enable **relaxation of the sample** after gap setting of the measuring system which may cause high internal stresses particularly when testing highly viscous and viscoelastic samples. Simultaneously, this period is suited to enable **temperature equilibration**.

2) **Pre-shear intervals,** presetting a constant low shear rate, to ensure a defined sample preparation, to distribute the sample in the shear gap homogeneously, and to equalize or to reduce pre-stresses deriving from diverse sample preparation steps.

Example: Testing lubricating grease (according to DIN 51810-1)
Test program consisting of five intervals
1st interval: rest phase (for t = 1min): at $\dot{\gamma} = 0$
2nd interval: pre-shear phase (for t = 1min): at $\dot{\gamma} = 100s^{-1}$ = const
3rd interval: rest phase (for t = 2min): at $\dot{\gamma} = 0$
4th interval: shear rate ramp upwards (in t = 1min): with $\dot{\gamma} = 0$ to $\dot{\gamma}_{max}$
with $\dot{\gamma}_{max} = 1000s^{-1}$ for greases exhibiting soft consistency (showing NLGI classes 000 to 1; NLGI is the National Lubrication Grease Institute in the USA),
and with $\dot{\gamma}_{max} = 500s^{-1}$ for greases exhibiting rigid consistency (i.e., NLGI class 2),
5th interval: high-shear phase (for t = 5min): at $\dot{\gamma} = \dot{\gamma}_{max}$ = const
Analysis: With the viscosity values η_1 at the beginning and η_2 at the end of the fifth interval, the "relative viscosity change" is calculated as follows:

$$\eta_{rel} = (\eta_1 - \eta_2) \cdot 100/\eta_1 \text{ (specification of } \eta_{rel} \text{ in [%])}$$

3.4.2.1 Structural decomposition and regeneration (thixotropy and rheopexy)

a) Thixotropic behavior
Experiment 3.4: Shaking bottles containing ketchup and paraffin oil
1) When shaking continuously, the ketchup is becoming thinner and thinner. During the subsequent period of rest, it is thickening more and more and after approximately 10 minutes it has returned to its original gel-like consistency.
2) The paraffin oil is solid when at rest, but immediately turns to a liquid state when shaken. It takes approximately eight hours until the oil has reached its initially solid state again.

The term "thixotropy" is a combination of Greek words and is not that easy to translate: "Thixis" means touching, in the sense of deranging when bringing in motion (e.g. when stirring or shaking), and "trepein" means turning, changing, transforming; it is a "property of some solidified colloids to temporarily turn into a liquid state under the influence of mechanical forces" [352, 377]. As a rheological term, "thixotropy" might be translated as "structural change or transition due to a mechanical load". Unfortunately in many industrial laboratories due to lack of knowledge, the term "thixotropy" is given various meanings, and correspondingly, there are a lot of more or less useful test procedures (see also Chapters 11.2.3b, 11.2.4a and 11.2.11a and c).

In order to perform useful rheological tests, in minimum two intervals have to be considered: "Thixotropic behavior" means reduction of structural strength during a **shear load phase and** a more or less rapid but complete structural regeneration during the subsequent **period of rest** [101]. **This cycle of decomposition and regeneration is a completely reversible process** (see Figure 3.40).

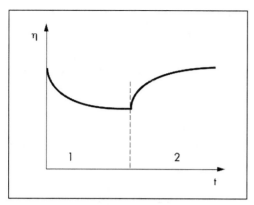

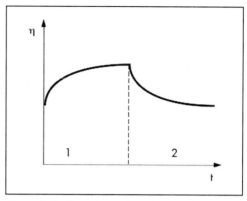

Figure 3.40: Time-dependent viscosity function of a thixotropic material
(1) structural decomposition when applying a constantly high shear load
(2) structural regeneration when at rest

Figure 3.41: Time-dependent viscosity function of a rheopectic material:
(1) increasing structural strength when applying a constantly high shear load
(2) decreasing structural strength when at rest

Thixotropic behavior is defined as time-dependent behavior, and is correctly determined in a scientific sense only if
1) **both decomposition and regeneration** of the structure are taken into consideration,
2) testing is performed at a **constant shear load** in each one of the test intervals.

Examples of thixotropic materials
Almost all dispersions (suspensions, emulsions, foams) such as pastes and creams; gels, ketchups, coatings, paints, printing inks, sealants, drilling fluids, plaster dispersions (e.g. in geo-technics), soap sols

b) Non-thixotropic behavior
Experiment 3.5: Stirring yogurt
After stirring, yogurt remains considerably thinner compared to the initial state, even when waiting a long period of time.

Each material displaying a certain degree of structural decomposition in the shear phase with time, of course is showing time-dependent behavior. However, such a material is not thixotropic if the initial structural strength does not completely return finally, even after an "infinitely" long period of time at rest. In this case a **permanently remaining structural change** has taken place. If structural regeneration does not occur completely (i.e. to 100%), it is sometimes referred to as "incomplete" or "false" thixotropy. In this case, instead of "thixotropy" it is better to speak of "partial regeneration", e.g. expressed in a percentage compared to the initial viscosity value.

Example of a meaningful specification, related to the viscosity value: After a high-shear phase, during a rest period of t = 120s, structural regeneration has taken place up to 70% compared to the initial structural strength-at-rest before shearing.

c) Rheopectic behavior
Rheopectic behavior means an increase in structural strength when performing a high-shear process which is followed by a more or less rapid but complete decomposition of the increased structural strength during a subsequent period of rest. **This cycle of generation and re-composition, of increase and decrease in structural strength is a completely reversible process** (see Figure 3.41). Rheopexy is sometimes called "anti-thixotropy" or "negative thixotropy" [18] which may lead to confusion. **Rheopectic behavior is defined as**

a time-dependent behavior (like thixotropic behavior). Testing of rheopectic behavior is similar to the thixotropy test, with appropriate modifications (see Chapter 3.4.2.2).

Materials showing rheopectic behavior tend to inhomogeneous flow. **Wall-slip effects and phase separation** should always be taken into account. In industrial practice, rheopectic behavior is much less common than thixotropic behavior.

Examples of rheopectic materials
When working with dispersions showing a high concentration of solid matter or gel-like particles such as ceramic casting slips, latex dispersions and plastisol pastes, rheopectic behavior should always be taken into account.

Note: Non-thixotropic behavior
Shear-induced and permanently remaining increase in structural strength

Example: Testing a dispersion under the following conditions
1st interval: for t = 1min at $\dot{\gamma}$ = 100s^{-1}; result: viscosity increase η = 0.1 to 1Pas
2nd interval: for t = 3min at $\dot{\gamma}$ = 0.1s^{-1}; result: reaching η = 10Pas after a short time, remaining on this high value afterwards

In this case, on the one hand an increase in viscosity in the high-shear interval can be observed. But on the other hand, there is no decrease of the viscosity value in the subsequently following low-shear interval. Therefore, **this is not rheopectic behavior, since here, a permanently remaining shear-induced structural change has taken place**. Of course when testing these kinds of materials, loss of solvent or drying effects must be excluded. More about shear-induced effects: See Chapter 9.2.2 and Figure 9.23.

3.4.2.2 Test methods for investigating thixotropic behavior

There are several possibilities to characterize thixotropic behavior, and correspondingly, different results may occur finally [101].

a) Step test consisting of three intervals
Preset
1) With controlled shear rate (CSR): profile $\dot{\gamma}$(t) with three intervals as a step function (see Figure 3.42)
2) With controlled shear stress (CSS): profile τ(t) with three intervals as a step function (similar to Figure 3.42)

Result: Time-dependent viscosity function η(t), see Figure 3.43

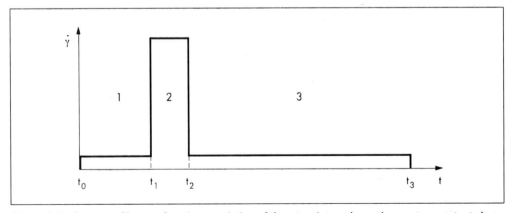

Figure 3.42: Preset profile: step function consisting of three test intervals, each one at a constant shear rate, at (1) low-shear, (2) high-shear, and (3) again at low-shear conditions

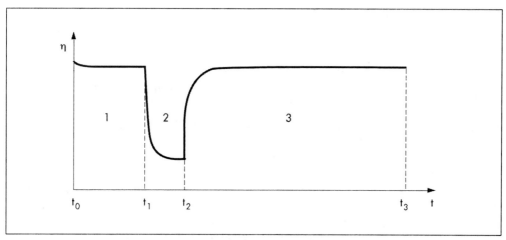

Figure 3.43: Time-dependent viscosity of a thixotropic material, (1) at low-shear conditions showing the "reference value of the viscosity-at-rest", (2) structural decomposition, and (3) structural regeneration

For measurements like this, the following three test intervals are preset.

1) Reference interval (low-shear)
Shear conditions "at rest", i.e. **low-shear conditions**, are preset in the period between t_0 and t_1. The aim is here to achieve a fairly constant η-value in the whole first interval, in terms of the **"reference value of the viscosity-at-rest"**. This value is used later as the reference value to be compared to the viscosity values obtained in the third test interval, showing structural regeneration if it occurs.

2) High-shear interval
High-shear conditions are preset in the period between t_1 and t_2 in order to break the internal structure of the sample. This interval is used to simulate the high-shear conditions occurring during an **application process**, for example, painting and coating using brush, roller or blade, or when spraying.

3) Regeneration interval (low-shear)
Shear conditions "at rest" are preset **again** in the period between t_2 and t_3, at the same shear conditions like in the first interval to facilitate **regeneration of the sample's structure**. This interval is used to simulate the **low-shear conditions** occurring directly after the coating process when the material is only slightly stressed by its own weight due to gravity.

For practical users, the crucial factor to evaluate structural regeneration is the behavior **in the time frame which is related to practice**. This period of time has to be defined by the user before the test according to the requirements, usually after a number of experiments performed (e.g. by the application department). Example A: For a wet coating, a regeneration time of $t = 60$ or $120s$ is desired in order to obtain good leveling behavior. Example B: For a drop of an adhesive or for a printing ink a time of only $t = 1$ or $2s$ is desired to achieve dot sharpness in a relatively short time. If the desired η-value has not been reached within this **"time related to practice"** (or "practice-relevant time"), then the sample is not considered to be thixotropic, related to this application.

Example 1: Presetting shear rates for all intervals
1st interval (for $t = 60s$, with 5 measuring points): at $\dot{\gamma} = 0.1s^{-1}$ = const
2nd interval (for $t = 30s$, with 5 points): at $\dot{\gamma} = 1000s^{-1}$ = const
3rd interval (for $t = 180s$, with 50 points or more): at $\dot{\gamma} = 0.1s^{-1}$ = const again

Exactly the same shear profile has to be preset for each individual test if thixotropy values of different tests are to be compared, and that counts for all parameters: shear rates, number of measuring points, and duration of the test intervals.

Example 2: Presetting shear stresses for all intervals
1st interval: at τ = 10Pa; 2nd interval: at τ = 1000Pa; 3rd interval: again at τ = 10Pa.
Interval times and number of measuring points should be selected like in Example 1.

Example 3: Presetting shear rate and shear stress combined in series
1st interval with controlled shear stress (at a low stress value, conditions at rest)
2nd interval with controlled shear rate (at a high shear rate, to simulate the application process)
3rd interval with controlled shear stress (at a low stress again, to simulate the low weight force of the applied wet coating layer)
Using this type of test, sometimes differences between samples may be observed which might hardly be observed when using an other test type. Often, the first interval is omitted here (as in Figure 3.40). On the other hand, as a disadvantage must be stated that pre-tests are required to find out a useful shear stress value to be selected for the first and third interval [101].

Note 1: Optimizing the test conditions
In order to get a useful "reference value of the viscosity-at-rest", the viscosity values should be as constant as possible in the first interval. If this condition is not met, the following actions can be taken:
1) If the $\eta(t)$-curve comes from above and shows constant viscosity values only after a certain period of time, the preset shear rate was too high for the sample to be still in a state of rest. Therefore, at these shear conditions a certain degree of structural decomposition is already taking place. **Action:** A lower shear rate should be selected.
2) If the $\eta(t)$-curve comes from below and shows constant viscosity values only after a certain period of time, then **transient behavior** is measured. Transient shear viscosity η^+ is a function of both shear rate and time, i.e. $\eta^+ = \eta(\dot\gamma, t)$. In this case, the period of time was too short for the sample to adapt evenly throughout the entire shear gap to the applied low-shear conditions. **Action:** The measuring point duration should be extended. **As a rule of thumb: The measuring point duration should be at least as long as the value of the reciprocal shear rate** $(1/\dot\gamma)$. After a first trial with this preset, the measuring point duration may have to be extended further until good test conditions are achieved. Sometimes, however, even shorter times as $t = 1/\dot\gamma$ are sufficient (see also the Note in Chapter 3.3.1b and Figure 2.9, no. 5: transient behavior).

Note 2: Which mode of testing is more useful – shear rate or shear stress control?
If the instrument is able to control very rapid changes in shear rates, then controlled shear rate (CSR) tests are often preferable to controlled shear stress (CSS) tests. The reason is that the process of structural decomposition and regeneration is directly dependent on the degree of the shear rate or deformation, respectively. Shear stress in fact is causing this deformation, but it is the deformation itself leading to the change in the structural strength. This is correct since the viscosity value is only changing if the acting shear force indeed produces a sufficiently high deformation. As a result of this interrelation, in most cases, measuring results obtained from CSR tests are more reproducible.

Optional methods to analyze structural regeneration
A number of options to analyze thixotropic behavior are given below, many users prefer evaluation according to method M4.

M1) "Thixotropy value" as a viscosity difference
The extent of thixotropic behavior is determined in terms of the viscosity change $\Delta\eta$, which is the difference between the maximum viscosity after structural regeneration and the mini-

mum viscosity after structural decomposition. With η_{min} at the time point t_2 and η_{max} at the time point t_3 the following holds (see Figure 3.43):

$\Delta\eta = \eta_{max} - \eta_{min}$

Note 3: Alternative analysis methods (to method M1) from industrial practice

Some users in industrial laboratories evaluate thixotropic behavior by the following simple methods to carry out quality control of coatings.

Preset: Test, consisiting of two measuring intervals: at first high-shear (HS), and then low-shear (LS) to enable structural regeneration of the sample. Analysis:

a) "Thixotropy index": $TI = (\eta_L - \eta_H)/t_R$

with η_H in mPas at the end of the HS interval; and η_L in mPas at the regeneration time t_R after the beginning of the LS interval (Examples: $t_R = 60s$; or 30 or 90 or 300s)

Calculation: **Viscosity change** in a previously defined time in the LS interval, in the form of the **slope ($\Delta\eta/\Delta t$) of the $\eta(t)$-curve**, with the unit (mPas/s). A straight line is therefore adapted between the last measuring point of the HS interval and the measuring point at time point t_R in the subsequent LS interval. Evaluation: The faster the regeneration and the higher the corresponding η-value obtained, the higher is the TI value. Example: With $\eta_H = 100$mPas and $\eta_L = 1000$mPas after $t_R = 60s$, then: $TI = (900\text{mPas}/60s) = 15\text{mPas/s}$.

b) "Structure recovery index": $SRI = \lg \eta_L - \lg \eta_H$

with ηH, η_L and t_R as above. Calculation: Viscosity difference in the form of logarithmic viscosity values. Evaluation: The stronger the obtained regeneration, the higher the SRI value. Example: With $\eta_H = 100$mPas and $\eta_L = 1000$mPas (e.g. after $t_R = 30s$), then: $SRI = (\lg \eta_L - \lg \eta_H) = 3 - 2 = 1$. Usually here, specifications are given without any unit.

M2) "Total thixotropy time"

The "total thixotropy time" is determined as the time difference between t_2 at the end of the second interval indicating structural decomposition in terms of η_{min}, and the time point in the third test interval when reaching the maximum value η_{max} after structural regeneration. Thus, the "total thixotropy time" is the period of time required for the complete (100%) regeneration, i.e when reaching again the reference value of the viscosity-at-rest which was determined in the first test interval. Of course, this period of time might be shorter than $(t_3 - t_2)$ if regeneration is finished already before time point t_3 is reached.

M3) "Relative thixotropy time" required to reach a certain percentage of regeneration

For QC tests, analysis of the "total thixotropy time" according to method M2 may take a too long time. Therefore then in the third test interval, the "relative thixotropy time" is determined as the period of time for the η-value to reach the relative value of, for example, 75 or 90% compared to the reference value obtained in the first interval (which counts as the "100% η-value" here).

Example (to method M3): Testing PVC plastisol pastes

Evaluation of plastisols as used in automotive industry as underseals or seam sealants. **Requirement:** After the application of the plastisol, the conveyor line transporting the car should not be set in motion before the structural strength of the plastisol has regenerated to 75% of the reference value at rest. Otherwise vibrations might cause sagging or even dripping of the coatings in an uncontrollable way. Previously performed laboratory tests resulted in a viscosity reference value $\eta(100\%) = 2000$mPas at time point t_1.

Determination: Which period of time is required after time point t_2 to reach 75% of the reference viscosity value, thus here, $\eta(75\%) = 1500$mPas?

M4) Percentage of regeneration within a previously defined time period

The "percentage of regeneration" taking place in the third interval is determined at certain time points which have been defined before the test by the user (e.g. after t = 30 and 60s). The

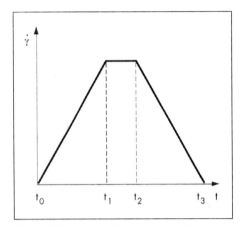

Figure 3.44: Preset of a time-dependent shear rate profile consisting of three test intervals: upward ramp, high-shear phase, and downward ramp

η-values are read off at these time points and the percentage is calculated in relation to the reference value of the first interval which counts as the "100% value" then.

Example (to method M4): Comparison of two coatings

Different behavior of two coatings in the regeneration phase is illustrated by Table 3.3.

Analysis: Coating 1 shows complete regeneration within 120s already (related to the viscosity value). This may facilitate to obtain the required wet layer thickness. Here, with 87% structural recovery attained after 30s the final structural strength has almost been reached already. Coating 2 displays a slower structural regeneration, showing a long lasting and therefore good leveling behavior. However, this coating may show a certain tendency to sagging on vertical areas, which may prevent to achieve the desired layer thickness.

For more information on tests to evaluate thixotropic behavior, see Chapter 8.5.2.2 (using oscillatory tests). Corresponding analysis can be performed in an adapted form for "rheopexy" and rheopectic behavior. However, this behavior is hardly occurring in industrial practice (see also the Note in Chapter 3.4.2.1c: shear-induced increase in viscosity).

b) Flow curves and hysteresis area (for evaluating thixotropic behavior)

This testing and analysis method for determining thixotropic and rheopectic behavior is now outdated (DIN 53214 of 1982, meanwhile withdrawn). Nevertheless, it is still used in many industrial laboratories to carry out simple QC tests.

Preset

1) With controlled shear rate (CSR): profile $\dot{\gamma}(t)$, see Figure 3.44
2) With controlled shear stress (CSS): profile $\tau(t)$, similar to Figure 3.44

Example 1: Preset of the shear rate

1st interval (shear rate ramp upwards, in t = 120s): with $\dot{\gamma}$ = 0 to 1000s⁻¹
2nd interval (high-shear phase, for t = 60s): at $\dot{\gamma}$ = 1000s⁻¹ = const

Table 3.3: Regeneration of two coatings in terms of η(t) and in %

	Coating 1		Coating 2	
	η [Pas]	Reg. [%]	η [Pas]	Reg. [%]
At the end of the first interval, at low-shear conditions; the reference value of the viscosity-at-rest	15	100%	30	100%
At the end of the second interval, at high-shear conditions	0.5	(3%)	1.0	(3%)
Regeneration in the third interval after t = 30s	13	87%	4	13%
after t = 60s	14	93%	10	33%
after t = 120s	15	100%	15	50%

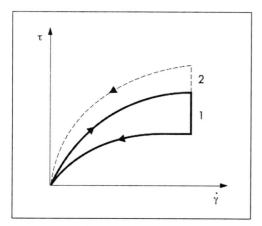

Figure 3.45: Flow curves obtained when controlling the shear rate, showing a hysteresis area: (1) with decreasing, and (2) with increasing structural strength when shearing

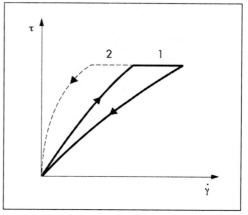

Figure 3.46: Flow curves obtained when controlling the shear stress, showing a hysteresis area: (1) with decreasing, and (2) with increasing structural strength when shearing

3rd interval (shear rate ramp downwards, in t = 120s): with $\dot{\gamma}$ = 1000 to 0s^{-1}
Therefore here, the total test duration is t = 300s.

Example 2: Preset of the shear stress
1st interval (shear stress ramp upwards, in t = 120s): with τ = 0 to 2000Pa
2nd interval (high-shear phase, for t = 60s): at τ = 2000Pa = const
3rd interval (shear stress ramp downwards, in t = 120s): with τ = 2000 to 0Pa

Measuring result
1) With CSR: Flow curves $\tau(\dot{\gamma})$ showing a hysteresis area, see Figure 3.45
2) With CSS: Flow curves $\dot{\gamma}(\tau)$ showing a hysteresis area, see Figure 3.46

The area between the upward and downward curves in Figures 3.45 and 3.46 is called **"hysteresis area"**. "Hysteros" is a Greek word meaning "later", and "hysteresis" literally is the "dependence of a (physical) state on previous states" [377]. In this case, we can interpret this term as a "time-dependent loop" or "a loop which is influenced by time-dependent behavior". In the past, the terms "thixotropic area" or "rheopectic area", respectively, were used. The "thixotropic value" was determined by the following procedure.
1) When presetting the shear rate, determination of the difference of the first area between the upward curve and the $\dot{\gamma}$-axis, and the second area between the downward curve and the $\dot{\gamma}$-axis (see Figure 3.45): Samples for which a positive value of resulting area is obtained are termed "thixotropic" (no. 1), and those with a negative value are called "rheopectic" (no. 2).
2) When presetting the shear stress, evaluation of the difference of the first area between the downward curve and the τ-axis, and the second area between the upward curve and the τ-axis (see Figure 3.46): Samples for which a positive area value is obtained are referred to as thixotropic" (no. 1), and those with a negative value are named "rheopectic" (no. 2).

Note: Hysteresis area and shearing power
The shearing power which corresponds to the thixotropic structural decomposition, can be calculated as the product of the total amount of the area between the two measured flow curves and the sheared volume of the sample [215]: $P = [(\tau \cdot \dot{\gamma})_1 - (\tau \cdot \dot{\gamma})_2] \cdot V$
with τ [Pa], $\dot{\gamma}$ [s^{-1}] and V [m^3]. The resulting unit is 1(Pa · m^3)/s = 1Nm/s = 1J/s
It is, however, very optimistic to assume that all parts of the whole sample are sheared homogeneously, and therefore at a constant shear load or shearing power.

Comment: No undisturbed structure regeneration

Using the "hysteresis area method", flow behavior is only determined in a state of motion, and thus, only in the phase of structural decomposition. This method does not reveal any information about structural regeneration at rest, since a state of rest is not part of the test method at all. This is why in current literature the analyzed hysteresis area no longer is referred to as the "thixotropic area" or "rheopectic area", respectively. However, in most cases it is just the phase of regeneration which is important for practical users if leveling and sagging behavior after an application process of a coating has to be evaluated. As a consequence, it is therefore better to use the step test here, as explained above in Chapter 3.4.2.2a to determine thixotropic behavior.

c) Very simple evaluation methods (for evaluating thixotropic behavior)

1) Time-dependent viscosity ratio or "thixotropy index" (using a single test interval only)

Some users evaluate thixotropic behavior by the following simple testing and analysis method: with a single test interval, presetting a constant medium or high rotational speed n = const (or shear rate). Afterwards the "thixotropy index" (TI) is calculated as follows

$TI = \eta_1(t_1)/\eta_2(t_2)$

with the time points t_1 and t_2 [s], (e.g., $t_1 = 30s$, and $t_2 = 600s$). For flow behavior independent of time TI = 1, for time-dependent shear-thinning TI > 1, and for time-dependent shear-thickening TI < 1

Comment: Here, the term "thixotropy index" is misleading since this ratio quantifies time-dependent structural decomposition of a material only. Thixotropic behavior, however, can only be quantified if – directly after the break of a material's superstructure – also the subsequent time-dependent structural recovery under low-shear condition is evaluated (to TI, see also Note 3 in Chapter 3.4.2.2a, and Note 2 in Chapter 3.3.2). Therefore instead of TI, this ratio should better be called "time-dependent viscosity ratio under a constant shear load" or similar.

2) "Bingham build-up" (BBU) and "rate of build-up" (RBU) after a "20-minute gelation test"

consisting of two test intervals: first high, then low shear load [99].

Some users still perform the following simple test and analysis method consisting of two intervals. In the first part, preset is a constantly high rotational speed n_H [min^{-1}] for a period of $t_{10} = 10min$, and in the second part a constantly low speed n_L for another 10min = t_{20} (e.g. for ceramic suspensions, with $n_L = n_H/10$, for example, at $n_H = 100min^{-1}$ and at $n_L = 10min^{-1}$). Please be aware that these η-values are relative viscosity values if the test is performed using a spindle (which is a relative measuring system; see also Chapter 10.6.2). Here, instead of the shear stress often is used dial reading DR (which is the relative torque value M_{rel} in %), and the viscosity values are calculated then simply as $\eta = DR/n$ (with the rotational speed n in min^{-1}). Usually here, all units are ignored.

"Bingham build-up" (BBU) indicates the change of the relative viscosity values between the end of the second, low-shear interval and the first, high-shear interval.

Calculation: $BBU = \eta_L(n_L, t_{20}) - \eta_H(n_H, t_{10}) = (DR_L/n_L) - (DR_H/n_H)$.
Example: with $n_H = 100$, $n_L = 10$, $DR_H = 50$, $DR_l = 40$,
then: $BBU = (40/10) - (50/100) = 4 - 0.5 = 3.5$

"Rate of build-up" (RBU) is the partial change over the first two minutes in the second, low-shear interval related to the total change over the full ten minutes in this interval.

Calculation: $RBU = [\eta(n_L, t_{12}) - \eta(n_H, t_{10})] / [\eta(n_L, t_{20}) - \eta(n_H, t_{10})]$
$= [(DR_{L,12}/n_L) - (DR_H/n_H)] / [(DR_L/n_L) - (DR_H/n_H)]$
with time point t_{12} after 12 minutes (or after two minutes in the second interval, resp.)

Example: with n_H, n_L, DR_H and DR_L as above, and $DR_{L,\,12}$ = 30, then:
RBU = [(30/10) – (50/100)] / [(40/10) – (50/100)] = (3 – 0.5)/(4 – 0.5) = 2.5/3.5
Thus: RBU = 0.71 = 71%

Comment: Using this simple method, usually a **too high rotational speed is applied in the "low-shear" phase** to enable regeneration conditions which are related to practice. At these conditions, which are not really simulating low-shear conditions or even the state-at-rest, any **regeneration – if at all – is merely possible to a partial, very limited degree**.

3.4.3 Time-dependent flow behavior of samples showing hardening

Rheological tests are physical tests and usually it is assumed that the chemical structure of the sample does not change during the measurement. However, if it is aimed to determine time-dependent flow behavior during a hardening process, e.g. during a chemical curing reaction, then both: shear conditions and measuring temperature should be kept constant (thus testing takes place at isothermal conditions). The shear load should be sufficiently low to ensure an undisturbed hardening or curing process.

a) Minimum viscosity, gel time and gelation time, gel point and gelation point
An η(t)-diagram of a time-dependent hardening material is presented in Figure 3.39 (no. 3); see also Figure 3.48 (although here, temperature T on the x-axis has to be replaced by time t). Sometimes, the viscosity curve shows a minimum value η_{min} after a certain period of time. Evaluating coatings, the following information is important for practical users: At this point a wet coating layer may show optimum flow, spreading and leveling behavior. However, if the value of η_{min} is too low, a wet layer may be too thin, or it may show edge creep, giving not enough edge protection finally. On the other hand, if η_{min} is too high, a layer may not level out smoothly enough, and de-aeration may be not sufficient to obtain a surface without so-called "pinholes", "craters" or air bubbles finally. All these effects may reduce the gloss of the surface after all. Often, as a definition is taken: The gel time or gel point is reached when the viscosity has increased to a certain value which was pre-defined by the user. The terms gel time, gel point, gelation time and gelation point are often used with the same meaning. (Examples showing gel formation of modified corn starches are e.g. presented in [251].)

Example 1: Gelation of an epoxy resin
1) Time point at the viscosity minimum: after t = 90s (e.g. showing η_{min} = 60Pas then)
2) Gel time (when reaching the pre-defined value of η = 100Pas): after t = 200s

Example 2: "Isothermal viscosity development" of reactive resin mixtures (acc. to DIN 16945)
After mixing the components for 10 minutes, gelation time is determined as the period required to reach a previously defined upper limiting value of the viscosity (termed η_2). The following values are specified for three different consistencies (with η_1 as the viscosity directly after mixing):
1) for $\eta_1 \leq 250$mPas,   then: η_2 =   1500mPas
2) for 250mPas < $\eta_1 \leq 1000$mPas,   then: η_2 =   7500mPas
3) for $\eta_1 > 1000$mPas,   then: η_2 =   15,000mPas
Some further, but often very simple, test methods to determine whatsoever gel times or gelation times are briefly presented in Chapters 11.2.1d/e/f, 11.2.8e, 11.2.11a/b and 11.2.12a4/5/6; and in Chapter 11.2.13 information can be found about the following terms: incubation time, vulcanization time, scorch time, rise time, cure time of uncured rubbers and elastomers.

Note 1: Pot life, open time, and gelation time of resins
The following terms are often used to analyze the period of time for the use of a resin in an open container at the processing temperature. The reason for the increasing gel formation may be drying, oxidation or air humidity. Typical evaluation by a rotational test is a viscosity measurement at a low constant shear rate, e.g. at $\dot{\gamma} = 1s^{-1}$. Possible criteria for analysis are:
1) The **pot life** is the period of time without any noticeable viscosity change.
2) The **open time** is the period of time for which the material is still able to flow. This time is over if a previously defined, high limiting value of the viscosity is reached; this time point is often also called the **gelation time**.
Tip: It is more meaningful to evaluate here the **viscoelastic properties** using oscillatory tests (see Note 1 in Chapter 8.5.3b).

b) Comparison of controlled shear rate (CSR) and controlled shear stress (CSS) tests
For hardening or curing processes, CSR tests are involved with the disadvantage that the preset constant shear rate remains unchanged even if the viscosity values are continuously increasing and the sample is becoming more and more solid and therefore more inflexible. This can lead to irreversible partial destruction of the structure, therefore decisively disturbing a homogeneous hardening process. **CSS tests offer an advantage here:** The torque (or shear stress, resp.) is kept constant and the hardening sample causes the resulting rotational speed (or shear rate, resp.) to decrease. Using the CSS mode, the continuing hardening process is disturbed less and less, and therefore, it is less influenced by the resulting decreasing degree of deformation. Finally, if the resistance force of the solid sample is larger than the shear force which is applicable by the test instrument, the rotational speed will be displayed as n = 0 (or as shear rate $\dot{\gamma} = 0$, respectively). This indicates that the measuring system has come to a standstill after all.

Note 2: Advantage of oscillatory tests when determining the gel point
Nowadays, rotational tests should no longer be used for accurate investigations of processes such as gel formation, hardening or chemical curing reactions, since here, the process kinetics and reaction development, and therefore the test results, are often strongly influenced by the test conditions. Instead, oscillatory tests in the linear viscoelastic (LVE) deformation range should be performed since in this range, the user can be sure that the sample's structure is strained only to a very limited extent, therefore remaining undestroyed during the whole test. Using this mode of testing, as well the onset of gel formation as well as the gel time can be exactly determined in the form of the sol/gel transition point (see Chapter 8.5.3b with Figure 8.35). A further advantage of oscillatory tests is the fact that the test can be still continued, even if the hardened material is already showing gel-like character, behavior of a soft or even of a rigid solid. In this case, viscosity values would be displayed as already "infinitely high", being therefore no longer detectable by rotational tests. Further information on gel formation can be found in [128, 217, 220].

3.5 Temperature-dependent flow behavior and viscosity function

This section informs about the temperature-dependence of viscosity. Here, the degree of the shear load is kept at a constant value for each single test interval.

3.5.1 Description of the test

Preset
1) With controlled shear rate (CSR): $\dot{\gamma}$ = const (see Figure 3.36), and time-dependent temperature profile T(t), e.g. as an upward or downward ramp

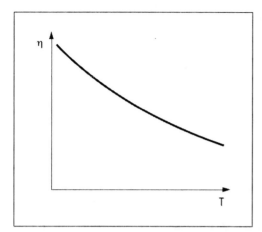

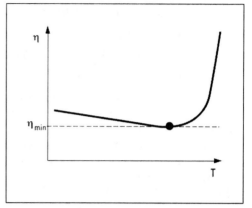

Figure 3.47: Temperature-dependent viscosity curve

Figure 3.48: Temperature-dependent viscosity curve of a material showing gel formation, hardening or curing

2) With controlled shear stress (CSS): τ = const (similar to Figure 3.36), and time-dependent temperature profile T(t), e.g. as an upward or downward ramp

Result
Function of τ (or $\dot{\gamma}$, respectively) and η dependent on the temperature (as in Figures 3.37 to 3.39, if time t is replaced by temperature T on the x-axis; see also Figures 3.47 and 3.48).

The shape of the curves of τ(T) and η(T) are similar because τ and η are proportional since $\tau = \eta \cdot \dot{\gamma}$ (here with $\dot{\gamma}$ = const). However, the curves of the functions of $\dot{\gamma}$(T) and η(T) show an inverse shape since $\dot{\gamma}$ and η are inversely proportional (since $\eta = \tau/\dot{\gamma}$, here with τ = const).

3.5.2 Temperature-dependent flow behavior of samples showing no hardening

Viscosity values always depend on the measuring temperature. In almost all cases, viscosity decreases if the sample is heated. Users mostly are interested in the **softening or melting temperature**. Highly viscous materials usually show greater temperature dependence compared to low-viscosity ones.

When cooling, the **solidification temperature** is determined. The term **crystallization temperature** is used for crystal-forming materials and **freezing point** for water, and sometimes, **congealing temperature** for gel-forming materials.

Usually, the temperature is presented on the x-axis of the η(T)-diagram on a linear scale, and viscosity on the y-axis, either on a linear scale, as shown in Figure 3.47, or on a logarithmic scale, if the values are covering a wide range.

Due to the strong dependence of all rheological parameters on the temperature, for the test protocol it is recommended to specify the measuring temperature exactly for each measuring point. **It is therefore essential to control the temperature carefully.** For information on **"freeze-thaw-cycle tests"** in order to evaluate temperature stability of emulsions, see Chapter 8.6.2.2b: oscillatory tests.

3.5.3 Temperature-dependent flow behavior of samples showing hardening

Figure 3.48 presents a temperature-dependent viscosity function of a material showing gel formation, hardening or curing.

Minimum viscosity, gel or gelation temperature, gel point or gelation point
The $\eta(T)$-curves mostly show a viscosity minimum η_{min}. This point is sometimes called the **softening** or **melting temperature**. For coatings, η_{min} decisively influences leveling, sagging, layer thickness, degassing, and edge creep behavior (e.g. of paints and powder coatings). The gel temperature or gel point is reached if the viscosity has increased to a certain limiting value which was pre-defined by the user. The terms gel temperature, gel point, gelation temperature and gelation point are often used with the same meaning.

Information given in Chapter 3.4.3 on the test conditions and on the **possible differences arising when using the different test modes controlled shear rate (CSR) or controlled shear stress (CSS)** also applies here. Among others, one of the **advantages of oscillatory tests** is the **accurate determination of the sol/gel transition temperature** after the onset of gel formation (see Chapter 8.6.3b).

Example 1: Gelation of an epoxy resin
1) Temperature at the viscosity minimum: at T = +165°C (e.g. showing η_{min} = 10Pas)
2) Gel temperature (when reaching the pre-defined value of η = 100Pas): at T = +175°C

Example 2: "Gelation Point" when cooling mineral oils (acc. to ASTM D5133 and D7110)
Here, oils are cooled down in the range of T = -5 to -40°C, at a constant cooling rate of $\Delta T/\Delta t$ = 1K/h, which corresponds to one degree Celsius per hour (according to D7110 with 3K/h). The "gelation point" is defined as the temperature value at which the oil viscosity is reaching η = 40,000mPas = 40Pas. Note: Here, a "Brookfield viscosity" is usually measured at the rotational speed of n = 0.3min⁻¹, and the shear rate is assumed to be $\dot\gamma$ = 0.2s⁻¹. In order to evaluate this relative viscosity values: see Chapter 10.6.2.

Note 1: Gelation index GI and GI temperature (according to ASTM D341 and D7110)
When cooling oils, the "gelation index" GI is determined in temperature steps of ΔT = 1K (Kelvin; acc. to ASTM D5133 in the range of T = -5 to -50°C) or of ΔT = 3K (acc. to ASTM D7110 in the range of T = -5 to -40 °C) between two measuring points. Calculation of the GI after each step as
GI = $(-1) \cdot [(\lg \lg \eta_1) - (\lg \lg \eta_2) / (\lg T_1 - \lg T_2)]$, with η in mPas and T in K,
until the maximum of the GI function is exceeded finally. See also ASTM D341: Empirical $\eta(T)$ equation by MacCoull, Walther and Wright [332].
Presentation in a diagram as [lg lg y(lg x)] curve, the so-called "gelation curve".
Analysis: The GI value is reached at the maximum of the GI curve, and the corresponding GI temperature (GIT) is determined. GI is therefore the value at the maximum viscosity increase.
Comment: Corresponding modern test are oscillatory tests to determine the sol/gel transition temperature (see Chapter 8.6.3b).

Note 2: Cloud point, pour point, freezing point, dropping point of petrochemicals
In order to analyze **cooling behavior of petrochemicals** there are many, and often very simple, measuring and analysis methods. Samples include **fuels** such as kerosene, gasoline, Diesel oils and heating oils, **lubricants** such as mineral oils and lubricating greases, **paraffins** and **waxes. Examples** (here in alphabetic order):
a) **Borderline pumping temperature** (BPT)
b) **Cloud point** (CP; ISO 3015; ASTM D2500, D5771 - stepped cooling method, D5772 - linear cooling rate method, D5773 – constant cooling rate method, D7397; DIN EN 23015)
c) **Cold filter plugging point** (CFP; ASTM D4539, D6371, D7467)

d) **Congealing point** (ISO 2207; ASTM D938; DIN ISO 2207)

e) **Crystallization point**

f) **Drop melting point** (ASTM D127)

g) **Dropping point** (ISO 2176; ASTM D566, D2265, D3954; DIN ISO 2176)

h) **Flocculation point**

i) **Freezing point** (ISO 3013; ASTM D2386)

k) **Gelation point**

l) **Melting point** (ASTM D87)

m) **No-flow point** (ASTM D7346)

n) **Pour point** (PP; ISO 3016; ASTM D97, D5950; DIN ISO 3016; IP 15)

o) **Solidification point**

p) **Wax appearance point** (ASTM D3117)

Here, some frequently used **methods of determination** and criteria are explained briefly.

M1) **Optical method:** When cooling, the **cloud point** is reached at the temperature at which the previously transparent sample becomes turbid and "cloudy" due to the precipitation of paraffin crystals.

M2) **Filter method:** When cooling, the **period of time** increases which is needed to pump a certain amount of sample through a fine-meshed filter (e.g. CFP point).

M3) **Sagging method:** When cooling, the **pour point** (PP) is reached at the temperature at which the sample still flows off a vertical surface. First the **solidification point** (SP) is determined as the point at which the sample is no longer able to flow at the transition from the liquid to the solid state. Afterwards is added $\Delta T = 3K$, thus: $PP = SP + 3K$.

M4) **Yield point method:** When cooling, the **yield point** occurs at the critical temperature when measuring with a simple **rotational viscometer**, for example, at a constant low shear rate or shear stress. This gives insight into problems with pumping processes such as start-up of pumping and continued pumping (e.g. BPT point, see also Chapter 11.2.11d to g).

M5) **Flow cup method:** When heating, the **dropping point** is reached at the transition temperature from the solid state to the liquid state, i.e., when the sample begins to flow through the orifice in the bottom of the flow cup.

Comment: All the above-mentioned simple methods are dependent on the test conditions and the skill of the tester. The methods therefore can only deliver relative values. Recommended methods for determining the crystallization or melting temperature are explained in Chapter 8.6.2.2a (oscillatory tests).

3.5.4 Fitting functions for curves of the temperature-dependent viscosity

The advantage of a fitting function is that it can be used to characterize the shape of a whole measuring curve using only a few model parameters, although the curve actually may consist of a large number of individual measuring points. A variety of viscosity/temperature fitting functions are mentioned in specialized literature, e.g. in ASTM D341, DIN 51563 and 53017 [215]. As an example, explained is below the Arrhenius relation which usually is used for low-viscosity liquids.

☞   For "Mr. and Ms. Cleverly"

The relations described here only apply to **thermo-rheologically simple materials**, i.e. materials, which do not change their structural character in the observed temperature range. Therefore, they do not change from the "sol state" to the "gel state" or vice versa (see also Chapter 8.2.4a, Note 1). In the temperature range which is related to practice, most **polymer solutions and polymer melts** are showing thermo-rheologically simple behavior. However, this applies usually **not to dispersions and gels**.

a) Arrhenius relation, flow activation energy E_A, and Arrhenius curve

An approximation model for kinetic activity in chemistry was developed in the general form by *Svante A. Arrhenius* (1859 to 1927) who introduced an "activation constant" [11]. The Arrhenius relation in the form of a $\eta(T)$ fitting function describes the change in viscosity for both increasing and decreasing temperatures:

Equation 3.6 $\eta(T) = c_1 \cdot \exp(-c_2/T) = c_1 \cdot \exp[(E_A/R_G)/T]$

with the temperature T in [K], (i.e. using the unit Kelvin), and the material constants c_1 [Pas] and c_2 [K] of the sample (where $c_2 = E_A/R_G$), the flow activation energy E_A [kJ/mol], and the gas constant $R_G = 8.314 \cdot 10^{-3}$ kJ/(mol $\cdot$ K)

Conversion between the temperature units:

Equation 3.7   T [K] = T [°C] + 273.15K

At a certain temperature, the **flow activation energy E_A** characterizes the energy needed by the molecules to be set in motion against the frictional forces of the neighboring molecules. This requires exceeding the internal flow resistance, with other words, a material-specific energy barrier, the so-called potential barrier [246].

The exponential curve function (Equation 3.6) occurs in a semi-logarithmic diagram as a straight line showing a constant curve slope if (1/T) is plotted on a linear scale on the x-axis (with the unit: 1/K), and η on a logarithmic scale on the y-axis. In this **lg η/(1/T) diagram**, the so-called **Arrhenius curve**, temperature-dependent behavior occurs as a downwardly or upwardly sloping straight line for a heating or a cooling process, respectively.

Note: Recommended temperature ranges for fitting functions (Arrhenius and WLF)

The Arrhenius relation is useful for low-viscosity liquids and polymer melts in the range of $T > T_g + 100K$ (with the glass-transition temperature T_g, see Chapter 8.6.2.1a) [233, 276]. For analysis of polymer behavior at temperatures closer to T_g, it is better to use the WLF relation. For more information on this time/temperature shift method TTS, see Chapter 8.7.1.

b) Viscosity/temperature shift factor a_T, and Arrhenius plot

The following holds for the viscosity/temperature shift factor a_T in general:

Equation 3.8 $a_T = \eta(T)/\eta(T_{ref})$

The dimensionless factor a_T is the ratio of the two viscosity values at the temperature T and at the reference temperature T_{ref}. For polymers, this relation is only valid for the values of the zero-shear viscosity η_0. The following holds for the **temperature shift factor according to Arrhenius** (with T in [K]):

Equation 3.9 $a_T = \exp\left[\dfrac{E_A}{R_G}\left(\dfrac{1}{T} - \dfrac{1}{T_{ref}}\right)\right]$

The semi-logarithmic, so-called **Arrhenius plot** presents the temperature shift factor a_T on the y-axis on a logarithmic scale versus the reciprocal temperature 1/T on the x-axis (with the unit: 1/K).

In order to estimate viscosity values at temperatures at which no measuring values are available, proceed as follows:
1) Select T_{ref} (e.g. a temperature, at which an η-value is available).
2) Calculate the shift factor a_T for another available $\eta(T)$-value (using Equation 3.8).
3) Calculate the flow activation energy value E_A (using Equation 3.9).
4) Calculate the shift factor a_T for the desired $\eta(T)$-value (using Equation 3.9).
5) Result: Calculate the desired $\eta(T)$-value (using Equation 3.8).

Table 3.4: Temperature-dependent viscosity values, see the example of Chapter 3.5.4b

T [°C]	50	60	70	80
T [K]	323	333	343	353
η [Pas]	2.00	1.55	1.26	1.00

Example: Calculation of flow activation energy and viscosity/temperature shift factor of a mineral oil, and determination of viscosity values at further temperatures

From a mineral oil is known: η_1 = 2.00Pas (at T_1 = +50°C = 323K), and η_2 = 1.00Pas (at T_2 = +80 °C = 353K). Desired is the viscosity value η_3 at T_3 = +60 °C (= 333K).

1) Here is selected: T_{ref} = T_1
2) a_T is calculated for T_2 (as a_{T2}): a_{T2} = η_2 $(T_2)/\eta_1$ (T_{ref}) = 1.00Pas/2.00Pas = 0.5
3) E_A is calculated with T_{ref} = 323K and T_2 = 353K:
$\ln (a_T)$ = $([E_A/R_G] \cdot [(T_2)^{-1} - (T_{ref})^{-1}])$, therefore: E_A = $([R_G \cdot \ln (a_T)] / [(T_2)^{-1} - (T_{ref})^{-1}])$
E_A = $8.314 \cdot 10^{-3}$ (kJ/mol · K) · $\ln(0.5)/(2.83 \cdot 10^{-3} - 3.10 \cdot 10^{-3})$ 1/K = 21.3kJ/mol
4) a_T is calculated for T_3 = 333K (as a_{T3}): a_{T3} = $\exp ([E_A/R_G] \cdot [(T_3)^{-1} - (T_{ref})^{-1}])$
a_{T3} = $\exp ([21.3/8.314 \cdot 10^{-3}] \cdot [3.00 \cdot 10^{-3} - 3.10 \cdot 10^{-3}])$ = $\exp (-0.256)$ = $1/e^{0.256}$ = 0.774
5) Result: η_3 (T_3) = $a_{T3} \cdot \eta_1(T_{ref})$ = 0.774 · 2.00Pas = 1.55Pas

The already mentioned and a further calculated temperature-dependent viscosity value are presented in Table 3.4.

𝒢𝒻   End of the Cleverly section

3.6 Pressure-dependent flow behavior and viscosity function

Conversion of pressure units:

Equation 3.10   1bar = 0.1MPa = 10^5Pa = 10^5N/m²

(and 1000psi = approx. 6.89MPa, however, psi is not an SI-unit; see also Chapter 14.3i)

In most cases, the viscosity values of fluids are increasing with increasing pressure (fluids are gases and liquids). However, **liquids** are influenced very little by the pressure applied since liquids, in contrast to gases, are almost **non-compressible** at low and medium pressures. In order to explain the reason for this, you can imagine a volume element filled with the following materials: **For gases** the degree of **space filling** is only **around 0.1%, for liquids** however it is **around 95%**, and **crystalline solids** are even **completely filling the space**. Fluids, in contrast to solids, are able to flow since their atoms and molecules can change places due to the free space in between them [198]. The following comparison illustrates this point: For most liquids, the fairly considerable change in pressure of Δp = 0.1 to 30MPa (= 1 to 300bar) causes about the same small change in viscosity like the small change in temperature of ΔT = 1K.

Even under the enormous pressure difference of Δp = 0.1 to 200MPa (= 1 to 2000bar), the viscosity increase of low-molecular liquids is usually showing just a factor of 3 to 7 only. However, for highly viscous mineral oils this factor can be up to 20,000 and for synthetic oils even up to 8 million. In technical application processes, very high peak pressures can occur, for example, for drilling fluids or muds in deep oil wells showing p = approx. 15 to 30MPa (= 150 to 300bar) [95], and for lubricants in cogwheels, in ball bearings and sliding bearings or in gears showing p > 1GPa (= 10^{10}bar) [18].

For most liquids, the viscosity values are increasing with increasing pressure since the amount of free volume within the internal structure is decreasing due to compression, and therefore, the molecules are more and more limited in their mobility. This increases the internal frictional forces and, as a consequence, also the flow resistance. The dependence of $\eta(p)$ is stronger if the molecules show a higher degree of branching.

Water does not behave like most other liquids. For T < +32 °C, at a pressure of up to p = 20MPa (= 200bar), above all the structure of the three-dimensional network of hydrogen bridges will be destroyed. In this temperature range, the mentioned network is relatively strong compared to the structures of other low molecular liquids. Therefore here, the viscosity values are decreasing with increasing pressure. For T > +32 °C however, water begins to behave like most other liquids, and then, viscosity increases with increasing pressure.

For further information on pressure-dependent rheological behavior: see [215, 276].

&⌐   For "Mr. and Ms. Cleverly"

Viscosity/pressure shift factor a_p

The following holds for the viscosity/pressure shift factor a_p in general:

Equation 3.11 $a_p = \eta(p)/\eta(p_{ref})$

The dimensionless factor a_p is the ratio of the two viscosity values at the pressure p and at the reference pressure p_{ref}. For polymers, this relation is only valid for the values of the zero-shear viscosity η_0. The **viscosity/pressure coefficient** α_p is defined as follows:

Equation 3.12 $a_p = \exp(\alpha_p \cdot \Delta p)$

with α_p in [1/MPa = MPa^{-1}], at the pressure difference $\Delta p = p - p_{ref}$ [MPa]

Typical values of α_p for liquids are in between +0.005 and +0.05MPa^{-1} (or $5 \cdot 10^{-3}$MPa$^{-1} \leq \alpha_p \leq 50 \cdot 10^{-3}$MPa^{-1}). For water at T < +32°C, the value of α_p is negative, since viscosity decreases with increasing pressure as explained above; for T > +32°C, α_p is positive, i.e. viscosity increases now with increasing pressure.

In order to estimate viscosity values at pressures at which no measuring values are available, proceed as follows:
1) Select p_{ref} (i.e. a pressure, at which an η-value is available), usually is selected:
p_{ref} = 0.1MPa (= 1bar).
2) Calculate the shift factor a_p for another available $\eta(p)$-value (using Equation 3.11).
3) Calculate the coefficient α_p (using Equation 3.12).
4) Calculate the shift factor a_p for the desired $\eta(p)$-value (using Equation 3.12).
5) Result: Calculate the desired $\eta(p)$-value (using Equation 3.11).

Example 1: Calculation of viscosity/pressure coefficient and shift factor of a mineral oil, and determination of viscosity values at further pressures
From a mineral oil is known: η_1 = 0.300Pas (at p_1 = 0.1MPa = 1bar), and η_2 = 2.22Pas (at p_2 = 100MPa = 1000bar). Desired is the viscosity value η_3 at p_3 = 75MPa (= 750bar).
1) Here is selected: $p_{ref} = p_1$ = 0.1MPa
2) a_p is calculated for p_2 (as a_{p2}): $a_{p2} = \eta_2(p_2)/\eta_1(p_{ref})$ = 2.22Pas/0.300Pas = 7.40
3) α_p is calculated, with $\Delta p = \Delta p_{21} = p_2 - p_{ref}$:
$\ln(a_p) = \alpha_p \cdot \Delta p$, and thus: $\alpha_p = \ln(a_p)/\Delta p$
$\alpha_p = \ln(7.40)/(100-0.1)$MPa = 2.00/99.9MPa = 0.02MPa^{-1} (= $20 \cdot 10^{-3}$MPa^{-1})
4) a_p is calculated for p_3 (as a_{p3}), with $\Delta p = \Delta p_{31} = p_3 - p_{ref}$:
$a_{p3} = \exp(\alpha_p \cdot \Delta p_{31}) = \exp[0.02MPa^{-1} \cdot (75 - 0.1)MPa] = e^{1.5}$ = 4.48
5) Result: $\eta_3(p_3) = a_{p3} \cdot \eta_1(p_{ref})$ = 4.48 · 0.300Pas = 1.34Pas

Table 3.5: Pressure-dependent viscosity values, see the example of Chapter 3.6

p [MPa]	0.1	1	10	25	50	75	100
η [Pas]	0.300	0.306	0.366	0.495	0.813	1.34	2.22

The already mentioned and some further pressure-dependent viscosity values are presented in Table 3.5.

Example 2: Calculation of the viscosity/pressure coefficient of a crude oil

From a crude oil is known: η_1 = 36mPas (at p_1 = 0.1MPa = 1bar),
and η_2 = 45mPas (at p_2 = 10MPa = 100bar). Desired is the α_p-coefficient.

1) p_{ref} = p_1 = 0.1MPa

2) a_p = $\eta_2 (p_2)/\eta_{ref} (p_1)$= 45/36 = 1.25

3) α_p = ln $(a_p)/\Delta p$ = ln $(1.25)/(10 - 0.1)$MPa = 0.0225MPa^{-1} (= 22.5 · 10^{-3} MPa^{-1})

෫ End of the Cleverly section

4   Elastic behavior and shear modulus

In this chapter are explained the following terms given in bold:

Liquids		Solids	
(ideal-) viscous flow behavior	viscoelasticflow behavior	viscoelastic deformation behavior	(ideal-) elastic deformation behavior
Newton's law	Maxwell's law	Kelvin/Voigt's law	Hooke's law
flow/viscosity curves	creep tests, relaxation tests, oscillatory tests		

4.1   Introduction

Before 1990, there were only few users of torsional and oscillatory rheometers performing tests on solid materials under scientific measuring conditions. Meanwhile however, the progress in technology of the corresponding instruments has enabled many users to characterize the elastic behavior of solid samples also in a range of very low deformations, and therefore, in a non-destructive range.

4.2   Definition of terms

The following section uses the **Two-Plates-Model** in order to define further rheological parameters (see also Chapter 2.2). The lower plate is stationary (deflection s = 0). The upper plate with the (shear) area A is deflected by the (shear) force F and the resulting deflection s is measured (see Figure 4.1). Between the plates there is the constant distance h, and the sample is sheared in this shear gap. It is assumed that the following **shear conditions** are given:
1) The sample shows **adhesion** to both plates **without any wall-slip effects**.
2) **The sample is deformed homogeneously** throughout the entire shear gap, i.e., no inhomogeneous "plastic deformation" is occurring (see also Chapter 3.3.4.3d and Figure 2.9).

Accurate calculation of the rheological parameters is only possible if both conditions are met.

The real geometric conditions in rheometer measuring systems are not as simple as in the Two-Plates-Model. However, if a shear gap is narrow enough, the necessary requirements are largely met and the definitions of the following rheological parameters can be used.

4.2.1   Deformation and strain

Definition of the **shear deformation**, also termed **shear strain**:

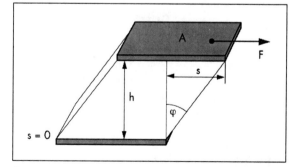

Equation 4.1 $\gamma = s/h$

γ (pronounced: "gamma"), with the deflection path s [m] and the distance

Figure 4.1: Two-Plates-Model for shear tests to illustrate deformation of a material in the shear gap

Thomas G. Mezger: The Rheology Handbook
© Copyright 2011 by Vincentz Network, Hanover, Germany
ISBN 978-3-86630-864-0

h [m] between the plates, see Figure 4.1. The following holds: $s/h = \tan\varphi$, with the deflection angle φ [°], (phi, pronounced: "fee" or "fi").

The unit of the shear deformation γ is [1], it is therefore dimensionless.

Most samples have to be measured at very low γ-values in order to remain in that limited part of the deformation range which can be analyzed scientifically. Therefore, in most cases, **it is useful to specify the γ-values in %** ($= 0.01 = 10^{-2}$).

Example: A deformation of $\gamma = s/h = 0.1 = 10\%$ occurs in a shear gap of h = 1mm if one of the plates is deflected by s = 0.1mm.

Note: Use of the terms deformation and strain
Sometimes, the terms "deformation" and "strain" are used as synonyms. In order to use a clear language, "strain" should be chosen if a **controlled shear strain** test is performed. And **"deformation"** should be selected to outline the consequences for the passively reacting sample if a **controlled shear stress** test is carried out. However, in many industrial laboratories people use both terms without making a difference. In most other languages besides English there are existing no different terms for γ when performing these two different test modes, and therefore then, in both cases the term "deformation" is used.

⌁⌐   For "Mr. and Ms. Cleverly"

The relation between deformation γ and shear rate $\dot{\gamma}$

The symbol $\dot{\gamma}$ for the shear rate is derived from γ. The following holds:

Equation 4.2 $\Delta\gamma/\Delta t = (\gamma_1 - \gamma_0) / (t_1 - t_0)$

with the deformation γ_0 [%] at the beginning and γ_1 [%] at the end of the test, with the time points t_0 [s] at the beginning and t_1 [s] at its end, with the change in deformation $\Delta\gamma$ [%] and the test period as time interval Δt [s]. Using the scientific notation for infinitesimal parameters:

Equation 4.3 $d\gamma/dt = \dot{\gamma}$

Therefore, a shear rate $\dot{\gamma}$ is an infinitely small change in deformation ($d\gamma$) which takes place in an infinitely short time period (dt). In other words (according to ASTM D4092): The shear rate $\dot{\gamma}$ is "the time rate of change of shear strain". Expressed mathematically: $\dot{\gamma}$ with the unit [s^{-1}] is the time derivative of γ. In other words: **The shear rate is the time-dependent rate of deformation, or briefly, strain rate.**

⌁⌐   End of the Cleverly section

4.2.2 Shear modulus

When measuring ideally elastic solids at a constant temperature, the ratio of the shear stress τ and corresponding deformation γ is a material constant if testing is performed within **the reversible-elastic deformation range**, the so-called **"linear-elastic range"**. This material specific value is referred to as the shear modulus G and reveals information about the rigidity of a material. Materials showing comparably stronger intermolecular or crystalline cohesive forces exhibit higher internal rigidity, and therefore, also a higher G-value. (Previously, the shear modulus was sometimes also called **"modulus of elasticity in shear"** or **"rigidity modulus"**). Definition of the shear modulus:

Equation 4.4 $G = \tau/\gamma$

The unit of the shear modulus is [Pa], ("pascal"), and $1Pa = 1N/m^2$
For rigid solids, the following units are also used:
1 kPa ("kilopascal") $= 1000Pa = 10^3Pa$

1 MPa ("megapascal") = 1000kPa = 1,000,000Pa = 10^6Pa (= $1N/mm^2$)
1 GPa ("gigapascal") = 1000MPa = 1,000,000,000Pa = 10^9Pa

A previously used unit was [dyne/cm²], then: 1dyne/cm² = 0.1Pa. However, this is not an SI-unit. In Table 4.1 are listed values of shear moduli of various materials.

$\mathcal{GO}$ For "Mr. and Ms. Cleverly"

Information on parameters obtained from tensile tests
Conversion of G- and E-values

Equation 4.5 $E = 2 \cdot G (1 + \mu)$

with the **tensile modulus E** [Pa], often called **"modulus of elasticity"** or **"Young's modulus"**, and **Poisson's ratio** μ with the unit [1], (my, pronounced: "mu" or "mew"). For a brief

Table 4.1: Values of G- and E-moduli, and of Poisson's ratio μ, at the temperature T = +20 °C; own data and from [112, 247, 268, 276, 318]

Material	G modulus	μ	E modulus
Very soft gel structures (examples: spray coatings, salad dressings)	5 to 10Pa		
Soft gel structures (example: brush coatings)	10 to 50Pa		
Viscoelastic gels (typical dispersions, lotions, creams, ointments, pastes of food, cosmetics, pharmaceuticals, medicals)	50 to 5,000Pa (often 100 to 500Pa)		
Puddings (containing 5 / 7.5 / 10 / 15% starch)	0.1 / 0.5 / 1 / 5kPa		
Adhesives before hardening – soft paste structure – strong structure, e.g. filled sealants	0.1 bis 10kPa 50 bis 500kPa		
Gummi bears, jelly babies	10 bis 500kPa		
Spread cheese / soft / semi-hard / hard / extra hard cheese	1 / 10kPa 0.1 / 0.5 / 1MPa		
Butter (example): at T = +10 / +23°C	2 MPa / 50kPa		
Soft natural gums unfilled gums filled gums eraser gum (India rubber) technical elastomers hard rubbers (e.g. car tires)	0.03 to 0.3MPa 0.3 to 5MPa 3 to 20MPa 1MPa 0.3 to 30MPa 10 to 100MPa	0.49 0.40 to 0.45 0.35 to 0.40 0.40 to 0.45 0.35 to 0.40	0.1 to 1MPa 1 to 10MPa 10 to 50MPa 1 to 100MPa 30 to 300MPa
PU coating, highly viscous/rigid one-pack PU adhesive two-pack PU reactive adhesive	30kPa / 1.0GPa 1 to 10MPa 200 to 600MPa		100kPa / 2.5GPa
Bitumen (example): at T = 0 / -10 / -30 / -50°C	10 / 50 / 200 / 500 MPa		
Thermoplastic polymers, unfilled, unlinked (usually)	0.1 to 2GPa	0.30 to 0.35	1 to 4GPa

Material	G modulus	μ	E modulus
PE-LD	70 to 200MPa	0.48	200 to 600MPa
PE-HD	300 to 800MPa	0.38	0.7 to 2GPa
PP	0.2 to 0.5GPa	0.35	0.5 to 1.3GPa
PP, filled	1 to 3GPa	0.25	1.8 to 6.5GPa
PVC-P (plasticized, flexible, T_g > +20°C)	0.5 to 5MPa	0.40	1.5 bis 15MPa
PVC-U (unplasticized, rigid)	0.3 to 1GPa	0.35	1 bis 3GPa
PVC, filled	1 bis 3GPa	0.25	3 bis 8GPa
PEEK-CF (with 40 / 65% carbon fibres)			up to 80 / 155GPa
Pure resins	1 to 2GPa	0.40	3 to 5GPa
filled and fiber-reinforced resins (dependent on the fiber orientation)	2 to 12 (24) GPa	0.25 to 0.35	5 to 30 (60) GPa
Wood (axial)			4 to 18GPa
wood (radial)			0.3 to 0.6GPa
Ice (at T = -4°C)	3.7GPa	0.33	9.9GPa
Bone			18 to 21 GPa
Ceramics, porcelain	15 to 35GPa, 25GPa	0.20	40 to 80GPa
Marble stone	28GPa	0.30	70GPa
(Window) Glass	30GPa	0.15	70GPa
Aluminum (Al 99.9%)	28GPa	0.34	72GPa
Gold (Au)	28GPa	0.42	81GPa
Brass (Cu-Zn)	36GPa	0.37	100GPa
Cast iron	40GPa	0.25	100GPa
Bronze (Cu-Sn)	43GPa	0.35	116GPa
Steel	80GPa	0.28	210GPa
Diamonds			1200GPa

information on *Thomas Young* (1773 to 1829 [402]) and *Siméon D. Poisson* (1781 to 1840, [292]), see Chapter 13.2. Poisson's ratio μ is the value of the ratio of the lateral (transversal) deformation to the corresponding axial deformation, resulting from uniformly distributed axial stress below the proportional limit of the material (according to ASTM D4092; by the way, in this standard instead of the sign μ the sign ν is used). The following holds (e.g. according to DIN 13316):

Equation 4.6 $0 \leq \mu \leq 0.5$

The higher the value of Poisson's ratio, the more ductile is a material; or: The lower the μ-value, the more brittle is its behavior when breaking. Cork, showing μ = 0, is a material with one of the two extreme values. Therefore here

Equation 4.7 $E = 2 \cdot G$

On the other hand, for viscoelastic liquids occurs the other extreme value of μ = 0.50. In this case, there is no volume change when stressing or straining these kinds of materials. Close to that value are soft and very flexible rubbers showing μ = 0.49. The same value occurs when testing polymers exhibiting behavior of viscoelastic liquids at temperatures above the glass-transition temperature $(T > T_g)$. This applies also to other incompressible and isotropic materials. Hence, for these kinds of materials counts:

Equation 4.8 $E = 3 \cdot G$

Note: Conversion of G- and E-modulus values

In general, calculation of G-values from E-values, and vice versa, is not recommended since there is evidence that suggests the Poisson's ratio varies from material to material in the same material class, and may vary from temperature to temperature for the same material (according to ASTM D1043). Therefore, these conversions must be regarded as rough estimates only.

Stress/strain diagrams of tensile tests

Performing tensile tests, E-modulus values are determined in the linear-elastic range, i.e. in a range of very low strain values. In this range of a σ/ε diagram, the curve function shows a constant slope. The following applies here:

Equation 4.9 $E = \sigma/\varepsilon$

with the **tensile stress** σ [Pa], (pronounced: "sigma"), and the **tensile strain or elongation** ε in [%], (pronounced: "epsilon"). Further information on tensile tests can be found in Chapters 10.8.4.1 and 11.2.14; ISO 6892; DIN EN ISO 6892, DIN 50125 and [104, 160, 198, 234, 246].

🖝   End of the Cleverly section

4.3   Shear load-dependent deformation behavior

Experiment 4.1: Playing with a spiral spring

When a constant tensile or compression load is applied to the spring, it immediately deforms to a constantly remaining degree of deflection. After removing the load, the spring immediately recoils to the initial position.

Experiment 4.2: Playing with a steel ball

When the steel ball is dropped onto a rigid ground, for example, on a thick plate of marble stone, it bounces back almost to the initial height.

4.3.1 Ideally elastic deformation behavior according to Hooke

a) Hooke's law

Formally, ideally elastic (or Hookean) deformation behavior is presented by Hooke's law:

Equation 4.10 $\tau = G \cdot \gamma$

Robert Hooke (1635 to 1703) wrote in 1676, in his textbook "de potentia restitutiva" [193]: Deformation of solids is proportional to the applied force ("ut tensio sic vis"; meaning: as the extension so is the force). Based on later works on solid-state physics and mechanics by *Jacob Bernoulli* (in 1689) and *Leonhard Euler* (in 1736/1765, "mechanica" [126]), finally *Augustin Louis Cauchy* (in 1827 [71]) stated the accurate and modern form of Hooke's hypothesis.

If the τ/γ-function is presented in a diagram, ideally elastic behavior is displayed in the form of a straight line coming from the origin point, showing a constant slope. This slope value corresponds to the value of G. When presented as a function of $G(\gamma)$ or $G(\tau)$, both curves indicate a constant plateau value for G if the linear-elastic range is not exceeded.

The values of the shear modulus of ideally elastic or Hookean solids are independent of the degree and duration of the shear load applied.

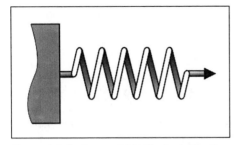

Figure 4.2: Spring model to illustrate ideally elastic behavior

🦯   For "Mr. and Ms. Cleverly"

For tensile tests, similarly to shear tests, Hooke's law applies in the following form (see also Chapter 4.2.2):

Equation 4.11 $\sigma = E \cdot \varepsilon$

b) The spring model
The spring model is used to illustrate the behavior of ideally elastic or Hookean solids (see Figure 4.2).

Ideally elastic behavior, explained by the behavior of a spring
1) When loading
Under a constant force, the spring immediately shows a corresponding deformation remaining at a constant value as long as the force is applied. When applying forces of differing strength to the spring, it can be observed in all cases: The resulting deflection is proportional to the force applied. The proportionality factor corresponds to the rigidity of the spring, and therefore, to the spring constant or elasticity, respectively.

2) When removing the load
As soon as the force is removed, the spring recoils elastically, and this means immediately, step-like and completely, returning to the initial state. No permanent deformation remains finally. Comparison: A Newtonian liquid shows no re-formation at all which can be illustrated by the behavior of the dashpot model, see Chapter 2.3.1b.

Summary: Behavior of the spring model
Under a constant load, the spring deforms immediately and remains deformed as long as the load is applied. After removing the load, the previously occurring deformation disappears immediately and completely. In other words: **After a load cycle, an ideally elastic material completely returns to the initial state.** These kinds of solids show absolutely no sign of viscous behavior.

Comparison: Metal spring and Hooke's law
For tension and compression springs, the force/deflection law holds (elasticity law according to Hooke):

Equation 4.12 $F = C_H \cdot s$

with the spring force F [N], the spring constant C_H [N/m] which is the "rigidity" of the spring and the index "H" is due to Hooke; and the deflection s [m] of the spring.

Here: F corresponds to the shear stress τ, C_H corresponds to the shear modulus G, and s corresponds to the shear deformation γ.

Note: Elastic behavior, and stored deformation energy
Deformation energy acting on an ideally elastic body during a shear process **will be completely stored** within the deformed material. When the load is removed, the stored energy can be recovered without any loss, enabling the complete reformation of the material. Therefore here, after deformation and reformation, **a completely reversible process** has taken place since the shape of the sample is unchanged after the experiment is finished.

For all materials showing ideally elastic deformation behavior, there are existing relatively strong interactive forces between their atoms or molecules. As examples for those very dense, stiff and rigid materials can be imagined stone and steel with crystalline structures at room temperature. If the linear-elastic range is exceeded, they show brittle fracture without any sign of time-dependent creep or creep recovery, e.g. in the form of very slow time-dependent deformation, partial reformation or stress relaxation, respectively. These kinds of materials do not show any visco-elastic behavior, since there is absolutely no viscous component available.

☞   End of the Cleverly section

5   Viscoelastic behavior

In this chapter are explained the following terms given in bold:

Liquids		Solids	
(ideal-) viscous flow behavior	**viscoelastic flow behavior**	**viscoelastic deformation behavior**	(ideal-) elastic deformation behavior
Newton's law	**Maxwell's law**	**Kelvin/Voigt's law**	Hooke's law
flow/viscosity curves	creep tests, relaxation tests, oscillatory tests		

5.1   Introduction

Viscoelastic (VE) materials are always showing viscous **and** elastic behavior simultaneously. The viscous portion behaves according to Newton's law which was presented in Chapter 2, and the elastic portion behaves according to Hooke's law which was explained in Chapter 4. Depending on their rheological behavior, VE liquids differ from VE solids. VE materials display a time-dependent, delayed response as well when a stress or a strain is applied as well as when it is removed.

5.2   Basic principles

5.2.1   Viscoelastic liquids according to Maxwell

Experiment 5.1: The thickened liquid in a glass beaker
After swirling the filled glass a few times and then stopping the motion immediately, the VE liquid still continues to swirl. When the liquid comes to rest, the surface is leveling again after a short period of time.

_GS_   For "Mr. and Ms. Cleverly"

5.2.1.1 Maxwell model

Behavior of a VE liquid can be illustrated using the combination of a spring and a dashpot in serial connection (see Figure 5.1). Both components can be deflected independently of each other. The spring model shows Hookean behavior as described in Chapter 4.3.1b, and the dashpot model displays Newtonian behavior as explained in Chapter 2.3.1b. The model was named in honor to _James C. Maxwell_ (1831 to 1879), who first presented the corresponding mathematical fundamentals (in 1867/1868, in the form of a differential equation) [245; 155].

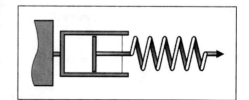

a) Viscoelastic flow behavior, illustrated by use of the Maxwell model (see Figure 5.2):
1) **Before loading:** Both components of the model exhibit no deformation.

Figure 5.1: Maxwell model

Thomas G. Mezger: The Rheology Handbook
© Copyright 2011 by Vincentz Network, Hanover, Germany
ISBN 978-3-86630-864-0

2) When loading

2a) When applying a constant force, only the spring displays an immediate deformation until it reaches a constant deflection value which is proportional to the value of the loading force. Therefore, immediately after the beginning of the load phase it is only the spring which is deformed.

2b) Afterwards, when still under the acting constant force, also the piston of the dashpot is beginning to move, and it is now continuously moving on as long as the force is applied. After a certain period of time under load, both components are showing a certain extent of deformation which corresponds as well to the degree of the force as well as to the time of loading.

As a result of the load phase, the time-dependent deformation function occurs in a $\gamma(t)$-diagram as an immediate deformation step, i.e. a step which is independent of time, followed by a straight line sloping upwards.

3) When removing the load

The spring recoils elastically, i.e. it moves back immediately and completely. The distance travelled by the dashpot, however, remains unchanged.

As a result of the phase after removing the load, the $\gamma(t)$-function occurs as an immediate reformation step, i.e. a step which is independent of time. Afterwards, the γ-value remains unchanged on a constant value.

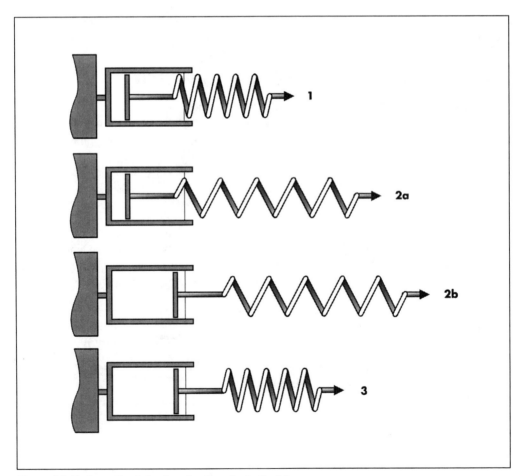

Figure 5.2: Deformation behavior of a viscoelastic liquid

Summary: Behavior of the Maxwell model

After a load cycle, these kinds of samples remain partially deformed. The extent of the reformation represents the elastic portion, and the extent of the permanently remaining deformation corresponds to the viscous portion. There is an **irreversible deformation process** taking place since such a sample occurs in a changed shape at the end of the process because its reformation is not complete, even after a long period of time at rest. These kinds of materials behave essentially like a liquid, and therefore, they are referred to as **viscoelastic liquids** or **Maxwell fluids**.

b) Differential equation according to the Maxwell model

In order to analyze Maxwellian behavior during a load cycle, the following differential equation is used (with the index "v" for the viscous portion and "e" for the elastic one):

Assumption 1: The total deformation is the sum of the individual deformations applied to the two model components.

$$\gamma = \gamma_v + \gamma_e$$

This applies also to the shear rates: $\dot{\gamma} = \dot{\gamma}_v + \dot{\gamma}_e$
since $\dot{\gamma} = d\gamma/dt$ (as explained in Chapter 4.2.1)

Assumption 2: The same shear stress is acting on each one of the two components.

$$\tau = \tau_v = \tau_e$$

Newton's law applies to the viscous element: $\quad \eta = \tau_v/\dot{\gamma}_v \quad$ or $\quad \dot{\gamma}_v = \tau_v/\eta$
Hooke's law applies to the elastic element: $\quad G = \tau_e/\gamma_e \quad$ or $\quad \gamma_e = \tau_e/G$
and $\dot{\gamma}_e = \dot{\tau}_e/G$ respectively, with the change of the shear stress over time as $\dot{\tau} = d\tau/dt$ [Pa/s], this is the time derivative of τ.

The sum of the shear rates results in the **differential equation according to Maxwell**:

Equation 5.1 $\qquad \dot{\gamma} = \dot{\gamma}_v + \dot{\gamma}_e = \tau_v/\eta + \dot{\tau}_e/G = \tau/\eta + \dot{\tau}/G$

The solution and use of this differential equation are described in Chapters 7.3.2c and 7.3.3.2 (relaxation tests) as well as in Chapter 8.4.2.1 (oscillatory tests/frequency sweeps).

&ℴ   End of the Cleverly section

5.2.1.2 Examples of the behavior of VE liquids in practice

a) Die swell, or post-extrusion swelling effect (see Figure 5.3)
Experiment 5.2: Extrusion, using a small toy extruder, producing spaghetti-like strands
1) Die swell does not occur when extruding **plasticine** (or wax). Immediately after the extrusion, the plasticine strands exhibit a stable form. Therefore, this material shows behavior of a solid.

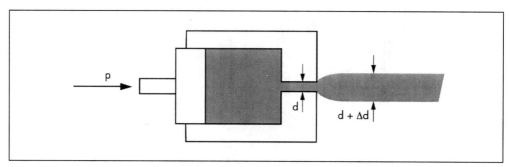

Figure 5.3: Die swell of a viscoelastic material after an extrusion (with pressure p, diameter d of the die, and diameter increase Δd of the extrudate)

2) Silicone polymer (unlinked PDMS) clearly displays **post-extrusion die swelling**. After a certain time when at rest, the extruded strands are slowly flowing and spreading, penetrating one another finally. Therefore, the silicone is showing behavior of a viscoelastic liquid.

Due to the high pressure in the extrusion die, the polymer molecules are deformed and oriented into shear direction. As soon as leaving the die, the molecules of a visco-elastic material are recoiling since they are in a stress-less state then. As a consequence, re-formation in shear direction occurs, and in order to compensate this, they are expanding into other directions, for example, right-angled to the direction of the extrusion. Therefore, the extrudate shows an increased diameter finally. See also Note 1 of Chapter 11.4.2.2b: high-pressure capillary viscometers; further information on die swell effects, including images, can be found in [45, 46, 203, 352].

Examples: Materials showing die swell
Polymers, when extruding rods, pipes or profiles (e.g. window frames), and when spinning synthetic fibers.

Sometimes, the following parameters are determined:
Swell ratio (or swell factor):
$SR = d_1/d = (d + \Delta d)/d$
Die swell (or relative swell):
$\Delta d/d = (d_1 - d)/d = (d_1/d) - 1$
with the diameters d of the die and d_1 of the extrudate, and their change Δd ($= d_1 - d$)

b) The Weissenberg effect when stirring
Experiment 5.3: The two stirrer vessels, containing water and a polymer solution
Water as an ideally viscous liquid exhibits the usual flow behavior during stirring, forming a vortex around the stirrer axis showing the lowest water level in the center of the vessel (see Figure 5.4: left side). The highest level occurs at the wall of the vessel. In contrast to that, the viscoelastic polymer solution creeps up the stirrer shaft when stirring; this is called the "Weissenberg effect" (see Figure 5.4: right side).

Karl Weissenberg (1893 to 1976) studied VE effects in detail [383]. In 1951, he was the first one who presented an instrument which really deserves to be called a "rheometer" (see also Chapter 13.4). When shearing VE liquids at certain conditions, such as sufficiently high rotational

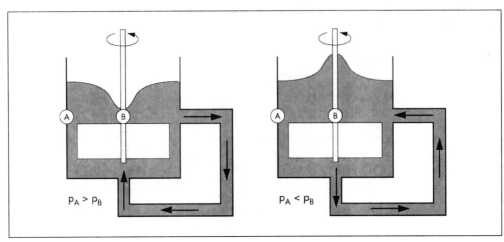

Figure 5.4: Liquids in two stirrer vessels, left: displaying ideally viscous flow behavior, and right: viscoelastic behavior showing the Weissenberg effect (with the pressures p_A at the inner wall and p_B in the center of the vessel)

speeds and at low temperatures, they are displaying the Weissenberg effect, which is also called the **"rod climbing effect"**. For these kinds of fluids, the pressure and flow conditions in the vessel are changing compared to the behavior of an ideally viscous fluid, and as a consequence, reversed flow direction of the VE liquid can be observed in a bypass pipeline as illustrated in Figure 5.4. Of course, these effects might have an enormous impact on the effectiveness of a stirring process, and the success in mixing corresponding liquids might be greatly limited. By the way, this effect also occurs if a magnetic stirrer is used, i.e. without using a stirrer shaft. In this case, a rising "mountain of liquid" can be seen in the center of the container above the rotating stirring rod. Further information on the Weissenberg effect, including images, can be found in [45, 46, 203, 352].

Examples: Materials showing the Weissenberg effect
Dough, emulsion paints, polymer melts, highly concentrated polymer solutions and dispersions when stirring at a sufficiently high rotational speed

c) Tack and stringiness when performing coating processes
Offset printing inks, adhesives, polymer-modified bitumen (PmB), grease and coatings often exhibit **tacky behavior** and **stringiness** with long filaments during application when printing, coating and painting with a brush or a roller, respectively (see Figure 5.5). See also Chapter 11.2.1g: finger test.

d) Mouth feeling
When sticky, tacky, stringy and ropy foodstuffs or pharmaceuticals containing thickening agents show a too high elastic portion, this may lead to unpleasant mouth sensation and might even cause problems when swallowing [382].

∽   For "Mr. and Ms. Cleverly"

Note 1: Explanation of the Weissenberg effect
For rotational systems, the **circumferential velocity** v increases with increasing distance from the axis of rotation: v [m/s] = $\omega \cdot r$, with the angular velocity ω [s^{-1}] = $(2\pi \cdot n)/60$, the radius r [m], and the rotational speed n [min^{-1}]. Since many polymer molecules cannot follow the rapid motion, e.g. at the edge of the stirrer vessel, they gather in the middle. This causes the "mountain of liquid" in the center of the container. As a simple **illustrative example** you can imagine: **On a roundabout** (merry-go-round) rotating at a constant speed, it is better to sit closer to the center than more outside. On the latter position, the "stress by motion" is increased due to the higher circumferential velocity compared to a place which is closer to the rotational axis.

Note 2: Tack and stringiness
Stringiness and tack can be tested and scientifically analyzed, adjusted and controlled via the **damping factor** tanδ obtained when performing oscillatory tests. See also the Notes and

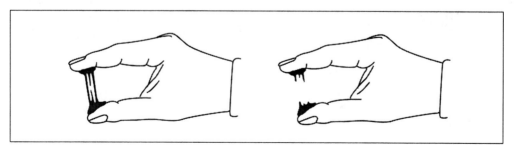

Figure 5.5: Comparison of two more or less tacky materials, one of them shows stringiness when strained at a high deformation [274]

Examples in Chapters 8.2.4a and 8.5.2.2d on automotive adhesives. A further testing mode are **tack tests**, see Chapter 10.8.4.2.

Note 3: Normal stresses
Extrudate swell, the Weissenberg effect, as well as all other viscoelastic effects are resulting from normal forces or normal stresses, respectively (see Chapter 5.3).

&ᶜ   End of the Cleverly section

Summary: Behavior of viscoelastic liquids
In order to get an optimal behavior for practical use, the viscous and elastic portions of visco-elastic materials should show a well-balanced ratio.

5.2.2 Viscoelastic solids according to Kelvin/Voigt

Experiment 5.4: Breaking a silicone rubber showing a low degree of cross-linking
When applying a high tensile load, this VE mass breaks into two pieces. Then, the edges are moving slowly, but only partially backwards. Finally, the edges of the two pieces remain stable in their dimension without exhibiting any further reformation.

Experiment 5.5: Comparison of two rubber balls
Two rubber balls having the same weight, but consisting of different base materials, are hit on a stone floor. The balls are bouncing back to different heights, clearly indicating different elastic behavior. One of the balls does not bounce back as high since obviously, this rubber consists of a viscoelastic material showing a comparably smaller elastic portion. See also Experiment 8.2 in Chapter 8.4.3a: Bouncing rubber balls (oscillatory tests, and damping factor tanδ).

&ᶜ   For "Mr. and Ms. Cleverly"

5.2.2.1 Kelvin/Voigt model

Behavior of a VE solid can be illustrated using the combination of a spring and a dashpot in parallel connection (see Figure 5.6). Both components are connected by a rigid frame. The mathematical fundamentals, in the form of a differential equation, were first presented by *Oskar E. Meyer* (in 1874 [250; 155]). However, due to later works of *William Thomson ("Kelvin",* in 1878 [359]) and *Woldemar Voigt* (in 1892 [376]), usually nowadays, the model is called the "Kelvin/Voigt model". Sometimes it is also named "Voigt/Kelvin model" (as in DIN 1342-1), or "Kelvin model" [18], or "Voigt model" [276].

a) Viscoelastic behavior, illustrated by use of the Kelvin/Voigt model (see Figure 5.7):
1) Before loading
Both components of the model exhibit no deformation.

2) **When loading**
Deformation is increasing continuously as long as a constant loading force is applied. The two components can only be deformed together, which means simultaneously and to the same extent, because they are connected by a rigid frame. Since its motion is slowed down by the presence of the dashpot, the spring cannot carry out an immediate, step-like deformation as it would do if it were independent of the dashpot.

As a result of the load phase, the deformation process occurs in the γ(t)-diagram as a curved,

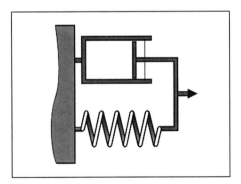

Figure 5.6: Kelvin/Voigt model

time-dependent exponential function (e-function) showing moderately increasing deformation values until a certain maximum value γ_{max} is reached finally.

3) When removing the load

The spring immediately aims to elastically step back to its initial shape, and this driving force is causing both components to reach their initial positions finally. However, this step occurs only after a certain period of time. Therefore, due to the presence of the dashpot there is a delayed process going on. But indeed, in the very end the model will show no longer any remaining deformation.

As a result of the phase after removing the load, the re-formation process occurs again as a curved, time-dependent e-function in the $\gamma(t)$-diagram, but now of course showing moderately decreasing deformation values. After a sufficiently long period of time the reformation will be completed, i.e. displaying $\gamma = 0$ again, like at the beginning of the test.

Summary: Behavior of the Kelvin/Voigt model

After a load cycle, these kinds of samples show indeed delayed but complete re-formation which fully compensates for the previously occurred deformation. There is a **reversible deformation process** taking place since such a sample occurs in an unchanged shape in the

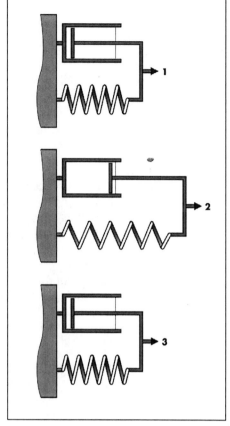

Figure 5.7: Deformation behavior of a viscoelastic solid

very end of the test. These kinds of materials behave essentially like a solid, and therefore, they are referred to as **viscoelastic solids** or **Kelvin/Voigt solids**.

b) Differential equation according to the Kelvin/Voigt model

In order to analyze Kelvin/Voigt behavior during a load cycle, the following differential equation is used (again with "v" for the viscous portion and "e" for the elastic one):

Assumption 1: The total shear stress applied will be distributed on the two model components.
$\tau = \tau_v + \tau_e$
Assumption 2: Deformation of both components occurs to the same extent, and this applies also to the shear rate.
$\gamma = \gamma_v = \gamma_e$ or $\dot{\gamma} = \dot{\gamma}_v = \dot{\gamma}_e$
with $\dot{\gamma} = d\gamma/dt$ (as explained in Chapter 4.2.1)
Newton's law applies to the viscous element: $\tau_v = \eta \cdot \dot{\gamma}_v$
Hooke's law applies to the elastic element: $\tau_e = G \cdot \gamma_e$
The sum of the shear stresses results in the **differential equation according to Kelvin/Voigt:**

Equation 5.2 $\tau = \tau_v + \tau_e = \eta \cdot \dot{\gamma}_v + G \cdot \gamma_e = \eta \cdot \dot{\gamma} + G \cdot \gamma$

The solution and use of this differential equation are described in Chapters 6.3.3 a/b and 6.3.4.3 (creep tests).

℘   End of the Cleverly section

5.2.2.2 Examples of the behavior of VE solids in practice

The following processes are taking place when a load is applied to a VE solid.

Elastic behavior

When performing a deformation process, a more or less large proportion of the **deformation energy** applied will be stored by the strained material. When removing the load, only the stored proportion of the deformation energy will be completely available afterwards. Therefore, this proportion of the deformation energy corresponds to the elastic behavior of a sample being the driving force for the re-formation process finally (see Chapter 4.3.1b).

Viscous behavior

In general, when a material is forced to flow, a certain **relative motion** occurs between its components (molecules, particles, superstructures). Therefore, a process is taking place which is always combined with **internal friction**, leading to the so-called **"viscous heating"**. The proportion of the deformation energy used up by the material during the shear process corresponds to the permanently remaining changes in the internal structure. Of course, this consumed portion is no longer available, e.g. for a complete reformation process after all. A part of the produced friction heat may heat up the sample itself, and another part may be conducted to the outside, being therefore lost (dissipated) for the sample, e.g. heating up the environment then (see also Chapter 2.3.1b).

Examples from daily practice:

a) Rubber buffers, and damping of mechanical vibrations

Viscoelastic behavior of a rubber buffer has to be balanced. If the viscous portion is too small, **absorption properties** are reduced, e.g. **to damp the effects of mechanical vibrations**. In this case, a buffer may be too rigid and inflexible, i.e., a large part of the deformation energy arising from mechanical oscillations cannot be used up in the energy-absorbing material components in order to render the consequences of this energy harmless.

Example: Insulating mats for damping vibrations of machines

b) Noise protection, and damping of sound waves

Acoustic absorption properties are reduced if viscoelastic **sound insulation materials** show a too small viscous portion. In this case, the material is unable to damp and absorb a sufficiently high portion of deformation energy which is occurring here in the form of the energy of motion produced by the molecules in the air. The goal is to transfer as much as possible of the mechanical energy of the sound waves into heat energy. Without sufficient absorption, this **sound energy** will be reflected by the material or it is directed through the material to the outside then. As a consequence, noise pollution arises. Specialists in acoustics are aiming to optimize the **damping factor** which is the crucial characteristic parameter to evaluate **absorption of structure-borne sound** of materials under dynamic, i.e., oscillating load [210]. This material-specific factor depends on the occurring frequency and on the temperature. See also Chapter 8.2.4a: damping factor, oscillatory tests.

Example: Noise protection materials for vehicles and other machines, walls of sound absorbing rooms, sound insulating mats. See also Example 3 of Chapter 8.7.1c: Acoustic damping behavior of technical rubbers.

c) Car tires, deformation energy and viscous heating

On the other hand, if a **damping material** has a too large viscous proportion, this can lead to excessive **viscous heating**, and therefore, to an even destructive degree of deformation finally. In this case, too much deformation energy would be transformed into heat, more than can be simultaneously stored or transported through the material to the outside. As a consequence, further supply of deformation energy might lead to partial or even complete destruction, i.e., to softening or even to the break of the material.

Example: Car tires, and lost deformation energy

Note by the way about car tires and deformation energy: Several million US dollars in fuel are wasted every day due to insufficient air pressure in tires. Tire air pressure that is 20% too low increases fuel consumption by 10% and reduces the working life of the tires by 15%[159].

d) Automotive bumpers

A bumper of a car should be able to absorb as much mechanical impact energy as possible, or alternatively, should lead it away to other components of the vehicle without any permanently remaining deformation or even destruction. For this purpose, material scientists, physicists and automotive engineers work closely together to develop design modifications and modern **VE composite materials** in order to optimize the spectrum of the mechanical and thermal properties. For example, integrated damping elements such as crushable bins or **crash-boxe**s are used which are deforming or folding in a defined, desired way when it comes to an accident, therefore absorbing most of the impact energy. Re-formable casings and facings are available which can resist collisions with a speed of up to 15km/h without any damage.

e) Shock absorbers, and the Kelvin/Voigt model

For a shock absorber consisting of a spring/dashpot combination, the spring should keep the deflection within a defined limit and the dashpot should absorb a part of the impact energy by the viscous dashpot fluid (e.g. oil). After the impact, the shock absorber should be pushed back to its initial position by the elastic behavior of the spring. Therefore here, it is aimed to perform a delayed but completely reversible process, comparable to the behavior of the Kelvin/Voigt model.

f) Equipment for sportive activities, and viscoelastic properties

Nowadays, success in sports also depends crucially on the VE properties of the equipment used which should neither be too rigid nor too flexible. Some games are only possible due to the equipment's VE behavior, and others are only attractive and tricky due to this. Sports goods, sports kit and sportswear etc. have become a paying playground for developers of VE composite materials.

Examples: The pole for pole vaulting; shoes when running and jumping (using "gel soles"); balls containing an air bubble or made of composite or solid material; the combination of the racket handle, racket strings, ball and net in tennis, squash, badminton, table tennis or ping-pong. See also Experiment 8.2 of Chapter 8.4.3a: Bouncing rubber balls (and tanδ).

g) Biological materials, and synthetic "bio-materials"

Life itself could not have developed in all its diversity, were it not for the complex interplay of natural VE liquids and VE solids.

Examples: Blood (if the elastic portion of this VE liquid is too high, there is the risk of a stroke), mucus, the synovial fluid (to lubricate the knee joint), the vitreous body of the eyes (the water content of this gel-like material is more than 98%); soft, flexible, swellable, taut or firm tissues (such as leaves, skin, cornea, body fat, cartilages); arteries and veins (if they are becoming too inflexible, this may lead to arteriosclerosis, which means "artery hardening"), twigs, tendons, barks, bones, wood, tree trunks.

The **spine** is a complex natural composite material consisting of disks (fibers with a gelatinous core), vertebrae (spongy bone containing bone marrow), ligaments, joints and connecting tissue. Its task is to give stability to the body, to absorb impacts, to show elasticity and also to enable us to make bending, stretching and torsional motions. These areas of medical technology/biotechnology are opening up new opportunities for designers and engineers who work on the modification and development of natural and artificial tissues, cartilages, bone cements and implants (such as TEMPs, tissue engineered medical products [206]).

When performing rheological tests on biological materials, it is important to take into consideration that these kinds of materials usually cannot be deformed evenly in the entire shear gap due to their mostly inhomogeneous structures and superstructures, related to macroscopic structure sizes in the range of around 0.1 to 1mm. This may lead to conditions at which the reproducibility of test results will be poor, and therefore, these kinds of materials often cannot be analyzed scientifically in terms of absolute values (see also Chapter 3.3.4.3d: inhomogeneous, "plastic" behavior).

Summary: Behavior of viscoelastic solids
Desired behavior of viscoelastic materials can be obtained only if the viscous and elastic portions are in a well-balanced ratio.

Example: Bamboo
Required behavior is given for bamboo bars only if both, flexibility (viscous behavior) **and** rigidity (elastic behavior) are present, showing the right mixture of both components. Only in this case, the material may show the optimal (viscoelastic) behavior. By the way: Also in East Asian nature philosophy, bamboo is an illustrative example to characterize balanced behavior or properties.

ᘒ⌒   For "Mr. and Ms. Cleverly"

5.3 Normal stresses

This chapter is intended for giving merely in a nutshell some basic information on normal stress tests and on the corresponding terminology in this field.

If a viscoelastic material is deformed, there are not only one-dimensional forces or stresses acting in the direction of the deformation. In fact, there is always a **state of three-dimensional deformation**. This can be illustrated using a **(3 · 3) tensor** (see also Chapter 13.2: *A.L.Cauchy*, 1827 [71], and *J.C. Maxwell*, 1855 [245]). It contains nine values, which can be displayed in a three-dimensional Cartesian coordinate system. In order to explain this, you can imagine the behavior of a cube-shaped volume element of our test material. The following stress values occur in a **stress tensor**:

1) Into x-direction, which is the shear or deformation direction: $\tau_{xx}, \tau_{yx}, \tau_{zx}$
In the Two-Plates-Model, the x-axis is pointing to the right (see Chapter 4.2, Figure 4.1).
2) Into y-direction, which is the direction of the shear gradient: $\tau_{xy}, \tau_{yy}, \tau_{zy}$
In the Two-Plates-Model, the y-axis is pointing upwards.
3) Into z-direction, which is the indifferent or neutral direction: $\tau_{xz}, \tau_{yz}, \tau_{zz}$

In the Two-Plates-Model, the z-axis is pointing out of the page, towards the reader.

The first index of the stress tensor values indicates the position of the area of the cube-shaped volume element on which the stress is acting. The term **"normal direction"** is used in mathematics and physics to determine the position of an area. For example here, the area with the x-direction as the normal direction means, it is the area on which the x-coordinate is standing in a right-angled position. The second index of the stress sensor values indicates the direction of the acting stress. Thus, for the x-direction counts here, it is pointing to the right side.

The tensor contains the **three normal stresses** τ_{xx}, τ_{yy} and τ_{zz}. For most viscoelastic liquids, the τ_{xx} component clearly shows the highest value of the three normal stresses. The other six tensor stresses are shear stresses. If a liquid is flowing, i.e. if viscous behavior clearly dominates, these six stress values can be reduced to the τ_{yx} component only, and the other five stress values can be ignored. This is the reason why in this textbook, beside of the chapter at hand, there is mentioned this one τ-value only, and therefore here, it can be written without any index.

The shear stress $\tau_{yx} = \tau$ is acting in x-direction on an area which is parallel to the plates of the Two-Plates-Model, and the normal direction of the corresponding area is the y-direction. We can imagine this area as an individual flowing layer. An **illustrative example** is a **stack of beer mats** (see Experiment 2.1 and Figure 2.2 of Chapter 2.2). Here, each individual beer mat represents an area with the y-direction as the normal direction, and each beer mat is moving into x-direction when applying a shear force F.

Usually, the following curve functions are presented in a diagram on a double-logarithmic scale.

1) The **1st normal stress difference**, as a function of the shear rate $N_1(\dot{\gamma})$

Equation 5.3 $N_1 \, [Pa] = \tau_{xx} - \tau_{yy}$

2) The **2nd normal stress difference**, as $N_2(\dot{\gamma})$

Equation 5.4 $N_2 \, [Pa] = \tau_{yy} - \tau_{zz}$

3) The **1st normal stress coefficient**, as $\Psi_1(\dot{\gamma})$; psi, pronounced: "psee" or "sy"

Equation 5.5 $\Psi_1 \, [Pa \cdot s^2] = N_1/\dot{\gamma}^2$

showing the **1st zero-normal stress coefficient** as a plateau value in the low-shear range

Equation 5.6 $\Psi_{1,0} \, [Pa \cdot s^2] = \lim_{\dot{\gamma} \to 0} \Psi_1(\dot{\gamma}) = const$

4) The **2nd normal stress coefficient**, as $\Psi_2(\dot{\gamma})$

Equation 5.7 $\Psi_2 \, [Pa \cdot s^2] = N_2/\dot{\gamma}^2$

showing the **2nd zero-normal stress coefficient** as a plateau value in the low-shear range

Equation 5.8 $\Psi_{2,0} \, [Pa \cdot s^2] = \lim_{\dot{\gamma} \to 0} \Psi_2(\dot{\gamma}) = const$

In order to determine the first normal stress difference, **the raw data measured by a rheometer are the values of the normal force F_N in [N] in axial direction** (y-direction). Normal forces of samples are forces acting into the direction of the shaft of the measuring bob, trying to push the upper plate or the cone upwards or the lower plate downwards, respectively (when using a parallel-plate or cone-and-plate measuring system). For Information on tests using the **normal force control** (NFC) option, see Chapters 9.4.6 and 9.7b.

Effects of normal forces can be found in industry, for example in the form of
a) **"post-extrusion swell"** or **"die swell"** effects (see Figure 5.3) and as **"melt fracture"** when extruding polymer melts
b) the **"Weissenberg effect"** or **"rod-climbing effect"** when stirring (see Figure 5.4)

When performing rotational tests with viscoelastic samples, the corresponding effects may occur in the form of streaks and similar defects on the surface of cylinder measuring systems or on the edges of cone-and-plate and parallel-plate systems. Further information on normal stresses can be found e.g. in DIN 13316 [18, 46].

Note: Lodge/Meissner relation
In 1976, *Arthur S. Lodge* (1942 to 2007) and *Joachim Meissner* (*1929) presented the following relation [229, 394]:

Equation 5.9 $N_{1,LM} = \gamma \cdot \tau$

Using this LM-relation, with the values of shear strain γ [1] and shear stress τ [Pa] which were preset or determined via relaxation tests (see Chapter 7), the 1st normal stress difference N_1

[Pa] can be calculated. Numerous tests with many standard polymers have confirmed this empirical rule. Comparisons have shown good correlation between

a) N_1 values which are measured directly by the normal stress sensor of a rheometer

b) $N_{1,LM}$ values, which are calculated by use of the LM relation from data which are measured by stress relaxation tests [222]

ᘗ   End of the Cleverly section

In the following Chapters 6 to 8, the most important types of tests are presented which can be performed to measure viscoelastic behavior: creep tests, relaxation tests, and oscillatory tests.

6   Creep tests

In this chapter are explained the following terms given in bold:

Liquids		Solids	
(ideal-) viscous flow behavior Newton's law	**viscoelastic flow behavior** Maxwell's law	**viscoelastic deformation behavior** Kelvin/Voigt's law	(ideal-) elastic deformation behavior Hooke's law
flow/viscosity curves	**creep tests**, relaxation tests, oscillatory tests		

6.1   Introduction

Creep and creep recovery tests are used to analyze the viscoelastic (VE) behavior performing two **shear stress steps**. This method is mostly used to examine chemically
unlinked and unfilled polymers (melts and solutions), but it is also suitable to evaluate the
behavior of chemically cross-linked polymers, gels and dispersions showing a physical-
chemical network of forces.

In industrial practice, however, creep tests have lost of importance since rheometers with
air bearings are meanwhile available enabling the user to directly preset very low rotational
speeds. Previously, creep tests have been the only way to produce very low shear rates, even if
indirectly, and therefore to obtain information about the behavior of polymers in the zero-shear
viscosity range, for example, to determine the average molar mass. Today however, with modern
rheometers measurements with direct shear rate control are possible as well in the low-shear
range. For these reasons since around 1995, creep tests are mostly carried out by scientists only,
especially to achieve results under the extreme low-shear conditions of $\dot{\gamma} < 10^{-3}\mathrm{s}^{-1}$.

6.2   Basic principles

Experiment 6.1: Creep and reverse creep of a hot-melt adhesive
At room temperature, the hot-melt adhesive at hand behaves like a flexible but tacky, soft solid.
Placed in a flat cylindrical container, the adhesive is sticking to both its base and its top cover.
Turning the cover manually with a constant force against the base causes the adhesive to
resist at first, but then yielding slowly, showing creeping motion with increasing deformation
until the cover reaches a twist angle of approximately 90° against the base of the container.
When releasing the force on the cover, external forces are no longer acting on the adhesive
and together with the cover it is turning back. This motion is faster at first, becoming slower
and slower then, reaching finally the initial position again.

6.2.1   Description of the test

Preset: Shear stress step function τ(t), see Figure 6.1

1) Immediate step in stress from $\tau = 0$ to τ_0, then keeping constantly τ_0 = const; this stress
interval lasts from the time point t_0 to t_2.

Thomas G. Mezger: The Rheology Handbook
© Copyright 2011 by Vincentz Network, Hanover, Germany
ISBN 978-3-86630-864-0

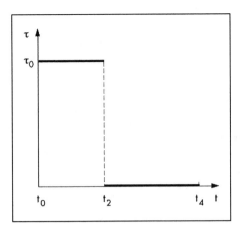

Figure 6.1: Preset of two intervals when performing creep tests:
1) stress phase, as an immediate step to a constant stress value, followed by the creep phase
2) rest phase, as an immediate step to zero-stress, followed by the creep recovery phase

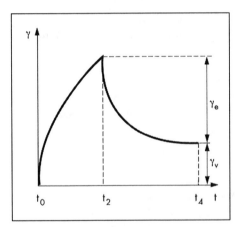

Figure 6.2: Creep and creep recovery curve, showing also the final values of the reformation γ_e and the permanently remaining deformation γ_v

2) Immediate step in stress from τ_0 back to $\tau = 0$, then remaining constantly at $\tau = 0$; this interval at rest lasts from t_2 to t_4.

Both steps should be performed as fast as possible. This requires the use of a highly dynamic rheometer drive.

The following values are found empirically for the **permissible maximum deformation** γ_{max}:
1) For polymers (unlinked and unfilled, solutions and melts):
$\gamma_{max} \leq 50\%$, however, sometimes even up to $\gamma_{max} = 100\%$
2) For most dispersions (i.e. emulsions, suspensions, foams), cross-linked polymers (such as elastomers, rubbers, thermosets), and gels:
$\gamma_{max} \leq 1\%$, however, sometimes counts $\gamma_{max} < 0.1\%$ only

When applying higher deformation values, there is the risk of exceeding the limiting value of the linear viscoelastic (LVE) range. In this case, the basic laws of rheology are no longer valid, i.e. the laws of Newton and Hooke and also the relations of Maxwell, Kelvin/Voigt and Burgers. More information about the LVE-range and further γ_{max} values, see Chapter 8.3.3.1: oscillatory tests, amplitude sweeps.

Example: Test preset for a polymer sample
1) Stress phase: step in stress to $\tau_0 = 500Pa$, this stress value is kept constantly for $t = 5min$
2) Rest phase: step to $\tau = 0Pa$, the sample remains unstressed for $t = 10min$

Measuring result: Creep curve and creep recovery curve $\gamma(t)$, see Figure 6.2

As a test result, the time-dependent deformation function $\gamma(t)$ is measured. In a diagram, the creep and recovery curve is displayed showing γ on the y-axis, and time t on the x-axis. Usually, both parameters are presented on a linear scale.

The first part of the curve in the time interval between t_0 and t_2 is termed **creep curve** (or deformation curve). The second part between t_2 and t_4 is referred to as **creep recovery curve** (or reformation curve). The reformation value γ_e indicates the elastic portion of the VE behavior, and the value of the finally remaining deformation γ_v represents the viscous portion.

6.2.2 Ideally elastic behavior

When presetting a step in stress, ideally elastic materials show an immediate, step-like deformation. After removing the load, the reformation takes place immediately and completely (see Figure 6.3). In the creep recovery phase, the complete deformation energy which was previously stored by the deformed material during the creep phase, can be recovered now to be used up for the reformation process.

Here, the following applies for the ratio of the finally occurring deformation values:
$\gamma_e/\gamma_{max} = 100\%$ and $\gamma_v/\gamma_{max} = 0$, with $\gamma_{max} = \gamma_v + \gamma_e$.
The terms are illustrated in Figure 6.2.

6.2.3 Ideally viscous behavior

As long as loaded by a constantly acting stress, ideally viscous materials are showing continuously increasing deformation. After removing the load, there is no reformation at all since these kinds of materials do not have any elastic portion (see Figure 6.4). Ideally viscous materials are not able to store any deformation energy during the load interval, and therefore after releasing the load, they finally remain deformed to the same extent as they showed in the end of the load phase.

Here, the following applies for the ratio of the finally occurring deformation values:
$\gamma_v/\gamma_{max} = 100\%$ and $\gamma_e/\gamma_{max} = 0$

6.2.4 Viscoelastic behavior

For VE materials under a constant stress, the resulting deformation can be imagined to be devided in two portions: the first one occurs immediately and the second one delayed. After removing the load, delayed reformation takes place either partially or completely. The $\gamma(t)$-diagram shows for both creep and creep recovery curve, the shape of an increasing or decreasing exponential function, respectively (e-function; see Figure 6.5). The degree of the reformation depends on the elastic portion of the sample.

For **VE liquids** (1) after releasing the load and even after a longer rest phase, a certain extent of deformation is remaining permanently. This value represents the viscous portion. This behavior is exhibited by concentrated polymer solutions and polymer melts. Here, a load cycle is an irreversible process, since the shape of the sample remains permanently changed after the experiment is finished.

For **VE solids** (2), delayed but complete reformation occurs if the period of testing is sufficiently

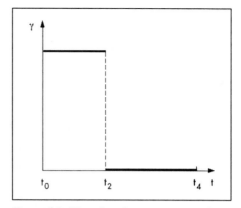

Figure 6.3: "Creep" and creep recovery of an ideally elastic solid (inverted commas, since there is no creep at all but only an immediate step in the stress phase)

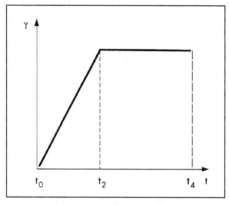

Figure 6.4: Creep and "recovery" curve of an ideally viscous liquid (inverted commas, since there is no recovery in the rest phase at all)

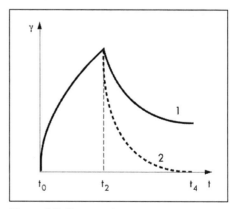

Figure 6.5: Creep and creep recovery curves of two different viscoelastic materials, both are showing delayed reformation: (1) partially for the VE liquid, and (2) completely for the VE solid

long. This behavior is displayed by chemically cross-linked materials, gels and concentrated dispersions with a gel-like structure at rest. Here, a load cycle is a reversible process since the shape of the test material will be the same again finally when compared to the initial shape.

6.3   Analysis

6.3.1 Behavior of the molecules

Behavior of a polymer sample can be illustrated by observing the process taking place in a conglomeration of polymer molecules, when regarding a single macromolecule only (see Figure 6.6).

a) At rest, before the step in stress
Still in a stress-less state, the chain of an individual polymer molecule occurs as a spherical molecule coil which may have many entanglements with neighboring chains. This is the state at rest requiring minimum expense of energy.

b) Under constant stress, in the creep phase
The observed macromolecule reacts to the constant shear stress showing slow, creeping motion. Forced by the constant load, the spherical coil begins to leave its state of rest exhibiting increasing deformation. It changes more and more to the shape of an ellipsoid whose axis is oriented lengthways between the shear direction and the direction of the shear gradient. More illustrative: An ellipsoid is shaped "like an American football". As a consequence, deformation is increasing now as well of the individual molecule, as well as of the superstructure of the whole polymer sample due to the entanglements between the molecule chains.

c) After releasing the stress, in the creep recovery phase
Immediately after the step-like removal of the stress, each one of the molecules is trying to return to a rest position bare of any deformation. Continuously, the molecules are recovering more and more from the previously occurred deformation, reducing it continuously by slow and compensating, creep motions within the superstructure. This is a delayed process which is also called "retarded". For unlinked polymers, the extent of reformation may tend to zero if the creep phase was long enough, i.e., if a correspondingly great number of disentanglements occurred in the stress interval. For samples showing a chemical or physical network, the extent of partial reformation corresponds to the elastic portion. Fully cross-linked polymers are reforming completely as shown in Figure 6.5 (no. 2) if the LVE range was not exceeded. Finally they will achieve the same shape as they displayed initially, before the stress was applied (as shown in Figure 6.6, no. 1).

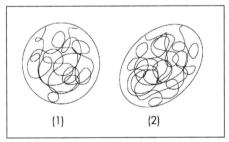

Figure 6.6: Deformation process of a polymer molecule when performing a creep test (Figure from [18])
(1) at rest, before applying the stress
(2) after a certain period of time under stress

In rheology, a delayed deformation or reformation process after applying or removing a stress is referred to as **retardation**. As a comparison: The term "relaxation" is used to describe the behavior at rest after applying strain (deformation), e.g. when performing relaxation tests (see Chapter 7).

⌒⌒   For "Mr. and Ms. Cleverly"

6.3.2 Burgers model

The following components are combined in series to analyze creep and creep recovery behavior:

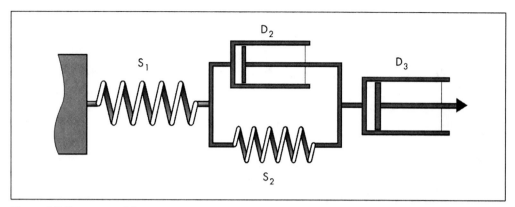

Figure 6.7: Burgers model

spring S_1; spring S_2 and dashpot D_2 (both in parallel); and dashpot D_3 (see Figure 6.7). Indeed, this is a combination of the Maxwell model (S_1 and D_3) **and** the Kelvin/Voigt model (S_2 and D_2). This combined model is called the "Burgers model", since in 1935 it was presented by *J.M. Burgers* (1895 to 1981) [60].

Analysis of the deformation behavior on the basis of the Burgers model is rather complex. A differential equation of the second order have to be solved using the following parameters: as well the shear stress τ, the first derivative of the shear stress $\dot{\tau}$ which is the time-dependent stress rate or "velocity of change in stress", and the second derivative $\ddot{\tau}$ which is the time-dependent change of the stress rate or "acceleration of stress"; as well as the deformation γ, the first derivative of the deformation $\dot{\gamma}$ as a function of time which is the rate of change in deformation or shear rate, and the second derivative $\ddot{\gamma}$ which is the "acceleration of deformation" [155, 237, 366].

Viscoelastic behavior, illustrated by use of the Burgers model:

a) Creep phase
When applying the force F, the following deformation behavior occurs:
1) Immediate, step-like deformation of spring S_1 at the beginning of the test
2) Delayed deformation of spring S_2 and dashpot D_2 (like the Kelvin/Voigt model)
3) Continuously increasing deformation of dashpot D_3 showing the constant rate of deformation (or shear rate) $d\gamma/dt = \dot{\gamma}_3$ = const now
After a sufficiently long test period, all springs and dashpots are deflected to a certain extent dependent on the test conditions, i.e. on the constant stress value and on the period of time of the stress applied.

b) Creep recovery phase
When removing the force F, the following reformation behavior occurs:
1) Immediate, step-like elastic reformation of spring S_1
2) Delayed reformation of spring S_2 and dashpot D_2 (like the Kelvin/Voigt model)
3) Dashpot D_3 remains completely deflected
If the sample completely returns to its initial position, then it is a VE material showing the character of a viscoelastic solid (see Figure 6.5, no. 2). In this case, the dashpot D_3 is without function in the Burgers model, and can therefore be ignored for analysis.

6.3.3 Curve discussion

In order to explain the analysis in detail, the creep curve (from t_0 to t_2) and the creep recovery curve (from t_2 to t_4) are divided into the following sectors, see Figure 6.8:

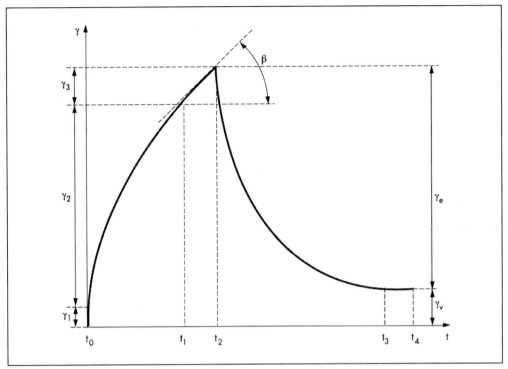

Figure 6.8: Creep and creep recovery curve γ(t), showing the parameters used for analysis

1) γ_1 as step-like, purely elastic deformation occurring immediately after the beginning of the test, idealized: "without any time delay, in zero-time" (behavior of spring S_1)

2) γ_2 as delayed viscoelastic deformation (spring S_2 together with dashpot D_2)

3) γ_3 as purely viscous deformation (dashpot D_3); after reaching steady-state behavior, i.e. a constant rate of deformation (or shear rate), the γ-curve is showing a constant slope angle β then

4) $\gamma_{max} = \gamma_1 + \gamma_2 + \gamma_3 (= \gamma_e + \gamma_v)$ is the maximum deformation, at the end of the stress phase

5) γ_e is the extent of reformation after the creep recovery phase (representing the elastic portion of the viscoelastic behavior)

6) γ_v is the extent of remaining deformation after the creep recovery phase (viscous portion)

a) Creep curve

Various denominations are used for the first test interval: creep curve, deformation curve, load(ing) phase, or stress phase.

Transient flow behavior with a non-constant rate of deformation (shear rate) $\dot{\gamma}$ occurs between the time points t_0 and t_1, then: $\gamma = f(\tau_0, t)$. Here, the slope value of the time-dependent deformation curve depends as well on the applied shear stress τ_0 as well as on the passing time. **Steady-state behavior** with $\dot{\gamma}$ = const occurs between the time points t_1 and t_2, when $\gamma = f(\tau_0)$. Here, the γ-curve shows a constant slope, now depending on the applied stress τ_0 only but no longer on the time. Steady-state creep is reached when the curve displays a constant slope angle β finally.

The creep function, describing the time-dependent deformation behavior during the stress phase, can be formulated as follows:

Equation 6.1 $\gamma(t) = \gamma_1 + \gamma_2(t) + \gamma_3(t) = (\tau_0/G_1) + (\tau_0/G_2) \cdot [1 - \exp(-t/\Lambda)] + (\tau_0 \cdot t)/\eta_0$

with the shear modulus G_1 [Pa] $= \tau_0/\gamma_1$ (corresponding to the spring constant of S_1 and visible in the creep curve as an immediate deformation step due to the purely elastic behavior); the retardation time Λ [s] $= \eta_2/G_2$ (pronounced: "lambda", see Chapter 6.3.4.3) with the shear modulus G_2

[Pa], (the spring constant of S_2) and the shear viscosity η_2 [Pas], (the dashpot constant of D_2); and the zero-shear viscosity η_0 [Pas], (the dashpot constant of D_3). The medium term of the formula is obtained from the differential equation according to Kelvin/Voigt (see Chapter 5.2.2.1b):

$\tau = G_2 \cdot \gamma + \eta_2 \cdot \dot\gamma$

The following applies for t = 0:

$\gamma(0) = (\tau_0/G_1) + (\tau_0/G_2) \cdot [1 - e^0)] + (\tau_0 \cdot 0)/\eta_0 = (\tau_0/G_1) + (\tau_0/G_2) \cdot (1 - 1) + 0$

thus: $\gamma(0) = (\tau_0/G_1)$

i.e., at the very beginning of the test the only element which is deflected is the spring S_1, which is deformed immediately, without any delay.

The following applies for t = ∞, or for practical users, after a "very long" time:

$\gamma(\infty) = (\tau_0/G_1) + (\tau_0/G_2) \cdot [1 - (1/e^\infty)] + C = (\tau_0/G1) + (\tau_0/G_2) \cdot (1 - 0) + C$

thus: $\gamma(\infty) = (\tau_0/G_1) + (\tau_0/G_2) + C$

i.e., S_1 and S_2 are fully deflected and therefore also D_2. The deformation of D_3 would be, strictly speaking, "infinitely" large ($\gamma_3 = \infty$), this is formulated here in terms of a C for a correspondingly large value which is reached in the very end of the creep phase.

The following applies for t = Λ, i.e., when reaching the retardation time:

$\gamma(\Lambda) = (\tau_0/G_1) + (\tau_0/G_2) \cdot [1 - (1/e)] + (\tau_0 \cdot \Lambda)/\eta_0$

thus: $\gamma(\Lambda) = (\tau_0/G_1) + 0.632 \cdot (\tau_0/G_2) + (\tau_0 \cdot \Lambda)/\eta_0$

i.e., S_1 is fully deflected, and up to this time point, S_2 and therefore also D_2 are deflected partially to an extent of 63.2%. D_3 however, is deflected to a very small and not significant extent only, since the third term has merely reached a relatively low value up to this point.

b) Creep recovery curve

Also for the second test interval various denominations are used: creep recovery curve, reformation curve, retardation curve, or rest phase.

Transient flow behavior during the reformation occurs between the time points t_2 and t_3. Finally, the deformation value reached is the maximum deformation γ_{max} reduced by γ_e. The latter represents the elastic portion of the viscoelastic behavior. For a test being sufficiently long, a VE liquid will show a certain, permanently remaining deformation value then (which is γ_v = const), and therefore **steady-state deformation behavior**. For a VE solid, steady-state (equilibrium state) is reached when the material displays complete reformation after all.

The creep recovery function, describing the time-dependent reformation behavior during the rest phase, can be formulated as follows:

Equation 6.2 $\quad \gamma(t) = \gamma_{max} - \gamma_1 - \gamma_2(t) = \gamma_{max} - (\tau_0/G_1) - (\tau_0/G_2) \cdot [1 - \exp(-t/\Lambda)]$

The following applies for t = 0:

$\gamma(0) = \gamma_{max} - (\tau_0/G_1) - (\tau_0/G_2) \cdot [1 - e^0] = \gamma_{max} - (\tau_0/G_1) - (\tau_0/G_2) \cdot (1 - 1)$

thus: $\gamma(0) = \gamma_{max} - (\tau_0/G_1)$

i.e., immediately after releasing the load, only the spring S1 recoils without any delay.

The following applies for t = ∞, or for practical users, after a "very long" time:

$\gamma(\infty) = \gamma_{max} - (\tau_0/G_1) - (\tau_0/G_2) \cdot [1 - (1/e^\infty)] = \gamma_{max} - (\tau_0/G_1) - (\tau_0/G_2) \cdot (1 - 0)$

thus: $\gamma(\infty) = \gamma_{max} - (\tau_0/G_1) - (\tau_0/G_2)$

i.e., S_1 and S_2, and therefore also D_2 are fully reset; D_3 however, remains deflected. At the very end of the creep recovery phase it is merely the deformation value γ_v which still occurs.

The following applies for t = Λ, i.e., when reaching the retardation time:

$\gamma(\Lambda) = \gamma_{max} - (\tau_0/G_1) - (\tau_0/G_2) \cdot [1 - (1/e)]$

thus: $\gamma(\Lambda) = \gamma_{max} - (\tau_0/G_1) - 0.632 \cdot (\tau_0/G_2)$

i.e., S_1 is fully recoiled; and up to this time point, S_2 and therefore also D_2 is reset by 63.2%. D_3 however, is still fully deflected as it was at the end of the creep phase.

6.3.4 Definition of terms

6.3.4.1 Zero-shear viscosity

Zero-shear viscosity η_0 is one of the most important rheological parameters used in polymer industry. The value of η_0 is determined at the end of the creep phase as

$\eta_0 = \tau_0/\dot{\gamma}_3 = \tau_0/\tan\beta$

Unit: [Pas]; with the constant shear rate $\dot{\gamma}_3 = \gamma_3/(t_2 - t_1) = \tan\beta$ which is independent of time (see Figure 6.8); $\dot{\gamma}_3$ is often referred to as $\dot{\gamma}_0$.

The value of zero-shear viscosity corresponds to the behavior of the dashpot D_3 in the Burgers model, resulting in a "creep rate" value, i.e. in a very low rate of deformation or shear rate. In order to avoid exceeding the limit of the LVE deformation range between the time points t_1 and t_2, the total deformation value should not increase too much in this period (see also Chapter 6.2.1: values of γ_{max}).

a) Determination of the value of zero-shear viscosity using creep curves
1) Manual or visual determination, respectively
In practice, the aim is to produce a creep curve showing a constant slope at the end of the creep phase. In order to get a useful basis for the analysis, as a rule of thumb, at least 10% of the measuring points of the entire creep curve should occur within this range of steady-state flow. Taking a ruler, it is useful to draw a straight line in this test interval to facilitate visual evaluation. The following results are obtained from this sector of the curve: The change in deformation $\Delta\gamma$ read-off on the γ-axis, and the corresponding period of time Δt read-off on the time axis. The first calculation step is: $\Delta\gamma/\Delta t$; this is the deformation change in the corresponding period of time, or time-dependent rate of deformation (shear rate) $\dot{\gamma}_0$. The second calculation step (see Figure 6.8):

Equation 6.3 $\eta_0 = \tau_0/(\Delta\gamma/\Delta t) = \tau_0/[\gamma_3/(t_2 - t_1)] = \tau_0/\dot{\gamma}_0$

2) Automatic determination using an analysis software
First of all, the user defines the part of the final sector of the creep curve to be used for the characterization of steady-state behavior, i.e. when $d\gamma/dt$ = const is reached. Like above, as a useful rule of thumb at least 10% of the measuring points of the whole creep curve should occur within this steady-state sector to outline the corresponding curve interval. Then, the straight analysis line is fitted in this curve sector by the software program, finally determining also the value of η_0.

b) Zero-shear viscosity and average molar mass
Zero-shear viscosity η_0 is a material constant for unlinked polymers and using this value, a relative specification can be given for the average molar mass M, since the following proportionality holds:

Equation 6.4 $\eta_0 \sim M^{3.4}$

Therefore, polymers with a higher molar mass are showing higher η_0-values (see also Chapter 3.3.2.1a3 with Equation 3.3). For non-flowing materials, e.g. for chemically cross-linked elastomers such as rubbers, the calculated η_0-values obtained would approach towards infinity. Therefore of course, it is not useful to determine any whatsoever η_0-value for these kinds of materials.

c) Comparison: different methods to determine η_0
At a selected temperature, the value of zero-shear viscosity is a material constant which can be determined using the following rheological test methods:
1) Rotational tests: η_0 as the limiting value of the shear viscosity function $\eta(\dot{\gamma})$ "at rest", i.e. for $\dot{\gamma} \to 0$ (see Chapter 3.3.2.1a)

2) Oscillatory tests (frequency sweeps): η_0 as the limiting value of the complex viscosity function $\eta^*(\omega)$ "at rest", i.e. for $\omega \to 0$ (see Chapter 8.4.2.1a)

3) Creep tests: η_0 at the end of the stress phase, when reaching steady-state flow behavior.

6.3.4.2 Creep compliance, and creep recovery compliance

Creep (shear) compliance J(t) can be calculated using the following parameters: the preset shear stress τ_0 and the resulting deformation function $\gamma(t)$ obtained in the creep phase. Definition:

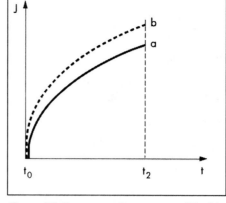

Figure 6.9: Creep compliance curves J(t), a) in the LVE range, b) outside the LVE range, showing comparatively higher J-values then

Equation 6.5 $\qquad$ $J(t) = \gamma(t)/\tau_0$

Unit: [1/Pa = Pa^{-1}]. The "shear compliance" is the reciprocal value of the shear modulus which can be imagined as "rigidity" then:

Equation 6.6 $\qquad$ $J = 1/G$

The shape of the J(t)-curve is similar to the $\gamma(t)$-curve since for creep tests, the shear stress τ_0 is preset as a constant value. Usually, J is presented on the y-axis, and time t on the x-axis (see Figure 6.9).

a) Instantaneous compliance

Definition of the instantaneous (shear) compliance J_0:

Equation 6.7 $\qquad$ $J_0 = 1/G_0 = \gamma_0/\tau_0$

Unit: [1/Pa = Pa^{-1}]. The following holds:

Equation 6.8 $\qquad$ $J_0 = \lim_{t \to 0} J(t)$

J_0 is the limiting value of the J(t)-function at the very beginning of the creep test, i.e. when t = 0.

G_0 [Pa] is the "instantaneous (shear) modulus" representing the sum of the elastic behavior of the springs S_1 and S_2, and γ_0 [%] is the "instantaneous deformation", i.e. the limiting value of the deformation at the time point t = 0.

The value of γ_0 is determined at the intersection of the γ-axis and the straight line which is fitted to the steady-state sector of the creep curve, showing the curve slope $\tan\beta$ (see Figure 6.8). It is the same straight line which is used to determine the value of zero-shear viscosity. To enable accurate determination of the values of J_0 (or G_0, respectively), at least 10% of the measuring points of the entire creep curve should occur within this steady-state sector.

Both parameters, J_0 and G_0, are coefficients to characterize the elastic behavior of a material. Polymers with similar structures but a higher molar mass M are showing lower J-values or higher G-values, respectively. For non-flowing materials, e.g. for chemically cross-linked elastomers such as rubbers, the calculated J-values obtained would approach towards zero or the G-value towards infinity, respectively. Therefore, these kinds of materials would be absolutely rigid, showing no yielding at all.

b) Creep recovery, and equilibrium compliance

Sometimes, the J(t)-function in the rest phase is referred to as the **(creep shear) recovery compliance J_r**. Definition:

Equation 6.9 $J_r(t) = \gamma_r(t)/\tau_0$

unit: [1/Pa = Pa⁻¹], with the reformation γ_r [%]. The following holds for the equilibrium (shear) compliance J_e:

Equation 6.10 $J_e = \lim_{t\to\infty} J_r(t)\ (= J\infty)$

Je [1/Pa = Pa⁻¹] is the limiting value of the $J_r(t)$-function after an "infinitely long" period of time, i.e. when t = ∞. The following holds: $J_e = J_0$

Equilibrium of the recovery compliance shows the same value if determined on the one hand via J(t) in an "infinitely short" period of time as above (in terms of J_0), or, on the other hand via $J_r(t)$ after an "infinitely long" period (in terms of J_e or J_∞), i.e. at the very end of the test when finally reaching again steady-state behavior or equilibrium of forces, respectively. However, when using the function of J(t) or $J_r(t)$, the same value for J_0 and J_e may only be achieved if indeed the limiting value of the LVE deformation range has never been exceeded.

c) Determination of the limiting value of the LVE range
Evaluating creep curves, many users prefer the J(t)-function to the γ(t)-function. Since J(t) = γ(t)/τ_0, all J(t)-curves essentially overlap with one another, independent of the preset stress τ_0 as long as the limit of the LVE range is not exceeded. Since Hooke's law applies to each time point, the preset stress results in the corresponding proportional deformation value. Therefore, the corresponding ratio value of J(t) is independent of the preset stress value when still measuring in the LVE range.

It is easy to check whether the limit of the LVE range has been exceeded by presenting in the same diagram all of the individual J(t)-curves resulting from several creep tests which are performed by presetting a different constant τ_0-value for each individual test. Those J(t)-curves, which are not overlapping with one another but deviate significantly upwards, have obviously been measured under conditions outside the LVE range (see curve b in Figure 6.9). In this case, the internal structure has been deformed already too much by the preset stress, which was too high in this case. As a consequence, the partially destroyed structure of the material is yielding more, i.e. it will be deformed to a greater extent as corresponding to Hooke's law (τ/γ = const). In other words: The compliance of the structure increases, and then, higher J-values are obtained.

Therefore, tests which are nevertheless carried out under these conditions reveal information on **non-linear behavior** outside the LVE range. For the above reasons, J(t)-diagrams are well suited to check whether the LVE range has been exceeded or not. For detailed information on the LVE range: see Chapters 8.3.2 and 8.3.3, and about non-linear behavior: see also Chapter 8.3.6 (LAOS).

Note: Master curve of J(t)-functions via time/temperature shift (WLF method)
For all thermo-rheologically simple materials tested in the LVE deformation range, a **temperature-invariant master curve of the J(t)-function** can be determined from several individual J(t)-functions of which each one was measured at a different temperature. The corresponding master curve is generated using the WLF relation and the time/temperature shift (TTS) method (see Chapter 8.7.1). This method is useful to produce J(t)-values also in a time-frame for which no measuring data are available, e.g. in the short-term range.

6.3.4.3 Retardation time

Ideally elastic materials are showing as well immediate deformation after applying a step-like stress, as well as immediate reformation after removing the stress afterwards in the step-like form again. For all VE samples, this elastic behavior occurs with a certain time delay. To evaluate this time-dependent deformation behavior, two parameters have been defined: the relaxation time λ [s] and the retardation time Λ [s], (both are pronounced "lambda"; see also Chapter 14.2: Greek alphabet). The term "relaxation time" is explained in detail in Chapter 7. It is used in combination with tests when presetting the strain (deformation) or strain rate (shear rate), e.g. when performing relaxation tests. **Relaxation** is a process in the state at rest after a forced deflection or strain and can be described as "delayed elasticity" in the sense of "delayed stress decrease" (see Chapter 7.3.3.2). On the other hand, the term "retardation time" is used for tests when presetting the stress, e.g. when performing creep tests. **Retardation** is a delayed response to an applied force or stress and can be described as "delayed elasticity" in the sense of "delayed re-formation".

a) Retardation time Λ in the Kelvin/Voigt model
Behavior of VE solids becomes clear when using the differential equation according to Kelvin/Voigt (see Chapter 5.2.2.1b): $\tau = \eta \cdot \dot{\gamma} + G \cdot \gamma$

Using $\Lambda = \eta/G$ or $\eta = \Lambda \cdot G$ then:

Equation 6.11 $\tau = (\Lambda \cdot G) \cdot \dot{\gamma} + G \cdot \gamma$ or $\tau/G = \Lambda \cdot \dot{\gamma} + \gamma$

Here, the symbol Λ is taken for the retardation time. Some authors choose the symbol λ_K to show the correlation between the Kelvin/Voigt model – which is used to characterize the rheological behavior of VE solids – and this specific time [18]. The retardation time determines the time-dependent deformation and reformation behavior of the parallel connected components spring and dashpot of the Kelvin/Voigt model for both intervals, as well for the stress phase as well as for the rest phase. The solution of the differential equation leads to the following time-dependent exponential function:

Equation 6.12 $\gamma = (\tau/G) \cdot [1 - \exp(-t/\Lambda)]$

1) At the time point $t = \Lambda$, the following applies to the creep phase:
$\gamma(\Lambda) = (\tau/G) \cdot [1 - (1/e)] = 63.2\% \cdot (\tau/G) = 63.2\% \cdot \gamma_{max}$
Therefore counts for the creep phase: The retardation time Λ of the Kelvin/Voigt model is reached if the γ-value has increased to 63.2% of the maximum deformation γ_{max} which will finally occur at the end of the stress interval (see also Chapter 6.3.3a).
2) At the time point $t = \Lambda$, the following applies to the creep recovery phase:
$\gamma(\Lambda) = \gamma_{max} - (\tau/G) \cdot [1 - (1/e)] = \gamma_{max} - 63.2\% \cdot (\tau/G) = (100\% - 63.2\%) \cdot \gamma_{max}$
and thus: $\gamma(\Lambda) = 36.8\% \cdot \gamma_{max}$
Therefore counts for the creep recovery phase: The retardation time Λ of the Kelvin/Voigt model is reached if the γ-value has decreased to 36.8% of γ_{max} (see also Chapter 6.3.3b).

b) Retardation time in the Burgers model
Using the Burgers model, the calculation seems to be even more complex since here, the immediate reformation of the first spring S_1 in the recovery phase also has to be taken into account for the determination of the γ-values. However, since the spring recoils immediately, i.e. idealized "in zero-time", there is no influence on the value of Λ. Therefore, it has no effect on the time-dependent behavior.

6.3.4.4 Retardation time spectrum

When describing the behavior of real **polymers**, it is important to take into account that the molecules do not have a single molar mass only since they may have a more or less wide **molar mass distribution (MMD)**. Therefore a retardation time spectrum should be given preference, since here, a single retardation time only is not sufficient for an appropriate analysis.

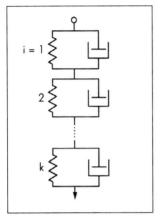

Figure 6.10: Generalized Kelvin/Voigt model

a) Generalized Kelvin/Voigt model

Generalized Kelvin/Voigt models are therefore used to analyze the creep and creep recovery (retardation) functions. A single Kelvin/Voigt model consists of a single spring and a single dashpot in parallel connection (see Figure 5.6). For a generalized Kelvin/Voigt model, however, several Kelvin/Voigt elements are connected in series (see Figure 6.10).

Each one of the individual Kelvin/Voigt elements displays the behavior of an individual polymer fraction having a specific molar mass and molecular structure. Each fraction is represented in the model by a spring and a dashpot which together produce the characteristic values of the viscoelastic behavior of this one fraction. This results in the corresponding individual retardation time Λ_i.

The following applies to each individual Kelvin/Voigt element:

$$\tau_i = \eta_i \cdot \dot{\gamma} + G_i \cdot \gamma$$

with the counting number i = 1 to k; and k is the total number of all Kelvin/Voigt elements available.

The following holds: $\Lambda_i = \eta_i / G_i$, with the individual retardation time Λ_i [s]

Thus: $\qquad\qquad\quad \gamma_i(t) = (\tau_0 / G_i) \cdot [1 - \exp(-t/\Lambda_i)]$

Dependent on the shear stress step and on its removal, each one of the Kelvin/Voigt elements is showing an individual time-dependent deformation or reformation behavior, respectively. The resulting total deformation value γ occurs as the sum of all individual deformation values γ_i:

Equation 6.13 $\qquad \gamma(t) = \sum_i \gamma_i(t) = \sum_i (\tau_0 / G_i) \cdot [1 - \exp(-t/\Lambda_i)]$

It is possible to use this calculation since here applies the **principle of superposition** according to *Ludwig Boltzmann* (1844 to 1906; [47] of 1874). According to this principle, for data of the linear viscoelastic range, the ratio of the value pairs of stress/deformation also applies to its multiples and sums (see also Chapter 8.3.2: LVE range) [114, 249]. Sometimes, an extra spring and an extra dashpot are connected in series as additional components to the generalized Kelvin/Voigt model to enable also analysis, on the one hand of reversible elastic behavior at very low deformations, and on the other hand of purely viscous flow behavior at high deformations (this model is comparable to the Burgers model, see Figure 6.7).

b) Discrete retardation time spectrum

A discrete retardation time spectrum consists of individual value pairs showing a limited total number k of Kelvin/Voigt models connected in series, e.g. with k = 5. In this case, the discrete retardation time spectrum consists of 5 individual value pairs, e.g. expressed in terms of the retardation time-dependent creep compliances J_i (Λ_i), here with i = 1 to 5. A corresponding example, but using the relaxation time-dependent relaxation moduli, is explained in Chapter 7.3.3.3b (see Table 7.1 and Figure 7.9).

The shape of both exponential functions, the creep and creep recovery curve, is determined by the spectrum of retardation times. For many polymers, the range of retardation times spreads over several decades, and frequently up to much more than 100s. Often Λ_i values up to 1000s (= approx. 17min) or 10,000s (= approx. 167min = almost 3h) can be found, and sometimes periods which are even longer. Therefore, for samples showing highly viscous or viscoelastic behavior, retardation times of at least half an hour should be taken into account.

Summing up the individual creep compliances $J_i(\Lambda_i)$, the function of the time-dependent creep compliance J(t) can be determined in the form of a fitting function:

Equation 6.14 $J(t) = \sum_i (1/G_i) \cdot [1 - \exp(-t/\Lambda_i)]$

$= (1/G_1) \cdot [1 - \exp(-t/\Lambda_1)] + (1/G_2) \cdot [1 - \exp(-t/\Lambda_2)] + (1/G_3) \cdot [1 - \exp(-t/\Lambda_3)] + ...$

The discrete retardation time spectrum can be illustrated in a diagram displaying an individual point for each individual value pair (Λ_i/J_i). Usually, the Λ_i values [s] are presented on the x-axis and the J_i values [Pa^{-1}] on the y-axis (similar to Figure 7.9). In the same diagram, the calculated fitting function J(t) may be displayed; here, time t [s] is shown on the x-axis and the J-values [Pa^{-1}] on the y-axis, using the same scale for Λ_i and t, and also the same scale for J_i and J on the other axis.

c) Continuous retardation time spectrum

The continuous retardation time spectrum $H(\Lambda)$ is produced from an "infinite" number of individual values i = 1 to k (and k $\rightarrow \infty$). H is called the "amount" or "intensity", and $H(\Lambda)$ is referred to as the "distribution function of the retardation times Λ". The sum of the continuous spectrum is usually presented in the form of an integral. The corresponding integral equations are calculated using special analysis programs.

Usually, $H(\Lambda)$ is presented with H on the y-axis and Λ on the x-axis. The data points of $H(\Lambda)$ at low or high Λ-values are indicating the number of molecules (or other components) with short or long retardation times, respectively. For information on the relaxation time spectrum $H(\lambda)$, including a diagram, see Chapter 7.3.3.3c and Figure 7.10; most of the information given there applies also to $H(\Lambda)$, either directly or in an adapted form.

6.3.5 Data conversion

When performing creep tests, data are measurable indeed at such low shear rates which would cause problems in many cases if determined when using other test methods, or the corresponding test would take an extremely long time then. Creep test data are often used to characterize **long-term behavior**. They may be **converted**, for example **from J(t)-values to frequency sweep values at very low frequencies**.

Example: Conversion of creep test data to frequency sweep data

Data of a creep test are available in terms of a $\gamma(t)$-function. The aim is to determine the corresponding frequency function. The following steps are performed:

Using the available creep function data $\gamma(t)$

$\rightarrow$ Calculation of the function of the creep compliance J(t)

$\rightarrow$ Calculation of the continuous retardation time spectrum $H(\Lambda)$

$\rightarrow$ Data conversion to determine the corresponding frequency sweep in terms of G' & G''(ω) (see Chapter 8.4: oscillatory tests, frequency sweeps, with storage modulus G', loss modulus G'' and angular frequency ω).

This method is particularly useful if it is aimed to produce **values in the zero-shear (low-shear) viscosity range**.

Information given in Chapter 7.3.4 on data conversion and on curve fitting methods also applies here, in the appropriate form, to creep test data. Data conversion can be performed using data of both the time-dependent relaxation modulus G(t) or creep compliance J(t). Relaxation spectrum $H(\lambda)$ and retardation spectrum $H(\Lambda)$ correspond to each other approximately. However, it should be taken into account that data which are depending on relaxation times λ or retardation times Λ, respectively, are measured at different shear conditions: for data related to λ with controlled strain (deformation) γ, and for data related to Λ with controlled stress τ. There might be a different response by the structure of the sample when subjected to these different shear conditions.

6.3.6 *Determination of the molar mass distribution*

Using H(Λ) data, special software analysis programs enable users to determine the molar mass distribution (MMD) of unlinked polymers. In a MMD diagram, usually the relative amount w [%] is presented on the y-axis, and the molar mass M [g/mol] on the x-axis (see Chapter 7.3.5 and Figure 7.11).

⤳   End of the Cleverly section

7 Relaxation tests

In this chapter are explained the following terms given in bold:

Liquids		Solids	
(ideal-) viscous flow behavior Newton's law	**viscoelastic flow behavior** Maxwell's law	**viscoelastic deformation behavior** Kelvin/Voigt's law	(ideal-) elastic deformation behavior Hooke's law
flow/viscosity curves	creep tests, **relaxation tests,** oscillatory tests		

7.1 Introduction

Relaxation tests are used to analyze the viscoelastic (VE) behavior performing a **strain step** (deformation step). The full name of this test is step strain or step deformation test emphasizing the set parameter, or **stress relaxation test** which urges the resulting parameter. This test method is mostly used to examine chemically unlinked and unfilled polymers (melts and solutions), but it is also suitable to evaluate the behavior of chemically cross-linked polymers, gels and dispersions showing a physical-chemical network of forces.

In industrial practice for quality assurance, relaxation tests are rarely carried out as routine tests. But they are used in several R & D departments in the polymer industry, e.g. to determine the molar mass distribution (MMD). Useful results in the short term range, however, can only be produced when using a highly dynamic rheometer which is indeed capable to perform the step in strain in a really very short time interval, i.e. in at least $\Delta t = 100ms = 0.1s$.

7.2 Basic principles

7.2.1 Description of the test

Preset: Shear strain step function $\gamma(t)$, see Figure 7.1

1) Preparation phase: Immediate step in strain from $\gamma = 0$ to γ_1, then keeping constantly this low pre-strain value from time point t_1 to t_2 (i.e. $\gamma_1 =$ const). The purpose of this interval is to level out possible effects of pre-stresses which might be still present due to previously performed measures of sample preparation, and therefore, this preparation interval is intended to improve reproducibility of the test results.

2) Immediate step in strain from γ_1 to $\gamma_0 =$ const.

3) The strain value is kept constantly at $\gamma_0 =$ const in the whole test interval between the time points t_2 to t_3.

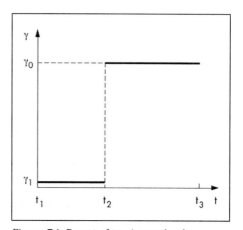

Figure 7.1: Preset of two intervals when performing relaxation tests:
1) pre-strain as a small step, followed by a rest phase to level out possible pre-stresses,
2) the effective step in strain, followed by the stress relaxation phase to be analyzed

Thomas G. Mezger: The Rheology Handbook
© Copyright 2011 by Vincentz Network, Hanover, Germany
ISBN 978-3-86630-864-0

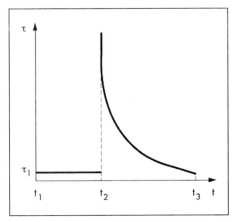

Figure 7.2: Stress relaxation curve (in the second test interval)

The linear viscoelastic (LVE) deformation range should not be exceeded. For various materials, information on values of the **permissible maximum deformation** γ_{max} can be found in the Chapters 6.2.1 and 8.3.3.1.

Example: Test preset for a polymer sample
1) Pre-strain phase: Step in strain to $\gamma_1 = 1\%$, this strain value remains unchanged for t = 5min
2) Step in strain to $\gamma_0 = 20\%$
3) Relaxation phase: This strain value remains unchanged for t = 20min

Discussing the first test interval: Why is it useful to apply a pre-strain?
There are users who prefer to perform the effective step in strain not directly from a non-deformed state, i.e. from $\gamma = 0$. When analyzing polymers, reproducibility can often be increased if a low, constant strain γ_1 has already been applied to the sample before performing the effective, large γ_0-step. This counts above all for the following reason: Gap setting (e.g. when using a parallel-plate system) may cause a high shear load on the sample leading to relatively large internal stresses. Therefore, it is advantageous to set a certain pre-strain γ_1 - showing of course clearly lower values compared to the effective strain step γ_0 which is applied subsequently - in order to compensate the effects of the stresses deriving from all previously performed measures in sample preparation. Also this "background" of a sample should be regarded as one of the test parameters, and indeed it counts for all kinds of rheological testing, but it is particularly true when performing relaxation tests. By the way, internal stresses may occur also when using the automatic gap setting (AGS) option. However, gap setting by use of the normal force control (NFC) option may improve the reproducibility of this process and its consequences (for more information on AGS and NFC see Chapter 10.4.6).

The first test interval should result in relatively low τ-values only. Afterwards, the effective γ_0-step should not be performed before the finally occurring τ_1-value has reached indeed a comparatively negligible low level.

Example: Typical recovery times of polymers
For highly viscous and viscoelastic polymer melts like a silicone polymers (unlinked PDMS), at least half an hour should be taken into account until the extent of the internal stresses, resulting from sample preparation, has reached such a low level which is no longer significant for the later test result and therefore can be ignored. Pre-stresses and recovery time may be reduced if setting a gap of H ≥ 1mm when using a parallel-plate measuring system.

Measuring result: Shear stress relaxation curve $\tau(t)$, see Figure 7.2.
As a test result, the time-dependent stress relaxation function $\tau(t)$ is measured between the time points t_2 and t_3. In a diagram, typical the relaxation curves are showing τ on the y-axis, and time t on the x-axis. Let us begin with the presentation of both parameters on a linear scale.

7.2.2 Ideally elastic behavior

Experiment 7.1: Twisting an eraser
An eraser gum is taken in both hands and twisted quickly at one end until a torsion angle of around 45 degrees is reached, and this deformation is kept constantly then. Since the

rigid internal structure of the gum is not able to reduce any internal stress, as a response to the forced strain there is a certain resistance force in the form of a torque, which is remaining on a constant value then, even after a longer time.

For ideally elastic solids after a step in strain, no recovery (stress relaxation) occurs at all. Therefore, the $\tau(t)$-curve remains constantly on the plateau value τ_2 (see Figure 7.3).

7.2.3 Ideally viscous behavior

Experiment 7.2: Rotating a wooden disk on a water surface

A wooden disk floating on a water surface is rotated quickly to an angle of about 90 degrees. Afterwards, this position is maintained constantly. Immediately after stopping the rotation of the disk, the water shows no more resistance force or torque, respectively.

For ideally viscous liquids after a step in strain, immediate and complete stress relaxation occurs. At the time point t_2, the $\tau(t)$-diagram indicates a very sharp maximum looking like a needle-shaped peak (see Figure 7.4).

7.2.4 Viscoelastic behavior

Experiment 7.3: Rotating a spatula in a silicone polymer

A spatula is rotated by approximately 45 degrees in a beaker containing a silicone polymer (unlinked PDMS). Afterwards, this position is kept constantly. At first, the silicone shows a strong elastic resetting force in the form of a torque, which is continuously decreasing then.

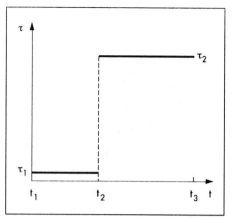

Figure 7.3: "Relaxation curve" of an ideally elastic solid (inverted commas, since there is no stress relaxation at all)

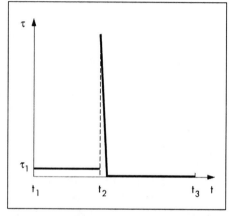

Figure 7.4: Relaxation "curve" of an ideally viscous liquid (inverted commas, since here, an immediate needle-shaped peak is appearing only)

For VE materials after a step in strain, delayed stress relaxation is occurring, either partially or completely. The $\tau(t)$-diagram presents the shape of an exponential function (e-function; see Figure 7.5). The degree of stress relaxation depends on the viscous portion of the test material.

For **VE liquids** (1), delayed but complete stress relaxation occurs if the period of testing is sufficiently long. This behavior is displayed by concentrated polymer solutions and polymer melts. A more steeply falling τ-curve in the short term range indicates the presence of a large fraction of smaller molecules, since these are showing shorter relaxation times. For stress curves falling more moderately a wider spectrum of relaxation times is present, i.e. there is a wider molar mass distribution (MMD). Comparatively higher stress values in the long term range are indicating longer relaxation times, and therefore, a greater number of larger molecules having correspondingly higher molar masses.

For **VE solids** (2) even after a very long period of time, the stress is relaxing to a certain extent only. The $\tau(t)$-curve asymptotically approaches the final value which is called the **equi-**

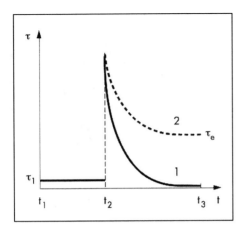

Figure 7.5: Stress relaxation curves of two different viscoelastic materials, both are showing delayed relaxation: (1) completely for the VE liquid, and (2) partially for the VE solid (τ_e is the equilibrium stress value)

librium stress τ_e. This indicates that at least a part of the molecules are chemically or physically linked, and therefore are not able to move freely. This behavior is exhibited by chemically cross-linked materials, gels and concentrated dispersions showing a gel-like structure at rest.

7.3 Analysis

7.3.1 Behavior of the molecules

Behavior of a polymer sample can be illustrated by observing the process taking place in a conglomeration of polymer molecules, when regarding a single macromolecule only.

Figure 6.6 illustrates the deformation process when performing a relaxation test:
(1) At rest, before applying the strain
(2) Immediately after the step in strain, or at the beginning of the relaxation phase, respectively

a) At rest, before the step in strain
Still in a non-deformed state, the chain of an individual polymer molecule occurs as a spherical molecule coil which may have many entanglements with neighboring chains. This is the state at rest requiring minimum expense of energy.

b) Performing the step in strain
Due to the shock-like step strain, the observed macromolecule is forced to leave its state of rest. As a result of the immediately applied strain, the spherical coil is deformed rapidly and takes on the shape of an ellipsoid whose axis is oriented lengthways between the shear direction and the direction of the shear gradient. More illustrative: It is shaped now "like an American football". As a consequence, internal stresses are occurring immediately now as well in the individual molecule, as well as within the superstructure of the whole polymer sample due to the entanglements between the molecule chains.

c) In the stress relaxation phase
Immediately after the step in strain, each one of the molecules is trying to return to a stress-free state. Continuously, the molecules are relaxing more and more from the effects of the preset strain by reducing the internal stresses by a slow and compensating motion within the superstructure. This relaxation occurs as a delayed process. At the end of the test, the molecules are in a recoiled state again, showing the shape of spheres again. However, the resulting configuration of the individual molecules is no longer the same as it was initially before the step in strain. Unlinked polymers may exhibit complete stress relaxation if the period of time was long enough, i.e., if a corresponding great number of disentanglements occurred up to then (as shown in Figure 7.5, no. 1). For samples showing a chemical or physical network, the extent of the partial stress reduction corresponds to the viscous portion. For fully cross-linked polymers such as thermosets there is no relaxation at all (as shown in Figure 7.3). Here, if the LVE range was not exceeded, the configuration of the molecules remains unchanged even after a long period of time (as shown in Figure 6.6, no. 2).

In rheology, the trial of a deformed material to reach the steady-state after applying a strain is referred to as **relaxation**. As a comparison: The term "retardation" is used to describe the delayed reaction after applying a stress, e.g. when performing creep tests (see Chapter 6).

Of course, smaller molecules are reaching the stress-free state-at-rest faster compared to larger molecules which need a longer relaxation time. Shorter relaxation times therefore characterize the rheological behavior of smaller, "faster" molecules. In contrast, longer relaxation times are representing the behavior of larger, "slower" macromolecules.

&⁀   For "Mr. and Ms. Cleverly"

7.3.2 Curve discussion

a) Before the effective strain step
The first test interval is only for the purpose to create reproducible test conditions before performing the effective step in strain. The user should wait until τ_1 has reached a relatively low value or until the rheometer used is displaying a torque (or stress) value close to zero, respectively (see Figure 7.2). Usually, this interval is not analyzed.

b) During the strain step
Since the strain step is carried out in an "infinitely short time" in the ideal case, this interval is not analyzed either.

c) Stress relaxation curve
The internal stresses of the sample resulting from the step in strain are measured in terms of shear stress values τ. For VE liquids, the $\tau(t)$-curve shows a decreasing curve slope over time until the internal stresses are reduced completely. In this case, the $\tau(t)$-function curve is tending asymptotically towards the value $\tau = 0$Pa finally. Of course, the limiting value $\tau = 0$ cannot really be detected by a rheometer, since at some point the lower resolution limit for determining the torque or shear stress is exceeded, respectively. This counts even for the most sensitive instrument available. For VE solids, stress reduction takes place only partially, finally reaching the value $\tau = \tau_e$.

Transient behavior occurs as long as the τ-value is changing with time, then $\tau = f(\gamma_0, t)$. Here, the shape of the $\tau(t)$-curve depends on both as well on the applied strain γ_0 as well as on the passing time. **Steady-state behavior** with τ = const is not reached before the relaxation phase is finished, then $\tau = 0$ or $\tau = \tau_e$, respectively. The shear stress shows a constant value now which is no longer depending on the time.

The time-dependent **stress relaxation function**, obtained from the differential equation according to Maxwell (see Chapter 5.2.1.1b), can be formulated as follows:

Equation 7.1 $\tau(t) = \gamma_0 \cdot G \cdot \exp(-t/\lambda)$

with the relaxation time λ [s] = η/G. The shear modulus G corresponds to the behavior of the spring, and the viscosity η to the behavior of the dashpot of the Maxwell model.

The following applies for t = 0:
$\tau(0) = \gamma_0 \cdot G \cdot e^0 = \gamma_0 \cdot G \cdot 1 = \gamma_0 \cdot G = \tau_{max}$
i.e., the maximum shear stress τ_{max} occurs directly after the step strain.

The following applies for t = ∞, or for practical users, after a "very long" time:
$\tau(\infty) = \gamma_0 \cdot G \cdot (1/e^\infty) = \gamma_0 \cdot G \cdot 0 = 0$
i.e., the shear stress is completely relaxed finally.

The following applies for t = λ, i.e., when reaching the relaxation time:
$\tau(\lambda) = \gamma_0 \cdot G \cdot (1/e) = 0.368 \cdot \gamma_0 \cdot G$
i.e., up to this time point, the stress has relaxed partially, and that is 36.8% of the τ_{max}-value.

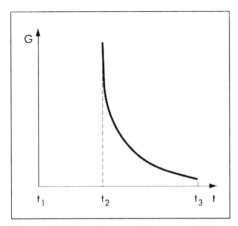

Figure 7.6: Relaxation modulus curve G(t), on a linear scale

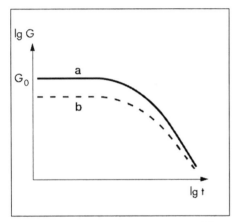

Figure 7.7: Relaxation modulus curves G(t), on a logarithmic scale,
a) in the LVE range, G_0 is the instantaneous shear modulus or "plateau value", b) outside the LVE range, showing comparatively lower G-values then

7.3.3 Definition of terms

7.3.3.1 Relaxation modulus

The (shear stress) relaxation modulus G(t) can be calculated using the following parameters: The preset shear strain γ_0 and the resulting stress relaxation function $\tau(t)$. Definition:

Equation 7.2 $G(t) = \tau(t)/\gamma_0$

unit: [Pa]

The shape of the G(t)-curve is similar to the $\tau(t)$-curve since for relaxation tests, the shear strain γ_0 is preset as a constant value. Usually, G is displayed on the y-axis, and time t on the x-axis (see Figure 7.6). Presentation on a logarithmic scale is advantageous when analyzing the short term range since then, low and even very low values can be exhibited clearly and visible (see Figure 7.7).

a) Instantaneous shear modulus
In the short term range, the limiting value of the G(t)-function shows a constant value which is called the "instantaneous (or initial) shear modulus G_0" or "plateau value G_0" (see Figure 7.7). The following holds:

Equation 7.3 $G_0 = \lim_{t \to 0} G(t)$

G_0 is the limiting value of the G(t)-function at the very beginning of the relaxation test, i.e. when t = 0.

b) Equilibrium shear modulus
Sometimes, also the equilibrium (shear) modulus G_e is determined:

Equation 7.4 $G_e = \lim_{t \to \infty} G(t) \ (= G_\infty)$

G_e [Pa] is the limiting value of the G(t)-function after an "infinitely long" period of time, i.e. when t = ∞. The sample structure is totally relaxed now, when finally reaching again steady-state behavior or equilibrium of forces, respectively. For viscoelastic liquids, G_e is tending to zero with time, and for viscoelastic solids the limiting value will be $G_e = \tau_e/\gamma_0$ in the very end of the test.

c) Determination of the limiting value of the LVE range
Since $G(t) = \tau(t)/\gamma_0$, all G(t)-curves essentially overlap with one another, independent of the preset strain γ_0 as long as the limit of the LVE range is not exceeded. Since Hooke's law applies to each time point, the preset strain results in the corresponding proportional stress value. Therefore, the corresponding ratio value of G(t) is independent of the preset strain value when still measuring in the LVE range.

It is easy to check whether the limit of the LVE range has been exceeded by presenting in the same diagram all of the individual G(t)-curves resulting from several relaxation tests which are performed by presetting a different constant γ_0 value for each individual test. Those G(t)-curves, which are not overlapping with one another but deviate significantly downwards, have obviously been measured under conditions outside the LVE range (see curve b in Figure 7.7). In this case, the internal structure has been deformed already too much by the preset strain, which was too high in this case. As a consequence, the partially destroyed structure of the material is showing less resistance force as corresponding to Hooke's law (τ/γ = const.). In other words: The rigidity of the structure decreases, and then, lower G-values are obtained.

Therefore, tests which are nevertheless carried out under these conditions reveal information on **non-linear behavior** outside the LVE range. For the above reasons, G(t)-diagrams are well suited to check whether the LVE range has been exceeded or not [222, 257]. For detailed information on the LVE range see Chapters 8.3.2 and 8.3.3, and about non-linear behavior see also Chapter 8.3.6 (LAOS).

Note: Master curve of G(t)-functions via time/temperature shift (WLF method)
For all thermo-rheologically simple materials tested in the LVE deformation range, a **temperature-invariant master curve of the G(t)-function** can be determined from several individual G(t)-functions of which each one was measured at a different temperature. The corresponding master curve is generated using the WLF relation and the time/temperature shift (TTS) method (see Chapter 8.7.1). This method is useful to produce G(t)-values also in a time-frame for which no measuring data are available, e.g. in the short-term range.

7.3.3.2 Relaxation time

As above, the term "relaxation time" is used when performing measurements when controlling the strain or strain rate (shear rate). On the other hand, "retardation time" is used when performing tests with controlled stress (see also Chapter 6.3.4.3).

Relaxation time λ in the Maxwell model

Behavior of VE liquids becomes clear when using the differential equation according to Maxwell (see Chapter 5.2.1.1b):

$\dot{\gamma} = \tau/\eta + \dot{\tau}/G$

using $\qquad \lambda = \eta/G \qquad$ or $\qquad G = \eta/\lambda \qquad$ then:

Equation 7.5 $\qquad \dot{\gamma} = \tau/\eta + (\lambda \cdot \dot{\tau})/\eta \qquad$ or $\qquad \dot{\gamma} \cdot \eta = \tau + \lambda \cdot \dot{\tau}$

Here, the symbol λ is taken for the relaxation time. Some authors choose the symbol λ_M to show the correlation between the Maxwell model – which is used to characterize the rheological behavior of VE liquids – and this specific time [18]. The relaxation time determines the time-dependent behavior of the serially connected components spring and dashpot of the Maxwell model. The solution of the differential equation leads to the following time-dependent exponential function:

Equation 7.6 $\qquad \tau = \gamma \cdot G \cdot \exp(-t/\lambda)$

At the time point t = λ, the following applies to the relaxation phase:

$\tau(\lambda) = \gamma \cdot G \cdot (1/e) = 0.368 \cdot \gamma \cdot G = 36.8\% \cdot \gamma \cdot G = 36.8\% \cdot \tau_{max}$

Therefore counts for relaxation phase: The relaxation time λ of the Maxwell model is reached if the τ-value has decreased to 36.8% of the maximum shear stress τ_{max}, which occurred immediately after the step in strain (see also Chapter 7.3.2c). In other words: In this period of time, the stress has decreased already by 63.2% from the initial value τ_{max}.

In literature, information can be found on behavior of very small molecules showing extremely short relaxation times: Water with $\lambda = 10^{-12}$s and mineral oils with $\lambda = 10^{-9}$s. Of course

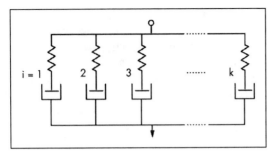

Figure 7.8: Generalized Maxwell model

these are not measured values, they are derived from assumed molecule dimensions and calculated using appropriate model concepts (see also Chapter 10.8.2.5: Characteristic times of motion) [352].

7.3.3.3 Relaxation time spectrum

When describing the behavior of real polymers, it is important to take into account that the molecules do not have a single molar mass only since they may have a more or less wide **molar mass distribution (MMD)**. Therefore a relaxation time spectrum should be given preference, since here, a single relaxation time only is not sufficient for an appropriate analysis.

a) Generalized Maxwell model
Generalized Maxwell models are therefore used to analyze the stress relaxation functions. A single Maxwell model consists of a single spring and a single dashpot in serial connection (see Figure 5.1). For a generalized Maxwell model, however, several Maxwell elements are connected in parallel (Figure 7.8).

Each one of the individual Maxwell elements displays the behavior of an individual polymer fraction having a specific molar mass and molecular structure. Each fraction is represented in the model by a spring and a dashpot which together produce the characteristic values of the viscoelastic behavior of this one fraction. This results in the corresponding individual relaxation time λ_i.

The following applies to each individual Maxwell element:
$\dot{\gamma}_i = \tau/\eta_i + \dot{\tau}/G_i$
with the counting number i = 1 to k; and k is the total number of all Maxwell elements available.
The following holds: $\lambda_i = \eta_i/G_i$, with the individual relaxation time λ_i [s]
Thus: $\tau_i(t) = \gamma_0 \cdot G_i \cdot \exp(-t/\lambda_i)$

After the step in strain, each one of the Maxwell elements is showing an individual time-dependent stress relaxation behavior. The resulting total stress value τ occurs as the sum of all individual stress values τ_i:

Equation 7.7 $\tau(t) = \sum_i \tau_i(t) = \gamma_0 \cdot \sum_i G_i \cdot \exp(-t/\lambda_i)$

When analyzing data from the LVE range, it is possible to use this calculation since here, the **principle of superposition** according to L. Boltzmann is valid (see also Chapter 6.3.4.4a).

Sometimes, an extra spring and an extra dashpot are connected in series as additional components to the generalized Maxwell model to enable also analysis, on the one hand of a reversible elastic behavior at very low strains (deformations), and on the other hand of purely viscous flow behavior at high-strain (deformation) conditions.

b) Discrete relaxation time spectrum
A discrete relaxation time spectrum consists of individual value pairs showing a limited total number k of Maxwell models connected in parallel, e.g. with k = 5. In this case, the discrete relaxation time spectrum consists of 5 individual value pairs, e.g. expressed in terms of the relaxation time-dependent relaxation moduli $G_i(\lambda_i)$, here with i = 1 to 5.

Example: The discrete relaxation time spectrum of a silicone polymer (unlinked PDMS, poly-di-methyl-siloxane) is presented in Table 7.1 (see also Figure 7.9). The G-values are decreasing more and more with time. After t = 0.1s, the G-modulus is already reduced to the half, and after t = 100s merely $5 \cdot 10^{-5}$ = 0.005% of the initial "rigidity" is still present.

It is the relaxation time spectrum which controls the time period required to reach the final equilibrium stress value or even the complete stress relaxation, respectively. Also the shape of the exponential relaxation curve is determined by the spectrum of relaxation times. For many polymers, the range of relaxation times spreads over several decades, and frequently up to much more than 100s. Often λ_i values up to 1000s (= approx. 17min) or 10,000s (= approx. 167min = almost 3h) can be found, and sometimes periods which are even longer. Therefore, for materials showing highly viscous or viscoelastic behavior, relaxation times of at least half an hour should be taken into account.

Table 7.1: Discrete relaxation time spectrum of a PDMS polymer at T = 25 °C[233]; see also the corresponding diagram in Figure 7.9

counting number of the Maxwell elements	individual relaxation time λ_i [s]	individual relaxation modulus G_i [Pa]
i = 1	0.01	200,000
i = 2	0.1	100,000
i = 3	1.0	10,000
i = 4	10	100
i = 5	100	10

Summing up the individual relaxation moduli $G_i(\lambda_i)$, the function of the time-dependent relaxation modulus G(t) can be determined in the form of a fitting function:

Equation 7.8 $G(t) = \sum_i G_i \cdot \exp(-t/\lambda_i) = G_1 \cdot \exp(-t/\lambda_1) + G_2 \cdot \exp(-t/\lambda_2) + G_3 \cdot \exp(-t/\lambda_3) + ...$

The discrete relaxation time spectrum can be illustrated in a diagram displaying an individual point for each individual value pair (λ_i/G_i). Usually, the λ_i values [s] are presented on the x-axis and the G_i values [Pa] on the y-axis (Figure 7.9). In the same diagram, the calculated fitting function G(t) may be displayed; here, time t [s] is shown on the x-axis and the G-values [Pa] on the y-axis,

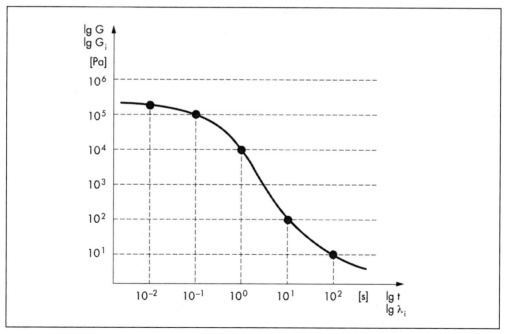

Figure 7.9: Testing a PDMS polymer: Function of the calculated time-dependent relaxation modulus G(t) determined from the five individual values of the discrete relaxation time spectrum $G_i(\lambda_i)$; see also Table 7.1

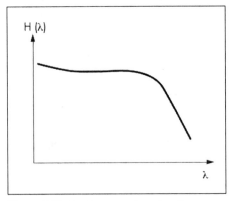

Figure 7.10: Continuous relaxation time spectrum H(λ)

using the same scale for λ_i and t, and also the same scale for G_i and G on the other axis. Note: The curve of the G(t)-function does not have to cover directly each one of the individual points of the discrete spectrum as it is illustrated in Figure 7.9.

c) Continuous relaxation time spectrum

The continuous relaxation time spectrum H(λ) is produced from an "infinite" number of individual values i = 1 to k (and k → ∞). H is called the "amount" or "intensity", and H(λ) is referred to as the "distribution function of the relaxation times λ". The sum of the continuous spectrum is usually represented in the form of an integral. The corresponding integral equations are calculated using special analysis programs.

Usually, H($\dot{\lambda}$) is presented with H on the y-axis and λ on the x-axis (see Figure 7.10). The data points of H(λ) at low or high λ-values are indicating the number of molecules (or other components) with short or long relaxation times, respectively.

7.3.4 Data conversion

Using this analysis tool, various rheological parameters of viscoelastic functions might be calculated from one another. Software programs using corresponding algorithms are available to carry out these calculations, enabling the user to convert data of relaxation tests, creep tests and oscillatory tests, e.g. from G(t) into G' & G''(ω), or from J(t) into G' & G''(ω), or vice versa (see Chapter 8.4: Oscillatory tests, frequency sweeps, with storage modulus G', loss modulus G'' and angular frequency ω). It should be taken care that these conversions are only useful for data which are measured in the LVE range. The aim is to extend the time-frame considerably for these parameters beyond the measured data available.

Examples

1) Relaxation modulus G(t) → frequency functions G' & G''(ω)

Data of a relaxation test are available in terms of a τ(t)-function. The aim is to determine the corresponding frequency sweep data.

The following steps are performed:

Using the available stress relaxation data τ(t)

→ calculation of the function of the relaxation modulus G(t)

→ calculation of the continuous relaxation time spectrum H(λ)

→ data conversion to determine the corresponding frequency sweep in terms of G' & G''(ω)

This method is particularly useful if it is aimed to produce **data at low frequencies**, e.g. in the **zero-shear viscosity range**. These data are often used to characterize **long-term behavior**. Alternatively via frequency sweeps, it would be extremely time-consuming to produce the concerning data in the corresponding frequency range.

2) Frequency functions G' & G''(ω) → relaxation modulus G(t)

Data from a frequency sweep are available in terms of G' & G''(ω).

The aim is to **determine** the corresponding **continuous relaxation time spectrum and** the function of the **relaxation modulus**:

Using the available frequency sweep data G' & G''(ω)

→ optimal λ_i values have to be defined by the user, or are selected automatically by the software program used (e.g. like in Table 7.1, selecting a single relaxation time per decade:

$\lambda_1 = 0.01s / \lambda_2 = 0.1s / \lambda_3 = 1s / \lambda_4 = 10s / \lambda_5 = 100s$, etc.)

→ determination of the individual pairs of $G_i(\lambda_i)$-values of the discrete relaxation time spectrum by fitting to the frequency sweep data (e.g. via linear regression and iterative error reduction using the method of quadratic deviation)

→ calculation of $H(\lambda)$

→ determination of G(t) via data conversion using approximating regularizing methods

3) Frequency functions G' & G" at high ω-values → G(t) or J(t)

A frequency master curve is available, produced from **frequency sweep data** which have been **shifted into the high-frequency range** using the WLF relation and time-temperature shift method TTS (see Chapter 8.7.1). By data conversion, **short-term values of a relaxation modulus function G(t)** or other rheological functions might be produced which usually cannot be obtained by relaxation tests or by other tests in this time range, or which cannot be obtained with the desired degree of accuracy.

4) Frequency functions G' & G" at low ω-values → relaxation modulus G(t)

By conversion, it is also possible **to produce long-term values of a G(t)-function from frequency sweep data** which have been **shifted** by use of the TTS (WLF) method **into the low-frequency range using the master curve** concept. However, this method is not useful in practice since there is more time needed to measure the corresponding frequency sweep data than time to carry out the appropriate relaxation tests.

5) Creep compliance J(t) → relaxation modulus G(t)

Performing creep tests, data can be measured at very low deformation rates (shear rates), which are not attainable when using other test methods, or which are indeed attainable, but only under extremely time-consuming conditions. These data in terms of the creep function $\gamma(t)$ or as creep compliance J(t), respectively, are representing long-term behavior and might be converted also into G(t)-values (see Chapter 6.3.5).

In the future, these types of analysis methods will probably play an even greater role than is the case today. Tests and analysis may be carried out automatically and on-line. At first, a short pre-test will identify the general type of the sample, as a "rheological fingerprint" (e.g. like in Example 2 of Chapter 8.9.1). Afterwards, a software program will decide the mode of testing to be used and in which measuring range. For polymers for example, the program may calculate the average molar mass M and molar mass distribution MMD finally, using data such as zero-shear viscosity or relaxation or retardation time spectrum. In order to obtain the desired results, the software program will be interacting with appropriate data banks to get the information required on all imaginable polymer types.

As already mentioned, there are several methods to convert the measured rheological material functions, such as G(t), J(t), G' & G"(ω), into the relaxation or retardation time spectrum. Here, approximation procedures are used to solve for example "fuzzy problems", e.g. using regularizing methods. Since the 1950s a lot of studies were published on this subject [155], in order to name a few authors, for example: J.D. Ferry [134], M. Baumgärtel [24] and H.H. Winter [392], (using their IRIS software program), A. Schausberger, A.Y. Malkin and V.V. Kuznetsov [237], N.W. Tschoegl [367], F.R. Schwarzl [327], M.L. Williams [389], C. Friedrich and B. Hoffmann [143], J. Honerkamp and J. Weese [192].

Information given in Chapter 6.3.5 on data conversion also applies here, in the appropriate form, to relaxation test data. Data conversion can be performed using data from both the time-dependent relaxation modulus G(t) or creep compliance J(t). Relaxation spectrum $H(\lambda)$ and retardation spectrum $H(\Lambda)$ correspond to each other approximately. However, it should be taken into account that data which are depending on relaxation times λ or retardation times Λ, respectively, are measured at different shear conditions: for data related to λ with controlled

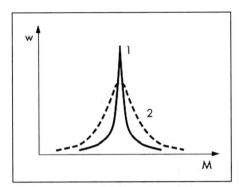

Figure 7.11: Both curves indicate the same average value of the molar mass, but there are different molar mass distributions (MMD): (1) narrow MMD, and (2) wide MMD

strain (deformation) γ, and for data related to Λ with controlled stress τ. There might be a different response by the structure of the sample to these different shear conditions.

7.3.5 Determination of the molar mass distribution

Using $H(\lambda)$ or $H(\Lambda)$ data, special software analysis programs enable users to determine the molar mass distribution (MMD) of unlinked polymers. In a MMD diagram, usually the relative amount w [%] is presented on the y-axis, and the molar mass M [g/mol] on the x-axis (see Figure 7.11; and also Chapter 6.3.6).

&⌢   End of the Cleverly section

8 Oscillatory tests

In this chapter are explained the following terms given in bold:

Liquids		Solids	
(ideal-) viscous flow behavior Newton's law	**viscoelastic flow behavior** Maxwell's law	viscoelastic deformation behavior Kelvin/Voigt's law	(ideal-) elastic deformation behavior Hooke's law
flow/viscosity curves	creep tests, relaxation tests, **oscillatory tests**		

8.1 Introduction

Oscillatory tests are used to examine all kinds of viscoelastic materials, from low-viscosity liquids to polymer solutions and melts, dispersions (i.e. suspensions, emulsions and foams), pastes, gels, elastomers, and even rigid solids. This mode of testing is also referred to as **dynamic mechanical analysis (DMA)**.

Probably the first ones who applied periodic oscillatory measurements for scientific purposes were R. Eisenschitz and W. Philippoff in 1933 [116, 288]. H. Roelig in 1938 [308], A.P. Aleksandrov and Y.S. Lazurkin in 1939 [7] designed apparatuses to preset oscillatory mechanical forces using a combination of a pre-stressed spring and an eccentrically rotating mass or by means of a cam compressing a plate spring, respectively, in order to determinate the resulting deformation amplitude by an optical system [237, 384]. With modern instruments, the excitation of oscillatory motion is usually controlled electronically. Here for example, the torque is measured via the required operating current of the drive, and the deflection angle is detected by an opto-electronic incremental position sensor (see also Chapter 11.6).

For most of the rheometers used in industrial laboratories, the bob is the oscillating part of the measuring system (Searle method, see Chapter 10.2.1.2a). But there are also types of instruments, where the cup (Couette method) or the lower plate, respectively, is set in oscillatory motion (see Chapters 10.2.1.2b, and 11.6.1 with Figure 11.6).

8.2 Basic principles

In order to explain oscillatory tests, the **Two-Plates-Model** is used (see also Chapter 2.2 with Figures 2.1 and 2.9, and Chapter 4.2 with Figure 4.1). Figure 8.1 illustrates how oscillatory motion of the upper plate might mechanically be produced using a drive wheel. A pushing rod is connected at the one end eccentrically to the driving wheel, and at the other end it is fixed to the upper oscillating plate.

The bottom plate is stationary (deflection s = 0). The distance h between the plates is the shear gap dimension. When the wheel is rotating, the upper plate with the shear area A is pushed back and forth by the shear force ±F (see Figure 8.2). The motion of the upper plate causes shearing of the sample which is placed between the two plates, showing the deflection path ±s and the deflection angle ±φ. It is assumed that the following **shear conditions** are given, since an accurate calculation of the rheological parameters is only possible if both conditions are met.

Thomas G. Mezger: The Rheology Handbook
© Copyright 2011 by Vincentz Network, Hanover, Germany
ISBN 978-3-86630-864-0

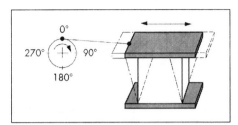

Figure 8.1: Two-Plates-Model to illustrate oscillatory shear tests

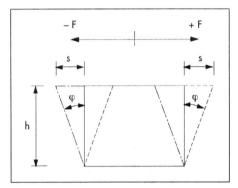

Figure 8.2: Oscillatory tests: shear force ±F, deflection path ±s and deflection angle ±φ in the shear gap h

1) The sample shows **adhesion** to both plates **without any wall-slip effects**.
2) **The sample is deformed homogeneously** throughout the entire shear gap, i.e., there is no inhomogeneous "plastic deformation" (see also Chapter 3.3.4.3d, and Figure 2.9).

In this case applies: The shear stress
$\pm\tau$ [Pa] = $\pm$F/A,
and the shear strain (or deformation)
$\pm\gamma$ = $\pm$s/h = $\pm$tanφ.

8.2.1 Ideally elastic behavior

When performing oscillatory tests on ideally elastic materials, i.e. completely inflexible, stiff and rigid solids, Hooke's law applies as follows:

Equation 8.1 $\tau(t) = G^* \cdot \gamma(t)$

with the **complex shear modulus G*** (pronounced: "G-star"), and the time-dependent values of the sine functions of τ and γ. For practical use, G^* can be imagined as the rigidity of the test material, i.e. as the resistance against deformation. A number of G- and G*-values, respectively, are listed in Table 4.1 of Chapter 4.2.2.

As illustrated in Figure 8.3, the motion of the upper plate is caused by the rotation of the drive wheel. In this example, the resulting force is measured at the bottom plate. When performing a full rotation, the wheel is turning over a rotation angle of 360°. This corresponds to a complete oscillation cycle of the time-dependent functions $\tau(t)$, and $\gamma(t)$ or $\dot\gamma$ (t), respectively.

Ever then during continuous rotation, when the wheel is passing the angle positions of 0° or 180°, the upper plate is showing the zero position and therefore, $\gamma(t) = 0$ and also $\tau(t) = 0$. The velocity however, is at a maximum here, i.e. $\dot\gamma$ (t) = $\dot\gamma_{max}$. At the angle position of 90°, the upper plate shows the maximum deflection to the right, and correspondingly at 270° occurs the maximum deflection to the left. Therefore here, $\gamma(t) = \gamma_{max}$ and $\tau(t) = \tau_{max}$, or $\gamma(t) = -\gamma_{max}$ and $\tau(t) = -\tau_{max}$, respectively. The velocity is zero at these two positions, i.e. $\dot\gamma$ (t) = 0, since at this point, the sense of motion is reversing.

Since $G^* = \tau(t)/\gamma(t)$ = const, the $\tau(t)$-curve is always "in phase" with the $\gamma(t)$-curve, i.e. both curves are occurring without any delay between preset and response in the form of **sine curves**. Therefore, they are showing the same frequency, and also the zero-transitions of the curves are appearing at the same time points. Formally, the **sinusoidal strain (or deformation) function** is presented by

Equation 8.2 $\gamma(t) = \gamma_A \cdot \sin\omega t$

with the shear strain (deformation) amplitude γ_A [%], and the angular frequency ω (pronounced: "omega") in [rad/s] or in [s⁻¹]. The **amplitude** is measured from basis line (zero) to peak on one side of the sine curve only.

Note: Oscillation in terms of the frequency f in Hertz, or as angular frequency ω
The oscillation frequency can be specified in two ways: as the angular frequency ω in [rad/s] or in [s⁻¹], or as the frequency f in [Hz], (in honor to *Heinrich Rudolph Hertz*, 1857 to 1894 [183]).

The disadvantage when using the frequency f is that "Hz" is not an SI-unit. The unit rad/s or s^{-1} of the angular frequency ω, however, is an SI-unit. When relating the concerning calculations to the so-called "unity circle" as it is used in mathematics and physics, instead of the unit rad/s also s^{-1} can be used. Calculations are facilitated and often made possible only when working with ω in [rad/s] or [s^{-1}], respectively, instead of using f in [Hz]. For more information on SI-units see Chapter 13.4 (1960), and Chapter 14.3: conversion of units.

Conversion of the two frequencies:

Equation 8.3 $\omega = 2\pi \cdot f$ (with the circle constant π = 3.141 ...)

Example: f = 10Hz corresponds to ω = 62.8rad/s or $62.8s^{-1}$

For samples showing ideally elastic behavior, there is no delay between the $\gamma(t)$- and the $\tau(t)$-curve; or using a scientific term: there is no phase shift. There is no phase shift angle δ (pronounced: "delta") between the two curves, thus: $\delta = 0°$ (or, if specified in rad: $\delta = 0$). Here, the $\dot{\gamma}$ (t)-curve, the shear rate function, is shifted by 90° compared to the $\gamma(t)$-curve, i.e. the $\dot{\gamma}$ (t)-curve occurs as a **cosine curve** in relation to the $\gamma(t)$-curve, if the latter is presented as a sine curve. Further information on δ is given in Chapter 8.2.3.

For "Mr. and Ms. Cleverly"

Since $\dot{\gamma} = d\gamma/dt$ (see Equation 4.3 of Chapter 4.2.1), the following holds for oscillatory tests: The time derivative of the sinusoidal strain (or deformation) function $\gamma(t)$ results in the strain rate (or shear rate) function $\dot{\gamma}$ (t) which occurs in the form of a cosine function then. Therefore:

Equation 8.2 $\gamma(t) = \gamma_A \cdot \sin\omega t$
Equation 8.4 $\dot{\gamma}(t) = \gamma_A \cdot \omega \cdot \cos\omega t$

End of the Cleverly section

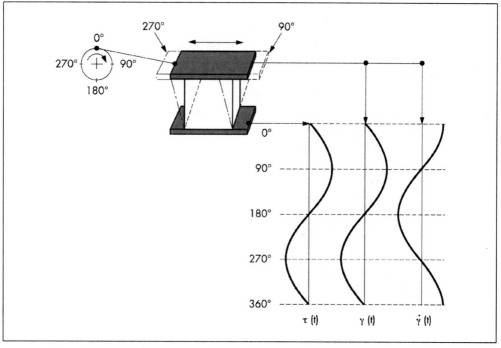

Figure 8.3: For ideally elastic behavior: time-dependent functions of $\tau(t)$, $\gamma(t)$ and $\dot{\gamma}(t)$ in the form of sine and cosine curves

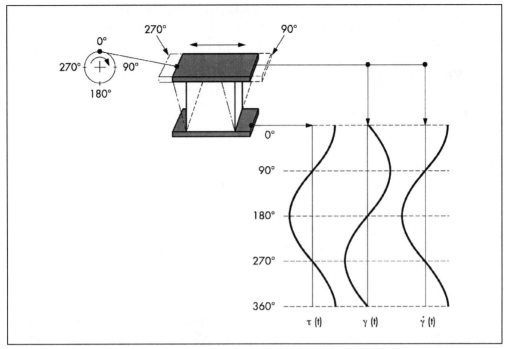

Figure 8.4: For ideally viscous behavior: time-dependent functions of $\tau(t)$, $\gamma(t)$ and $\dot{\gamma}(t)$ in the form of sine and cosine curves

8.2.2 Ideally viscous behavior

When performing oscillatory tests on ideally viscous fluids, Newton's law applies as follows:

Equation 8.5 $\tau(t) = \eta^* \cdot \dot{\gamma}(t)$

with the **complex viscosity** η^* (pronounced: "eta-star"), and the time-dependent values of the sine functions of τ and $\dot{\gamma}$. For practical use, η^* can be imagined as the viscoelastic flow resistance of a sample (see also Chapter 8.2.4b). A number of η- and η^*-values, respectively, are listed in Table 2.3 of Chapter 2.2.3.

Since $\eta^* = \tau(t)/\dot{\gamma}(t) = $ const, the $\tau(t)$-curve is always "in phase" with the $\dot{\gamma}(t)$-curve, i.e. both curves are appearing without any delay between preset and response, showing the same frequency. If the $\gamma(t)$-curve is presented as a sine curve, both the $\tau(t)$-curve and the $\dot{\gamma}(t)$-curve are occurring as cosine curves then (see Figure 8.4). For samples showing ideally viscous behavior, there is a delay between the τ-curve and the γ-curve with the phase shift angle of $\delta = 90°$ (or, if specified in rad: $\delta = \pi/2$ rad).

8.2.3 Viscoelastic behavior

Preset
1) With controlled shear strain as a sine function: $\gamma(t) = \gamma_A \cdot \sin\omega t$ (see Figure 8.5)
2) With controlled shear stress as a sine function: $\tau(t) = \tau_A \cdot \sin\omega t$, with the shear stress amplitude τ_A [Pa]

Measuring result
1) When presetting the strain γ: the τ-curve as a phase-shifted sine function

Equation 8.6 $\tau(t) = \tau_A \cdot \sin(\omega t + \delta)$

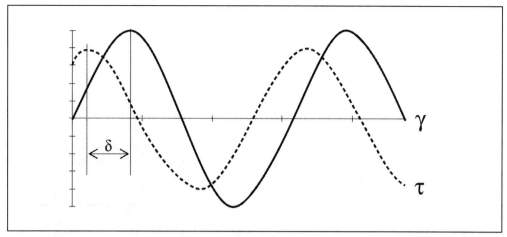

Figure 8.5: For viscoelastic behavior: time-dependent functions of γ(t) and τ(t), showing indeed the same frequency, but here, a phase shift angle δ occurs between the two sine curves

with the **phase shift angle δ between the preset and the resulting curve** (see Figure 8.5), which is usually specified in degrees [°], or alternatively but rarely used, in rad. Sometimes, δ is referred to as the **loss angle**.

2) When presetting the stress τ: the γ-curve as a phase shifted sine function

Equation 8.7 $\gamma(t) = \gamma_A \cdot \sin(\omega t + \delta)$

Also here, the resulting sine curve is shifted by the angle δ compared to the preset sine curve. The phase shift angle occurs always between 0° and 90°. Thus:

Equation 8.8 $0° \leq \delta \leq 90°$ (or, if specified in rad: $0 \leq \delta \leq \pi/2$ rad)

As already mentioned, for ideally elastic behavior δ = 0°, for ideally viscous behavior δ = 90°, and for viscoelastic behavior 0° < δ < 90°. For practical users: Testing viscoelastic materials, the resulting sine curve always shows a certain delay compared to the preset sine curve. In rheology, however, this shift is usually not expressed in terms of a time but of an angle. This is due to the use of "complex mathematics" which is required for some calculation steps when analyzing the test results (see for example Figures 8.6 and 8.7).

8.2.4 Definition of terms

a) Complex shear modulus, storage modulus, loss modulus, and loss factor
Hooke's law applies, here in the complex form:

Equation 8.9 $G^* = \tau(t)/\gamma(t)$

with the **complex shear modulus G*** [Pa], and the values of the sinusoidal functions of τ(t) in [Pa] and γ(t) with the unit [1] or in [%]; in ASTM D4092 is stated: G* = "complex modulus, measured in shear", to distinguish it from E* = "complex modulus, measured in tension or flexure"

Parameters, resulting from harmonic-periodic processes, meaning sinusoidal oscillatory shearing, always should be written in the complex form, i.e. marked with a star. Therefore, complex shear moduli G* always should be marked with a star, to distinguish them from "common" shear moduli which are not measured by oscillatory tests and analyzed correspondingly. Analysis is performed afterwards using "complex mathematics". As a comparison:

Shear moduli G are carrying no additional sign if they are determined under constant, steady-state shear conditions, i.e. when detecting each one of the individual measuring points by applying a constant shear strain or a constant shear stress within the linear elastic range; as explained in Chapter 4 and illustrated by Figure 4.1.

The storage modulus G', with the unit [Pa], (pronounced: "G-prime")
The G'-value is a measure of the **deformation energy stored** by the sample during the shear process. After the load is removed, this energy is completely available, now acting as the driving force for the reformation process which will compensate partially or completely the previously obtained deformation of the structure. Materials which are storing the whole deformation energy applied are showing completely **reversible deformation behavior** since after a load cycle, they occur with an unchanged shape finally. Thus, G' represents the **elastic behavior** of a material. (In ASTM D4092 is stated: G' = "storage modulus, measured in shear", to distinguish it from E' = "storage modulus, measured in tension or flexure".)

The loss modulus G'' with the unit [Pa], (pronounced: "G-double-prime")
The G''-value is a measure of the **deformation energy** used up by the sample during the shear process and therefore afterwards, it is **lost** for the sample. This energy is spent during the process of changing the material's structure, i.e. when the sample is flowing partially or altogether. Flow, and also viscoelastic flow, means: There is **relative motion between the molecules, clusters, particles, aggregates** or other components of the superstructure such as "domains" or crystals. Then, there are frictional forces between these components, and as a consequence, **frictional heat** occurs. This process is also called **"viscous heating"**. Scientists say: **Energy is consumed** during this friction process; it is **"dissipated"**, meaning: It is lost. A part of this energy may heat up the sample, and another part may be lost in the form of heat also to the surrounding environment. Energy losing materials are showing **irreversible deformation behavior** since after a load cycle, they occur with a changed shape finally. Thus, G'' represents the **viscous behavior** of a test material. (In ASTM D4092 is stated: G'' = "loss modulus, measured in shear", to distinguish it from E'' = "loss modulus, measured in tension or flexure".)

The loss factor or damping factor tanδ with the unit [1], (pronounced: "tangent delta" or "tan delta"). Definition:

Equation 8.10 $\tan\delta = G''/G'$

The loss factor is calculated as the quotient of the lost and the stored deformation energy. It therefore reveals the **ratio of the viscous and the elastic portion** of the viscoelastic deformation behavior. The following holds in general:

Equation 8.11 $0 \leq \tan\delta \leq \infty$ (since $0° \leq \delta \leq 90°$, see Equation 8.8)

Ideally elastic behavior might be specified in terms of $\delta = 0°$ or as $\tan\delta = 0$ since here, G' completely dominates G''. Ideally viscous behavior might be expressed as $\delta = 90°$ or as $\tan\delta = \infty$ since here, G'' completely dominates G'. If viscous and elastic behavior are exactly balanced, i.e. G' = G'', then $\tan\delta = 1$ or $\delta = 45°$ (or, if specified in rad: $\delta = \pi/4$ rad).

Note 1: The sol/gel transition point ("gel point"), and tanδ
Reaching the value of $\tan\delta = 1$ is an important analysis criteria for **gel formation, hardening and curing processes**. In this case, the so-called sol/gel transition point is reached.

For the fluid or liquid state ("sol state") holds: $\tan\delta > 1$ (since G'' > G')
For the gel-like state or solid state holds: $\tan\delta < 1$ (since G' > G'')
At the sol/gel transition point holds: $\tan\delta = 1$ (since G' = G'')

For more information on the "gel point", i.e. on the sol/gel transition in terms of a time point or as a temperature, see the Chapters 8.5.3b and 8.6.3b. See also Experiment 8.2 in Chapter 8.4.3a: bouncing rubber balls and tanδ.

Note 2: Tack and stringiness, and tanδ

Tacky behavior can be evaluated and controlled using the loss factor tanδ. When dividing into pieces, a sample will show stringiness only if the tanδ-value occurs in a certain medium range between the extremes "ideally viscous" and "ideally elastic". When reaching tanδ-values below or above these values of the medium range, tack can be either reduced or even prevented completely. Practical users can imagine the following two extreme materials without any stringiness when they are devided: On the one hand pure water showing an "infinitely high" tanδ-value since it will just flow apart; and on the other hand a stone

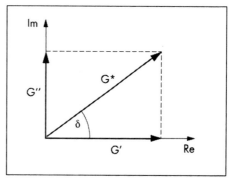

Figure 8.6: Vector diagram showing G', G" and the resulting vector G*

showing tanδ = 0 when breaking with a brittle fracture. See also Chapter 10.8.4.2: tack test.

Example: Stringiness of an adhesive, and tanδ

After a coating process using a robot, an automotive adhesive showed stringy behavior between the robot's nozzle and the substrate, therefore causing an impurity problem as well for the cars to be coated, as well as for the coating equipment and also for the floor. The problem was finally solved by adding an inorganic rheological additive to the adhesive such as silica or clay. By doing this, the loss factor was shifted from initially tanδ = 0.3 to merely 0.15 which was a useful value in this case. By this action, **the tanδ-value was shifted from a medium range towards a lower value which resulted in a more gel-like or paste-like, solid and brittle character of the material used**. See also the Example of Chapter 8.5.2.2d: stringiness of uncured adhesives, evaluation and classification using tanδ-values.

&ᷧ   For "Mr. and Ms. Cleverly"

1) Notation of G' and G" in terms of sine and cosine functions

Equation 8.12 $G' = (\tau_A/\gamma_A) \cdot \cos\delta$

Equation 8.13 $G'' = (\tau_A/\gamma_A) \cdot \sin\delta$

2) Vector diagrams illustrating G*, G', G" and tanδ

Viscoelastic behavior of all kinds of materials consists as well of a viscous as well as of an elastic portion. This sum can be illustrated by a vector diagram with G' on the x-axis and G" on the y-axis (see Figure 8.6). The lengths of the vectors correspond to the total amount of each one of the displayed parameters. G* is the vector sum, i.e. the resultant of the two components G' and G", therefore characterizing the complete **viscoelastic behavior which is composed of both an elastic and a viscous portion**.

Based on the **theorem of Pythagoras** (570 to 496 BC) concerning the lengths of the sides of a rectangular triangle, the relation between G*, G' and G" reads:

Equation 8.14 $|G^\star| = \sqrt{(G')^2 + (G'')^2}$

By the way, this relation was known at least 1000 years before Pythagoras and it was already used by the Babylonians, Egypts and Chinese; but it was Pythagoras himself who proved the validity of this theorem finally by application of mathematical methods [322, 337]. Based on this trigonometric relation, the loss factor occurs in Figure 8.6 as the side opposite to δ divided by the side adjacent to δ, which is the definition of the tangent of the angle δ, or briefly, tanδ.

Equation 8.10 $\tan\delta = G''/G'$

&ᷧ   End of the Cleverly section

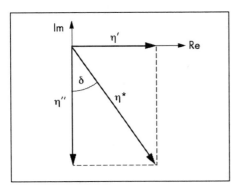

Figure 8.7: Vector diagram showing η', η" and the resulting vector η*

b) The complex viscosity, its real and imaginary parts

Newton's law applies, here in the form:

Equation 8.15 $\eta^* = \tau(t)/\dot{\gamma}(t)$

with the **complex viscosity** η* **[Pas]**, and the values of the sinusoidal functions of τ(t) in [Pa] and $\dot{\gamma}$(t) in [s⁻¹]

Complex viscosities η* always should be marked with a star to distinguish them from "common" shear viscosities which are not measured by oscillatory tests and analyzed correspondingly. Analysis is performed afterwards using "complex mathematics". As a comparison: Shear viscosities η are carrying no additional sign if they are determined under constant, steady-state shear conditions, i.e. when detecting each one of the individual measuring points by applying a constant shear rate; as explained in Chapter 2.2.3a and illustrated by Figure 2.9 (no. 2).

Diagrams and tables usually are showing η* in terms of the total amount |η*|, i.e. presenting also the two strips to indicate the total amount. This is recommended due to formal reasons since "complex numbers" are used for the calculation (see Chapter 8.2.4c).

The following holds for the total amounts:

Equation 8.16 $|G^*| = \omega \cdot |\eta^*|$

⌇⌇   For "Mr. and Ms. Cleverly"

1) Notation of η' and η" in terms of sine and cosine functions
Real part of the complex viscosity

Equation 8.17 $\eta' = G''/\omega = (\tau_A \cdot \sin\delta)/(\gamma_A \cdot \omega)$

Unit: [Pas], η' represents the viscous behavior.

Imaginary part of the complex viscosity

Equation 8.18 $\eta'' = G'/\omega = (\tau_A \cdot \cos\delta)/(\gamma_A \cdot \omega)$

Unit: [Pas], η" represents the elastic behavior.

2) Vector diagrams showing η*, η', η" and tanδ
Complex viscosity η* can be presented in the form of the vector sum of the two components η' and η" (see Figure 8.7). Using the **theorem of Pythagoras**, for the total amount of η* counts:

Equation 8.19 $|\eta^*| = \sqrt{(\eta')^2 + (\eta'')^2}$

The loss factor is illustrated in Figure 8.7 as the side opposite to δ divided by the side adjacent to δ.

Equation 8.20 $\tan\delta = \eta'/\eta''$

Note 1: Elastic behavior in terms of G' or η", and viscous behavior as G" or η'
The elastic portion can be specified as both G' or η", and the viscous portion as both G" or η'. Please note: The number of primes of G and η is different for each one of the two behaviors. Usually, G-values are used when discussing oscillatory tests in industrial practice, and η-values can be found rarely there. Therefore, η' and η" are not mentioned further here.

Note 2: The term "(absolute) dynamic viscosity"

Sometimes, the term "dynamic viscosity" or "absolute dynamic viscosity" is used for η^* [327]. Some authors use this term to refer to the real part of the complex viscosity η' only [18, 233]. To avoid confusion and in agreement with the majority of the current international authors, here, the term "complex viscosity" will be used for η^* (see also Chapter 2.2.3a).

Note 3: Unusual terms used for η' and η''

Some authors have tried to give individual names to each of the above viscosity parameters. Similar to circuits in electrical engineering, η' is then referred to as the "in-phase component of the complex viscosity", and η'' as the "out-of-phase component" (or according to DIN 13343, in German: "Wirkviskosität" for η', meaning "effective viscosity", and "Blindviskosität" for η'', meaning "blind viscosity", here in the sense of "ineffective"). However, these terms are rarely used and therefore not used here. For more information on the analogy between electric and mechanical oscillation circuits see [237, 367].

c) Presentation of the parameters in complex notation

For people interested in mathematics, some additional notes are given here, using the "complex plane" or "Gaussian plane", named in honor to the mathematician *Carl Friedrich Gauss* (1777 to 1855) [150]. This plane is spread out using the "real axis" (x-axis) and the "imaginary axis" (y-axis), in order to illustrate the concept of the **complex numbers** (see Figures 8.6 and 8.7). Here, the imaginary unit $i = \sqrt{-1}$, the "negative root", is used. G^* in complex notation:

Equation 8.21 $G^* = G' + iG''$

with the real component G' and the imaginary component G''. In Figure 8.6, G^* can be seen as the sum of G' and G'' in the form of the vector resultant. η^* in complex notation:

Equation 8.22 $\eta^* = \eta' - i\eta''$

In Figure 8.7, η^* is shown as the sum of the real part η' and the imaginary part $(-\eta'')$. The following counts:

Equation 8.23 $G^* = i\omega \cdot \eta^*$

And, as already mentioned, for the total amounts holds (see Equation 8.16): $|G^*| = \omega \cdot |\eta^*|$

When using the parameters defined in the Chapter at hand, the introducing chart which is shown at the beginning of Chapters 2 to 8 could be extended as follows:

ideally viscous flow behavior	behavior of a viscoelastic liquid	viscoelastic behavior showing 50/50 ratio of the viscous and elastic portions	behavior of a viscoelastic gel or solid	ideally elastic deformation behavior
$\delta = 90°$	$90° > \delta > 45°$	$\delta = 45°$	$45° > \delta > 0°$	$\delta = 0°$
$\tan\delta \rightarrow \infty$	$\tan\delta > 1$	$\tan\delta = 1$	$\tan\delta < 1$	$\tan\delta \rightarrow 0$
$(G' \rightarrow 0)$	$G'' > G'$	$G' = G''$	$G' > G''$	$(G'' \rightarrow 0)$

d) Conversion between shear deformation and shear rate

Shear rate values can be calculated from strain or deformation values, which are preset or obtained when performing oscillatory tests, as follows:

Equation 8.24 $\dot{\gamma} = \gamma_A \cdot \omega$

with the shear rate $\dot{\gamma}$ [s^{-1}], the angular frequency ω [s^{-1}], and the deformation (strain) amplitude γ_A [1 = 100%]. Caution: Here, the unit of γ_A is not taken in percent. See also the conversion examples for amplitude sweeps in Note 4 of Chapter 8.3.3.1, and for frequency sweeps in Note 1 of Chapter 8.4.1.

ᏜᏜ   End of the Cleverly section

8.2.5 The test modes controlled shear strain and controlled shear stress, raw data and rheological parameters

The following eight parameters have been explained above in the Chapter at hand:

G^*, G', G'', η^*, η', η'', δ, $\tan\delta$

The great number of different parameters which are often used when performing oscillatory tests might be confusing or discouraging for practical users who are not so familiar with rheology. However, when performing oscillatory tests, a rheometer produces not more than two different raw data (see Tables 8.1 and 8.2).

a) Tests with controlled shear strain, also called "controlled shear deformation" CSD
When performing controlled shear strain tests, deflection angles φ or shear strains γ, respectively, are preset and controlled by the rheometer (see Table 8.1). Sometimes, this test method is also referred to as a controlled shear rate test (CSR), since both the shear strain γ (or deformation) and the shear rate $\dot{\gamma}$ are closely related physically and mathematically (see Chapters 4.2.1 and 8.2.1). Presetting is made in the form of the following sine curves (index "A" for amplitude):

$\varphi(t) = \varphi_A \cdot \sin\omega t$ or $\gamma(t) = \gamma_A \cdot \sin\omega t$

The results are obtained as sine curves which are shifted by the phase shift angle δ:

$M(t) = M_A \cdot \sin(\omega t + \delta)$ or $\tau(t) = \tau_A \cdot \sin(\omega t + \delta)$

b) Tests with controlled shear stress (CSS tests)
When performing CSS tests, torques M or shear stresses τ, respectively, are preset and controlled by the rheometer (see Table 8.2). Presetting is made in the form of the following sine curves:

$M(t) = M_A \cdot \sin\omega t$ or $\tau(t) = \tau_A \cdot \sin\omega t$

The results are obtained as sine curves which are shifted by the angle δ:

$\varphi(t) = \varphi_A \cdot \sin(\omega t + \delta)$ or $\gamma(t) = \gamma_A \cdot \sin(\omega t + \delta)$

Table 8.1: Raw data and rheological parameters of oscillatory tests with controlled shear strain

Oscillation CSD	Preset	Results
raw data	deflection angle φ(t) [mrad]	torque M(t) [mNm] and phase shift angle δ [°]
rheological parameters	strain (deformation) γ(t) [%]	shear stress τ(t) [Pa] and δ [°]
calculation of the complex shear modulus		$G^* = \tau(t) / \gamma(t)$

Table 8.2: Raw data and rheological parameters of oscillatory tests with controlled shear stress

Oscillation CSS	Preset	Results
raw data	torque M(t) [mNm]	deflection angle φ(t) [mrad] and phase shift angle δ [°]
rheological parameters	shear stress τ(t) [Pa]	deformation γ(t) [%] and δ [°]
calculation of the complex shear modulus		$G^* = \tau(t) / \gamma(t)$

Summary: With oscillatory tests, there are measured only two independent raw data.
When performing oscillatory tests, a rheometer measures not more than two raw data as can be seen in Tables 8.1 and 8.2: M(t) and δ when presetting the strain, or φ(t) and δ when presetting the stress, respectively. From these two independent variables are determined the viscous (e.g. as G") and the elastic portion (e.g. as G') of the viscoelastic behavior. If further parameters are presented they should be regarded only as another possibility to show the obtained test results. However, these values do not contain any additional independent information. For reasons of clarity, the presentation of measured data in diagrams should be limited to the main parameters (e.g. to G' and G"). Again, rheometers can only produce two independent raw data values which – even after many steps of complex mathematical conversions – finally cannot result in more than two independent rheological parameters, independent of in whichever terms these results are specified after all.

c) Recommendation: Scaling of diagrams showing measuring curves
For oscillatory tests, presentation of diagrams is recommended in any case on a **double-logarithmic scale** for the following reasons:
1) Frequently, that part of the curves which is most important for analysis is just the range showing very low values.
2) Often, the numeral values of the curve functions are stretching over several decades.
Examples: Amplitude sweeps as lg G'(lg γ) and frequency sweeps as lg G'(lg ω), i.e. both in the form of log/log diagrams

Time and temperature are usually presented on a **linear scale**.
Examples: Time-dependent lg G'(t), or temperature-dependent lg G'(T) as semi-logarithmic (log/lin) diagrams.

Mostly it is useful to show the curves of G' and G" together in the same diagram. Sometimes these parameters are displayed together with the following ones:
1) Loss factor tanδ: For example when presenting processes like melting, phase modifications, sol/gel transitions, gel-formation or curing (like the curves shown in Figures 8.37, 8.39, 8.41 and 8.43). For the loss factor both a lin or a log scale can be selected, a log scale should be preferred if a wide range of values should be displayed.
2) Complex viscosity $|\eta^*|$: Especially when presenting frequency sweeps of polymer samples to illustrate whether the plateau of the zero-shear viscosity occurs or not (like in Figure 8.17, and by way of intimation in Figure 8.21). Usually, the function of $|\eta^*|$ is shown on a log scale.

d) Recommendation: Specification of parameters in test protocols (data tables)
For practical laboratory work it is recommended to present the following parameters in data tables of the test protocol, preferably using the units shown in brackets:
G' [Pa], G" [Pa], tanδ [1], γ [%], τ [Pa], ω in [rad/s] or in [s⁻¹], T [°C]
Only when testing liquids, polymer solutions and melts, there should be presented additionally: $|\eta^*|$ [Pas]

It is useless to show η^* for other kinds of materials like gels, semi-solid and stable dispersions or solids.

In any case, the following raw data should appear in the test report:
φ [mrad], M [mNm], δ [°]

Only the raw data can give an unfailing insight whether a test has been carried out in a useful measuring range, since only they **are really measured by a rheometer**. Therefore, also the raw data should always be presented in data tables of test protocols, because often just they are the crucial information when it comes to discussions about measuring results.

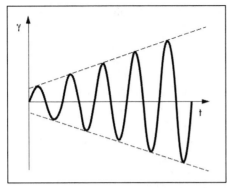

Figure 8.8: Preset of a shear strain amplitude sweep, or briefly, strain sweep

8.3 Amplitude sweeps

Amplitude sweeps are oscillatory tests performed at variable amplitudes, while keeping the frequency at a constant value (and also the measuring temperature).

8.3.1 Description of the test

Preset
1) For tests with controlled shear strain:
$\gamma(t) = \gamma_A \cdot \sin\omega t$
with a constant angular frequency ω = const, and a variable strain amplitude increasing with time $\gamma_A = \gamma_A(t)$, see Figure 8.8. This kind of test is also called a shear strain amplitude sweep, or briefly, **strain sweep**. Here, the period of time for each one of the oscillation cycles of the measuring bob is kept constant (frequency), only the maximum value of the bob's deflection angle is increasing continuously (amplitude).

2) For tests with controlled shear stress: $\tau(t) = \tau_A \cdot \sin\omega t$
with ω = const and a variable stress amplitude $\tau_A = \tau_A(t)$. This kind of test is also a called shear stress amplitude sweep, or briefly, **stress sweep**. In this case, the oscillatory motion of the measuring bob is caused by a torque whose maximum value is increasing continuously (amplitude). The period of time for each one of the oscillation cycles, however, is kept constant (frequency).

Note 1: Selection of the (angular) frequency value
When performing amplitude sweeps, **many users are selecting the angular frequency** ω **= 10rad/s**. Since $\omega = 2\pi \cdot f$, this value corresponds to a frequency of f = 1.6Hz approximately. Users who still prefer to work with the "old units" like Hz, often choose f = 1Hz which corresponds to ω = 6.28rad/s approximately. However, Hz is not an SI-unit. See also the Note in Chapter 8.2.1: f or ω; and Chapter 8.3.5: frequency-dependence of amplitude sweeps.

Measuring result
Usually for strain sweeps, lg γ is presented on the x-axis (and lg τ for stress sweeps, resp.), and lg G' and lg G'' are shown on the y-axis, i.e., logarithmic scales are used on both axis (see Figures 8.9 to 8.11).

Note 2: Three-dimensional diagrams of amplitude sweeps at different temperatures
The results of several amplitude sweeps which have been measured each one at a different but constant temperature (i.e. at isothermal conditions) can be presented in the form of a three-dimensional (3D) diagram, for example, with lg γ on the x-axis, lg G' and lg G'' on the y-axis, and the temperature T on the z-axis.

Note 3: Test preparation: Thermal-stability time of polymer melts
Before testing samples of polymer melts, it is recommended to check the thermal-stability time at the desired test temperature, for example, to avoid molecular degradation (according to ISO 6721-10; see also the Notes of Chapters 8.5.2 and 8.6.2).

8.3.2 Structural character of a sample

At low amplitude values, in the so-called **linear viscoelastic (LVE) range**, both the G'(γ)- and G''(γ)-curve display constant plateau values, mostly on different levels. The term linear LVE range is derived from the proportionality of the preset and measured parameters, and as a consequence, the resulting curve occurs as a straight line in the diagram, as follows:

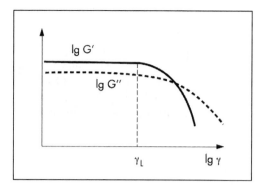

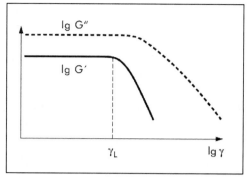

Figure 8.9: Strain amplitude sweep of a sample showing gel-like character in the LVE range, i.e. G' > G" (with the limiting value γ_L)

Figure 8.10: Strain amplitude sweep of a sample showing the character of a viscoelastic liquid in the LVE range, i.e. G" > G'

Linear-elastic behavior is found whenever Hooke's law is valid: For the raw data in terms of M_A/φ_A = const, or for the rheological parameters as τ_A/γ_A = const (= G^*), respectively.

Linear-viscous behavior is found whenever Newton's law applies: for $M_A/\dot{\varphi}_A$ = const or for $\tau_A/\dot{\gamma}_A$ = const (= η^*), respectively. For unlinked and unfilled polymers, the following holds in the LVE range: $|\eta^*| = \eta_0$

Analyzing the behavior in the LVE range, the following relations can be used to describe the "viscoelastic character" of a sample.

a) The character of a gel or solid, if G' > G"

Here, the elastic behavior dominates the viscous one (see Figures 8.9, 8.11 and 8.12). The sample exhibits a certain rigidity and this is no surprise for solids or stable pastes. However, there are many dispersions such as coatings, pharmaceutical lotions or foodstuffs which are showing low-viscosity flow behavior at medium and high shear rates but G' > G" in the LVE range. In this case they indicate gel-like consistency in the low-shear range. "Low-shear range" means at shear rates $\dot{\gamma} \leq 1s^{-1}$, and "at rest" – for practical users – means at $\dot{\gamma} \leq 0.01s^{-1}$. Nevertheless, when at rest, for these kinds of gels, "semi-solid matters" or viscoelastic solids, a certain firmness and stability can be expected even if they show a weak gel structure only.

Examples: Cross-linked polymers, rubbers, solids, stable dispersions, pastes and gels, but also coatings (e.g. when containing inorganic gellants like silica or clay), stable w/o emulsions, salad dressings (e.g. when stabilized by hydrocolloids such as starch, carrageen, or locust bean gum); and all materials showing a flow point (as presented in Figure 8.12), and therefore, no plateau value of a zero-shear viscosity.

b) The character of a liquid or fluid or a sol, if G" > G'

In this case, the viscous behavior dominates the elastic one (see Figure 8.10). The sample exhibits the character of a liquid, and that counts also for the LVE range. Even highly viscous materials with entangled molecule chains but without a consistent chemical network or physical network-of-forces are showing this behavior. At rest, these kinds of materials usually are not stable since they are flowing with time. However, this flow process can occur with a very, very low flow velocity.

Examples: Self-leveling fluids (when at rest), polymer solutions and polymer melts consisting of unlinked polymers (e.g. silicone, PDMS), bitumen (if not "frozen"); and all materials showing a zero-shear viscosity. These kinds of viscoelastic materials indeed show a yield point, but cannot show a flow point in principle since they are flowing in the entire measuring range, and therefore, of course already in the LVE range (as shown in Figure 8.10).

c) At the gel point, if G' = G" in the LVE range

If the values of the two moduli are balanced, the behavior is frequently called **"at the gel point"**. This state occurs seldom with samples to be measured in the industrial lab and it merely happens by accident, but if indeed appearing, it shows the behavior of a material at the borderline between liquid and gel-like. In this case, it is useful to perform further amplitude sweeps by presetting also other frequency values. Additionally here, for a further characterization of the sample's structure it may be useful to carry out also frequency sweeps which are measured at several strain amplitudes.

8.3.3 Limiting value of the LVE range

Amplitude sweeps are mostly carried out for the sole purpose of determining the limit of the LVE range. As long as the strain amplitudes are still below the limiting value γ_L the curves of G' and G" are remaining on a constant plateau value, i.e. the structure of the sample shows no significant change at these low deformations (see Figures 8.9 to 8.11, with the index "L" for "limiting value"). When measuring in the LVE range, practical users sometimes speak of "non-destructive testing" NDT. Another synonym, used above all by scientists, is "small angle oscillatory shear" SAOS (see also Chapter 8.3.6: SAOS and LAOS tests). At strain amplitudes higher than γ_L the limit of the LVE range is exceeded. Here, the structure of the sample has been changed already irreversibly, or it is even completely destroyed.

There are several options to determine the limiting value of the LVE range. **In most cases the G'-function is taken for the analysis** since it is mostly the G'-curve which shows as the first one the tendency to leave the LVE range.

a) Visual or manual analysis

1) When observing the measuring curve, the user decides at which γ-value the G'-curve begins to deviate noticeably from the LVE plateau. In order to facilitate visual analysis, a straight line may be drawn as a **fitting line** along the plateau of the G'-function.

2) Using the data table of the test protocol, the last γ-value of the LVE range is determined at which the G'-values are just beginning to show a noticeable deviation from the previously constant values. This measuring point is taken as the limiting value γ_L. The range of the tolerated deviation has to be defined previously by the user, e.g. as 5% or 3% (or even 10%). In other words: If 10% was selected as the tolerance to be allowed, all those G'-values which are below 90% of the plateau value are considered to be outside of the LVE range.

b) Automatic analysis using a software analysis program

Using an appropriate software program, "smoothing" of the data can be performed before starting the analysis. Then, the user defines the desired range of tolerance. Afterwards, the γ_L-value will be determined by the analysis program. Also here, the adaption of a straight fitting line on the plateau of the G'-function is the basis to determine of the limiting value of the LVE range.

Note: Recommended limits of the LVE range for polymers (according to ISO 6721-10)

When testing polymer melts, ISO 6721-10 recommends to chose as the limit of the LVE range that strain value (or stress, respectively) at which a difference of 5% occurs in the values of any of the parameters G^*, G' or G" compared to their values in the LVE range.

8.3.3.1 Limiting value of the LVE range in terms of the shear strain

The following limiting values of γ_L may be considered as values of **the permissible maximum strain (deformation)**. These values are empirically found and proven in practice in many repeated tests. However, they should only be used as a rough estimation as they may be not valid in all cases. Therefore **first of all, with every unknown sample an amplitude sweep should always be carried out**. Only after performing this kind of testing with the sample to be measured, the user can be sure to be on the safe side to avoid irreversible structure changes when carrying out further measurements. This is of crucial importance since the laws of Hooke and Newton are only valid in the LVE range (see also Chapter 8.3.2). Therefore, any evaluation of measuring results is not based on the basic laws of rheology if the results are obtained from tests which are performed at higher strain or stress values outside the LVE range.

The following might be used as a rough guide to estimate limiting values of the strain γ_L (or deformation, respectively):

1) For neat, unmodified resins, unlinked and unfilled polymer solutions and melts:
$\gamma_L \leq 50\%$ (usually, according to ASTM D4473).
However here, in some cases γ_L-values of only 10% or even up to 100% can be found.

2) For materials with a network, e.g. showing a physical-chemical superstructure or chemical bonds such as **dispersions** (i.e. suspensions, emulsions, foams), **pastes, gels and cross-linked polymers above the glass-transition temperature** (i.e. if $T > T_g$): $\gamma_L \leq 1\%$
According to ASTM D4473, for **prepregs** is recommended: $\gamma_L \leq 2\%$. Comment: For filled polymers should be set $\gamma_L \leq 1\%$, and in most cases even $\gamma_L \leq 0.1\%$ should be preferred.

3) For solids, thermosets, polymers below the glass-transition temperature, i.e. if $T < T_g$, **highly concentrated dispersions,** and materials showing a very inflexible structure such as **waxes, vaselines, lubricating greases** and **chocolate melts**, the limit is at least 10 times lower, thus: $\gamma_L \leq 0.1\%$
According to ASTM D4065, for materials showing an elastic modulus in the range from 0.5MPa to 100GPa, there is recommended: $\gamma_L < 1\%$. Comment: Here, in any case $\gamma_L \leq 0.1\%$ is required, and often even $\gamma_L \leq 0.01\%$.

4) Reversible elasticity was measured on **flexible biological tissues** up to $\gamma_L = 1200\%$. With "super-elastic" elastomers consisting of for example **soft rubbers** and copolymers, values of $\gamma_L = 1000$ to 2000% are achieved today. Applications of these products are in bio-technology, e.g. membranes, prosthesis for blood vessels (stents), artificial tendons and muscles, and can be found also in other fields such as loudspeaker membranes and packaging material (see also Chapter 5.2.2.2g) [160, 381, 398].

Note 1: Inhomogeneous deformation behavior
At very low strain values already, parts of inflexible and rigid samples may tend to inhomogeneous deformation behavior in the form of "plastic" creep or flow. This means: Deformation behavior occurs within such a material in a not homogeneously distributed form related to the material as a whole, and then of course, the same applies to the shear gap of the measuring system used. Therefore, these kinds of samples cannot be analyzed exactly in a scientific sense. In this case, raw data only but not rheological parameters should be considered for analysis. See also Chapter 3.3.4.3d and Figure 2.9: plastic behavior, and Chapter 8.2: required shear conditions.

Note 2: Frequency-dependence of the limiting value of the LVE range
When performing amplitude sweeps, the determined limiting values of the LVE range are valid for the applied frequency only. At a higher frequency, many samples might show higher rigidity, and as a consequence, higher G'-values then. In this case, such a sample might be more inflexible and more brittle which again might lead to a correspondingly lower limiting value of the LVE range (see also Chapter 8.3.5).

Note 3: The limiting value of the LVE range via the G"-curve, and the G"-peak
When exceeding the limit of the LVE range, with many gels, highly concentrated dispersions showing a network of forces, chemically cross-linked materials and solids, the curves of

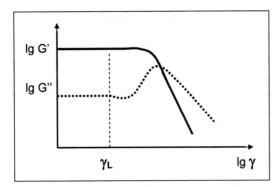

Figure 8.11: Strain amplitude sweep of a sample showing a G"-peak

G' and G" often do not slope downwards with increasing deformation. Then, the G"-curve in particular may show a more or less high peak on a more or less wide base, see Figure 8.11. However, peaks do not appear, or they are occurring only slightly, for polymers with unlinked linear molecules. The height of the peak usually increases both as well with increasing measuring frequency as well as with decreasing measuring temperature. An example of this are filled and cross-linked polymers, such as rubbers containing soot [398]. In rare cases several peaks may appear (see also Chapter 9.2.2: complex behavior and shear-banding).

Increasing G"-values indicate an increasing portion of deformation energy which is used up already before the final breakdown of the internal superstructure occurs, to irreversibly deform at first only parts of the latter. This may occur due to relative motion between the molecules, flexible end-pieces of chains and side chains, long network bridges, mobile single particles, agglomerates or superstructures which are not linked or otherwise fixed in the network. As a reminder: The loss modulus G" represents the deformation energy which is **lost (dissipated)** due to inner friction processes (see Chapter 8.2.4a).

Leaving the plateau value of the LVE range is therefore indicated not only by decreasing curves of G' and G", but also by a significant increase in the G"-curve. In both cases the **range of irreversible deformation** is entered. When a G"-peak appears, it can be assumed that initially **there was a network structure** when at rest, independent of whether this structure was built up via primary or secondary bonds. In this case, it can also be interpreted that this network does not suddenly collapse in the whole shear gap if the LVE range has been exceeded, since here at the beginning, micro-cracks are forming which grow into macro-cracks until the G"-peak is exceeded, until finally a large crack devides the entire shear gap. This event is displayed by the **crossover point G' = G"**, and means that the sample is flowing now as a whole. The maximum of the G"-curve and the crossover point of the curves of G' and G" are mostly occurring very close together. See also Chapter 8.3.6: LAOS tests at large deformations, and analysis of as well viscous shear-thinning or shear-thickening as well as elastically shear-softening or shear-stiffening behavior.

Example (to Note 3): The LVE range of a solid polymer, the G"-curve and micro-cracks

A designer studio was interested to get information about the behavior of a polymer at room temperature. Therefore, non-destructive tests are performed using polymer specimens in the form of torsion bars showing a rectangular cross-section (see also Chapter 10.7.1). With this very rigid samples, strain sweeps are carried out starting already at a strain value of $\gamma = 10^{-5} = 0.001\%$. In the LVE range, the test resulted in $G' = 10^9 Pa = 1GPa$ and $G" = 10^7 Pa$ (i.e. $\tan\delta = G"/G' = 0.01$). When reaching $\gamma = 10^{-4} = 0.01\%$, the G'-values still remained almost constantly on their high level, but the G"-values began to increase continuously.

Interpretation: Micro-cracks are appearing beyond this γ-value which marks the limiting value of the LVE range. The rising G"-curve is indicating an increase in lost energy. **When micro-cracks are developing, relative motion is occurring** along their initially very small fracture areas **causing internal friction** and, as a consequence, **frictional heat is**

produced during this irreversible deformation process. This process can be regarded as a slow-motion flow process, and therefore, scientists are speaking of "viscous heating". As a consequence, the corresponding portion of energy is lost for the sample afterwards. With **growing cracks**, also the friction areas are **increasing** and so does the **amount of the lost energy, and this process is indicated by the increasing values of the G"-curve**. Using a common rheometer, a polymer bar at room temperature usually cannot be tested until the final fracture occurs since the measurable maximum torque of these kinds of instruments is limited. But reaching that point is not necessary for the purpose of determining the limiting value of the LVE range, since the following holds: **The LVE range is exceeded at that point when the first curve of the functions of G' or G" is leaving the plateau value. Here**, in the example at hand, **it is the G"-curve** which is **sloping upwards**.

Note 4: Calculation of shear rate values occurring with amplitude sweeps

Using Equation 8.24, shear rate values can be calculated as follows:

Equation 8.24 $\qquad \dot{\gamma} = \gamma_A \cdot \omega$

Example (to Note 4): With the strain amplitude $\gamma_A = 10\% = 0.1$ and the angular frequency $\omega = 10s^{-1}$ the shear rate $\dot{\gamma} = 0.1 \cdot 10s^{-1} = 1s^{-1}$. This shows that the shear conditions occurring with amplitude sweeps performed in the LVE range, usually correspond to the low-shear range at $\dot{\gamma} \leq 1s^{-1}$. Important for practical users: Oscillatory tests are usually carried out in a completely different shear rate range compared to rotational tests, the two test types therefore are complementing each other.

8.3.3.2 Limiting value of the LVE range in terms of the shear stress

The stability of viscoelastic materials can also be evaluated by analyzing the dependence of G' and G" on the shear stress. The following applies here, similar to the explanations above for the strain or deformation: The LVE range in terms of the shear stress is exceeded if the first one of the stress-dependent values of G' or G" is beginning to deviate significantly from the plateau value.

Stress controlled tests are producing limiting values of the LVE range which are often **less frequency-dependent** than those obtained by strain controlled tests. Despite this, in industrial practice strain controlled tests are preferred. The reason is that the change of the structural strength is directly depending on the degree of deformation. Shear stresses, however, can influence the degree of deformation only indirectly (see also Note 2 of Chapter 3.4.2.2a: shear rate or stress control).

8.3.4 Determination of the yield point and the flow point by amplitude sweeps

Here, the measuring curves are presented versus the shear stress τ, and usually like above, the functions of G' and G" are selected for the evaluation, using logarithmic scales on both axis (see Figure 8.12).

8.3.4.1 Yield point or yield stress τ_y

Also here, it is the G'-function which is mostly used for analysis. The limiting value of the LVE range in terms of the shear stress τ is the "yield stress" or the "yield point" (see τ_y in Figure 8.12). No significant change of the internal structure occurs as long as stresses below the yield point are applied, and therefore, no yielding behavior can be observed in this range. Therefore, being still in the LVE range before reaching the yield stress, the sample is showing reversible viscoelastic behavior.

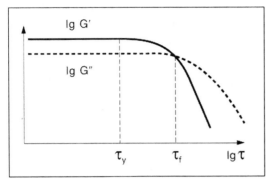

Figure 8.12: Stress amplitude sweep, showing the yield point τ_y at the limit of the LVE range, and the flow point τ_f when $G' = G''$ (pre-condition is $G' > G''$ in the LVE range)

a) Visual or manual analysis of the yield point

1) When observing the measuring curve, the yield point τ_y is the stress value at which the curve begins to deviate noticeably from the LVE plateau or from the corresponding **fitting straight line** used for analysis, respectively.

2) This value can also be determined from the data table of the test protocol, taking the last τ-value of the LVE range at which the G'-values are just beginning to show a noticeable deviation from the previously constant values. This measuring point is taken as the limiting value of the shear stress, or briefly, as the yield point τ_y. The range of the tolerated deviation has to be defined previously by the user, e.g. as 5% or 3% (or even as 10%). In other words: If the tolerance was selected as 10%, then all those values which are below 90% of the plateau value are considered to be outside of the LVE range.

b) Automatic analysis of the yield point using a software program

At first, the user defines the desired range of tolerance. Then, the τ_y-value will be determined by the analysis program. Also here, the basis of the determination is the straight fitting line on the plateau of the G'-function.

Note: Comparison of yield points obtained from rotational or oscillatory tests

Above analysis method often produces a yield point value which is comparable to that one obtained from a rotational test if the "method of the single straight fitting line" was taken as the basis of the determination, using the line which is adapted to the linear-elastic range (as explained in Chapter 3.3.4.2a). However, when using the "tangent crossover method" (as shown in Chapter 3.3.4.2b), the yield point value obtained is often not as comparable to the value achieved by the amplitude sweep method.

8.3.4.2 Flow point or flow stress τ_f

In order to determine the flow stress value, there has to be paid attention for the following: **Pre-condition is $G' > G''$**, and therefore the occurrence of a **gel-like character or solid state in the LVE range**. Then, when reaching the crossover point $G' = G''$, the gel-like character with $G' > G''$ changes to the liquid state showing $G'' > G'$ (see τ_f in Figure 8.12). Sometimes, also this point is called the "yield point" but in order to distinguish this point from the above τ_y-value, it is here called the "flow point". Indeed, in many languages besides English (and the Spanish: "valor de cedencia") instead of the "yield point" a term including "flow" is used: for example in German "Fliessgrenze", in French "seuil d'écoulement" [78], in Italian "soglia di scorrimento", in Dutch "vloeigrens", in Swedish "flytgräns", all meaning literally the "flow limit". However, one thing should be kept in mind: If the flow point is taken in terms of the point $G' = G''$, a measuring point will be determined which occurs already in the non-linear (NL) deformation range. Therefore, in a scientific sense, the flow point τ_f should be considered a relative value and not an absolute value. See also Chapter 8.3.6: LAOS tests in the NL range.

8.3.4.3 Yield zone between yield point and flow point

The range between τ_y and τ_f can be called the "yield zone" or "yield/flow transition range". Here still with $G' > G''$, the sample is showing gel-like behavior. However, the reversible-elastic deformation range is exceeded already, and therefore, **irreversible behavior** is present now, despite the still occurring dominance of elastic behavior. If the shear process was stopped within this range, G' and G" would show lower values then, compared to the initial plateau values of the LVE range. These decreased values might remain finally, even after the shear load is released, see also Chapter 3.3.4.3: Yield zone via rotational tests.

Example: The flow point and the flow transition index of grease (acc. to DIN 51810-2) [228] Recommended are the two measuring temperatures T = +25 and -40°C, the use of a parallel-plate measuring system (d = 25mm), and **trimming of the sample** (i.e. removing projecting sample material) at a distance of 1.025mm, before setting the shear gap of H = 1.000mm finally. The heating or cooling rate is $\Delta T/\Delta t$ = 0.4K/min, and after reaching the measuring temperature, there is an equilibration time of t = 15min if T = +25°C, or t = 30min if T = -40°C. The measurement is performed in terms of an amplitude sweep.

Preset: Strain γ = 0.01 to 100% and angular frequency ω = 10rad/s = const, with always 10 measuring points per decade. (Here, testing at controlled strain is called the measuring method A. Using controlled stress is the method B then, in the range of τ = 1 to 10,000Pa. However, the latter alternative is not recommended for these kinds of samples).

Analysis:
1) **Flow point τ_f at the crossover point G' = G"**
Possible further specifications are:
2) Values of G' and G", and as well of tanδ (= G"/G') in the LVE range
3) Yield stress τ_y and yield strain γ_y at the limit of the LVE range
4) Flow strain γ_f at the flow point (i.e. when G' = G")
5) **Flow transition index τ_f/τ_y** as the ratio value of the flow stress and the yield stress to characterize the breaking behavior of the inner structure. The closer this index to the value 1, the more likely the grease will show **brittle behavior**.

Note: When comparing this modern DIN measuring method to the methods when using torque testers according to ASTM D1478 and IP 186, the former provides information which is more relevant to practice. Additionally, torque testers are also considerably more expensive and testing needs clearly more time [110], see also Chapter 11.2.11i.

8.3.4.4 Evaluation of the two terms yield point and flow point

Of course, when using two different parameters instead of the only one in order to determine yielding, there are improved options available to characterize the behavior of a material. **Many scientists prefer to use the yield stress** in order to analyze materials and the structure strength, still being on the safe side against irreversible structure changes, since the limit of the LVE range is not exceeded yet. Additionally it can be taken for granted that this value is still an absolute measuring value.

Practical users, however, **often prefer to use the flow point** as the basis of evaluation since they want to know the point at which the internal structure is breaking to such an extent causing the material to flow finally. Frequently, the two values τ_y and τ_f deviate significantly from each other, sometimes more than a factor of 100. By the way, for comparison only: In terms of γ-values, these two analysis points mostly would show a difference of more than a factor of 10 then. Therefore, in order to cut back corresponding discussions, the analysis method and the evaluation criteria should be defined in the form of clear specifications already before performing test and analysis.

8.3.4.5 Measuring programs in combination with amplitude sweeps

Sometimes in industrial practice, combined test programs are useful which include besides an amplitude sweep also other test intervals.

Example: Amplitude sweep upwards and downwards for testing plastisols

In order to evaluate the behavior of uncured plastisols used as **underseals** and **seam sealants for automobiles**, the following **test method** is recommended by the "Working Group Plastisol Rheology" consisting of members from plastisol manufacturers, users (OEM, car producers) and measuring instrument manufacturers:

Measurement: Using an air-bearing rheometer, a parallel-plate measuring system (d = 25mm) and the shear gap of H = 1mm, at the measuring temperature T = +25°C (or upon consultation at another temperature between T = +20 and +38°C).

Preset: For all test intervals (T_0 to T_4) at the constant angular frequency ω = 10rad/s, and with the duration of t = 10s for each measuring point

T_0: Waiting time (without shear, interval of rest and temperature equilibration); for t = 5min

T_1: **Amplitude sweep** at γ = 5 to 100%; with 12 measuring points

T_2: **High-shear** interval, maintaining γ = 100% = const; with 30 measuring points

T_3: **"Reversed amplitude sweep"** at γ = 100 to 5%; with 12 measuring points

T_4: Optional, if in T_3 there is still occurring no crossover point G' = G": continued measurement at γ = 4 to 0.1%; with 14 measuring points

Analysis (A_1 to A_6), and **evaluation for practical use**:

Regarding T_1: Filling process of the dosing equipment, or sagging behavior in a storage tank

(A_1) τ-value (in Pa) at the crossover point G' = G", as the **flow point at the onset of pumping**

Regarding T_2: Shear stability when pumping, volume flow rate during application (e.g. when using robots with flat stream nozzles), overspray, die swell effect, seam width, turning behavior on edges

(A_2) $|\eta^*_2|$ value (in Pas) at the last measuring point, as the **flow behavior during application**

(A_3) ($\Delta|\eta^*|/|\eta^*_1|$) · 100% (in %), the relative viscosity difference between the first and the last measuring point, referring to the first measuring point, as the percentage **structural decomposition**

(A_4) δ-value (phase shift in degrees) at the last point, as the **viscoelastic behavior**

Regarding T_3 and T_4: structural regeneration, leveling, sagging, dripping, contour stability, layer thickness, brushability

(A_5) τ-value (in Pa) at the crossover point G' = G", as the **stability flow point**

(A_6) $|\eta^*|$ value (in Pas) at the last measuring point, as the **flow behavior after application**

See also the Example in Chapter 8.5.2.2d: testing adhesives in terms of a combination of an amplitude sweep and a step-test.

$\mathcal{GS}$ For "Mr. and Ms. Cleverly"

8.3.5 Frequency-dependence of amplitude sweeps

When performing amplitude sweeps at a lower angular frequency than ω = 10rad/s, many non-polymeric samples do not show a pronounced change of the G'- and G"-values compared to the values at ω = 10rad/s, neither for the values in the LVE range nor for the limiting value of this range.

Note 1: Testing at higher frequencies may result in higher values of G' and G"

A variety of materials are showing comparably higher values of G' and G" in the LVE range when measured **at a higher frequency value**, e.g. five or ten times higher, since then, **many structures are exhibiting less flexibility and therefore higher rigidity** at this faster motion. In this case, often a smaller LVE range can be observed and also a steeper fall of the G'-curve after exceeding the LVE range. This may indicate a **tendency to more brittle beha-**

vior of the internal superstructure. Particular for unlinked polymers, even the viscoelastic character might be changed if there is G" > G', thus liquid behavior at lower frequencies, but G' > G" and therefore gel-like behavior at higher ones.

Note 2: Amplitude sweeps usually do not simulate the behavior at rest

Usually when performing amplitude sweeps, the samples are not measured in the state-at-rest since the tests are mostly carried out at a constant angular frequency of ω = 10rad/s which is not really very slow motion and also not the state-at-rest (see also Note 4 of Chapter 8.3.3.1: calculation of shear rate values occurring with amplitude sweeps). However, **if it is aimed to evaluate a material at rest, a frequency sweep should be performed and analyzed at very low frequency values**, e.g. at ω = 0.01rad/s (see also Chapters 8.4.3 to 8.4.5). If amplitude sweeps were carried out at this low frequency, too much time for testing is required, at least for industrial users.

Note 3: Materials showing strong dependence on the frequency

Sometimes, unexpected and extraordinary results may be obtained when performing amplitude sweeps, see the following examples:

Example 1 (to Note 3): An adhesive, flowing at high frequencies but not at low ones

An uncured adhesive as it is used in automotive industry showed undesired flow immediately after the application, but it exhibited no flow after stirring it softly and very slowly in a small container. Amplitude sweeps displayed in the LVE range: G" > G' at ω = 10rad/s, but G" = G' at ω = 1rad/s and even G" < G' at ω = 0.1rad/s. A subsequent frequency sweep confirmed these results: The crossover point G" = G' occurred at ω = 1rad/s, for ω < 1rad/s there was G' < G", and for ω > 1rad/s then G" > G'. This indicated a gel-like superstructure at rest, and therefore, a certain dispersion stability there.

Example 2 (to Note 3): Printing inks, gel-like at low frequencies but liquid at higher ones

Many offset printing inks display a frequency-dependent rheological character when performing amplitude sweeps: **at low frequencies G' > G"**, i.e. having a gel-like character showing as well a yield point as well as a flow point here; however **at higher frequencies G" > G'**, i.e. having the character of a liquid showing indeed a yield point but no more a flow point now. However, the user should reflect whether this kind of testing is really simulating the situation during the practical use of an ink. For example: Is an ink really in a "state of rest" when evaluating the situation in an ink fountain of the printing machine if there is a stirrer revolving in this container? Thus, is the concept of yield point determination or of similar values really close to practice in this case? See also Example 3 of Chapter 8.4.4.

Comment: The Notes and Examples above may illustrate that indeed **the state of rest is not simulated by a typical amplitude sweep when performed at an angular frequency of ω = 10rad/s**. But, as explained e.g. in Example 2, measurements in the state-at-rest are not required in any case in order to simulate behavior in practice.

Note 4: Three-dimensional diagrams of amplitude sweeps at different frequencies

The results of several amplitude sweeps which have been measured each one at a different but constant frequency can be presented in the form of a three-dimensional (3D) diagram, for example, with lg γ on the x-axis, lg G' and lg G" on the y-axis, and the angular frequency ω on the z-axis.

8.3.6 *SAOS and LAOS tests, and Lissajous diagrams*

a) Linear behavior in the LVE range and SAOS tests

As an alternative name for the **linear viscoelastic (LVE) range** also the term **SAOS** is used, meaning **"small amplitude oscillatory shear"**. Here, the preset and measuring parameters

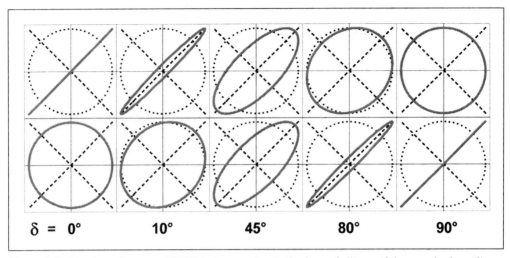

Figure 8.13: Lissajous diagram of SAOS tests occurring in the form of ellipses if the sample shows linear viscoelastic behavior: upper row as stress-strain diagrams τ(γ) to present the elastic behavior, lower row as stress-strain rate diagrams τ(γ̇) to display the viscous behavior (with the phase shift angle δ)

should behave linearly to one another. The preset amplitudes have to be selected correspondingly small in order to produce proportional response amplitudes. If this occurs, the measurement is performed within the LVE range of the sample.

Preset is a constant frequency and a strictly sinusoidal well-known amplitude as shear strain γ_A or shear stress τ_A respectively. When still remaining in the LVE range, the response which occurs as τ_A or γ_A will be also strictly sinusoidal and the amplitude will be proportional to the preset then. With viscoelastic behavior, the resulting sine curve will show a phase shift angle δ when compared to the preset sine curve. Usually, the results of these kinds of oscillatory tests with the focus on the LVE range, are presented in the form of the functions of the storage modulus G' and the loss modulus G". This is sufficient in typical standard rheology aiming analysis in the LVE range.

One method for presentation and analysis is the use of **Lissajous diagrams** [227; 154, 237, 276, 384]. In order to evaluate the elastic behavior, the γ-values are shown on the x-axis and the τ-values on the y-axis, see Figure 8.13: upper row. Usually, both measuring parameters are related to their maximum values. Then as (γ/γ_{max}) and (τ/τ_{max}), the scaled range on both axis only includes values between (-1) and (+1).

If the plotted functions are strictly sinusoidal, ellipses are appearing in the diagram. Using this kind of diagrams, for each rheological measuring point all values of both sine curves are presented, from $(-\gamma_{max})$ via γ = 0 to $(+\gamma_{max})$ and also from $(-\tau_{max})$ via τ = 0 to $(+\tau_{max})$. The difference to the usual presentation of amplitude sweeps at SAOS measuring conditions is that with Lissajous diagrams there are not only occurring just two points in the form of G' and G" for each individual measuring point. However, here for each individual measuring point, the diagram shows the whole function of both sine curves of the preset and the response in the form of an ellipse curve.

The area of the ellipse τ(γ) represents the dissipated energy density E/V [J/m³], which is the deformation energy related to the sheared volume of the sample. As a comparison: The unit of this area is calculated via $(\tau \cdot \gamma)$ as [Pa · 1] = [N/m²] = [J/m³].
Purely elastic behavior occurs in the τ(γ)-diagram as a straight line, because here, γ and τ are proportional (Hooke's law). This extreme case of the "ellipse" has no area, since here, no energy is dissipated.

Purely viscous behavior occurs in the $\tau(\gamma)$**-diagram as a perfect circle**, since here, the whole deformation energy is dissipated. To evaluate the viscous behavior, it is useful to present a Lissajous diagram in this case with the shear rate $\dot{\gamma}$ on the x-axis and, as above, τ on the y-axis; see Figure 8.13: lower row. In the $\tau(\dot{\gamma})$-diagramm, a straight line is resulting for a purely viscous liquid, because here, $\dot{\gamma}$ and τ are proportional (Newton's law).

Viscoelastic behavior in the LVE range occurs in the $\tau(\gamma)$**-diagram as an ellipse**, since here, part of the deformation energy is dissipated.

b) Non-linear behavior in the NL range and LAOS tests

Outside the LVE range, the **non-linear (NL) range** is entered. Alternatively the term **LAOS** is also used, meaning **"large amplitude oscillatory shear"**. Please note that the terms SAOS and LAOS are strictly limited to shear load, whereas the terms LVE range and NL range are not. The latter terms can also be used when testing under extensional load or under compression.

As described for SAOS also with LAOS, preset is a strictly sinusoidal well-known amplitude at a constant frequency, e.g. as strain γ_A. In the NL range, more and more of the measuring response will occur which is no longer strictly sinusoidal, for example, when observing the resulting τ-values. The response at low preset values being still sinusoidal, now the response sine curve is occurring increasingly in a changed shape. Scientists say: Here, the measured parameter responds indeed still periodically, however, it is non-harmonically now.

LAOS can be analyzed in many ways. The historically established method is the **Fourier transform** (FT analysis, or Fast Fourier Transform FFT, "FT rheology"). The response wave, the deformed sine curve, is analyzed by first being mathematically divided into many sine curves showing different frequencies (higher harmonics): One sine curve is related to the exciting frequency (basic oscillation, first harmonic), plus a sine curve for the double frequency (second harmonic), plus a sine curve for the third frequency, and so on. Each sine curve provides rheological parameters: There is a storage modulus related to the exciting frequency, one for the double frequency, and one for the triple frequency, and so on. This analysis provides correspondingly many parameters and is therefore often less illustrative. In standard rheology with SAOS tests, however, only the first wave is occurring.

With some experience, the **Lissajous curves** immediately provide qualitative insights on the rheological behavior. When exceeding the LVE range, both the viscous and the elastic behavior of a material are changing. Viscosity may decrease or increase, thus, **viscous flow behavior may become shear-thinning or shear-thickening**. Independent of this, elasticity may decrease or increase, thus, **elastic deformation behavior may become shear-softening or shear-stiffening**. The change in the elastic behavior is evaluated using the stress-strain Lissajous diagrams $\tau(\gamma)$, and the change in the viscous behavior is evaluated using the stress-strain rate Lissajous diagrams $\tau(\dot{\gamma})$. Typical Lissajous figures for LAOS differ significantly from all shapes which may occur for SAOS. Even when the values of G' and G" remain almost the same when increasing the strain (or stress), therefore pretending to measure still within the LVE range, a Lissajous diagram reliably shows clearly whether testing takes place in the NL range already.

In the $\tau(\gamma)$-diagram, **viscoelastic behavior in the NL range** increasingly leads to deviations from the ellipsoidal shape. The Lissajous curves at first may show dents or bulges and then perhaps even the form of an "S-shaped banana". In extreme cases this may even lead to multiple crossings of the circumferential line of the Lissajous curves, also known as "secondary loops". To get a measure for the changes, the following rheological parameters are recommended [129]:

The tangent modulus G'_1 at zero-strain is the slope value of the tangent on the $\tau(\gamma)$-curve at $\gamma = 0$ (or "zero-elasticity"), and the **secant modulus** G'_2 at the maximum strain value is the slope value of the straight line between the zero-point of the $\tau(\gamma)$-diagram and the curve

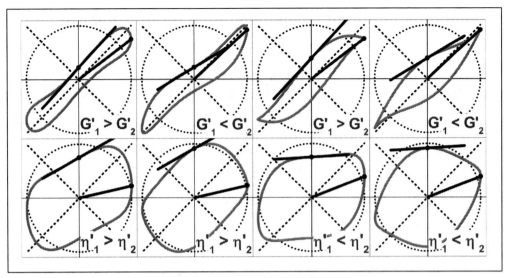

Figure 8.14: Lissajous diagrams of LAOS tests showing non-linear behavior of a sample: upper row as τ(γ)-diagrams to exhibit the elastic behavior, and lower row as τ(γ̇)-diagrams to illustrate the viscous behavior

point at γ_{max} (see Figure 8.14: upper row). Only for the LVE range applies: The storage modulus $G' = G'_1 = G'_2$. Correspondingly, the **tangent viscosity** η'_1 at the minimum shear rate is the slope value of the tangent on the τ(γ̇)-curve at $\dot{\gamma} = 0$ (or "zero-viscosity"), **and the secant viscosity** η'_2 at the maximum shear rate is the slope value of the straight line between the zero-point of the τ(γ̇)-diagram and the curve point at $\dot{\gamma}_{max}$ (see Figure 8.14: lower row). Only for the LVE range applies: The real part of the complex viscosity $\eta' = \eta'_1 = \eta'_2$.

Note: In other publications, both parameters are also referred to as G'_M for minimum-strain or zero-strain elasticity and G'_L for large-strain elasticity, and to as η'_M for minimum-shear rate viscosity and η'_L for large-shear rate viscosity [129].

Summary: SAOS tests in the LVE range measure both the elastic and the viscous behavior of a sample. These tests illustrate clearly whether a sample is predominantly viscous (fluid-like) or elastic (gel-like, solid). If the measurements are extended to the NL range it is also possible to determine the **changes of both the elastic and viscous behavior** within a single measuring point. **With viscoelastic materials, four qualitative behaviors under LAOS conditions** can be distinguished:
1) viscous shear-thinning and elastic shear-softening (with $\eta'_1 > \eta'_2$ and $G'_1 > G'_2$)
2) viscous shear-thinning and elastic shear-stiffening (with $\eta'_1 > \eta'_2$ and $G'_2 > G'_1$)
3) viscous shear-thickening and elastic shear-softening (with $\eta'_2 > \eta'_1$ and $G'_1 > G'_2$)
4) viscous shear-thickening and elastic shear-stiffening (with $\eta'_2 > \eta'_1$ and $G'_2 > G'_1$)

Examples of applications: complex materials
Concentrated dispersions (i.e. suspensions, emulsions, foams), colloidal gels and hydrogels, semi-solid foods, fats, biopolymers, mucus, viscoelastic solutions and dispersions showing worm-like micelles (as "nano-composites"), biological materials such as tissue, soft and semi-solid materials ("soft matter")

Note 1: Complex rheological behavior in the non-linear range
There are also gel-like samples which firstly show elastic softening in response to increasing deformation, before they finally exhibit elastic stiffening. Examples of natural products: snail slime, spider silk, networks made of keratin threads.

Note 2: Presetting of sine waves or triangle functions, also in the NL range
In order to carry out useful analysis in the non-linear range, it is essential to preset an ideal sine wave. This can be achieved via **real-time position control** (see Chapter 11.6.2b: DSO, direct strain oscillation). Without presetting an ideal sine wave, a part of the non-linearity would be included already in the preset, and therefore, it would also occur as a part of the measurement response. Measurements of complex samples under LAOS conditions also show in general: **When performing controlled stress tests, different results are obtained compared to controlled strain tests.** This becomes clear when comparing the corresponding Lissajous curves, since these kinds of curves will show different details in the shape of the curve for these two modes of testing. And this is also confirmed if **triangular functions are preset instead of sine waves** [222].

Note 3: Wall-slip, phase separation, and inhomogeneous deformation behavior
Considering dispersions and gels, inhomogeneous deformation behavior and therefore effects like wall-slip and phase separation may occur when exceeding the LVE range (see Chapter 3.3.4.3d and Figure 2.9: "Plastic behavior").

⌒⌒   End of the Cleverly section

8.4   Frequency sweeps

Frequency sweeps are oscillatory tests performed at variable frequencies, keeping the amplitude at a constant value (and also the measuring temperature). Sometimes, the term **"dynamic oscillation"** is used as a synonym for "variable frequency" (as in ASTM D4440).

Frequency sweeps are used to investigate time-dependent deformation behavior since the frequency is the inverse value of time. Short-term behavior is simulated by rapid motion, i.e. at high frequencies, and long-term behavior by slow motion, i.e. at low frequencies.

Experiment 8.1: Bouncing or spreading of PDMS
Time-dependent rheological behavior of a polymer can be observed when playing with a silicone polymer (unlinked PDMS).

1) Short-term behavior
Hit on the ground with a short and rapid impact, the silicone bounces like a rubber ball, and therefore, like an elastic solid. Correspondingly, G' > G" is measured at appropriately high frequencies which are simulating short-term behavior.

2) Long-term behavior
When exposed to very slow motion, e.g. if left at rest, the silicone flows very, very slowly like a highly viscous liquid, gradually spreading under its own weight. Correspondingly, G" > G' is measured at appropriately low frequencies which are simulating long-term behavior.

Note 1: Optional specification as frequency in Hertz or as angular frequency
There are two options to specify frequency values:
1) **Frequency f** using the unit [Hz], ("Hertz")
2) **Angular frequency** ω using the unit [rad/s] or [s^{-1}]. The following holds:

Equation 8.3 $\omega = 2\pi \cdot f$ (with the circle constant π = 3.141...)

In rheology and in the field of physics in general, the use of the angular frequency ω instead of the frequency f is becoming increasingly widespread. To avoid confusion, ω should never be referred to simply as "frequency", but always as "angular frequency" (see also the Note in Chapter 8.2.1).

Note 2: Units of the angular frequency and angular velocity (using rad/s or s^{-1})
For both the angular frequency of oscillatory tests and the angular velocity of rotational tests, the same symbol ω and the same unit [rad/s] or [s^{-1}] is used. Sometimes in daily prac-

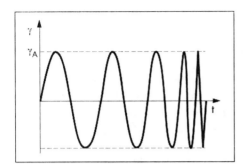

Figure 8.15: Preset of a frequency sweep, here with controlled shear strain

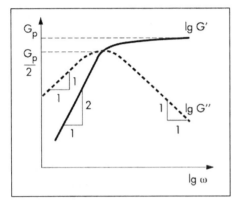

Figure 8.16: Frequency sweep of a polymer solution or melt showing the behavior of a Maxwell fluid, here in terms of the curve functions of G' and G", with the plateau value G_P

tice, "rad" is ignored, and therefore, only [s⁻¹] is used, which is also an SI-unit. This is possible when relating the concerning calculations to the so-called "unity circle", which is often used in mathematics and physics.

8.4.1 Description of the test

Preset

1) For tests with controlled shear strain:
$\gamma(t) = \gamma_A \cdot \sin\omega t$ with γ_A = const and a variable angular frequency $\omega = \omega(t)$, see Figure 8.15.
Here, the maximum value of the deflection angle of the measuring bob is kept constant (amplitude), only the period of time for each one of the oscillation cycles is increasing or decreasing continuously, respectively (frequency).

2) For tests with controlled shear stress:
$\tau(t) = \tau_A \cdot \sin\omega t$, with τ_A = const and $\omega = \omega(t)$
In this case, the oscillatory motion of the measuring bob is caused by a torque whose maximum value is kept constant (amplitude). The period of time for each one of the oscillation cycles however, is increasing or decreasing continuously, respectively (frequency).

Information on values of the permissible maximum strain (deformation) is given in Chapter 8.3.3.1. For each new and unknown sample before performing a frequency sweep, the limiting value of the LVE range has to be determined (in terms of the strain limit γ_L or the yield stress τ_y, respectively; see Chapters 8.3.3.1 and 8.3.4.1). Therefore, **first of all always an amplitude sweep has to be carried out**. Only in this case, useful test conditions for frequency sweeps can be selected to ensure that the measurement is really performed in the **LVE range**.

Measuring result

Usually, lg ω (or alternatively, lg f) is presented on the x-axis, and lg G' and lg G" are shown on the y-axis, i.e., logarithmic scales are used on both axis (see Figure 8.16). Sometimes, additional parameters are displayed on a second y-axis such as $|\eta^*|$ when testing liquid samples such as polymer solutions and melts (see Figures 8.17 and 8.21).

⌒⌒   For "Mr. and Ms. Cleverly"

Note 1: Calculation of shear rate values occurring with frequency sweeps

Based on Equation 8.24, shear rate values can be calculated as follows:

Equation 8.24 $\dot{\gamma} = \gamma_A \cdot \omega$

Example (to Note 1): With the strain amplitude γ_A = 1% = 0.01 and the angular frequency ω = 100s⁻¹ the shear rate $\dot{\gamma}$ = 0.01 · 100s⁻¹ = 1s⁻¹. This shows that the shear conditions occurring with frequency sweeps performed in the LVE range usually correspond to the low-shear range at $\dot{\gamma} \leq$ 1s⁻¹. Only in rare cases, e.g. with unlinked and unfilled polymers, a value of $\dot{\gamma}$ = 50s⁻¹ is exceeded; for example showing $\dot{\gamma}$ = 62.8s⁻¹ if γ_A = 10% = 0.1 and ω = 628s⁻¹ (or f = 100Hz, respectively). Important for practical users: Oscillatory tests are usually carried out in a completely different shear rate range compared to rotational tests, the two test types therefore are complementing each other.

Note 2: Three-dimensional diagrams of frequency sweeps at different temperatures
The results of several frequency sweeps which have been measured each one at a different but constant temperature (i.e. at isothermal conditions) can be presented in the form of a three-dimensional (3D) diagram, for example, with lg ω on the x-axis, lg G' and lg G'' on the y-axis, and the temperature T on the z-axis.

Note 3: Three-dimensional diagrams of frequency sweeps at different amplitudes
The results of several frequency sweeps which have been measured each one at a different but constant amplitude can be displayed in a three-dimensional (3D) diagram, for example, with lg ω on the x-axis, lg G' and lg G'' on the y-axis, and the strain as lg γ on the z-axis. Here, even amplitude sweeps and frequency sweeps can be shown together in the same 3D diagram.

Note 4: Test preparation - thermal-stability time of polymer melts
Before testing samples of polymer melts, it is recommended to check the thermal-stability time at the desired test temperature, for example, to avoid molecular degradation (according to ISO 6721-10; see also the Notes of Chapters 8.5.2 and 8.6.2).

⚬⚭   End of the Cleverly section

8.4.2 Behavior of unlinked polymers (solutions and melts)

For unlinked polymers holds: The average molar mass is M > 10,000g/mol, and typical technical polymers are showing 50,000 < M < 500,000g/mol (see also the Note in Chapter 3.3.2.1a: values of the critical molar mass M_{crit} and polymers). If the polymer concentration is high enough, the unlinked polymers are entangled with each other (see also Chapter 3.3.2.1: the entanglement model; and Figure 3.11: dependence of viscosity on the polymer concentration). However here, besides the purely mechanical interactions such as the frictional forces and the elasticity caused by the entanglements, it is assumed that neither chemical bonds such as primary valency bonds of a chemical network nor physical-chemical bonds as secondary bonds of a network-of-forces appear between the macromolecules. Since there are absolutely no attractive forces between the macromolecules, they are able to move slowly even under the lowest shear force, gliding along each other, showing partial or even complete disentanglement. Here, the most effective forces occurring are solely friction forces between the molecule chains which are neither based on chemical nor on physical-chemical effects but exclusively on mechanical ones. More about structures and the behavior of polymers can also be found e.g. in [114, 233, 249, 321, 352].

⚬⚭   For "Mr. and Ms. Cleverly"

8.4.2.1 Single Maxwell model for unlinked polymers showing a narrow MMD

In a simplified and idealized case, all macromolecules of a polymer are assumed to be linear, all displaying the same length and the same molar mass. "Linear molecules" means, they do not have any side branches. In this case, the polymers show an absolutely narrow **molar mass distribution (MMD)**. As an **illustrative example**: Imagine a conglomerate of a number of cooked and therefore flexible **spaghetti**, all of the same length.

The behavior of simple viscoelastic liquids can be characterized using a single Maxwell model which is a combination of a spring and a dashpot in serial connection, see Figure 5.1 of Chapter 5.2.1.1. For the frequency-dependence of Maxwell fluids holds:

Equations 8.25 and 8.26 $$G'(\omega) = G_P \cdot \frac{\omega^2 \cdot \lambda^2}{(1 + \omega^2 \cdot \lambda^2)} \quad \text{and} \quad G''(\omega) = G_P \cdot \frac{\omega \cdot \lambda}{(1 + \omega^2 \cdot \lambda^2)}$$

with the plateau value G_P [Pa], and the relaxation time λ [s]; for λ see Chapter 7.3.3.2.

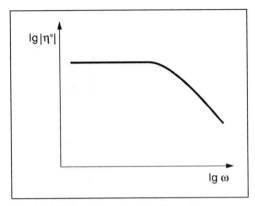

Figure 8.17: Frequency sweep of a polymer solution or melt showing the behavior of a Maxwell fluid, here in terms of the curve function of the complex viscosity $|\eta^*|$

a) Discussion of curves of unlinked polymers showing a narrow MMD
Shape of the curves of the frequency dependent functions of G' and G'' when presented in a diagram on a logarithmic scale (see Figure 8.16):
1) At low frequencies, the G'-curve shows the slope 2:1 since G' and ω^2 are proportional for $\omega \to 0$.
2) At high frequencies (with $\omega \to \infty$), the G'-curve is reaching the constant plateau value G_P.
Summary: For the frequency-dependent limiting values of the storage modulus applies:
$\lim_{\omega \to 0} G'(\omega) = 0$ and $\lim_{\omega \to \infty} G'(\omega) = G_P$
Maxwell fluids are flexible and liquid at very slow motion ("at rest") or at low frequencies, respectively, but they are inflexible and rigid at very rapid motion or at high frequencies, respectively, and the parameter G_P represents the maximum rigidity.
3) At low frequencies, the G''-curve shows the slope 1:1 since G'' and ω are proportional for $\omega \to 0$.
4) At high frequencies, the G''-curve shows the slope (-1):1 since G'' and $1/\omega$ are proportional for $\omega \to \infty$.
5) At the point when $\omega \cdot \lambda = 1$, the G''-curve reaches its maximum value $G''_{max} = G_P/2$.

If an oscillation cycle is exactly as long as λ (or if $1/\omega = \lambda$, resp.), the viscous behavior and the elastic behavior are balanced, showing G' = G'' ($= G_P/2$) then.

Summary: For the frequency-dependent limiting values of the loss modulus applies:
$\lim_{\omega \to 0} G''(\omega) = 0$ and $\lim_{\omega \to \infty} G''(\omega) = 0$

Figure 8.17 presents the shape of a **frequency-dependent function of the complex viscosity**. At low ω-values, Maxwell fluids are indicating the range of the **zero-shear viscosity** η_0 where the viscosity function shows a constant plateau value. The following applies to the limiting value of the complex viscosity:

Equation 8.27 $\eta_0 = \lim_{\omega \to 0} |\eta^*(\omega)|$

For people working with polymers, the zero-shear viscosity value is a very important parameter since it is proportional to the average molar mass M. Further information on this relation can be found in Chapter 3.3.2.1a (rotational tests: shear viscosity function, Figure 3.10 and Equation 3.1) and in Chapter 6.3.4.1 (Creep tests).

b) Interpretation of frequency sweeps of unlinked polymers
Sometimes it is not easy to interpret the frequency-dependent functions of G*, G', G'' and $|\eta^*|$. On the one hand, the G*-values are increasing with the frequency values, i.e. the polymer seems to show increasing rigidity. On the other hand, the η^*-values are decreasing here, i.e. the flow resistance of the polymer melt seems to decrease. The following explanations may be helpful to understand this apparent contradiction:
1) Formally, the following applies (see Equation 8.16):
$|G^*| = |\eta^*| \cdot \omega$ or $|\eta^*| = |G^*|/\omega$
Due to this formal calculation, with higher angular frequencies higher G*-values are obtained, but lower η^*-values. Indeed, this formality may be accepted by mathematicians, but for physicists and engineers it might be not sufficient.

2) Energy balance (deformation and heat): With increasing frequency and therefore faster motion, the polymer structure of the temporary network-of-entanglements is showing more and more inflexibility and rigidity. In this state of strain, more deformation energy can be stored now as is lost by friction between the molecules due to the increasingly reduced relative motion between their chains. As a consequence, elastic behavior is showing increasingly dominance, and therefore, G' and |G*| are increasing. At the same time, viscous behavior is losing more and more importance, and therefore, the portion of the lost deformation energy will be correspondingly reduced, thus, G" and |η*| are decreasing. On the other hand with decreasing frequencies, the structure of the temporary network-of-entanglements shows more and more flexibility then, due to an increasing number of disentanglements. In this state of motion, even when it is very slow, more deformation energy will be transferred into frictional heat, and therefore is lost for the polymer. Now, the fluid structure obtained is no longer able to store as much energy as in the entangled state. Due to the increasing relative motion between the macromolecules, viscous behavior is showing increasingly dominance, and therefore counts G" > G' and the values of |η*| are increasing, respectively.

3) Relaxation time (or relaxation time spectrum): Polymers showing relatively short average relaxation times are exhibiting entanglements only in the range of very fast motion, i.e. at high frequencies. Therefore, they are indicating only there dominating elastic behavior with G' > G" or high values of G*, respectively. Polymers having relatively long average relaxation times, however, are only able to flow in the range of very slow motion, i.e. at low frequencies. Therefore, they indicate only here dominating viscous behavior and G" > G' or high values of |η*|, respectively. For polymers, this low-shear range is the range of the plateau showing the zero-shear viscosity η_0.

Note 1: Specification of viscosity is not useful if G' > G"

If there is dominant elastic behavior showing G' > G", there is no reference to practice if viscosity values are used in analysis. According to Newton's law, useful viscosity values can only be determined if the shear rate is not zero but $\dot{\gamma} > 0$; see Equation 2.8: $\eta = \tau/\dot{\gamma}$ and Equation 8.15: $\eta^* = \tau(t)/\dot{\gamma}(t)$. In other words: Viscosity values are only useful if the material is actually flowing, and therefore, if it is not showing an infinitely high viscosity value. However if G' > G", a sample does not flow at all and there is obviously a dominant elastic gel-like or solid behavior. In this case in principle, rheological behavior can only be characterized in a useful way by Hooke's law or when using the Kelvin/Voigt model, but not by Newton's law or by the parameters "shear viscosity" or "complex viscosity". However here, the sample can usefully be characterized by the parameters G*, G' and G" or tanδ. This means, concerning the range in which samples are showing G' > G", most of the discussions about the topic "interpretation of viscosity curves in frequency sweeps" are redundant in this case.

Example (see Figure 8.21): On the left, presented is the frequency curve of a polymer solution showing G" > G' in the range of low frequencies. Thus, at rest it is a viscoelastic liquid. Here, it may therefore be useful to show also the viscosity curve and this counts for practical purposes at least until reaching the crossover point G' = G". In Figure 8.21 on the right, displayed is a measurement of a gel-like polymer **showing G' > G"** across the whole frequency range. **Here, it is not useful to present the viscosity curve since there is no reference to practical use** because a gel-like structure is not able to flow under these shear conditions in the LVE range.

Note 2: The Cole/Cole plot in the form of a semicircle

The Cole/Cole plot is another method to illustrate whether a material shows Maxwellian behavior, i.e. G" > G' in the low-frequency range, and a crossover point at G' = G" in the medium range, and G' > G" in the high-frequency range, finally reaching G' = G_P which is the so-called plateau value (as illustrated in Figure 8.16). Using a Cole/Cole plot, G' is presented on the x-axis and G'' on the y-axis, both **on a linear scale**. Sometimes for a modified diagram, G'/G_P and G''/G_P are displayed "to normalize" data, which is the so-called reduced Cole/Cole plot. Here, the values of G' and G'' are related to the plateau value G_P which occurs in the range of high frequencies.

For ideally Maxwellian behavior, the Cole/Cole curve shows the shape of a semicircle.
1) The semicircle begins at the origin point x = y = 0, since here, the values of G' and G'' are "invisibly small" due to the presentation on a linear scale.
2) The highest point of the semicircle occurs when G' = G'', or for x = y = 0.5 when using the normalized plot.
3) The semicircle is ending on the x-axis again showing G' = G_P, and G'' will be "invisible small" again due to the linear scaling, or x = 1 and y = 0 when using the normalized plot.

Deviations from the semicircular shape indicate deviations from the Maxwellian behavior. Since this plot is presented on a linear scale, particularly deviations in the high-frequency range are highlighted.

Example 1 [220]: Cole/Cole plots of viscoelastic surfactants (VES)
VES, showing a superstructure of linear chain-like or "worm-like" micelles, displayed a concentration-dependent deviation from the Maxwellian behavior in the range of high frequencies, and therefore, an unstable superstructure there. Summary: This is an indication to instabilities of the super-structure in this shear range (see also Chapter 9.1.1d: Fluctuation of surfactant molecules in micellar superstructures).

Example 2, [242]: Presentation of **Cole/Cole plots of a polystyrene melt** at different temperatures with the **real part of the complex viscosity η' on the x-axis and the imaginary part η'' on the y-axis**. At the value of the zero-shear viscosity η_0 the semicircular arch meets the x-axis. Here, with η'' at around zero, there is no significant portion of the elastic behavior available, and therefore, η' represents the purely viscous behavior in the range of η_0.

Comment: Nowadays, it is usual and recommended to present diagrams of frequency sweeps on both axis on a logarithmic scale as shown in Figure 8.16, since here, the viscoelastic behavior is illustrated clearly in the whole frequency range, i.e., it can be seen in particular also at very low values. Therefore, if deviations from the Maxwellian behavior might occur this can be observed in the whole time frame then, thus, from the short-term to the long-term range. For these reasons, **data presentation in the form of Cole/Cole plots is not recommended for practical users**.

Note 3: The van Gurp/Palmen plot [168, 364, 398]
The van Gurp/Palmen plot (vGP plot) is a method to present frequency sweep data of unlinked polymers giving in particular insight into their degree of branching. As already mentioned, the function of the storage modulus G'(ω) approaches at high frequencies the constant value of the plateau modulus G_P as the limiting value, and this is also valid for the complex shear modulus G*. (Sometimes G_P is also called G_N^0). Here, the following is defined: **The range of the G_P plateau begins** at the minimum of the frequency-dependent function of the loss factor tanδ(ω). In other words, after a conversion: The G_P value is reached at the minimum of the phase angle-dependent function of the complex shear modulus G*. Formally expressed: $G_P = \lim_{\delta \to 0} G^*(\delta)$. In the vGP plot, this minimum is clearly shown, with G* (in Pa) on the x-axis on a logarithmic scale and δ (in degrees) on the y-axis on a linear scale. Sometimes in a modified diagram, G^*/G_P is displayed in order to "normalize" these data. In this so-called **reduced vGP plot**, the G*-values are therefore related to G_P. A comparison of the curve shapes of different samples may give information on the molar mass, molar mass distribution (MMD) and, in particular, the **degree of branching of the macromolecules. Linear polymers** show a constant decrease and increase of the curve of lg G*(δ) on both sides of the curve "val-

ley" around the curve minimum. **Polymers with long-chain branching** however, display irregularities in the downwards sloping part of the curve, and sometimes they may even show step-like intervals there.

8.4.2.2 Generalized Maxwell model for unlinked polymers showing a wide MMD

Many viscoelastic liquids and polymers without a physical or chemical network can be characterized quite well for the purpose of a simple analysis when using a single Maxwell model only as the basis of calculation, i.e. the serial connection of a single spring and a single dashpot. This is acceptable for samples showing a narrow MMD since here, it can be assumed a single relaxation time λ to be sufficient for the characterization of the behavior. However, for unlinked and unfilled polymers with a wide MMD it is better to use a generalized Maxwell model which contains several Maxwell elements in parallel connection (see Figure 7.8 in Chapter 7.3.3.3a). As a consequence, several relaxation times are taken as the basis of calculation, and therefore, a **"spectrum of relaxation times"**. Then, for a polymer with a wide MMV each individual molar mass fraction can be characterized via the corresponding individual relaxation time. Afterwards, a software program calculates the values of G' and G" by summing up all of the individual G-values of all the corresponding individual Maxwell elements.

$\mathcal{6\!\!\smile}$ End of the Cleverly section

a) Curve discussion for unlinked polymers showing a wide MMD (see Figure 8.18):
Usually, the G'-function is taken to analyze frequency sweeps. For polymers exhibiting two levels of the G'-curve, this one at high frequencies is referred to as the **plateau** showing the value G_P, and the other one at medium frequencies is called the **rubber-elastic plateau** showing the value G_{RP}. In order to obtain information on polymer structures, it is useful to devide frequency sweeps into the following four intervals:

1) **Initial range** (or flow zone 1), when $\omega \leq \omega_1$
In the range of low frequencies, the behavior can be described by use of a single Maxwell model. Here, the curves of G' and G" show the slopes 2:1 and 1:1, respectively (see Figure 8.16). The curves indicate the dominance of the viscous portion, i.e. G" > G', and therefore,

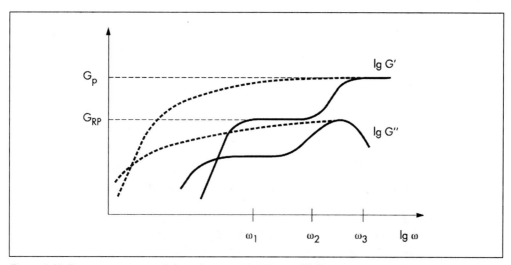

Figure 8.18: Frequency sweeps of two polymers with different MMDs, showing the plateau value G_P and the value of the rubber-elastic plateau G_{RP}. Uninterrupted lines: polymer with two fractions, each one with a narrow MMD and therefore showing two plateau values; interrupted lines: polymer with a wide MMD

the polymer exhibits the behavior of a viscoelastic liquid. Even when indicating a very high viscosity level, the polymer might be able to flow, displaying here the plateau value of the zero-shear viscosity η_0 then.

2) **Rubber-elastic range** with the plateau value G_{RP} between ω_1 and ω_2

In the range of medium frequencies occurs the so-called "rubber-elastic plateau" showing the more or less constant value of G_{RP}. Here, for the longer molecules it is no more possible to glide along each other, and therefore, their entanglements are beginning more and more to form a temporary network. Now the curves are indicating a slight dominance of the elastic portion, i.e. G' > G". The polymer exhibits viscoelastic behavior like a weak gel with a low structural strength or like a very soft rubber with an unfinished network or with a very low degree of cross-linking. Sometimes in this range there is approximately G' = G", and therefore, the behavior like of caoutchouc (i.e. a non-vulcanized rubber) or of dough can be imagined.

3) **Transition range** (or flow zone 2) between ω_2 and ω_3

In this range, the curves of G' and G" are steeply sloping up since here, only the smaller and therefore more mobile molecules are still deformable. But now, besides the long molecules also those with a medium length are showing more and more inflexibility. The curves are indicating increasingly the dominance of the elastic portion with G' > G", and therefore, the polymer exhibits more and more the viscoelastic behavior of a soft solid.

4) **Glassy range** showing the plateau value G_P when $\omega \geq \omega_3$

In the range of high frequencies, the G'-curve reaches the high and constant value of G_P, while the G"-value is decreasing continuously. Motion between the entangled molecule chains is very limited, since here, almost all parts of the macromolecules are almost immobilized now, showing the behavior of a temporary network which can be imagined like being in a "frozen state". Parts of the molecule chains still may vibrate a little but the main chains are no longer able to follow the rapid oscillatory motion as a whole. Now with G' ≫ G", the curves are indicating clearly the dominance of the elastic portion, and therefore, the polymer exhibits behavior of a rigid solid. In order to describe this situation, the term "glassy state" was selected, assuming that glass at room temperature was a frozen liquid. (More about the "glassy state of glass": see Note 2 of Chapter 8.6.2.2a.)

Polymers with a wide MMD contain a mixture of macromolecules showing clearly differing lengths. As an **illustrative example**: Imagine a conglomerate of a number of cooked, flexible **spaghetti of different lengths**. In a diagram, all the individual rubber-elastic plateaus, each representing an individual molecule fraction, can no longer be recognized as individual steps as shown in Figure 8.18 (uninterrupted lines) since here, all the individual plateaus are overlapping one another. As a consequence, at increasing frequencies the G'-curve slopes upwards becoming gradually flatter and flatter until reaching the plateau value G_P finally, as shown in Figure 8.18 (interrupted lines). The change in the slope during the transition from the sloping part to the plateau value of G_P is more pronounced for polymers showing a narrower MMD, i.e., the transition of the G'-curve from the slope 2:1 to the constant plateau value at high frequencies is occurring more abruptly then.

b) Comparison of unlinked polymers concerning molar mass and MMD

Many polymer melts are a mixture of different kinds of macromolecules, since besides linear long-chained ones there may be linear short-chained ones, and often also branched ones. Under a shear load indeed, all entanglements between the molecule chains can be loosened. However, this requires a higher shear force for the longer and the more branched molecules, or it takes a longer period of time under shear load, respectively. Due to the different behaviors, it is useful to compare only samples with a similar molecular structure, e.g. either polymers showing a linear molecule structure or polymers with a comparable degree of branching. And of course, that counts even more for polymers with a different chemical constitution.

c) Optional methods to analyze frequency sweeps of unlinked polymers
In order to compare polymer solutions and melts, practical users mostly select method M5.

M1) The plateau value G_P and the average molar mass
A comparatively higher plateau value of G_P indicates a higher degree of the structural strength of the temporary network when exposed to rapid motion, i.e. in the range of high frequencies. Interpretation: There is a greater amount of entanglements indicating longer molecules, and therefore usually, the polymer shows a higher average molar mass then.

M2) The shift of the curves of G' and G", and the average molar mass
If the curved part of the functions of G' and G" occurs at a comparatively lower frequency, this indicates the presence of longer or more branched molecules showing a longer relaxation time, and therefore, a higher average molar mass. At very low frequencies they are indeed able to move along each other, but at higher frequencies they are increasingly beginning to block each other's relative motion. The curves of G' and G" are rising then. Smaller molecules, however, are blocking each other not before relatively high frequencies are reached. Therefore, if the curved part of the curves occurs at a comparatively higher frequency, this indicates the presence of smaller or less branched molecules with a shorter relaxation time, and therefore, with a lower average molar mass.

M3) The slope of the curves of G' and G", and MMD
A comparatively steeper slope in the curves indicates a narrower MMD. Correspondingly, a wider MMD results in a more moderate slope of the G'-curve and a more gradual transition to the constant plateau value of G_P (see Figure 8.22). The G"-curve shows a less pronounced peak then.

M4) Vertical shift of the crossover point G' = G", and MMD
When comparing two polymers with the same average molar mass but different MMDs, the following applies: For the polymer showing the wider MMD, the crossover point G' = G" occurs at a lower G-value. Hence, a vertical shift of the crossover point depends on the MMD. Vertical means in parallel to the y-axis.

M5) Horizontal shift of the crossover point G' = G", and average molar mass
When comparing two polymers with the same MMD but different average molar masses, the following applies: **For the polymer showing the higher average molar mass, the crossover point G' = G" occurs at a lower angular frequency** ω_{co} (with the index "co" for crossover point). Since its longer molecules are less flexible and less mobile, they cannot follow the fast motion in the range of high frequencies, and then: G' > G" (see Figure 8.19). Hence, a horizontal shift of the crossover point depends on the value of M. Horizontal means in parallel to the x-axis.

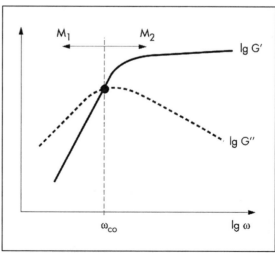

Figure 8.19: Frequency sweep of a polymer showing the crossover point G' = G" and its dependence on the average molar mass M: The crossover point occurs at a lower frequency for $M_1 > M_2$

捷   For "Mr. and Ms. Cleverly"

Note: Extension of the frequency range

The following two examples may show how the frequency range can be extended largely when using the concept of the master curve (see Chapter 8.7.1: WLF time/temperature shift method).

1) Shear behavior at ω = 10,000rad/s (corresponding to a frequency of f = 1600Hz approximately) might be presented in the form of a master curve obtained from frequency sweep data after shifting into the high-frequency range. These data might simulate ultra short-term behavior, corresponding to impacts in a time interval of t = 10^{-4}s = 0.1ms (= 1/ω).

2) Behavior at ω = 0.0001rad/s = 10^{-4}rad/s might be presented using master curve data achieved from frequency sweep data after shifting into the low-frequency range. These data are simulating long-term behavior, corresponding to a period of t = 10,000s (= 1/ω) or almost 3 hours. Alternatively, a very long test time had to be taken into account for frequency sweeps performed at these very low frequencies, when performing a full oscillation cycle for each individual measuring point. (However, this is no longer required today; see the Note in Chapter 11.6.2b: Direct strain oscillation DSO).

捷   End of the Cleverly section

8.4.3 Behavior of cross-linked polymers

Cross-linked polymers show a network of bridges between the macromolecules which are chemically linked by primary bonds. This makes it impossible for most of the molecule chains to glide along each other without destruction of the chemical network. The more or less limited degree of internal deformation depends on the length of these linkages. Higher maximum deformation is possible in a network showing a structure which is not consistent or a low degree of cross-linking, e.g. of flexible elastomers such as soft rubbers. For a network with a high degree of cross-linking, e.g. of rigid thermosets, hard rubbers or highly esterified gels, the permissible maximum deformation is clearly limited to avoid exceeding the limit of the LVE range. For soft rubbers, cross-links may occur at an average distance of about 1000 atoms, and for thermosets at about 20 atoms only [114]. Typical elastomers are showing bridge lengths between 200 and 400 C-atoms [398].

a) Polymers showing a high degree of cross-linking

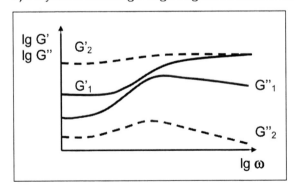

Highly cross-linked polymers can be deformed only in a very small range before irreversible deformation or even destruction occurs. In this case, the test results in a G'-value which is almost independent of the frequency (see Figure 8.20). Here, the curves of G' and G" display almost parallel lines throughout the whole frequency range showing a very slight curve slope only. For these kinds of materials, the value of G' is relatively high compared to G".

Figure 8.20: Frequency sweeps of two cross-linked polymers; uninterrupted lines: polymer with a low degree of cross-linking showing a certain flexibility at rest like elastomers; interrupted lines: polymer with a high degree of cross-linking showing almost constant rigidity throughout the whole frequency range like thermosets

Typically, the following ratio values occur for these kinds of stiff and rigid solids:

G': G" = 100:1 to 1000:1, or tanδ

(= G"/G') = 0.01 to 0.001, respectively

Examples: Thermosets such as reactive resins after curing, hard rubbers such as rigid eraser gums or car tires, hard cheese, but also very stiff food gels like "jelly babies" or "gummy bears", or other highly concentrated gelatin gels

Experiment 8.2: Bouncing rubber balls, and the damping factor tanδ
When rubber balls made of different materials bounce off the floor they reach different heights (see also Experiment 5.5 in Chapter 5.2.2). The following list presents with decreasing bounce height: polybutadiene (BR), polyisoprene (NR), polystyrene-butadiene (SBR), butyl. The balls' contact time with the floor is estimated to be t_c = 2 to 5ms.

Afterwards, a rheological evaluation was performed via frequency sweeps [398]. The value of the damping factor tanδ = G''/G' is read off at the angular frequency ω_c = $(1/t_c)$ which corresponds to the inverse contact time. Here $(1/0.005s) < \omega_c < (1/0.002s)$, therefore ω_c = 200 to 500s^{-1}. For the balls made of BR/NR/SBR/butyl this results in the values of tanδ = 0.04 to 0.85 (= G''/G'). The higher this value, the more the material is damping the bounce. Maybe, this becomes clearer when looking at the inverse values, showing the ratio of the elastic and the viscous portion in terms of G': G'' = 25:1 to 1.2:1 (= G'/G'' = 1/tanδ).

Note: The influence of the network density on the agglomeration of filler particles
Elastomers containing **fillers such as silica or soot** sometimes show lower structural strength despite having a higher network density of the polymer. This apparent contradiction can be explained [398].

1) In **unlinked filled polymers**, the filler aggregates are building up increasingly large, continuously growing flocculates and clusters. This process is accelerated at a higher temperature.

2) In **cross-linked filled polymers**, an agglomeration is of course only possible to a certain extent. Here, only aggregates can grow which are smaller than the length of the network bridges. That is because larger agglomerates are held by the network and cannot grow any further.

Since the larger agglomerates are causing structural strengthening, in this case, the result will be: **A denser network is leading to lower rigidity.** The occurring size of the agglomerates can therefore be controlled via the type and duration of the cross-linking process. At a first glance of contradictory nature, this is an illustrative example of how complex may be the interactive influence of different components in a mixture.

b) Polymers showing a low degree of cross-linking
For polymers showing a low degree of cross-linking, with increasing frequencies the G'-curve displays a gradual slope upwards to reach the constant plateau value G_P finally. The G''-curve slopes up to a maximum before it falls afterwards. The flexibility of the sample decreases continuously until reaching the plateau value G_P, exhibiting the maximum rigidity then. Typically, the following ratio values occur for these kinds of samples:
G': G'' = 10:1 up to 100:1, or tanδ (= G''/G') = 0.1 to 0.01, respectively.

The value of the **loss factor tanδ** = G''/G' shows the ratio of the sol to the gel component. The **sol component** represents the part of the **unlinked** and therefore **mobile molecule chains and chain ends**, which are using up **deformation energy when performing friction processes** with their environment. The **gel component** reflects the behavior of the **molecules which are integrated in the network**, and therefore cannot glide along one another. The corresponding part of the **deformation energy is stored** in the partially deformed network.

Examples: Elastomers such as cross-linked soft silicone rubbers (however, unlinked silicones are viscoelastic liquids showing G'' > G' in the range of low frequencies); flexible sealants such

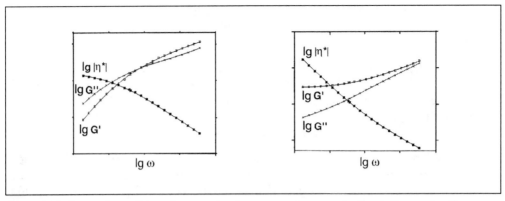

Figure 8.21: Comparison of two polymers using frequency sweeps, evaluation in the range of low frequencies.
Left: For unlinked molecules G" > G', the |η|-function approximates to the plateau value of the zero-shear viscosity, indicating the behavior of a viscoelastic liquid at rest.*
Right: For cross-linked molecules G' > G", the |η|-function slopes up towards an "infinitely high" value, indicating a "gel-like state" and therefore stability at rest. Comment: Since here in the entire range G' > G", the presentation and evaluation of the viscosity curve is not recommended for practical use*

as seam sealants used in automotive industry after curing; soft food gels such as corn starch (e.g. for dairy products), flexible biological tissues such as earlaps

c) Comparison of cross-linked and unlinked polymers

Frequency sweeps are a good tool to differentiate clearly between unlinked and cross-linked samples (see Figures 8.21 and 8.22):

1) Unlinked polymers

Towards lower frequencies, for unlinked polymers the values of both G' and G" are falling constantly, showing the maximum curve slopes of 2:1 for G' and 1:1 for G" then, when presenting on a logarithmic scale (see also Chapter 8.4.2.1 and Figure 8.16: behavior of a Maxwell fluid).

2) Cross-linked polymers

For polymers showing cross-linking, towards lower frequencies the G'-curve – and mostly, but less clearly, also the G"-curve – are approximating constant limiting values, displaying G' > G" finally. The **structural strength** or "rigidity" of a sample at rest might be expressed by use of this limiting value of G'. Sometimes, users call it the **"consistency-at-rest"**. When comparing two samples, those one with the higher degree of cross-linking will also show the higher plateau value of G' in the range of low frequencies, as shown in Figure 8.20.

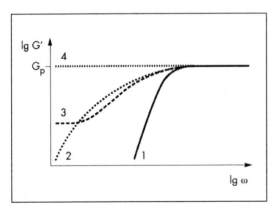

Figure 8.22: Comparison of polymers in principle, using G'-curves of frequency sweeps:
(1) Unlinked polymer showing a narrow MMD
(2) Unlinked polymer showing a wide MMD
(3) Polymer with a low degree of cross-linking, a flexible gel or a dispersion showing a weak structure at rest
(4) Polymer with a high degree of cross-linking, a rigid gel or a dispersion showing a strong structure at rest

☞   For "Mr. and Ms. Cleverly"

Formal notation of the limiting values of the frequency-dependent functions of G' and G"

1) **For unlinked polymers** (see Figure 8.16: **Maxwellian behavior**, and Figures 8.19, 8.21: left, and 8.22: no. 1 and 2):

$\lim_{\omega \to 0} G'(\omega) = 0Pa$, $G'(\omega)$ **shows the slope 2:1** on a logarithmic scale

$\lim_{\omega \to 0} G''(\omega) = 0Pa$, $G''(\omega)$ **shows the slope 1:1**

2) **For cross-linked polymers** (see Figures 8.20, 8.21: right, 8.22: no. 3 and 4):

$\lim_{\omega \to 0} G'(\omega) \neq 0Pa$ and $\lim_{\omega \to 0} G''(\omega) \neq 0Pa$, showing lim G' > lim G''

᷍   End of the Cleverly section

8.4.4 Behavior of dispersions and gels

In stable dispersions and gels, intermolecular interaction forces are building up a three-dimensional network of forces (see also Chapter 3.3.4.3c). These kinds of materials show G' > G'' in the whole frequency range, see Figure 8.23. Therefore, elastic behavior dominates the viscous one (see also Chapter 8.3.2a: Gel-like character). The curves of G' and G'' often occur in the form of almost parallel straight lines throughout the entire frequency range showing a slight slope only. Typically, for these kinds of samples the following ratio values can be found: G': G'' = 10:1 to 100:1, or tanδ (= G''/G') = 0.1 to 0.01, respectively.

The shape of the curves is approximately comparable to that one indicated by cross-linked polymers, see Chapter 8.4.3 and Figures 8.20 and 8.21: right. This is not surprising since both material systems are showing a **network structure: Dispersions and gels in the form of a physical network, and cross-linked polymers in the form of a chemical network**. When performing rheological tests, indeed the structural strength can be analyzed but not the type of the network. As long as the LVE range is not exceeded, both kinds of network structures are exhibiting a relatively constant structural strength in the whole frequency range.

a) Structural strength at rest, dimension stability, yield point, and frequency sweeps
When performing simple QC tests, rotational tests are often used to determine the "yield point" in order to characterize the internal structural strength-at-rest. In Chapter 3.3.4 is explained why this kind of determination of a "yield point" has become a subject of discussion among rheologists. For this reason, more and more users prefer to perform frequency sweeps in order to evaluate the **consistency-at-rest and long-term storage stability of dispersions** and to control **sedimentation, settling, flotation, syneresis and phase separation**. This new sight of view applies not only to people working in R & D but also increasingly to practical users working in QC laboratories.

Note 1: Long-term stability of dispersions at rest, and analysis at a low frequency value
In order to analyze behavior "at rest", for scientific investigations it is recommended to determine the G'-value – and if desired additionally also the G''-value – at the angular frequency $\omega = 0.01 rad/s$. For QC tests often higher ω-values are taken, e.g. $\omega = 0.1$ or 0.5 or even $1 rad/s$. In this case, the G'-curve should show only a very small slope towards lower frequencies, or even better: The shape of the curve should be as much as possible in parallel to the x-axis (as shown in Figure 8.23: Dispersion 1 on top). But please be

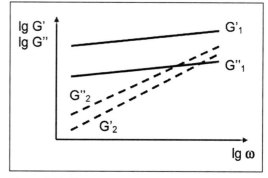

Figure 8.23: Frequency sweep of two dispersions; dispersion 1 shows G' > G'' and therefore a gel-like structure and physical stability at rest; dispersion 2 shows G'' > G' and therefore behavior of a liquid which may tend to phase separation

aware: Any G'-value "at rest" is not the "yield point" which is a shear stress value, even when both values are showing the same unit Pa.

As a rule of thumb based on a large number of experiments, **for many dispersions and gels** can be stated **for G'-values determined at ω = 0.01rad/s** if the following two pre-conditions are met: a) testing is performed in the LVE range, and b) occurrence of a gel-like character showing G' > G'':

1) **For G' ≥ 10Pa**, there is a **certain dispersion stability or gel stability** (see also Table 4.1 of Chapter 4.2.2).
2) For G' ≤ 1Pa however, there is hardly sufficient stability for any practical application.
3) For G'-values in between, further tests should be carried out, perhaps by determination of the "yield stress" **and** the "flow stress" via amplitude sweeps; see also Chapter 8.3.4. Examples from practice such as spray coatings, salad dressings and cosmetic lotions can be found in Table 4.1.

↶↷   For "Mr. and Ms. Cleverly"

Example 1: Stability of uncured PU adhesives, and evaluation via frequency sweeps
In order to evaluate the stability of highly filled, still uncured polyurethane (PU) adhesives, sometimes the **"dog tail test"** or **"nose test"** is performed. Extruded from a cartridge by use of a pistol, the adhesive shows the shape of a round strand and is stuck with its cross-section to a vertical wall in a way that it protrudes from the wall horizontally in the form of a cylinder (showing around L = 60mm, d = $2R$ = 40mm). After an observation time of t_1 = 2min, the reached inclination of the material is evaluated using five graduation steps [231]. In the following, explained is how the "dog tail test" can be simulated by a rheological test.
Preparatory calculation: With the material density of ρ = approx. 1000kg/m^3, the weight force F_G = approx. 0.75N (since F_G = $m \cdot g$ = $\rho \cdot V \cdot g$ = $\rho \cdot A \cdot L \cdot g$ = $\rho \cdot \pi \cdot R^2 \cdot L \cdot g$). With the sticking area A = $\pi \cdot R^2$, for the occurring shear stress counts: τ = F_G/A = $\rho \cdot L \cdot g$ = around 600Pa.
Measurement preset: Frequency sweep, stress amplitude τ_A = 600Pa and ω = 100 to 0.01rad/s.
Analysis: The value of the storage modulus G' is read off at ω = 0.01rad/s; this corresponds roughly to the inverse value of the observation time $1/t_1$ = (1/120)s for evaluating the stability (structural strength, material rigidity).
Summary: Classification of adhesives by their structural strength using a "dog tail test" depends on the observer, and therefore is subjective. In order to evaluate the stability of PU adhesives, practical application experiments have shown that they can be divided into the following classes when using their G'-values at ω = 0.01rad/s:

(1) G' < 1kPa: **very soft**/(2) G' = 1 to 10kPa: **soft**/(3) G' = 10 to 100kPa: **semi-soft**/(4) G' = 100 to 200kPa: **hard**/(5) G' > 200kPa = 0.2MPa: **very rigid**

See also Example 1 in Chapter 8.5.2.2d: Combination of an amplitude sweep and a step test for testing adhesives

↶↷   End of the Cleverly section

Note 2: Comparison of yield point values obtained from rotational and oscillatory tests
When performing oscillatory tests in the LVE range, evaluation of a sample's consistency is usually carried out at low frequencies. This kind of testing comprises many advantages over conventional methods of yield point analysis via rotational tests:

1) When determining the **yield point by a rotational test**, the measurement **results in a single value only**. Reproducibility is often unsatisfactory, especially for visco-**elastic** samples. Performing a **frequency sweep**, however, a complete G'-function is measured, and

therefore, **several values are achieved** at very low frequencies representing the structure-at-rest. This method improves the reliability of data evaluation.

2) G' and G" are **two resulting parameters**, based on the measurement of two independent raw data, therefore revealing more information as achieved by a single parameter only, as it is obtained when determining the yield point by a flow curve. For two samples showing the same **G'-value** at rest, the **additional information** obtained by the **values of G" or tanδ**, respectively, may be the decisive factor to enable the user to make a differentiation.

3) Often, the yield point value is clearly time-dependent (see Chapter 3.3.4.3b). On the other hand, when performing **frequency sweeps** each user measures at each individual measuring point at a defined frequency, and therefore, **at the same defined time conditions**.

Example 2: Specification of dispersion stability in terms of G' instead of a yield point
Possible specification in a test protocol when testing dispersion stability: Instead of the yield point (e.g. τ_y = 500Pa, determined at the shear rate $\dot{\gamma}_{min}$ = 0.5s^{-1}, by a rotational test), the following frequency sweep data are specified: G' (e.g. G' = 450Pa, determined at ω_{min} = 0.5rad/s), and if desired also G" (e.g. G" = 200Pa). Of course, also here holds: This G'-value is not the "yield point".

Example 3: Low-shear test for varnishes, instead of yield point tests via rotation
Preset: Use of an air-bearing rheometer is required, and a parallel-plate measuring system (with a gap H = 0.5mm), measuring temperature T = +23°C. Test program [127]:
1st interval (for t = 180s): period of rest (time for temperature equilibration)
2nd interval: Frequency sweep (6 measuring points) at γ = 10% = const, with increasing f = 0.5 to 20Hz or with decreasing f = 20 to 0.5Hz, respectively
Evaluation at f = 1Hz: using the values of $|\eta^*|$ and tanδ (or alternatively, of G' and G")
This test is recommended by Eurocommit (European Committee for Ink Testing Methods, Working Group Rheology), **to determine the viscoelastic low-shear behavior of varnishes or polymer solutions** used as raw materials of **offset printing inks**. The aim is **to replace previously performed yield point tests or flow point tests**, respectively, since at low-shear conditions, those varnishes and also finished offset printing inks, in most cases do not show a flow point at all but G" > G', and therefore, the character of a liquid, even when it is a high-viscosity liquid then. (See also Note 3 and Example 2 of Chapter 8.3.5: Offset printing inks)

Note 3: Syneresis of dispersions and gels
Gels and dispersions showing a gel-like structure, sometimes are becoming increasingly stiffer with time. As a consequence, with increasing structure density, water or other solvents might be pressed out of the more and more inflexible gel body. This process may be a reason for syneresis effects. In this case with time, on the surface of these kinds of materials often appears a liquid layer then (sometimes called "serum"). In order to prevent the occurrence of this process, **a higher value of the factor tanδ = G"/G' might be helpful**. Perhaps by adding rheological additives, it can be achieved to change the value of the loss factor tanδ from 0.01 to 0.1 (or even to 0.5), or the relation of (G': G") from 100:1 to 10:1 (or even to 2:1), respectively. This measure might lead to **more flexibility and deformability of the gel structure**, and finally as a consequence, in many cases to the desired reduction of syneresis effects.

b) Dispersion stability outside the LVE range
Example: Transport stability of dispersions
Vibration occurring during transportation by cargo trucks might cause sedimentation of particles in dispersions. For a suspension, G' > G" was found when measuring in the LVE range and evaluating in the range of low frequencies, and therefore, sedimentation stability was expected. By and by however, after several long-distance transports with cargo trucks there was settling of pigment particles. In this case, the simulation of the transport procedure

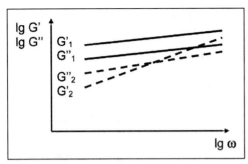

Figure 8.24: Frequency sweeps to evaluate transport stability of dispersions: Test 1, performed at a strain within the LVE range results in G' > G'' in the whole frequency range, and therefore, there is stability also at rest. Test 2, at an increased strain value reveals G'' > G' at low frequencies, and therefore, there is no dispersion stability when simulating the situation which may appear due to vibrations occurring in a container at transport conditions

with the occurring vibrations was improved by measurements which were performed under shear conditions a little bit outside the LVE range. Transport behavior was simulated now by applying shear conditions at an increased "vibration amplitude"; e.g. using **strain values between the yield strain at the limiting value of the LVE range γ_L (or γ_y) and the flow strain (γ_f) at the flow point** when G' = G''. Example: Instead of γ = 1% = const – as it was determined by an amplitude sweep for the limit of the LVE range – there was chosen γ = 10% = const then; see also the explanation given in Figure 8.24. However, the following has always to be taken into account: Using this method, relative values but not absolute values are obtained since the LVE range is already exceeded then, which happened in this case of course knowingly. See also the Example in Chapter 8.9.2: Superposition of rotation and oscillation.

8.4.5 Comparison of superstructures using frequency sweeps

Usually, the G'-curve is taken to describe the **"structural strength"** or **"consistency"** of a sample. In most cases, the different kinds of structures can clearly be distinguished **in the range of low frequencies**, i.e. below ω = 1rad/s, and for scientific demands usually at ω = 0.01rad/s. The following questions can be answered then (see also Figures 8.21 and 8.22):
1) **Polymers:** Are the molecules unlinked, showing G'' > G' then, or are they cross-linked, showing G' > G'' in this case? Which is the degree of cross-linking, when the corresponding G'-values are compared at the lowest frequency measured?
2) **Dispersions and gels:** Is there a "gel-like character", showing G' > G'' then, or is it the "character of a liquid", showing G'' > G' in this case? Which is the structural strength-at-rest or the gel strength, in terms of the value of G' at the lowest frequency measured?

☞   For "Mr. and Ms. Cleverly"

8.4.6 Multiwave test

Usually when performing frequency sweeps, one measuring point is measured after the other, each one at a single frequency from the lower to the higher ω-values or vice versa, until the complete curve is recorded. However, the test period might greatly be reduced if the rheometer is set to measure several frequencies at once. This results in a multiwave function then, which is a **multiple wave** produced from several superimposed single oscillations. Thus, a multiwave test is a multiple frequency test. Using this kind of test, great advantages might be achieved due to the option to save a lot of test time especially when testing time-dependent and temperature-dependent behavior of samples during a hardening, gel formation or curing process (see also Chapters 8.5.3 and 8.6.3).

Example 1: Design of a multiwave test, presetting nine different frequencies all at once
Simultaneously nine oscillatory curves are produced, each one showing a different individual angular frequency ω_i by use of a basic wave with the angular frequency ω = 0.1rad/s, which

Table 8.3: Multiwave test: Frequency factors f_i and the resulting angular frequencies ω_i (with i = 1 to 9)

frequency factor f_i	1	2	4	7	10	20	40	70	100
angular frequency ω_i [rad/s]	0.1	0.2	0.4	0.7	1	2	4	7	10

is multiplied by the nine individual frequency factors f_i (see Table 8.3; here, the individual counting numbers i = 1 to 9 are used for f_i and ω_i).

The complete frequency test covering two decades from ω = 0.1 to 10rad/s might be performed in the same time now which is usually required to determine only a single measuring point at ω = 0.1rad/s. Using this method, frequency sweeps might be carried out over six decades from ω = 0.001 to 1000rad/s in a comparatively short time performing just three measurements, when selecting the three basic waves at ω = 0.001/0.1/10rad/s and using again the mentioned nine frequency factors.

Example 2: Multiwave strain function, produced from three basic waves
Figure 8.25 presents three sinusoidal basic waves in form of fine lines, all showing the same strain amplitude of γ_A = 2% but each one of them with another constant angular frequency: the first one at ω = 1rad/s, the second one at ω = 2rad/s, and the third one at ω = 3rad/s. In order to produce the resulting multiwave curve, at each time point the three γ-values of the individual waves are summed up. As a result, the multiwave function occurs as the sum curve of the three individual strain functions available.

Note: Presetting of a multiwave strain function, and the LVE range
Presetting a multiwave function, the time-dependent strain amplitude values of several individual basic waves are summed up. At certain time points, the sum of these amplitudes may add up to a relatively high value; see for example in Figure 8.25: For each single sine curve counts indeed γ_A = 2% only, but for the sum curve holds $\gamma_A = \gamma_{max}$ = 5%. When performing strain

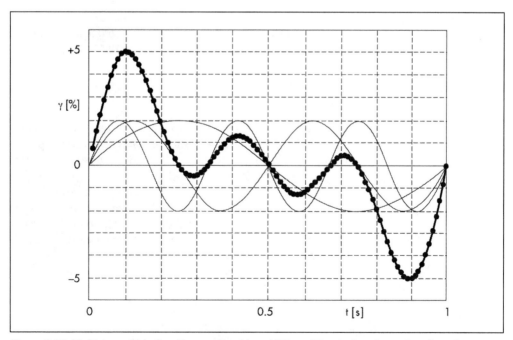

Figure 8.25: Multiwave strain function produced by addition of the strain values taken from three individual basic sine waves showing the angular frequencies ω = 1 and 2 and 3rad/s

controlled tests, **the resulting value of the total strain may exceed the limit of the LVE range**, therefore causing an irreversible deformation of the sample's structure. Appropriate software programs are available analyzing automatically and indicating whether the LVE range is exceeded. If desired, these programs are starting measures which prevent this occurring, e.g. using interactive control options.

8.4.7 Data conversion

In general when measuring in the LVE range, the rheological behavior of each viscoelastic material is depending on the relaxation or retardation time spectrum, respectively. Therefore, it is not that important which kind of test is performed since all **data from creep tests, relaxation tests and oscillatory tests can be converted from and into one another**. When performing oscillatory tests for this purpose, usually data of frequency sweeps are selected since using this mode of testing, the time-dependent rheological behavior can be characterized in the form of short-term and long-term behavior within a certain time frame selected by the user, therefore mirroring the corresponding spectrum of relaxation or retardation times, respectively. Other kinds of oscillatory tests such as amplitude sweeps or tests at constant dynamic-mechanical conditions (i.e. as well at a constant strain or stress amplitude as well as at a constant frequency), however, are all carried out at a single constant frequency only. In this case, as a time constant the effect of a single relaxation or retardation time might be analyzed only. More information and examples on conversion of rheological data see Chapters 6.3.5, 7.3.4 and 8.7.1e.

      ↝   End of the Cleverly section

8.5   Time-dependent behavior at constant dynamic-mechanical and isothermal conditions

Using this kind of oscillatory testing, both the frequency and the amplitude are kept at a constant value in each test interval. Therefore, constant dynamic-mechanical shear conditions are preset. Also the measuring temperature is kept constant, i.e. testing takes place at isothermal conditions. This kind of test is also referred to as **dynamic-mechanical analysis or DMA test**.

8.5.1 Description of the test

Preset
1) For tests with controlled shear strain: $\gamma(t) = \gamma_A \cdot \sin\omega t$, with both γ_A = const and ω = const (see Figure 8.26). Here, the period of time for each one of the oscillation cycles is kept constant (frequency), and also the maximum value of the bob's deflection angle is kept constant (amplitude).

2) For tests with controlled shear stress: $\tau(t) = \tau_A \cdot \sin\omega t$, with both τ_A = const and ω = const. In this case, the oscillatory motion of the measuring bob is caused by a torque whose amplitude is kept constant, and also the period of time for each one of the oscillation cycles is kept constant (frequency).

Here, many users select ω = 10rad/s = const; see also the Note in Chapter 8.3.1. Tests at a higher frequency may influence the result significantly; see also Chapter 8.3.5.

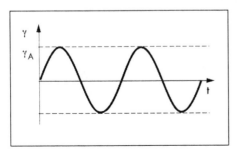

Figure 8.26: Preset of an oscillatory test with a constant amplitude and a constant frequency, here with controlled strain

Measuring result

Usually, the measuring curves are presented with time t on the x-axis on a linear scale, and G' and G" are shown on the y-axis both usually on the same logarithmic scale (as in Figures 8.31 and 8.35). Sometimes, also tanδ is displayed on a second y-axis, and there are users who like to present additionally $|\eta^*|$ on a third y-axis, usually also on a logarithmic scale. All of these presentations are therefore semi-logarithmic lin/log diagrams.

8.5.2 Time-dependent behavior of samples showing no hardening

This kind of testing is used to investigate time-dependent behavior of samples under the precondition to show **no chemical modification** during the measurement. The aim of these tests is to determine the behavior at constant dynamic-mechanical low-shear conditions, for example, by observing the change in the structural strength in terms of the time-dependent function of G' (see Figure 8.27). The structural strength may remain unchanged, increase or decrease with time.

Increase in the structural strength may occur for polymers due to an increasing number of molecular entanglements, and for dispersions and gels due to the formation of a gel structure based on an increasing number of physical interactions such as hydrogen bridges, or it may occur due to drying effects.

Decrease in the structural strength may occur for polymers due to an increasing number of molecular disentanglements, and for dispersions and gels due to softening or a decreasing number of physical interactions. A falling curve may also result when polymer molecules are destroyed due to degradation which reduces the average molar mass, but since the assumption for this chapter was "no chemical modification", these and other changes of the chemical structure should not appear when performing common rheological tests, unless these effects are clearly outlined in the test protocol.

Example: Testing bitumen and asphalt according to the SHRP specifications

In order to analyze polymer-modified bitumen and asphalt as materials used for road construction, oscillatory tests at constant dynamic-mechanical conditions are performed using the SHRP method according to the Strategic Highway Research Program of the USA and ASTM D7175, respectively, see also [1, 28, 351]. Here, three test methods are recommended:

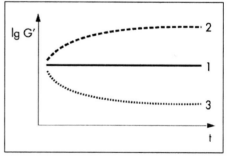

Figure 8.27: Time-dependence of the G'-function of materials without chemical modifications:
(1) independent of time, stable structure showing a constant structural strength
(2) increasing structural strength, e.g. of polymers due to additional molecular entanglements, or of dispersions and gels due to gel formation, or due to drying effects
(3) decreasing structural strength, e.g. of polymers due to molecular disentanglements, or of dispersions and gels due to decreasing interaction forces

1) For fresh original binders:
Preset: ω = 10rad/s = const, and γ = 12%;
specification: $(|G^*|/\sin\delta) \geq 1kPa$
Asphalt is too compliant, too flexible or too "soft" if showing a lower value. For heavily travelled road surfaces, this may lead to the formation of ruttings.
2) For binders which have been aged to a low degree using the Rolling Thin-Film Oven Test (RTFOT, according to ASTM D2872):
Preset: ω = 10rad/s = const, and γ = 10%;
specification: $(|G^*|/\sin\delta) \geq 2.2kPa$
Otherwise asphalt is too compliant, with the same consequences as above for test method 1.

3) For binders which have been aged to a high degree using the Pressure Aging Vessel Test (PAVT, according to ASTM D6521):
Preset: $\omega = 10\text{rad/s} = \text{const}$, and $\gamma = 1\%$;
specification: $(|G^*| \cdot \sin\delta) \leq 5\text{MPa}$
Asphalt is too inflexible or too "stiff" if showing a higher value. For road pavements exposed to heavy traffic and climatic temperature variations, this may lead to material fatigue and thermal cracking.

Note: Test preparation – thermal-stability time of polymer melts (acc. to ISO 6721-10)
Before testing polymer melts, it is recommended to check the thermal-stability time at the desired measuring temperature, for example, to avoid molecular degradation. Using a parallel-plate measuring system, testing is performed at the constant measuring temperature and at constant dynamic mechanical conditions (e.g. at $\omega = 10\text{rad/s}$ and $\gamma = \text{const}$, presetting a strain in the LVE range). The resulting time interval is defined as the time taken from the start of the test to the time point at which any of the measured values of G^*, G' or G'' have changed by 5% from their initial value.

Example: For an unfilled polymer at $T = 250°C$ was found a thermal-stability time of $t = 480\text{s} = 8\text{min}$ (at the shear conditions of $\omega = 10\text{rad/s}$ and $\gamma = 10\%$).
Sometimes, it may be necessary to carry out runs at more than one frequency of oscillation. Using new sample material then, subsequent measurements performed at the measuring temperature should be finished in a time shorter than the thermal-stability time.

8.5.2.1 Structural decomposition and regeneration (thixotropy and rheopexy)

a) Thixotropic behavior
"Thixotropic behavior" means reduction of the initial structural strength when performing a high-shear process, followed by a more or less rapid but complete structural regeneration in the subsequent period of rest (see Figure 8.28). **A sample shows thixotropic behavior only if the cycle of decrease and increase occurs as a completely reversible process.** For more information on thixotropic behavior and corresponding materials: see Chapter 3.4.2.1a.

This test is performed at constant shear conditions in each test interval, and this applies for both the amplitude and the frequency. For dispersions showing stability at rest and therefore a gel-like character with $G' > G''$, the aim of the **high-shear interval** is to break the internal superstructure at least to such an extent to obtain the character of a liquid with $G'' > G'$. Finally, an equilibrium of the shear forces and the flow resistance forces of the sample is reached. The steady-state behavior occurs not immediately in this case and the whole process is as well time-dependent as well as dependent on the preset high-shear conditions.

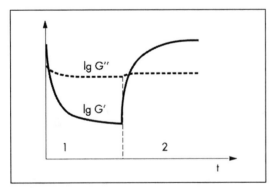

In the subsequent **state of rest**, a material exhibits thixotropic behavior only if the structural regeneration has made sufficient progress after a certain period of time, therefore showing the initial values of G' and G'' again, i.e., in this case again $G' > G''$. For most dispersions, the change in the G''-value is small in both the shear interval and the rest interval if compared to the G'-value.

Figure 8.28: Time-dependent functions of G' and G'' of a thixotropic material
(1) decreasing structural strength when applying a constantly high shear load, caused by a partial structural decomposition
(2) increasing structural strength when at rest, caused by structural regeneration

Note: Thixo-forming and thixo-forging of metals

"Thixo-forming" is a molding techno-logy which is an attractive alternative to casting and forging of metals. The materials are heated until they show a **dough-like consistency**. For **aluminium**, temperatures of around T = 500°C are required, and for steel bet-ween 1350 and 1450°C. In this partially liquid, viscoelastic state between on the one hand crystallized and solid behavior and on the other hand melted and liquid behavior, the metal can be shaped easily since it is flowing better than it does when cold forging. The forming process is comparable to **kneading** dough. A

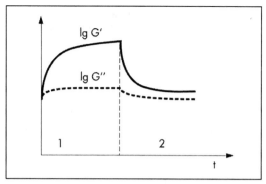

Figure 8.29: Time-dependent functions of G' and G" of a rheopectic material
(1) increasing structural strength when applying a constantly high shear load
(2) decreasing structural strength when at rest

large advantage is that it involves lower production costs. This is due to the considerably **lower mechanical load to be applied on the forming tools** such as presses, because the materials to be formed are considerably softer in this intermedium state. In addition, there are considerably **lower processing temperatures** compared to a melting process, and therefore, an enormous amount of **energy can be saved** when using this kind of molding.

Examples of applications: composites made of steel and aluminum used in the transporta-tion industry (cars and trucks, trams and railway, airplanes and ships), architecture, housings made of magnesium for cameras and portable computers, highly resilient thin-walled struc-tural components, foamed lightweight components

b) Rheopectic behavior

Rheopectic behavior means an increase of the initial structural strength when performing a high-shear process, which is followed by a more or less rapid but complete structural rege-neration in the subsequent period of rest, therefore showing decreasing structural strength then (see Figure 8.29). For more information on rheopectic behavior see Chapter 3.4.2.1c; and on shear-induced increase of structural strength see the Note in Chapter 8.5.2.2b and Figure 9.24 in Chapter 9.2.2.

8.5.2.2 Test methods for investigating thixotropic behavior

a) Step test consisting of three intervals
Preset
Oscillatory test showing three intervals, each one at constant dynamic-mechanical conditions (see Figure 8.30).

Result: Time-dependent functions of G' and G" (see Figure 8.31)

For measurements like this, the following three test intervals are preset:

1) The reference interval (low-shear)
Shear conditions "at rest", and therefore at low-shear conditions **in the LVE range**, are preset in the period between t_0 and t_1. The aim is here to achieve fairly constant values of G' and G" in the whole first interval. Usually, the G'-value at the end of the first test interval is selected to represent the sample's state-at-rest. This value is used later as the **"reference value of G'-at-rest"** to be compared to the G'-values occurring in the third test interval.

2) **The high-shear interval**

High-shear conditions, and therefore at shear conditions outside the LVE range are preset in the period between t_1 and t_2. The aim is here to break the internal structure of the sample. This interval is used **to simulate** the high-shear conditions occurring during an **application process**, when painting and coating with a brush, roller or blade, or when spraying.

3) **The regeneration interval (low-shear)**

Shear conditions "at rest", and therefore again at low-shear conditions in the LVE range, are preset in the period between t_2 and t_3. Here, the same shear conditions like in the first interval are chosen to facilitate **regeneration of the sample's structure**. This interval is used to simulate the low-shear conditions occurring directly after the coatings process when the material is only slightly stressed by its own weight due to gravity.

For practical users, the crucial factor **to evaluate structural regeneration** is the **behavior in the time frame which is related to practice**. This period of time has to be defined by the user before the test according to the requirements, usually after a number of experiments performed (e.g. by the application department). For example for a coating, desired is a regeneration time of $t = 60$ or 120s. If the crossover point $G' = G''$ has not been reached within this **"time related to practice"** (or "practice-relevant time"), then the sample is not considered to be thixotropic, related to this application.

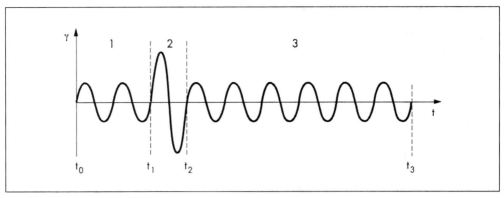

Figure 8.30: Preset profile: step function consisting of three intervals, each one at a constant strain amplitude, (1) at low-shear, (2) at high-shear, and (3) again at low-shear conditions

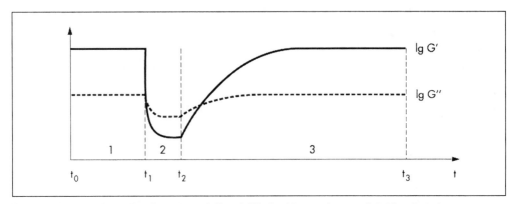

Figure 8.31: Time-dependent functions of G' and G'' of a thixotropic material, (1) at low-shear conditions showing the "reference value of G'-at-rest", (2) structural decomposition, and (3) structural regeneration

Example 1:

1st interval (with 5 measuring points, in t = 60s): γ = 1% and ω = 10rad/s
2nd Interval (with 5 measuring points, in t = 30s): γ = 100% and ω = 10rad/s
3rd interval (with 50 or more measuring points, in t = 180s): γ = 1% and ω = 10rad/s again

Exactly the same shear profile has to be preset for each individual test if thixotropy values of different tests are to be compared, and that counts for all parameters: strain amplitudes, (angular) frequency, number of measuring points, and duration of the test intervals.

Note 1: Optimizing the step test conditions

In order to get a useful "reference value of G'-at-rest", the G'-value should be as constant as possible in the first interval. If this condition is not met, the following actions can be performed:
1) If the G'(t)-curve comes from above showing constant values only after a certain period of time, then the preset strain was too high for the sample to be still in a state of rest, i.e. the LVE range is exceeded already. Therefore, at these shear conditions a certain degree of structural decomposition is already taking place.
Action: A lower strain amplitude should be selected, e.g. half as large or even lower. Here, it would be helpful to perform first an amplitude sweep to determine the limiting strain value of the LVE deformation range. Sometimes it is helpful to reduce also the frequency (see also Chapter 8.3.5: frequency-dependence).
2) If the G'(t)-curve comes from below showing constant values only after a certain period of time, then **transient behavior** is measured. In this case, the measuring time was too short for the sample to adapt evenly throughout the whole shear gap to the applied low-shear conditions.
Action: The measuring point duration must be extended. **As a rule of thumb: The measuring point duration should be at least as long as the value of the reciprocal shear rate** $\dot{\gamma} = \gamma_A \cdot \omega$ (see also Chapter 8.2.4d and Equation 8.24). After a first test with this preset, the measuring point duration may have to be extended further until steady-state test conditions are achieved. Sometimes, however, even shorter times than t = $1/\dot{\gamma}$ are sufficient. See also the Note in Chapter 3.3.1b: transient behavior, and Figure 2.9 (no. 5).

Note 2: Advantages of oscillatory tests over rotational tests

When performing rotational tests, only a single function is obtained which is the time-dependent shear viscosity η(t), as shown in Figure 3.43 (see Chapter 3.4.2.2). However, when presenting the results of **oscillatory tests**, for example, in terms of the functions of G'(t) and G''(t), additional information is achieved about the structural character since results from oscillatory tests are based on **two independent raw data**. A sample exhibits the character of a fluid as long as G'' > G' or tanδ > 1, and it shows gel-like behavior with stability if G' > G'' or tanδ < 1, respectively. But the greatest advantage of oscillatory tests is that a measurement can be still continued without any internal structural decomposition, even when the sample's character has meanwhile changed to the gel-like state. Of course also here, precondition is as ever: Testing is performed in the LVE range.

Optional methods to analyze structural regeneration

A number of options to analyze thixotropic behavior are given below, many users prefer the evaluation according to method M4, using the G'-values in order to characterize the structural strength.

M1) The "thixotropy value" in terms of the difference between G'-values

The extent of the thixotropic behavior is determined in terms of the change ΔG' in the values of the storage modulus. This is the difference between the maximum of G' after the structural regeneration and the minimum of G' after the structural decomposition. With G'_{min} at the time point t_2 and G'_{max} at t_3 the following holds (see Figure 8.31):

$$\Delta G' = G'_{max} - G'_{min}$$

M2) The "total thixotropy time"

The "total thixotropy time" is determined as time difference between the time point t_2 at the end of the second interval, indicating the structural decomposition in terms of G'_{min}, and the time point in the third test interval when reaching the maximum value G'_{max} after the complete structural regeneration. Therefore, the "total thixotropy time" is the period required for the complete (100%) regeneration of the structure, i.e. when reaching again the reference value of G'-at-rest which was determined in the first test interval. Of course, this period of time might be shorter than $(t_3 - t_2)$ if the regeneration is finished already before the time point t_3 is reached.

M3) "Relative thixotropy time" to reach a certain percentage of regeneration

For QC tests, analysis of the "total thixotropy time" according to M2 may take too long. Therefore then, in the third test interval the "relative thixotropy time" may be determined as the period for the G'-value to reach a previously defined relative value, for example, 75 or 90% compared to the reference value of G'-at-rest of the first interval (the latter counts as the "100% G'-value" here).

M4) The thixotropy time required to reach the crossover point G' = G"

Here, in the third interval the "thixotropy time" is determined as the period of time to reach the crossover point G' = G". Of course, as a pre-condition is required G' > G" in the first interval, and G" > G' in the second one. One advantage of this method is that the crossover point and the corresponding time point can be observed and read off easily.

M5) The percentage of regeneration within a previously defined period of time

The "percentage of regeneration" taking place in the third interval is determined at certain time points which have been defined by the user before the test (e.g. after t = 30 and 60s). The G'-values are read off at these time points and the percentage is calculated in relation to the reference value of G'-at-rest of the first interval which is taken as the "100% value" then.

Example (to method M5): Comparison of two coatings

Different behaviors of two coatings in the regeneration phase is presented in Table 8.4.
Analysis: Coating 1 shows complete regeneration of the initial structure strength within 120s (related to the G'-value). This may facilitate to obtain the desired wet layer thickness. Here, with 75% structural recovery attained after 30s, a relatively high value has been reached already. Coating 2 displays a slower structural regeneration, showing long lasting and therefore good leveling behavior. However, this coating may show a certain tendency to sagging on vertical areas, which may prevent to achieve the desired layer thickness.

Table 8.4: Regeneration of two coatings in terms of G'(t) and in %

	Coating 1		Coating 2	
	G' [Pa]	Reg. [%]	G' [Pa]	Reg. [%]
at the end of the first interval, at low-shear conditions; the reference value of G'-at-rest	10	100%	4	100%
at the end of the second interval, at high-shear conditions	1	(10%)	0.5	(12.5%)
regeneration in the third interval after t = 30s	7.5	75%	2.4	60%
after t = 60s	9	90%	2.8	70%
after t = 120s	10	100%	3	75%

b) Step test consisting of three intervals: oscillation/rotation/oscillation (ORO-test)

More and more users perform step tests as a combination of alternating intervals showing the test modes rotation and oscillation directly connected in series (see also Chapter 8.9.1: ORO-test). In this case, the intervals 1 and 3 are carried out as oscillation in the LVE range to simulate the state-at-rest, and interval 2 is performed as rotation to simulate the high-shear conditions of the application (see Figure 8.32). The advantage of this method is to have the option to apply a considerably higher shear load in the second interval as is the case when measuring in an oscillatory mode also here, and this even holds if a high strain amplitude was selected from outside the LVE range. This kind of test is performed as follows (ORO-test):

1st interval: Oscillation in the LVE range
2nd interval: Rotation at a high shear rate $\dot{\gamma}$
3rd interval: Oscillation in the LVE range again

Example of a preset

1st interval (with 5 measuring points, in t = 60s): at γ = 1% and ω = 10rad/s
2nd interval (with 5 measuring points, in t = 15s): at $\dot{\gamma}$ = 100s^{-1} (or even more)
3rd interval (with 50 or more measuring points, in t = 180s): at γ = 1% and ω = 10rad/s again

Two possible kinds of presentation of the measuring results are shown in Figure 8.33 in terms of G' and G'' (oscillation) and the shear viscosity η (rotation), as well as in Figure 8.34 in terms of the complex viscosity η^* (oscillation) and η (rotation). An application example of ORO-tests are metallic-effect automotive coatings, in order to optimize their leveling and sagging behavior after the application [57].

However, if a sample shows visco-**elastic** necking effects in the high-shear interval, for example in the form of edge failure effects when using cone-and-plate or parallel-plate measuring systems, then also here, the test mode oscillation should be selected like above.

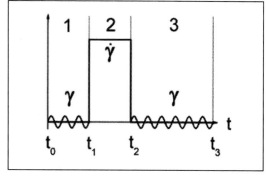

Figure 8.32: Preset profile: step function consisting of three intervals; (1) oscillation in the LVE range at a constant strain amplitude, (2) rotation at a high shear rate, (3) again like in the first interval

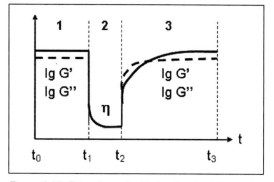

Figure 8.33: Time-dependent behavior of a thixotropic material: (1) low-shear oscillation showing the "state-at-rest" in terms of G' and G'', (2) high-shear rotation causing structural decomposition in terms of the shear viscosity η, and (3) again low-shear oscillation showing the structural regeneration as G' and G''

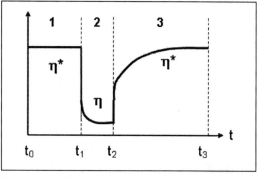

Figure 8.34: Time-dependent behavior of a thixotropic material: (1) low-shear oscillation showing the "state-at-rest" in terms of the complex viscosity η^*, (2) high-shear rotation causing structural decomposition in terms of the shear viscosity η, and (3) again low-shear oscillation showing the structural regeneration as η^*

c) Evaluating rheopectic behavior: step test consisting of three intervals in oscillation
Corresponding analysis can be used in an adapted form for **"rheopexy"** and the **"rheopectic behavior"**. However, this behavior occurs rarely in industrial practice. See also Chapter 3.4.2.1c: Rotational tests and rheopectic behavior.

Note: Non-rheopectic behavior
Shear-induced and permanently remaining increase of the structural strength
Example: A dispersion was tested under the following conditions.
1st interval (oscillation in the LVE range): with t = 2min, at γ = 0.1% and ω =10rad/s
Result: G" = 10Pa and G' = 7Pa, therefore showing the character of a fluid, since G" > G'
2nd interval (rotation): with t = 30s, at $\dot{\gamma}$ = 100s^{-1}
Result: viscosity increase from η = 0.5 to 1Pas
3rd interval (oscillation, like in the first interval): with t = 5min, at γ = 0.1% and ω = 10rad/s again
Result: Already after a short time, the values are increasing, remaining on G' = 300Pa and G" = 100Pa, indicating gel-like character now, since G' > G".

In this case, on the one hand an increase in viscosity in the high-shear interval can be observed. But on the other hand there is no decrease of the values of G' and G" in the subsequently following low-shear interval. Therefore, **this is not rheopectic behavior since here, a permanently remaining shear-induced structural change has taken place**. Of course when testing these kinds of materials, loss of solvent or drying effects must be excluded. More about corresponding results obtained with rotational tests: see Note of Chapter 3.4.2.1c; and to shear-induced increase of the structural strength: see Figure 9.24 of Chapter 9.2.2.

d) Measuring programs as a combination of step tests in oscillation and other test types
Sometimes in industrial practice it is useful to combine measuring programs in which one or several test intervals are carried out in the form of steps.

Example: Combination of an amplitude sweep and a step test for testing adhesives
The following is required when applying uncured construction adhesives in the automotive industry: desired flow behavior, reduced stringiness (target: "short"), stability of the adhesive bead, for example, when using an automated pump and application system. The following method is recommended for evaluating the behavior of highly filled, pasty, but still uncured adhesive systems, showing a maximum particle size of d = 0.1mm (acc. to DIN 54458, status of 2010).
Measurement: Using an air bearing rheometer, a parallel-plate measuring system (with d = 25mm) and the shear gap of H = 0.5mm, at the measuring temperature T = +23°C.

Explained is here the combination of the measuring methods A (amplitude sweep) and B (step test: first at high shear, then at low shear conditions).
Preset: For all test intervals (T_0 to T_3) at the constant frequency of f = 10Hz (or angular frequency ω = 62.8rad/s, resp.) to simulate shear rates in the range of 10 to 100s^{-1} (see also Equation 8.24: $\dot{\gamma} = \gamma_A \cdot \omega$).
T_0: Waiting time (without shear, interval of rest and temperature equilibration); for t = 5min
T_1: **Amplitude sweep** at γ = 0.01 to 100%
T_2: **High shear interval** maintaining γ = 100% = const (oscillation); the duration depends on the processing procedure to be simulated, e.g. t = 60s (or 120s, or only 4s)
T_3: **Low shear interval** at γ = 0.1% = const (oscillation); duration t = 60s

Analysis (A_1 to A_8), **and evaluation for practical use** (acc. to DIN 54458):
Regarding T_1: **Behavior at rest**
(A_1) G'-value (in Pa) in the LVE range, as the **stability at rest**

(A_2) tanδ-value in the LVE range, as the **tendency to stringiness at rest**
(A_3) τ-value (in Pa) at the crossover point G' = G'', as the **flow point at the onset of pumping**

Regarding T_2: **Behavior during the structural decomposition**
(A_4) $\eta^*{}_1$ and $\eta^*{}_2$ (in Pas), at the first and the last measuring point, as the **flow behavior during the application**
(A_5) ($\eta^*{}_2/\eta^*{}_1$), (dimensionless), as the **shear stability when pumping**
(A_6) $\eta^*{}_S = (\eta^*{}_1 + \eta^*{}_2)/2$, as the pumpability; mean value between beginning and end

Regarding T_3: Behavior at rest after the application, **structural regeneration** (analysis after t = 30s)
(A_7) G'_R (in Pa), as the **stability after the regeneration**
(A_8) tanδ$_R$, as the **tendency to stringiness after the regeneration**

Summary: Classification of uncured adhesives
a) **Stability** in terms of the value of G'_R: (1) < 5kPa: **very low**; (2) 5 to 10kPa: **low**; (3) 10 to 25kPa: **medium**; (4) 25 to 50kPa: **fairly high**; (5) 50 to 100kPa: **high**; (6) > 100kPa: **very high**

b) **Stringiness** in terms of the value of tanδ$_R$: 1) < 0.2: **hardly visible**; 2) 0.2 to 0.4: **very short**; 3) 0.4 to 0.8: **short**; 4) 0.8 to 1.5: **medium**; 5) 1.5 to 3: **long**; 6) > 3: **very long**

This is valid for a typical triangular geometry of the adhesive bead showing a height between 12 and 15mm and a base width of 7 to 8mm. Comparable tests: See as well the Note in Chapter 8.4.4a on the stability of uncured PU adhesives by amplitude sweeps and by the "dog tail test", as well as the Example in Chapter 8.3.4.5 on a testing method for plastisols used in automotive industry.

8.5.3 Time-dependent behavior of samples showing hardening

This kind of testing is used to investigate time-dependent behavior of materials showing hardening, chemical curing, or formation of a physical-chemical network during the measurement. Here, both the shear conditions and the measuring temperature are kept at constant values, i.e., testing takes place at isothermal conditions. These tests are performed at constant dynamic-mechanical low-shear conditions in the LVE range, in order not to influence reaction kinetics and to ensure an undisturbed cross-linking process .

a) Onset of curing, and development of a chemical cross-linking reaction
The **time point** t_{CR} **at the onset** of a curing process or of a chemical cross-linking reaction can be observed more or less clearly in a corresponding test diagram, since from this point on, the curves of G', G'' and |η*| are sloping up with time (see t_{CR} in Figure 8.35). Finally, the curves are asymptotically reaching constant values showing G' > G''.

Note: The onset of gelation according to ASTM D4473

"Onset of gelation" is the time point at the intersection of the following two tangents: the first one through the point of the minimum viscosity, and the second

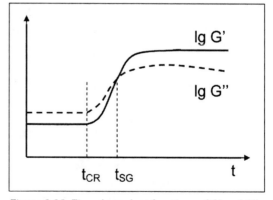

Figure 8.35: Time-dependent functions of G' and G'' of a curing polymer, showing the time point t_{CR} at the onset of the chemical reaction, and the time point t_{SG} at the sol/gel transition in terms of the crossover point G' = G''

one showing the maximum slope of the viscosity curve during the developing curing reaction, i.e., this one tangent is adapted to the turning point of the viscosity curve. For this purpose, meanwhile more and more users are selecting the G'-curve instead of the $|\eta^*|$-curve.

b) Sol/gel transition point, gel time, and gel point

Initially before the gel formation, samples mostly show the behavior of a liquid with $G'' > G'$, therefore being still in the **sol state**. Afterwards occurs the **gel state** with $G' > G''$ and gel-like or solid behavior. The time point t_{SG} at the intersection of the curves of G' and G'' indicates the sol/gel transition point, or briefly, the "gel point" (see Figure 8.35). At this time point **G' = G''** or $\tan\delta = G''/G' = 1$, respectively. Example: gel formation of gelatin solutions [251].

Note 1: Pot life, open time and gelation of resins

For resins in open containers at the application temperature, there are many terms used to evaluate the period of time in which they are still usable. The curing process may occur due to oxidation or air humidity. Typical oscillatory tests result in the time-dependent functions of G'(t) and G''(t) under constant dynamic-mechanical conditions in the LVE range, for example, at $\gamma = 1\%$ and $\omega = 10$ rad/s. Possible criteria for an evaluation are:

1) **Pot life** is the period of time in which the material is still showing no recognizable change in the values of G' and G''or $\tan\delta$ (= G''/G').

2) **Open time** is the period of time in which the material is still able to flow, thus, as long as $G'' > G'$ or $\tan\delta > 1$, respectively. This time is over when reaching $\tan\delta = 1$; and the corresponding time point is often called the **gelation time**. Application example: wood adhesive [395].

Note 2: Gel time of neat thermosetting resins and gel point of composites, acc. to ASTM D4473

For the cure behavior of reinforced thermosetting resins at rest, the "gel time" is reached at the time point when $\tan\delta$ is at the maximum, and G' levels out after a strong increase. For uncured, reinforced composites such as filled resins, prepregs or impregnated materials, the "gel point" is the time point where $G' = G''$. Starting an isothermal curing experiment at a temperature at which the uncured resin is still in a liquid state, the sample may form a gel structure which is build up by already branched molecules. This development is connected with a strong increase in viscosity, before finally transferring to the "glassy state" of a "glassy solid". In such a case, **two peaks in the tanδ-curve** may be observed (in ASTM D4473 called the "damping curve"). Then, the first peak should be associated with **gelation** and the second one with **vitrification**. If only a single peak occurs, it can be interpreted as either gelation or vitrification.

Note 3: Dynamic gel point of neat resins, according to ASTM D4473

The "dynamic gel point" of neat, unmodified resins is reached – after initial melting, flow at the viscosity minimum, and the subsequent onset of curing – at the time point when the complex viscosity has increased to $|\eta^*| = 100$ Pas.

Note 4: Sol/gel process in glass chemistry and ceramics technology

The chemical reaction of silanes to form a three-dimensional network is referred to as the sol/gel reaction in silane chemistry [49, 236, 335]. This reaction usually occurs in two steps via hydrolysis and subsequent condensation of **silanes** to **siloxanes**. After the first reaction step, a **silicate sol** occurs showing liquid character (siliceous sol, SiO_2 sol). Finally, there is a **gel structure forming as a three-dimensional network consisting of this inorganic glass-like material**. For example, the following chemical structural unit can be used to build up network bridges. It is shown here in a simplified form:

[-Si-O-Si-O-].

The synthesis can be controlled via several process parameters. Besides silicates (SiO_2), also other basic materials from ceramics can be used for the sol/gel process to form such **net-**

works, e.g. titanate (TiO_2), zirconate (ZrO_2) and aluminate (Al_2O_3). **Application examples: xerogels** ("dry gels" without any liquid, e.g. after sucking out, evaporating or pressing out the liquid) and **aerogels** ("air gels" with air pores) as **lightweight materials**, for **very fine filters** and membranes, or for catalysts.

One **drawback** of these inorganic sol/gel materials is their **brittleness**, which makes them too inflexible for many applications. From a rheological point of view they are very stiff and rigid, i.e. they can barely be deformed, and correspondingly, they are unstable under load. There is therefore great interest in silane-based inorganic-organic composites; see also Note 2 of Chapter 8.6.3b.

c) The minimum viscosity

Sometimes, the curve of time-dependent complex viscosity $|\eta^*|$ displays a minimum value η^*_{min}. The following is important for the practical use of coatings: At this point a wet coating layer may show optimum flow, spreading and leveling behavior. However, if the value of η^*_{min} is too low, a wet layer may be too thin finally, and it may show edge creep resulting in an insufficient degree of edge protection finally. On the other hand, if η^*_{min} is too high, the coating may not level out smoothly enough and de-aeration may be a problem, resulting in so-called "pinholes", "craters" or air bubbles finally. All these effects may lead to not sufficient surface gloss finally. See also Note 1 of Chapter 8.4.2.1b: Viscosity specification is not useful if $G' > G''$.

d) Further Notes and Examples
Example 1: Testing an epoxy resin
1) "Gel time" (acc. to ASTM D4473, when reaching the maximum of $\tan\delta$): after t = 90s (e.g. showing $\tan\delta_{max}$ = 10 then)
2) Time point at the viscosity minimum: after t = 90s (e.g. showing $|\eta^*_{min}|$ = 60Pas then)
3) "Dynamic gel point" (acc. to ASTM D4473, when reaching $|\eta^*|$ = 100Pas): after t = 200s
4) Gel point (when reaching $G' = G''$): after t = 210s (e.g. showing $G' = G''$ = 750Pa then)
5) Final values of G' and G'' after the curing reaction is finished: G' = 1GPa and G'' = 0.1GPa

Comment: In order to evaluate time-dependent curing behavior, most users select the following two time points only: The first one at the viscosity minimum (if it occurs), and the second one at the gel point when $G' = G''$. And frequently, determined are also the final values of G' and G'' when the curing process is finished.

Example 2: Testing a solid coating layer after curing
Preset: ω = 10rad/s = const and γ = 1% = const
Analysis after a previously defined time point (e.g. t = 15min) using the following specifications:
1) $G' \geq$ 1kPa, otherwise the solid coating layer is evaluated to be too flexible or too "soft"
2) $G' \leq$ 5kPa, otherwise the coating layer is regarded to be too inflexible or too "rigid"

Note 1: Frequency recommendation for testing curing resins (according to ASTM D4473) When evaluating curing behavior of thermosetting resins, according to ASTM D4473 it is recommended to set the frequency at f < 1.5Hz (or ω < 9.4rad/s, respectively).

Comment: Experience shows, also ω = 10rad/s (or f = 1.6Hz, resp.) is a well approved preset.

Note 2: The cure time
Some users determine the "cure time" at the intersection of the following two tangents: the first one on the high, constant plateau value obtained after curing, and the second one showing the maximum slope of the viscosity curve during the developing curing reaction, i.e., this one tangent is adapted to the turning point of the viscosity curve. For this purpose, meanwhile more and more users are selecting the G'-curve instead of the $|\eta^*|$-curve.

Note 3: Dual cure adhesives and UV radiation
Dual cure adhesives are coming increasingly in use, e.g. for bookbinding. The degree of cross-linking of hot-melt adhesives (e.g. polyurethane, PU) can be controlled by the following

two-stage reaction: Using **UV radiation**, the **first step** takes place within seconds after the application leading merely to a sufficient **initial cohesion**, e.g. to endure the rounding process of the spine. This may occur as an "onset of polymerization" combined with a limited degree of molar mass increase, or as a cross-linking reaction resulting initially in a network showing a low degree of cross-linking only. **Criterion:** Here, **the gel point with G' > G"** **should be just exceeded** (or tanδ < 1, resp.).

The second step takes place, for example, **due to air humidity** in the form of a continuing **chemical cross-linking reaction**, increasingly stabilizing the adhesive film. Then, the values of G' and G" are increasing continuously and the tanδ-value is decreasing, until the final degree of cross-linking is reached [54, 247].

Note 4: Advantages of oscillatory tests over rotational tests

1) When performing rotational tests on hardening materials, only a single curve function is obtained, usually showing the time-dependent shear viscosity $\eta(t)$, see Chapter 3.4.3 and Figure 3.48 (however here, on the x-axis should be imagined time t instead of the temperature T). With oscillatory tests, which are based on two independent raw data for each measuring point, resulting for example in the two functions of G'(t) and G"(t), however, correspondingly additional information is achieved about the structural character.

A sample exhibits liquid behavior as long as G" > G', and therefore, it can be deformed easily. Afterwards, gel-like behavior occurs with G' > G" indicating stability of the sample. **The greatest advantage of oscillatory tests compared to rotational tests is that measurements can be still continued without a significant structural destruction even if the sample's character has changed already to a gel-like state or even to the state of a cured and rigid solid.** Precondition is also here: Testing takes place in the LVE range which may be of course a very limited region for cured and therefore very inflexible materials.

2) As a further advantage counts: **In order to perform non-destructive tests**, there is **a better control** to keep the test conditions within the permitted deformation limits compared to rotational tests, since **oscillatory testing** is usually carried out **in the LVE range**. With increasing stiffness of the sample it may be necessary to adapt the strain by presetting smaller amplitude, to prevent any significant influence on reaction kinetics or even the fracture of the curing polymer structure during network formation. Therefore it may be useful to determine previously the appropriate strain values of the test material in an already cured state by performing a pre-test in the form of an amplitude sweep if a second sample is available. See also Chapter 10.7: Tests on solid torsion bars, and the Example to Note 3 of Chapter 8.3.3.1: Amplitude sweeps and the occurrence of micro-cracks in solid bars.

Note 5: Multiwave tests

Performing multiwave tests might be the fastest way to measure curing samples; see also Chapter 8.4.6.

Some further, but often very simple test methods to determine whatsoever gel times or gelation times are briefly presented in Chapters 11.2.1d/e/f, 11.2.8e, 11.2.11a/b and 11.2.12a4/5/6; and in Chapter 11.2.13 information can be found about the following terms: incubation time, vulcanization time, scorch time, rise time, cure time of uncured rubbers and elastomers.

8.6   Temperature-dependent behavior at constant dynamic mechanical conditions

Using this kind of oscillatory testing, both the frequency and the amplitude are kept at a constant value in each test interval, the only variable parameter is the temperature. Therefore, besides the constant dynamic-mechanical shear conditions a profile of the measuring temperature is preset. This kind of testing is also referred to as **dynamic-mechanical thermo-analysis or DMTA test**; or sometimes to as "dynamic thermo-mechanical analysis" DTMA (see also ISO 6721 and ASTM D4065).

Note: Temperature units
It is recommended to use SI-units (international standard units), and therefore, for the temperature should be used the unit K (Kelvin). Here however - as it is used in most industrial laboratories - temperature values are specified in °C (degrees Celsius, sometimes also called "centigrades"). Only for temperature differences ΔT is chosen the unit K, since for the latter it makes no difference if the values are given in K or in °C. For the conversion see also Chapter 14.3p.

Equation 3.7 $T [K] = T [°C] + 273.15$

Example 1: The temperature value of T = 20°C corresponds to T = 293K approximately.
Example 2: For the temperature difference $\Delta T = T_1 - T_2 = 303K - 283K = 20K$ the same numerical value is achieved like for $\Delta T = 30°C - 10°C = 20K$

Temperature program
There are two procedures in use to preset time-dependent temperature profiles:
1) **Ramps**, showing a linear heating or cooling rate in the form of a **ramp upwards or downwards**. Sometimes, this procedure is called a "(dynamic) temperature sweep".
2) **Steps** in the form of several discrete temperature steps, each one showing a constant temperature for a defined period of time. Sometimes, this procedure is called a **"step-and-hold function"**.

Example 1: According to DIN 53445 **for polymers**
Here, the heating or cooling rate of $\Delta T/\Delta t = 1K/min$ is recommended (or 1°C/min, resp.)

Example 2: According to ISO 6721 and ASTM D4065 **for solid materials**
Here, heating rates of $\Delta T/\Delta t = 1$ to 2K/min are specified, and alternatively, steps of $\Delta T = 2$ to 5K in time intervals of $\Delta t = 3$ to 5min.

Example 3: According to ASTM D4473 **for polymer melts and thermosetting resins** at high temperatures, heating rates of $\Delta T/\Delta t = 2$ to 5K/min are specified, and 0.5K/min as the minimum heating rate.

For practical use, the **heating or cooling rate** of $\Delta T/\Delta t = 1K/min$ (and as a maximum 2K/min) have been found suitable in most cases [398]. The procedure to control the temperature should be slow enough to ensure temperature equilibration throughout the whole sample including the measuring system. In any case, it must be guaranteed that the sample really shows the required and reported measuring temperature without a significant deviation and time lag. If temperature is preset in discrete steps instead of linear ramps, comparatively lower heating rates should be selected.

8.6.1 Description of the test

Preset
1) For test with controlled shear strain: $\gamma(t) = \gamma_A \cdot \sin\omega t$, with both γ_A = const and ω = const (see Figure 8.26)
2) For test with controlled shear stress: $\tau(t) = \tau_A \cdot \sin\omega t$, with both τ_A = const and ω = const
Here, many users select $\omega = 10rad/s$ = const (see also Note 1 of Chapter 8.3.1). Measurements performed at higher frequencies may influence the test results (see also Chapter 8.3.5).

Measuring result
Usually, the measuring diagram is presented with the temperature T on the x-axis on a linear scale, and G' and G" are shown on the y-axis both usually on the same logarithmic scale (as in

Figure 8.37). Sometimes, also tanδ is displayed on a second y-axis, and there are some users who like to show additionally |η*| on a third y-axis, usually also on a logarithmic scale. All of these presentations are therefore semi-logarithmic lin/log diagrams.

8.6.2 Temperature-dependent behavior of samples showing no hardening

This kind of testing is used to investigate temperature-dependent behavior of samples under the precondition to show **no chemical modification** during the measurement. For practical users, the focus is on softening and melting behavior, and scientists are trying to get additional information on the type of the structure of the investigated material. The aim of these tests is to determine the influence of temperature on physical properties like phase changes or other structural modifications occurring at constant dynamic-mechanical low-shear conditions in the LVE range.

Note: Test preparation – thermal-stability time of polymer melts (acc. to ISO 6721-10)
Before testing polymer melts, it is recommended to check the thermal-stability time at the desired measuring temperatures, for example, to avoid degradation of the macromolecules (for detailed information see also the Note in Chapter 8.5.2). It is useful to perform this test as well at the minimum as well as at the maximum measuring temperature, and if desired, somewhere in between to get a useful idea of an appropriate test period which is related to practice.

Figure 8.36: Configuration of macromolecules of an amorphous polymer

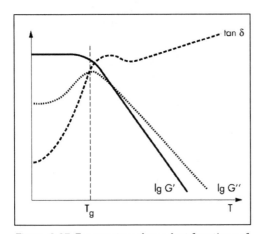

Figure 8.37: Temperature-dependent functions of G', G'' and tanδ of an amorphous polymer

8.6.2.1 Temperature curves and structures of polymers

Polymers can be divided into three groups corresponding to the configuration of the macromolecules: amorphous, partially crystalline and cross-linked. For more information on temperature-dependent behavior of polymers, see also e.g. [114, 233, 249, 268, 321, 379, 398].

a) Amorphous polymers
Here, **the molecule chains are chemically unlinked showing no homogeneous superstructure**, comparable to a structure of "felt-like fibers" (see Figure 8.36). The Greek term "a-morph" means "shapeless", "disordered", "inhomogeneous".

Examples
Polyvinylchloride (PVC-U, "unplasticized", "hard"), polystyrene (PS), polycarbonate (PC), polymethylmethacrylate (PMMA), polydimethyl-siloxane (PDMS, unlinked silicone)

1) **Temperature curves of amorphous polymers** (see Figure 8.37)
At very low temperatures, the polymer occurs in a "glassy state" (see also the Note to "glassy state" in Chapter 8.4.2.2a). Here, the macromolecules are almost immobile since they are "frozen". Since here G' > G'', the polymer shows the **consistency of a rigid and often brittle solid.**

With increasing temperature, from a technical point of view it is the **glass transition range** which is particularly important. This transition is mostly specified in terms of just one temperature value only which is referred to as the **glass transition temperature T_g**. The value of T_g is assumed to be approximately in the middle of the temperature range in which the glass transition is taking place (according to ASTM D4092). In this range the molecules are showing more and more mobility, and the polymer occurs in a **soft-elastic, rubber-like state**; like dough, for example. Important for practical users: The softening process takes place gradually only, covering a **wider temperature range** if the polymer shows a **wider molar mass distribution** (MMD).

With further increasing temperature, the rheological behavior is changing fundamentally, showing G" > G' finally. Therefore then, the molecules are able to move along one another which results in an increasing number of disentanglements. From now the polymer is melting more and more, displaying correspondingly the behavior of a **flowing polymer melt**, thus, it shows the character of a **viscoelastic liquid**.

When the previously rigid structure is becoming more and more flexible due to the temperature increase, there is an initial **rise of the G"-curve in the "glassy range"**, resulting from the increased amount of lost deformation energy which is used up by the sample. For the previously under-cooled and rigid structure it was impossible to perform any internal relative motion, or this was possible in a very limited degree only. With increasing temperature however, the structure is becoming weaker and weaker. As a consequence, more and more **motion occurs between the molecules** which again cause an increase in the amount of the **frictional forces**. Therefore, more and more **frictional heat** is produced, and afterwards, this kind of **energy is lost** for the sample, and it is lost **in the form of thermal energy**. This process can be observed in the diagram as an increasing G"-value (see also Chapter 8.2.4a: lost energy and loss modulus G").

Subsequently, when the first molecules are beginning to move freely, becoming increasingly flexible and mobile, the internal frictional forces are decreasing at first slightly and then more and more pronounced. This process can be observed as a decreasing G"-value then. Therefore, the **maximum of the G"-curve** indicates a temperature point at which the previously existing solidity of the superstructure is yielding more and more, breaking increasingly into smaller parts, turning into a state of flexibility.

Examples: Glass transition temperatures of amorphous polymers (approximate specs)
PVC-U (rigid): T_g = 80°C, PS: T_g = 106°C, PC: T_g = 150°C, PMMA: T_g = 105°C, PDMS: T_g = -125°C

2) Optional analysis methods to determine the glass transition temperature T_g
Several analysis methods are existing since there is no general agreement about the determination of T_g in the glass transition range [114, 407] (see also: ISO 11357-2, ASTM D3418, ASTM E1640, DIN 29971 and DIN 65583). Nowadays due to facilitated analysis by software programs, most users select the methods M1 or M2 as stated below, while previously above all methods M3 to M5 were taken, requiring a ruler only, of course leading to results which were obtained on a less profound scientific basis then. For all methods holds that G' and G" are presented on a logarithmic scale, and T on a linear scale.

M1) Maximum of the G"-curve, also called the G"-peak
The peak of the G"-curve indicates the point of the **maximum energy consumption**, and therefore, the maximum of the lost deformation energy. Many material scientists and designers prefer this analysis method to determine this first significant change in a sample's structure when heating, since the maximum of the G"-curve is always occurring at a lower temperature compared to the maximum of the tanδ-curve (see also ASTM D4065, D4092 and D5279).

M2) Maximum of the tanδ-curve, also called the tanδ-peak

This peak of the tanδ-curve indicates the maximum of the relation $G''/G' = \tan\delta$, and therefore, the maximum of the ratio value of the lost and the stored deformation energy. In other words: At the corresponding temperature, the **maximum relative energy consumption** takes place, and this relative value is related to the value of the stored energy. The tanδ-peak often occurs as a local peak or as an intermediate value of the curve in the glass transition range. Frequently for amorphous materials, a peak is not clearly visible, and then, the curve merely may show a decreasing slope of the rising curve or something like a plateau.

M3) Inflection point of the G'-curve

Here, T_g is determined at the inflection point of the G'-curve which is the point showing the highest slope value, and therefore, the largest change in the G'-values.

M4) Extrapolated initial T_g or T_{eig} or end-point of the freezing process

When cooling, the end point of the "freezing process" is indicated at the intersection of the following two tangents: the first one showing the slope at the inflection point of the G'-curve and the second one on the plateau value in the low-temperature range ("glassy state"). Important for designers: When heating, this is the temperature point of the first significant change of the material's rigidity, and therefore, the beginning of softening.

M5) Extrapolated final T_g or T_{efg} or beginning of the freezing process

When cooling, the onset of the "freezing process" is indicated at the intersection of the following two tangents: the first one showing the slope at the inflection point of the G'-curve and the second one representing the slight curve slope in the high-temperature range.

Notes on the determination of T_g

The T_g-value can vary considerably, depending on the details of the experimental technique used. Therefore, the observed T_g should only be considered as an estimate (according to ASTM D4092). Some explanations to this statement:

1) **The glass transition range is not a sharp temperature point**. Many polymers show softening in a **rather wide temperature range** which easily can cover $\Delta T = 30$ to $60K$

2) The T_g-value depends on the heating or cooling rate. Usually, a higher T_g-value occurs when presetting a higher heating rate and lower T_g-value results from a higher cooling rate.

3) The T_g-value depends on the applied frequency or strain rate (shear rate), respectively. Often, higher T_g-value and wider glass transition range occur, if higher frequencies are preset (see also Chapter 8.3.5: frequency-dependence).

As an extreme **Example**: Measurements at different frequencies resulted in different T_g-values of a sheeting which was produced from an emulsion coating. On the one hand when using an oscillatory rheometer, by a DMTA test at f = 1Hz, T_g = +5°C was obtained. On the other hand when using an ultrasonic rheometer at f = 5MHz, T_g = +35°C was achieved. Therefore here, a shift of $\Delta T = 30K$ towards higher temperature values was determined when applying a clearly higher frequency [270].

Note: Thermoplastic elastomers (TPE)

TPE consisting of amorphous polymers, even in the molten state might show a certain gel-like structure which is build up by chemical-physical interactions: see also the Note of Chapter 8.6.2.1c.

b) Partially crystalline polymers

Here, the mostly linear **molecule chains are chemically unlinked** showing a **partially homogeneous superstructure**. Therefore, for complete macromolecules or for parts of them there is the possibility to form tightly packed areas, so-called **crystalline zones** (see Figure 8.38). Since there are shorter distances between the molecules in these zones, there may occur an increased number of interaction forces in the form of physical secondary bonds. As a consequence, these regions usually exhibit as well more mechanical rigidity as well as higher heat stability. In between the crystalline zones there are amorphous regions, i.e. less densely packed areas. The more rigid zones are embedded in the softer regions if the polymer shows a lower degree of crystallization. Partially crystalline polymers are often also called **semi-crystalline polymers**.

Examples
Polyethylene (PE): PE-LD (low density) with 40 to 50% crystallinity (density ρ = 0.915 to 0.935g/cm³), PE-LLD (linear low density; ρ = 0.917 to 0.939g/cm³), PE-HD (high density) with 60 to 80% crystallinity (ρ = 0.942 to 0.965g/cm³); polypropylene (PP), polytetrafluorethylene (PTFE), polyamide (PA) with up to 60% crystallinity, polyethylenterephthalate (PET), polyetheretherketone (PEEK), amylose (which is a component of starch), many proteins (e.g. gelatin), DNA (desoxyribo-nucleic acid)

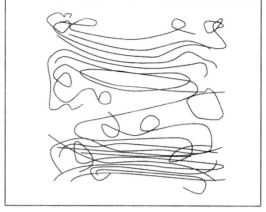

Figure 8.38: Configuration of macromolecules of a partially crystalline polymer

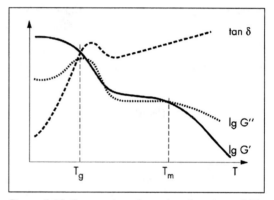

Figure 8.39: Temperature-dependent functions of G', G'' and tanδ of a partially crystalline polymer

Temperature curves of partially crystalline polymers (see Figure 8.39)
Below the glass transition temperature T_g, the molecules are almost immobile since they are "frozen", and therefore, in a "glassy state", and this counts as well for the amorphous regions as well as for the crystalline zones. Since here G' > G'', the polymer shows the **consistency of a rigid and often brittle solid**.

With increasing temperature, in the range **between T_g and the melting temperature T_m** more and more **macromolecules of the amorphous region are melting**, therefore showing increasingly mobility, and as a consequence, the polymer exhibits a softer consistency now. However, the still strong interaction forces between the molecules in the **crystalline zones** prevent the occurrence of any significant relative motion between the molecules in this area. Therefore in this temperature range, above all the amorphous regions are taking up the deformation energy. As an **illustrative example** the **currant cake**: The more solid currants correspond to the crystalline zones, and the more flexible **dough** to the amorphous regions.

For partially crystalline polymers showing a higher degree of crystallization, comparatively higher G'-values are obtained between T_g and T_m. Sometimes there is an "intermediate plateau value". In this temperature range, the so-called **"rubber-elastic region"**, the polymer displays the **behavior of an un-vulcanized or pre-vulcanized soft rubber**. G' and G'' are showing here about the same value, or tanδ = 1 approximately, since there is no homogeneous structure available throughout the entire sample (or shear gap, respectively).

When **reaching the melting temperature T_m**, the crystalline zones are melting too. From now, also the molecules of the crystalline zones are able to glide along one another. Finally

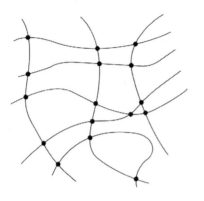

Figure 8.40: Configuration of macromolecules of a cross-linked polymer showing a chemical network

when G" > G', the **polymer is in a molten state** exhibiting the **behavior of a viscoelastic liquid**. However, the mostly spherical (spherolithic) crystallites do not "disappear" all at once but moderately in a certain period of time and temperature range. For this reason when heating, the curves of G' and G" do not fall steeply but gradually and sometimes even in steps if there are areas of the polymer structure which are softening or melting at different temperatures.

Clear curve steps can be observed particularly with polymer blends of which the different components cannot be mixed homogeneously like in solutions. Therefore here, the individual components are forming separated areas, so-called "domains". More or less clearly devided curve intervals are occurring then which represent the behavior of the individual components.

Examples: Glass-transition temperatures T_g and melting temperatures T_m of partially crystalline polymers (approximate specifications)

PE-LD:	$T_g = -125°C$ and	$T_m = +105$ to $118°C$,
PE-LLD:	$T_g = -125°C$ and	$T_m = +122$ to $126°C$
PE-HD:	$T_g = -125°C$ and	$T_m = +126$ to $135°C$,
PP:	$T_g = -3°C$ and	$T_m = +162$ to $168°C$ (ZN i-PP $161°C$),
PTFE:	$T_{g1} = -110°C$, $\quad T_{g2} = +130°C$ and	$T_m = +330°C$,
PA 6:	$T_g = +53°C$ and	$T_m = +220$ to $225°C$
PA 66:	$T_g = +70°C$ and	$T_m = +255$ to $260°C$
PET:	$T_g = +80°C$ and	$T_m = +250$ to $260°C$
PEEK:	$T_g = +141°C$ and	$T_m = +335°C$

The following relation often holds: $T_g/T_m = 0.5$ to 0.7,   with T in Kelvin

Note 1: Polyolefins dominate the market for standard polymers
Data on the distribution of different polymer types: **Polyolefins (PE and PP)** dominate the market for standard polymers, as shown by the statistics from 2006 on the subject of German plastics production, [399]. In many countries this dominance is even greater.
a) PP (20%), PE-LD and PE-LLD (18%), PE-HD (13%); **total 51%**
b) PVC (18%), PS (6%), PET (5%); total 29%
c) Technical plastics: PU (8%), PA (4%), ABS & SAN & ASA (3%); PC, PMMA, PBT, POM, and others (all each 1%); total 20%

Note 2: Thermoplastic elastomers (TPE)
TPE consisting of partially crystalline polymers, even in the molten state might show a certain gel-like structure, build up by chemical-physical interactions: see also the Note of the following Chapter 8.6.2.1c.

c) Cross-linked polymers
Here, **the molecule chains are connected by chemical primary bonds** (see Figure 8.40). Polymers showing a low degree of cross-linking are referred to as elastomers. Polymers showing a high degree of cross-linking, and therefore short network bridges, are termed thermosets (or duroplastics). As an example, for very soft rubbers links may occur between the main chains at an average distance of about 1000 atoms, and for thermosets at about 20 atoms only [114]. Typical elastomers show bridge lengths between 200 and 400 C-atoms [398].

Examples of elastomers

Silicone rubbers (SIR, cross-linked silicones), natural rubbers (NR), styrene/butadiene rubbers (SBR), polyurethanes with a low degree of cross-linking (PU), polyvinylchloride (PVC-P, "plasticized", with 20 to 50% softener amount), chewing gums (e.g. produced from the vegetable juice gutta-percha or using synthetic raw materials such as butadiene copolymers)

Examples of thermosets

Epoxy resins (EP), unsaturated polyesters (UP), phenolic resins (PF), polyurethanes with a high degree of cross-linking (PU), cross-linked polyethylene (PEX)

Information given on elastomers also applies to **gels with a low degree of**

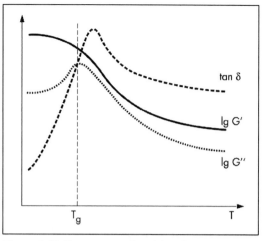

Figure 8.41: Temperature-dependent functions of G', G" and tanδ of a cross-linked polymer

cross-linking, e.g. amylopectin which is a component of starch showing a low degree of esterification. Correspondingly, information given on thermosets also applies to **gels with a high degree of cross-linking**, e.g. amylopectin showing a high degree of esterification.

Temperature curves of cross-linked polymers (see Figure 8.41)

Below the glass transition temperature T_g, the molecule network is "in a frozen state" ("glassy state"), and therefore, the cross-linked macromolecules are stiff and almost immobile. Since here G' > G", the polymer shows the **consistency of a rigid and brittle solid.**

Above T_g, only a limited deformation of the network is possible depending on the length and density of the network bridges. However here in principle, in the whole temperature range remains G' > G", and therefore, the structure cannot reach a liquid state even at higher temperatures. Due to the network consisting of strong chemical bonds, the polymer is not able to melt, therefore exhibiting the character of a **more or less flexible, soft or** still relatively **rigid solid**. See also Chapter 3.3.4.3c: primary bonds and secondary interaction forces.

At T_g, **polymers with a high degree of cross-linking** such as thermosets, often show a hardly visible decrease in the G'-curve; in this case – if at all – only a very small step downstairs may be observed. However, the G"-curve often rises steeply until T_g is reached, indicating a certain increase of consumed and therefore lost deformation energy. This shows a tendency to more and more flexibility and relative motion within the components of the superstructure which of course remains on a limited level. Therefore, more and more internal friction occurs, for example, if network bridges are gliding along other parts of the molecular network or along freely moveable chain ends of molecules. Or it occurs along some of the few still unlinked molecules which might be also present somewhere in between. Nevertheless, there is always clearly remaining G' > G" and this counts also for the temperature range T > T_g.

When reaching T_g, polymers **with a low degree of cross-linking** such as elastomers are showing a clearer step in the G'-curve downwards compared to those ones with a low degree of cross-linking (see Figure 8.42: no. 3a and 3b). Also here, the G"-curve rises until T_g is reached, but always applies: G' > G" or tanδ = G"/G' < 1, respectively. Thermosets exhibit lower tanδ-values compared to elastomers, for which even tanδ ≤ 1 may occur for very flexible samples, therefore here just touching the "gel point". Samples which are completely cross-linked throughout the whole bulk, however, are not able to flow.

Examples: Glass transition temperatures T_g of cross-linked polymers (approximate specs)

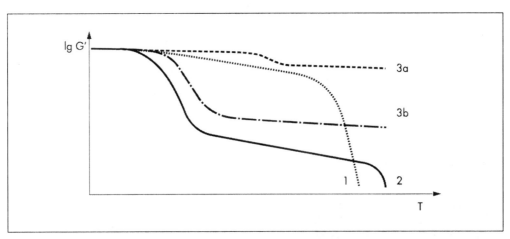

Figure 8.42: Comparison of temperature-dependent G'-curves for different configurations of macromolecules: (1) amorphous, (2) partially crystalline, (3a) with a high degree of cross-linking (thermosets), and (3b) with a low degree of cross-linking (elastomers)

Elastomers (soft)
SIR: T_g = -80°C, NR: T_g = -60°C, SBR: T_g = -52°C, PU: $T_g \geq$ -30°C (with a low degree of cross-linking), PVC-P (soft): T_g = -20 to +50°C (depending on the kind and the amount of the softener)

Thermosets (rigid at room temperature; here, T_g is often not clearly pronounced)
EP: $T_g >$ 45°C, UP: $T_g >$ 60°C, PF: T_g = 75°C, PU: $T_g \leq$ 90°C (with a high degree of cross-linking); PEX: T_g between +60 and 150°C (depending on the degree of cross-linking)

Note 1: Plasticizer content of polymers and the glass transition temperature T_g
Compared to polymers, plasticizers show a considerably lower molar mass. Adding them to polymers, usually leads to a decrease in the T_g value. When adding a large amount of plasticizers, sometimes **two local maxima** are occurring **in the curves of G'' and tanδ**. This may indicate the existence of **two separate phases containing different plasticizer content** [398].

Note 2: Thermoplastic elastomers (TPE)
Thermoplastic elastomers (TPE) are a special case. They belong to the group of **chemically unlinked polymers**, and therefore, they have an amorphous or a partially crystalline structure since their **network** is not build up by chemical primary bonds but by **secondary bonds**. With TPE, the curves of G' and G'' are similar in principle to those of the two unlinked polymer types mentioned. The inner structure may be a polymer blend which consists of a mixture of an unlinked matrix and a linked or as well unlinked inner phase. Besides these structures, there are many other possibilities for diverse structural designs. The bonds between the macromolecules are based on physical-chemical, **thermally reversible interactions** which may be disconnected partially or completely at sufficiently high temperatures $T > T_g$ or $T > T_m$, respectively. They are temporary, and therefore **not permanently effective**, in contrast to chemically linked elastomers. At $T < T_g$ for TPE as for all polymers, $G' > G''$. At $T > T_g$ or $T > T_m$ however, there is $G'' > G'$ and therefore occurs a highly viscous, viscoelastic, dough-like and deformable melt, which can be applied, for example, by injection and deep-draw molding, or by extrusion and extrusion blow molding [268]. (For information on the bond strength of primary and secondary bonds, see Chapter 3.3.4.3c.)

Examples: Glass transition temperatures T_g of TPE (approximate specifications)
TPA (with polyamide; PA6 or PA12, resp.): T_g = +80 or 160/220°C, resp.; **TPC** (with co-polyester): T_g = +160°C; **TPO** (with an olefin, e.g. propylene; PP-NBR): T_g = +160°C, **TBS** (with styrene; SBS): T_g = +95°C; **TPU** (with urethane): T_g = + 130 to 200°C

8.6.2.2 Temperature-curves of dispersions and gels

a) Softening, melting, solidification, crystallization, and freezing temperature

When performing a heating process, most users are interested in the **softening** or **melting temperature**. At low temperatures, stable dispersions and gels are showing G' > G", and therefore, gel-like behavior. The rheological character might be reversed at high temperatures if the material is becoming liquid, showing G" > G' then. Highly viscous materials usually have a higher temperature-dependence compared to low-viscosity fluids.

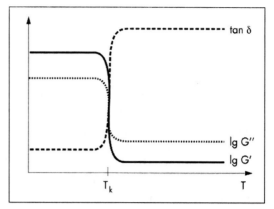

Figure 8.43: Temperature-dependent functions of G', G" and tanδ of a crystallizing material; with the crystallization temperature T_k

When performing a **cooling process** with a liquid, the **solidification temperature** is reached if the sample exhibits a rigid and solid consistency finally, showing G' > G" or tanδ < 1, respectively. Crystallizing samples build up a rigid crystal lattice then. Above the **crystallization temperature T_k**, these kinds of materials are in a molten and therefore liquid state, showing G" > G' or tanδ > 1, respectively (see Figure 8.43). A crystallization process will take place in a comparatively narrow temperature range if the individual components of the sample are dispersed more homogeneously. **For water**, this transition temperature is called the **freezing point**. **Examples:** water, fat of foodstuffs, ice cream, chocolate (i.e. cocoa butter); waxes, paraffins, vaseline, lubricating greases; see also [251, 387, 390].

When heating amorphous, and this means **non-crystallizing materials** such as bitumen or glass, also here, a transition takes place from a rigid, solid and brittle consistency with G' > G" to a liquid state with G" > G'. In this case however, a wider temperature transition range is covered as is the case for crystallizing samples. Also here, the glass transition temperature T_g is reached at the maximum of the G"-curve, and the melting temperature T_m occurs at the intersection of the curves of G' and G" or when tanδ = 1, respectively.

With non-crystallizing samples, the slopes of the curves of G' and G" are comparable to those of amorphous polymers (see Figure 8.37), since when heating, these curves are also decreasing more gradually, if compared to the more steeply falling curves of crystallizing materials. Frequently with amorphous materials at increasing temperature, however, the tanδ-curve is rising continuously, often showing merely something like a plateau in the intermedium range, but without displaying a clearly visible local maximum there.

Note 1: The filler content of polymers and the glass transition temperature T_g
Usually, the T_g value is only slightly influenced by the addition of fillers. This is true, independent of whether the particles are embedded in the polymer matrix via covalent primary bonds or via physical-chemical secondary bonds. However, the **shape of the curves of G" and tanδ, and therefore, the height of their maxima and the width of their peaks are often highly dependent on the filler content**. With a higher particle content or volume fraction of the particles, the curves are usually showing lower maxima and wider bases below the peaks [398].

Note 2: The "glassy state of glass" below the glass transition temperature T_g
There are different concepts to explain the state of cold glass below T_g [66, 398]. See also Chapter 13.3 (1938/39): the long-term experiment of stressing a glass sample by *Rayleigh*.

a) **From a dynamic-mechanical perspective**, the "glass-like" solid structure can be seen as an additional aggregation state besides the solid, liquid and gaseous state, namely as the state of an **amorphous solid**. In contrast, most materials such as stones (e.g. basalte) or metals (e.g. steel) solidify in the form of crystals. The solid state of cold glass is clearly characterized by rheological measurements showing G' > G", e.g. via oscillatory tests or DMTA (dynamic-mechanical thermal analysis). As an **example**, measured in the LVE range: At T = 500°C, G' = 800MPa and G" = 5MPa (here T_g = 610°C, read off at the maximum of G").

b) **From a thermodynamic** or **kinetic perspective**, this structural state can be seen as a very highly viscous, **"super-cooled" fluid**. This liquid is undergoing an extremely slow relaxation process, whose time constant is so high that it cannot be observed within the measuring time. Brownian motion behaves as it were "frozen" under these conditions.

Typically, technically oriented people working in practice and engineers tend to the first sight of view, whereas many chemists and theoretically oriented scientists prefer the second perspective.

Note 3: Phase change materials (PCM)

Example 1 (to Note 3): **Micro-encapsulated paraffin wax** as an encapsulated droplet showing a stable acrylic shell with a particle size of a few micrometers, absorbs a certain amount of heat during a phase transition if it liquefies at e.g. T = +23 or 26°C. As part of **building materials such as plaster or cement**, this functional material is therefore a latent heat reservoir which slows down a further increase in temperature, for example in a living-room. The melting point T_m of the paraffin can be controlled in the range between T = +6 and +60°C via the chain length of the molecules: For example, with 16 (or 18) C-atoms T_m is around +20°C (or +28°C) [21].

Example 2 (to Note 3): A mattress or bed cover showing 3 million microcapsules per cm^2 is designed to provide improved sleeping comfort due to its "dynamic temperature regulation", which adjusts itself to the body temperature of the sleeping person. At the **critical temperature** T_{crit} = +37°C the encapsulated PCM which is positioned on the cover, absorbs heat and **changes from the solid to the liquid state**. **Vice versa**, on the other hand it is able to give off heat when changing from the liquid to the solid state. PCM was originally developed as "functional fabric" for astronauts, as a coating for textiles to ensure "actively self-regulation" of the temperature [275].

Note 4: Pour point, solidification point, and dropping point of petrochemicals

Oscillatory tests at a constant low deformation and a constant frequency is a modern method for determining the solidification point (SP) or the pour point (PP) of petrochemicals **when cooling or** the dropping point **when heating**. Samples are vehicle fuels and heating fuels, lubricants, paraffins and waxes. Many users determine the phase transition from the liquid to the solid state and vice versa using the crossover point G' = G" or when tanδ = 1, respectively. For these kinds of measurements the conditions during sample preparation and temperature control, i.e. the cooling or heating gradient, are playing a crucial role. In addition, **shear-induced crystallization** may occur, and therefore, anisotropic behavior which is dependent on the shear direction (see also Chapter 9.2.2). With oscillatory tests, however, these effects are considerably smaller compared to rotational tests. For further specialist terms and further simple tests for evaluating the cooling behavior of petrochemicals see Note 2 in Chapter 3.5.3.

b) Freeze-thaw-cycle tests for testing temperature stability of emulsions

of testing is carried out to analyze temperature stability of semi-solid dispersions mulsions, and it is used for example as well for cosmetic, pharmaceutical and medical as well as for foodstuffs. Since it is aimed to produce thermo-stable emulsions, no phase n should occur, and therefore, also no significant change in the values of G' and G" or tanδ, respectively, when performing the following temperature program. During the whole test, the oscillatory shear conditions are kept constantly, e.g. at ω = 10rad/s and γ = 0.1%. Also here, as a precondition counts: Testing takes place within the LVE range.

Example 1 [56]: Freeze-thaw-cycle test in the temperature range of T = -10 to +50°C

Preset: Time-dependent temperature profile in the form of $3\frac{1}{2}$ cycles, each cycle consisting of four linear ramp-like intervals.

Ramp 1 (heating) in t = 10min from T = +20 to +50°C
Ramp 2 (cooling) in t = 10min from T = +50 to +20°C
Ramp 3 (cooling) in t = 10min from T = +20 to -10°C
Ramp 4 (heating) in t = 10min from T = -10 to +20°C

Therefore, for each individual cycle a period of t = 40min is required. Each cycle shows a temperature maximum at T = +50°C and a minimum at T = -10°C. Here, the total test time is t = (3.5 · 40min =) 140min, and there are four maxima and three minima. The heating and cooling rate is ΔT/Δt = 30K/10min = 3K/min.

Possible presentation of the results in the form of the following diagrams:
1) G' and G" (on a log scale) versus time t (on a linear scale); usually resulting in G' > G"
2) tanδ versus temperature T (both on a linear scale), usually resulting in tanδ < 1 (e.g. 0.2)

Analysis: An emulsion is stable if the values of G', G" and tanδ remain constantly over the whole period of time or temperature range, respectively. Under low-temperature conditions near to the freezing point, unstable products may show increasing values of G' and G", or even tanδ-curves showing loops ("hysteresis"). At high temperatures, however, unstable samples may show decreasing values of G' and G", and tanδ-curves with pronounced spikes which may occur due to irregular effects when softening. Instability occurs clearly visible if the measuring values vary from cycle to cycle.

Example 2 [36]: Freeze-thaw-cycle tests in the temperature range of T = +5 to +45°C

Preset: 15 cycles with the heating and cooling rate of ΔT/Δt = 5K/min

Calculation of the relative structural change ΔG' [1] in terms of the change in the G'-values compared to the initial value:

Equation 8.28 $\Delta G' = G'_{max,i}/G'_{max,1}$

with $G'_{max,1}$ [Pa] and $G'_{max,i}$ [Pa] which are the values of the maxima of the storage modulus in the first and i-th cycle (here with i = 1 to 15)

Presentation of the results is possible, for example, in the form of the following diagram: On the y-axis the value of ΔG' on a linear scale showing one value per cycle, for example in the range between 1 and 2; and on the x-axis the cycle number plotted in regular distances. In this case, 15 calculated single values would be displayed. Without any structural change, ΔG' would remain constantly on the value 1. With thermally unstable samples, the values of ΔG' are for example increasing with the number of cycles, and the more unstable the sample the more pronounced will be the increase in the ΔG'-values then.

After the emulsions have been produced, they are tested with regards to their temperature stability, for example after the following time intervals: after 48 hours, after one week, after four and eight weeks; for example after storage at T = +40 or +45°C over three or six months. To simulate different climate zones, different ranges for the test temperatures are chosen. For example, two further temperature ranges from industrial practice are: T = -12 to +60°C, and -25 to +20°C.

This method can be used to evaluate long-term stability of emulsions, often even after only 48 hours of testing. However, freeze-thaw-cycle tests cannot completely replace conventional long-term storage stability tests, for example, if behavior of emulsions should be predicted over a period of two years.

8.6.3 Temperature-dependent behavior of samples showing hardening

This kind of testing is used to investigate temperature-dependent behavior of materials showing hardening, chemical curing, or formation of a physical-chemical network during the measurement. Here, the shear conditions are kept at constant values, and therefore, these tests are performed at constant dynamic-mechanical low-shear conditions to ensure an undisturbed cross-linking process.

a) Onset of curing, and development of a chemical cross-linking reaction
Mostly, the **onset temperature** of a chemical cross-linking reaction or of a curing process can be observed clearly in a corresponding test diagram, since here, the curves of G', G'' and $|\eta^*|$ usually are showing a pronounced minimum and those of $\tan\delta$ a maximum, respectively. See T_{CR} in Figure 8.44 at the onset of the chemical reaction.
With increasing temperatures, the curves of G', G'' and $|\eta^*|$ are rising due to the developing curing reaction, showing G' > G'' finally. At even higher temperatures, even a fully cured sample with a completely developed network might show a certain softening effect, and then, G' and G'' might show decreasing values. However, the relation G' > G'' remains if the sample is not overheated, showing thermal destruction in this case.

b) Melting temperature, sol/gel transition temperature, gel temperature and gel point
Mostly, before the gel formation or curing, samples initially show the behavior of a liquid with G'' > G', therefore being still in the **sol state**. This state corresponds for resins to the state, when the **melting temperature T_m** is exceeded. When heating and after passing the curve minima, the sample's structure changes to the **gel-like state** with G' > G'', and therefore, to solid behavior finally. The temperature T_{SG} at the intersection of the curves of G' and G'' indicates **the sol/gel transition temperature** or **"gel temperature"**, or briefly, the **"gel point"** (see Figure 8.44). At this temperature **G' = G''** or $\tan\delta = G''/G' = 1$, respectively.

Note 1: Dynamic gel temperature DGT of neat resins according to ASTM D4473
The "dynamic gel temperature" or the "dynamic gel point" of neat, unmodified resins is reached – after initial heating and melting, flow at the viscosity minimum, and the subsequent onset of curing – at the temperature at which the complex viscosity has increased to $|\eta^*|$ = 100Pas.

Note 2: Inorganic-organic composites, hybrid materials
In the silane chemistry it is possible to combine inorganic and organic structural units with one another (see also Note 4 in Chapter 8.5.3b). For example, the following chemical structural unit can be used to build network bridges. It is presented here in a simplified form: [-C-C-Si-O-Si-C-C-].

Organosilanes as modified silanes showing organic chains, can build up a network in the form of an **inorganic-organic matrix**. Using these kinds of **nanocomposite materials** it is possible to connect polymer components and

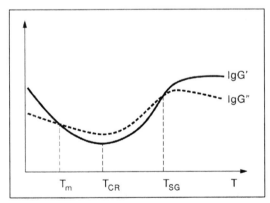

Figure 8.44: Temperature-dependent functions of G' and G'' of a curing polymer, with
- *T_m the melting temperature at the first crossover point G' = G''*
- *T_{CR} the temperature at the onset of the chemical reaction at the minimum of the G'-curve*
- *T_{SG} the sol/gel transition temperature at the second crossover point G' = G''*

mineral nanoparticles (e.g. showing a particle size between 5 and 50nm) or solid ceramic particles (e.g. showing a particle size between 0.2 and 1μm) if the particles are equipped with modified surfaces. These connections are build-up via covalent chemical primary bonds.

This opens up numerous possibilities for the design of materials between inorganic glass-like structures and these mixed composite structures, either with a dominant inorganic or a dominant organic character. Examples: "Ormosil", which are **organically modified silanes** or organo-functional silane systems; "ormocer", which are organically modified ceramic materials; PEX as a cross-linked polyethylene. It is also possible to produce an **organic polymer network within the inorganic network** of the base matrix. Sometimes, these kinds of superstructures are called "interpenetrating **network-in-the-network structures**".

One **advantage** of these hybrid materials over purely inorganic, glass-like ceramic systems is their increased flexibility. From a rheological point of view they are not as stiff and rigid but actually more deformable and less brittle. They can therefore withstand a certain degree of mechanical load without being damaged. The division of tasks among the different individual components can be adapted by material developers to the requirements at hand to achieve a balanced **viscoelastic system**: On the one hand, the **inorganic network** is responsible for the **rigidity and elasticity** since it should be able to store a certain part of the **deformation energy**; this can be evaluated via the G'-values. On the other hand, the **organic polymer network** is responsible for the **deformability and flexibility**, since it should absorb the corresponding part of the **deformation energy and make it harmless without destroying the structure**; this can be analyzed using the values of G'' or tanδ. This latter portion of the energy is mainly converted into heat, it is dissipated. Therefore, this lost energy is used up due to internal friction processes, it is absorbed or lost to the surroundings. Due to the variety of possibilities for developers and designers and the numerous potential fields of applications, the importance of inorganic-organic composites (hybrids) will also increase in the future [49, 236, 335].

c) The minimum viscosity
Sometimes, after a certain temperature increase the curve of the temperature-dependent complex viscosity $|\eta^*|$ shows the minimum value η^*_{min}. Information on the corresponding evaluation of coatings can be found in Chapter 8.5.3c. See also Note 1 of Chapter 8.4.2.1b: Viscosity specification is not useful if G' > G''.

d) Further Notes and Examples
Example: Testing an epoxy resin
1) Temperature at the viscosity minimum: at T = +165°C (e.g. showing $|\eta^*_{min}|$ = 10Pas then)
2) "Dynamic gel temperature" (according to ASTM D4473, when reaching $|\eta^*|$ = 100Pas): at T = +175°C
3) Gel temperature or gel point (when reaching G' = G''): at T = +180°C
(e.g. showing G' = G'' = 750Pa then)
4) Final values of G' and G'' after the curing reaction is finished: G' = 1GPa and G'' = 0.1GPa

Comment: In order to evaluate temperature-dependent curing behavior, most users select the following two time points only: The first one at the viscosity minimum, and the second one at the gel point when G' = G''. And frequently, determined are also the final values of G' and G'' when the curing process is finished.

Notes: Details to the following applications can be found in Chapter 8.5.3d since the explanations given there are also valid for the section at hand.
a) see Note 1: **Frequency recommendation for testing curing resins** (acc. to ASTM 4473)
b) see Note 2: **"Cure time"**

c) see Note 4: **Advantages of oscillatory tests over rotational tests**
Of course here, using in the adapted form, by replacing time t by the temperature T (e.g. t_{SG} by T_{SG}), and also replacing the mentioned Chapter 3.4.3 by Chapter 3.5.3.
d) see Note 5: **Multiwave tests**

    🖎   For "Mr. and Ms. Cleverly"

8.6.4 Thermoanalysis (TA)

Here, a short overview is given on several TA test methods. The terms **"thermal analysis"** or **"thermoanalysis" (TA)** are used when testing temperature-dependent properties of materials, and when determining characteristic material-specific temperatures (such as T_g, T_m, T_k, T_{CR}, T_{SG}); see for example ASTM E473 and DIN 51005. The following tests are TA tests:

a) DSC (Differential scanning calorimetry)
Presetting a defined heating or cooling rate ($\Delta T/\Delta t$) while the sample is mechanically unloaded, typical test diagrams show the mass-specific heat flow rate $\dot{Q}/m$ in [J/s · g] or in [W/g] versus the temperature T. Further results are endothermal and exothermal effects, change in the (mass) specific heat capacity Δc_p [J/g · K], and change in the mass-specific enthalpy $\Delta H/m$ [J/g].
See also ISO 11357, ISO 11409; ASTM D3417 and D3418, ASTM E793 and E794, ASTM E928, ASTM E967 and E968, ASTM E1356; DIN 51007, DIN 53765 and DIN 65467.

b) OIT (Oxidative induction time)
Here, also the DSC method is used. Presetting a defined atmosphere (air or oxygen), measured are thermal effects due to the oxidation of the sample such as thermo-oxidative degradation or destabilization. "Static OIT": Presetting a constant test temperature (isothermal conditions), typical test diagrams display $\dot{Q}$ versus time t. "Dynamic OIT": Presetting a defined heating rate, in typical test diagram is shown $\dot{Q}$ versus the temperature T.
See also ASTM D3895, ASTM D4565, and DIN EN 728.

c) TG (Thermogravimetry)
Presetting a defined heating or cooling rate while the sample is mechanically unloaded, typical test diagrams present the mass m [mg] or the relative mass [%] = [mg/0.1g] versus the temperature T or time t. Further results are the mass change Δm during evaporation, decomposition and chemical reactions; percentage of moisture; concentration of softener, filler and fibers.
See also ISO 11358, ASTM D6382, ASTM E1131, ASTM E1582, ASTM E1868, and DIN 51006.

d) TMA (Thermomechanical analysis)
Presetting a defined heating or cooling rate while the sample is mechanically unloaded or under a constant low compression or tensile load, typical test diagrams display the change in length $\Delta L/L_0$ [μm/m], or the coefficient of thermal expansion α_{TE} [10^{-6}/K] = [μm/m · K] versus the temperature T. Further results are shrinkage due to inhomogeneous structures and post-curing effects.
See also ISO 11359, ASTM E831, ASTM E1363, DIN 51045 and DIN 53752.

e) DMA (Dynamic mechanical analysis)
Sometimes, this type of test is called **DMTA**, dynamic-mechanical thermoanalysis, or **DTMA**, dynamic thermomechanical analysis. The sample is strained under a low mechanical load in the form of a harmonic oscillation. Presetting a constant test temperature, i.e. testing takes place at isothermal conditions (see also Chapter 8.5: DMA) or at a defined heating or cooling rate (see also Chapter 8.6: DMTA).
Testing can be performed as a **tensile test**, as **compression**, as **bending**, as a **linear-shear test** or as a **torsion test**, like it is carried out when using an oscillatory rheometer. Variants

of bending tests are loading on three-points, and the single or dual cantilever beam method. Typical test diagrams are presenting:

1) Tensile or bending tests: the complex tensile elasticity modulus E* (complex Young's modulus), the tensile storage modulus E', the tensile loss modulus E'', and the loss factor tanδ = E''/E' (see also Chapters 10.8. 4.1 and 11.2.14).

2) Compression tests: the corresponding compression moduli (see also Chapter 11.2.15)

3) Linear-shear tests and torsional shear tests: G*, G', G'' and tanδ as explained as well in Chapter 8 at hand, as well as in Chapter 10.7 for solid torsion bars (see also Chapter 11.2.16 to 18). See also ISO 6721, ASTM D4065, ASTM D4092, ASTM D4473, ASTM D5023, ASTM D5024, ASTM D5026, ASTM D5279, ASTM D5418, ASTM E1640, ASTM E1867, DIN 29971, DIN 53513, DIN 53545 and DIN 65583.

8.7 Time/temperature shift

Using the time/temperature shift method, and subsequently, by conversion of the available measuring data, the range of presentation of various rheological parameters can be extended beyond the really measured range, for example, as follows:

a) Based on a number of data measured at different temperatures, the aim is to obtain further values also for other temperatures of which no measuring data are available.

b) Based on a number of data measured in a certain time frame or frequency range, respectively, the aim is to obtain further values also for an enlarged time frame or frequency range of which no measuring data are available.

In this chapter, the **time/temperature shift** is abbreviated to **TTS**. For frequency-dependent parameters, the term **frequency/temperature shift** is also used, but this FTS frequently is also referred to as TTS.

The following relations only apply to **thermo-rheologically simple materials**, i.e. for materials which do not change their structural character in the temperature range considered. Usually, the following holds only **for melts and solutions of unlinked and unfilled polymers**. In order to achieve appropriate results, however, the measuring temperature should show enough distance to the glass transition temperature T_g.

Polymers are not thermo-rheologically simple but are called "thermo-rheologically complex" if they show one of the following structural modifications when heated: formation or destruction of superstructures with participation of interactive forces, gel formation, cross-linking reactions via chemical or physical links such as primary and secondary bonds. Therefore usually, **dispersions** (i.e. suspensions, emulsions, foams), **gels and surfactant systems** are not thermo-rheologically simple, and **it is not useful to apply the TTS method** to these kinds of materials.

The TTS relation is based on the following: velocity or time of shearing, respectively, and temperature have a comparable effect on a sample's rheological behavior. As an example, for typical polymer melts holds: **Heating** results in softening and thus in decreasing G'-values, **and shearing at very slow motion** is leading to **the same effect**. This can be illustrated when performing frequency sweeps at low frequencies, simulating long-term behavior of the sample. On the other hand, **cooling** results in increasing rigidity and thus in increasing G'-values, **and testing at high-speed conditions** is resulting in **the same effect**. This can be observed when performing frequency sweeps at high frequencies, simulating short-term behavior of the sample. Therefore, the following **correlations** can be stated:

1) **High temperatures and low strain rates (shear rates)**, i.e. when performing a limited degree of deformation in a relatively long period of time, both are resulting in increased molecular mobility, leading to more flexibility and therefore to softer behavior.

2) **Low temperatures and high strain rates (shear rates)**, i.e. when applying large deformation in a relatively short time interval, both are resulting in limited molecular mobility, leading to less flexibility and therefore to more rigid behavior.

8.7.1 Temperature shift factor according to the WLF method

Rising temperature results in a shorter average relaxation time λ since the mobility of the molecules is increasing then. The following applies for the **temperature-related shift factor**:

Equation 8.29 $a_T = \lambda(T)/\lambda(T_{ref})$

The shift factor a_T is the ratio of the two relaxation times at the temperature T [K] and at the reference temperature T_{ref} [K]. Correspondingly, for ideally viscous liquids the **shift factor for viscosity values** can be calculated in terms of the following viscosity/temperature ratio:

Equation 8.30 $a_T = \eta(T)/\eta(T_{ref})$

For polymer solutions and melts, this applies for the values of the zero-shear viscosity η_0 only.

a) Horizontal shift factor a_T and the master curve
The TTS relation can be used for all data points of parameters which are dependent on a relaxation time or retardation time, respectively. In other words: Using the **horizontal shift** factor a_T, single data points or even complete function curves can be shifted in a diagram in **parallel to the x-axis** if the frequency or time is presented on this axis. This means that a curve which was measured at a certain temperature T can be shifted, for example, onto another curve which was obtained at the reference temperature T_{ref} that was selected previously by the user.

The so-called **mastercurve** is produced by the following procedure: First of all several frequency sweeps are measured at different temperatures, and each curve is shifted then according to its individual shift factor towards the so-called reference curve. When all these individual shifted curves are superposed, showing afterwards the desired overlapping, as a result, a single curve function only will occur. This function is referred to as the **"temperature-invariant mastercurve with reduced variables"**. Mostly however, it is termed briefly the **"mastercurve"**.

The **frequency mastercurve** is presented with $(\omega \cdot a_T)$ on the x-axis, and for example, with G' and G" (or, if desired, also $|\eta^*|$) on the y-axis, both usually on a logarithmic scale. The following applies for each individual frequency curve to be shifted: $\omega_r = a_T \cdot \omega$
and in the logarithmic form: $\lg \omega_r = \lg a_T + \lg \omega$
with the **reduced angular frequency** ω_r which corresponds point by point to the angular frequency of the frequency sweep which was measured at the reference temperature T_{ref}, and with the ω-values of the curve which was measured at the temperature T. The latter curve will be shifted towards the reference curve then, using the shift factor a_T.
A positive value of the logarithmic shift factor ($\lg a_T$) indicates the shift direction of the frequency curve to the right side in order to meet the mastercurve, and correspondingly, a negative value of ($\lg a_T$) indicates shifting to the left.

Example 1: Shifting the frequency-dependent parameters G', G", G*, and $|\eta^*|$
In order to meet the master curve, initially available values of the complex shear modulus are shifted from $G^*(\omega,T)$ to $G^*(a_T \cdot \omega, T_{ref})$, and the same relation holds for the functions of G', G" and $|\eta^*|$.

Example 2: Shifting shear rate-dependent parameters
The same relation holds also for shear rate-dependent parameters. In this case, on the x-axis is presented the **reduced shear rate** $(a_T \cdot \dot{\gamma})$. Here, initially available viscosity values are shifted from $\eta(\dot{\gamma}, T)$ to $\eta(a_T \cdot \dot{\gamma}, T_{ref})$.

Example 3: Shifting time-dependent parameters

For time-dependent parameters, however, the shift factor is $(1/a_T)$, and therefore, their values are divided by a_T. Here, on the x-axis is presented the **reduced time** $t_r = t/a_T$. For example, initially available values of the relaxation modulus are shifted from $G(t,T)$ to $G(t/a_T,T_{ref})$, and initially available values of the creep compliance from $J(t,T)$ to $J(t/a_T,T_{ref})$ then.

b) Time/temperature shift using the WLF relation

This relation was published by *M.L. Williams, R.F. Landel and J.D. Ferry* in 1955, hence briefly, WLF[389]. For more information on this subject, see e.g. [233, 237, 249].

The WLF shift factor

Equation 8.31 $a_T = \exp([-c_1\,(T - T_{ref})]/[c_2 + (T - T_{ref})])$

with the two material specific coefficients c_1 and c_2, and T in [K]. The complete formula of the shift factor, including the density ρ [kg/m³] reads:

Equation 8.32 $a_T = \dfrac{\eta(T) \cdot T_{ref} \cdot \rho_{ref}}{\eta(T_{ref}) \cdot T \cdot \rho} = \exp\left[\dfrac{-c_1\,(T - T_{ref})}{c_2 + (T - T_{ref})}\right]$

Since the term $(T_{ref} \cdot \rho_{ref})/(T \cdot \rho) \approx 1$, it can be ignored in most cases by practical users.

Note: The relations of WLF and Arrhenius, and the temperature-related shift factor

Information on the temperature shift factor according to Arrhenius can be found in Chapter 3.5.4. However, the Arrhenius relation should only be used for low-viscosity liquids and for polymers in the range of $T > T_g + 100K$ (with the glass transition temperature T_g, see Chapter 8.6.2.1a). As a matter of fact, for polymer data at measuring temperatures closer to T_g better results are achieved when using the WLF relation.

c) Determination of the individual shift factors a_{Ti}
1) Manual determination using the WLF diagram

In the following, determination and calculation of a_T is explained briefly. Before computers were available, the following graphical procedure had to be carried out manually. The WLF diagram is produced by presentation of the following factors: temperature difference $(T - T_{ref})$ on the x-axis, and $[-(T - T_{ref})]/(lg\ a_T)$ on the y-axis. Using frequency sweep data of polymer melts, the following two examples may illustrate some options how to perform this analysis.

Example 1 (first section):
Determination of the shift factors of the available measuring curves of G' and G"

1) Data of three frequency sweeps are available, each one measured at the temperatures T_{M1}, T_{M2} or T_{M3} in the range of $\omega = 0.1$ to 100rad/s (i.e. over three decades; see Table 8.5 and Figure 8.45).
2) $T_{M3} = 483K\ (= 210°C)$ is selected as the reference temperature T_{ref}.
3) The individual values of a_T are determined for each available measuring curve (using the alternative 3a or 3b):
3a) Manual determination: The measuring curves of G' and G" are shifted in the direction of the x-axis (horizontally) one after the other, e.g. using transparent sheets, until each individual curve overlaps a part of the reference curve which is the curve at T_{ref}. In practice, this can be performed best by placing one curve upon another exactly at the individual crossover points G' = G". The shift factor for each individual curve is determined using the ω-values on the x-axis.
3b) Calculation of $a_{Ti} = \omega_{co}(T_{ref})/\omega_{co}(T_{Mi})$. Here, for each individual curve is taken the value of the angular frequency ω_{co} at the crossover point G' = G" (using the index "co" for "crossover point"). The achieved individual shift factors a_{Ti} are listed in Table 8.5. Of course, the a_T factor at T_{ref} is 1 since this curve is not shifted. The mastercurve occurs, after shifting all measuring curves onto the reference curve which results in partial overlapping of all of the individual frequency curves. In this case, the master curve covers five decades now: from $\omega = 0.1$ to 10,000rad/s (see Figure 8.46).

Example 1 (second section):
Determination of the shift factors of the curves of G' and G" at selected temperatures
4) First of all, a table is produced as a basis for the WLF diagram. To do this, the following individual values are manually determined or calculated for each of the three measuring temperatures T_{Mi} (with i = 1 to 3): 4a) $(T_{Mi} - T_{ref})$, 4b) a_{Ti}, 4c) lg a_{Ti}, 4d) $[- (T_{Mi} - T_{ref})]/(lg\ a_{Ti})$
5) Preparation of the **WLF diagram** with $(T - T_{ref})$ on the x-axis and $[- (T - T_{ref})]/(lg\ a_T)$ on the y-axis.
6) The measuring points at T_{M1} and T_{M2} are plotted into the WLF diagram.
7) A straight line is drawn through these measuring points: this is the so-called **WLF regression line**. Usually, this line is fitted optimally to all individual measuring points available; of course, the analysis has a better basis if there is a greater number of measuring points available.

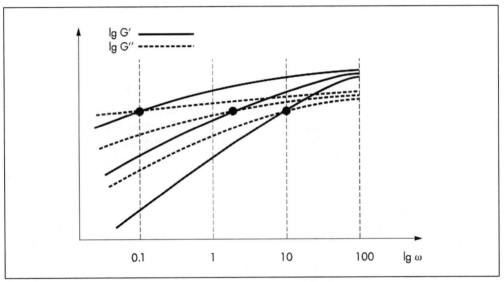

Figure 8.45: Three frequency sweeps of G' and G" covering about three frequency decades, each one measured at a different temperature indicating the crossover point G' = G" at another value of the angular frequency

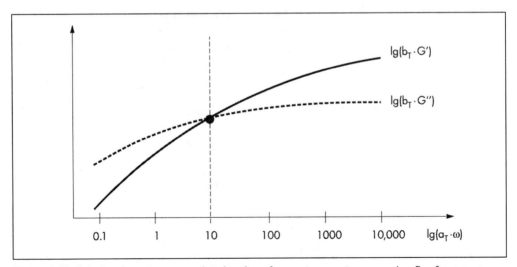

Figure 8.46: Calculated mastercurve, related to the reference temperature, covering five frequency decades now

8) The coefficients c_1 and c_2 of the formula of the WLF shift factor are determined graphically from the WLF diagram as follows:

8a) The slope value c_3 of the regression line is read off (here: $c_3 = 0.427$), then calculating: $c_1 = \ln10/c_3 = 2.30/c_3$ (result: $c_1 = 5.39$).

8b) The y-value of the WLF line at the point $(T - T_{ref}) = 0$ is read off and called c_4 (here: $c_4 = 55.6$), then calculating: $c_2 = [(c_1 \cdot c_4)/\ln10]$, (result: $c_2 = 130$).

9) Using the WLF relation with the determined coefficients c_1 and c_2, the shift factors a_T might be calculated for further temperatures which are selected by the user, e.g. for $T_4 = 473K$ (= 200°C), $T_5 = 493K$ (= 220°C) and $T_6 = 523K$ (= 250°C); see Table 8.5.

As an alternative to this calculation, the shift factors can be determined directly from the WLF diagram. To do this, first are calculated the x-values as $(T_i - T_{ref})$, and then, the corresponding resulting y-values can be read off using the WLF line (here for i = 4 to 6). From this, the a_{Ti}-values are calculated.

10) The results are occurring most illustratively at the crossover points G' = G" in terms of the ω_{co}-values; see Table 8.5.

Example 2 (first section):
Determination of the shift factors of the available measuring data of η_0

1) Data of frequency sweeps in terms of the zero-shear viscosities η_0 are available, measured at the five temperatures T_{Mi} (see Table 8.6; here with i = 1 to 5).

2) $T_{M3} = 483K$ (= 210°C) is selected as T_{ref}.

3) The individual values of a_T are calculated using $a_{Ti} = \eta_0(T_{Mi})/\eta_0(T_{ref})$; see Table 8.6.

Example 2 (second section):
Determination of the shift factors of η_0-curves at selected temperatures

4) The WLF table is produced, similar to Example 1.

5) The WLF diagram is prepared, with the same x- and y-axis like in Example 1.

6) The four measuring points are plotted into the diagram (without the one at T_{ref}).

7) The regression line is fitted, adapted to these four measuring points.

8) The coefficients c_1 and c_2 are determined graphically from the WLF diagram, similar to Example 1.

8a) The slope value c_3 of the regression line is read off (here: $c_3 = 0.31$), then calculating: $c_1 = \ln10/c_3$ (result: $c_1 = 7.43$).

8b) The y-value, called c_4, of the WLF line at the point $(T - T_{ref}) = 0$ is read off (here: $c_4 = 26.5$), then calculating: $c_2 = [(c_1 \cdot c_4)/\ln10]$, (result: $c_2 = 85.5$).

9) Now, using the WLF relation, the individual a_T-factors might be calculated for the other selected temperatures T_i, e.g. for $T_6 = 493K$ (= 220°C) and for $T_7 = 533K$ (= 260°C); see Table 8.6.

10) The corresponding η_0-values at the temperatures T_{Mi} are calculated using the formula shown in step 3; see Table 8.6.

Note: Vertical shift factor
The individual factor b_T for the shift of each individual frequency sweep curve in vertical direction, i.e. in parallel to the y-axis, should only be taken into consideration for very exact

Table 8.5: Temperature-dependent shift factors and crossover points of the individual frequency sweeps (see Chapter 8.7.1, Example 1), measured data under i = 1 to 3, and calculated values under i = 4 to 6

	i = 1	i = 2	i = 3	i = 4	i = 5	i = 6
T_{Mi} or T [K]	423	453	483	473	493	523
T_{Mi} or T [°C]	150	180	210	200	220	250
a_{Ti}	100	5	1	1.57	0.680	0.281
ω_{co} [rad/s]	0.1	2	10	6.37	14.7	35.6

investigations. Usually, the vertical shift factor is very small in comparison to the horizontal shift factor a_T since the former one is based above all on the relatively small temperature-dependence of the sample's density.

2) Automatic determination of a_T and b_T using an analysis software

In order to determine the temperature-invariant mastercurve with (G'/b_T) and (G''/b_T) presented on the y-axis versus $(\omega \cdot a_T)$ on the x-axis, it is easier to use an appropriate software program. However, for the reason mentioned above most users are working without calculation and presentation of the vertical shift factor b_T, and therefore, just G' and G'' are shown on the y-axis.

Of course, when working with an extended frequency spectrum obtained by use of the WLF method and a subsequently performed data shifting into the range of very low or very high frequencies, the options for an enhanced sample characterization will be increased. However, the user should always be realistic when estimating the lower and upper limiting values in order to obtain useful data after all.

Example 3: Acoustic damping behavior of technical rubber

Damping behavior of an elastomer was aimed to characterize in **the frequency range of f = 20Hz to 20kHz** which corresponds to angular frequencies between ω = 125 and 125,000rad/s. These three decades approximately cover the human range of hearing. Frequency sweeps are carried out in the range of ω = 0.05 to 500rad/s, i.e. over four decades, at the three temperatures T = 0/+30/+100°C. Using the measuring data available and **applying the WLF time/ temperature shifting method**, a **master curve** was produced covering now also a range of very high frequencies, for example, over eleven decades from ω = 0.05 to $5 \cdot 10^9$rad/s. This corresponds to frequencies from f = $8 \cdot 10^{-3}$ to $8 \cdot 10^8$Hz = 800MHz [398]. For more on damping behavior, see also Chapter 5.2.2.2b.

d) Inverse mastercurve

Using an appropriate software program, for a mastercurve which was achieved for a certain reference temperature, frequency curves in an extended frequency range might be produced now, related to any other temperature which may be selected by the user. This method is called the "inverse mastercurve".

Example: A mastercurve over seven ω-decades is available using data produced from several individual frequency sweeps, each one measured at a different temperature over four ω-decades. By determining the "inverse mastercurve", a frequency curve over seven decades may be obtained finally for any desired temperature, which also might be one of the measuring temperatures of the initially performed tests. Therefore, a frequency curve might be produced over seven decades now, although the initially measured curves covered four decades only.

e) Data conversion

Often, the mastercurve is produced over a wide frequency range for the sole purpose to use these data afterwards for a conversion, in order to extend the time range outside the measurable limits also for other functions of rheological parameters.

Table 8.6: Temperature-dependent shift factors and zero-shear viscosities of the individual frequency sweeps (see Chapter 8.7.1, Example 2), measured data under i = 1 to 5, and calculated values under i = 6 and 7

	i = 1	i = 2	i = 3	i = 4	i = 5	i = 6	i = 7
T_{Mi} or T [K]	463	473	483	503	523	493	533
T_{Mi} or T [°C]	190	200	210	230	250	220	260
a_{Ti}	10	2.4	1	0.2	0.1	0.46	0.065
η_0 [Pas]	50	12	5	1	0.5	2.30	0.33

Examples: Data from individual frequency sweeps are available at four different measuring temperatures, each sweep performed over four frequency decades. First of all, a mastercurve over seven frequency decades is produced using the TTS method according to the WLF-relation. The $H(\lambda)$-spectrum with relaxation times over seven decades might be determined via data conversion, i.e., covering a clearly wider time range now compared to the one which was measured initially. Further examples of data conversion are presented in Chapters 6.3.5 and 7.3.4; see also Chapter 8.4.7.

8.8   The Cox/Merz relation

In 1958, *W. P. Cox and E. H. Merz* introduced the following relation, which they found empirically for solutions and melts of many unlinked and unfilled polymers [82]:

Equation 8.33 $\eta(\dot{\gamma}) = |\eta^*(\omega)|$

This relation applies if the values of $\dot{\gamma}$ [s^{-1}] and ω in [s^{-1}] or in [rad/s] are equal in size.

This means: The shear rate-dependent viscosity function $\eta(\dot{\gamma})$ which is determined via rotational tests, and the (angular) frequency-dependent function of the complex viscosity $|\eta^*(\omega)|$ which is measured via oscillatory tests, are showing an identical shape of the curve over a wide range when presented in the same diagram (see Figures 8.47, as well as 3.10 and 8.17). In other words, η **and $|\eta^*|$ are displaying the same value** with the same unit Pas at the corresponding shear rate $\dot{\gamma}$ or angular frequency ω, respectively, which again have the same unit: s^{-1}. This applies for measurements performed in the LVE range, therefore showing for both functions the plateau value of the zero-shear viscosity η_0 there [222]. However, for various polymers this holds also beyond the η_0-range, sometimes up to $\dot{\gamma}$ or ω = 100s^{-1}, respectively, and rarely even up to 1000s^{-1} [379].

Scope of the Cox/Merz relation
Meanwhile, many tests have confirmed the usefulness of the Cox/Merz relation for the majority of **unfilled polymer solutions and melts**, however, not for dispersions and gels. Using this relation, time-consuming series of tests may be shortened since the viscosity function no longer has to be measured in both rotational and oscillatory mode but might be combined by parts which are measured either in the rotational or in the oscillatory mode. The relation is useful if only mechanical interactions are responsible for the rheological behavior, such as frictional forces between the macromolecules when deformation, disentanglements and flow occurs (see also Chapter 3.3.2.1). This assumption is true **for Maxwellian fluids**, i.e. liquids showing the slopes of 2:1 of the G'-function and 1:1 of the G''-function in the range of low frequencies if the frequency sweeps are presented on a logarithmic scale (see Chapter 8.4.2.1a and Figure 8.16).

However, any kinds of physical and/or chemical interactions are leading to a certain deviation from the Cox/Merz relation, and therefore: **This relation is not useful for materials showing G' > G''**, thus, **gel-like character** in the low-shear range as it is the case for **stable dispersions (i.e. suspensions, emulsions, foams), pastes, gels**, and of course, **solids**.

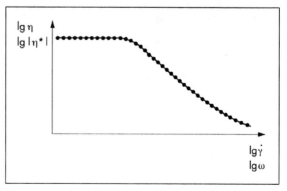

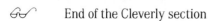   End of the Cleverly section

Figure 8.47: Cox/Merz relation: The same shape of the curve occurs for viscosity functions obtained as well from rotational tests as well as from oscillatory tests

8.9   Combined rotational and oscillatory tests

8.9.1 Presetting rotation and oscillation in series

With most rheometers it is possible to perform rotational and oscillatory tests (R & O) directly one after another in the form of a combined test program. This test method will be explained by the following two examples:

Example 1: Test consisting of three intervals in the form of an O-R-O series

Three measuring intervals are combined in series to examine the structural decomposition and regeneration of a sample in terms of the time-dependent G'-function. This is the same test method as presented in Chapter 8.5.2.2b (with Figures 8.32 to 8.34: Thixotropic behavior, including a numerical example).

1st test interval: Oscillation (at γ = const and ω = const, in the LVE range)

2nd test interval: Rotation (at high-shear conditions, at $\dot{\gamma}$ = const)

3rd test interval: Oscillation (at low-shear conditions, like in the first interval).

Example 2: Test with seven intervals in series to get a fast overview ("fingerprint")

A test consisting of seven intervals is performed to obtain a first and fast overview on the rheological behavior of a dispersion, here for example, in a total test duration of about 10min only.

1) Pre-shearing (rotation): at $\dot{\gamma}$ = $5s^{-1}$ = const, for t = 30s, to produce uniform start conditions

2) Rest phase without shearing, for t = 60s

3) Amplitude sweep (oscillation) to evaluate the viscoelastic character of the sample in the LVE range, to determine the limit of the LVE range related to the deformation, and possibly if G' > G'', to determine the "yield stress" (at the limit of the LVE range) and also the "flow stress" (at the crossover point G' = G'').

4) Rest phase without shearing, for t = 60s

5) Flow and viscosity curve (rotation): ramp from $\dot{\gamma}$ = 0.1 to $1000s^{-1}$, in t = 90s, to analyze the flow behavior and the viscosity function, respectively.

6) Step test (rotation, at high-shear/low-shear conditions):

6a) at $\dot{\gamma}$ = $1000s^{-1}$ = const (or higher), for t = 30s, to evaluate the structural decomposition

6b) at $\dot{\gamma}$ = $0.1s^{-1}$ = const (or lower), for t = 120s, to evaluate the structural regeneration, and the "thixotropic behavior".

Comment: No intermediate strain and no delay when performing tests in series

These combined measuring programs are only useful if the rheometer used is capable to continue the test program without a significant time delay and without changing the angle position of the measuring system in between the individual test intervals. Some rheometers are forced to return first to always the same zero position of the deflection angle before starting the next oscillatory test interval. Of course, if the angle position is changed between the test intervals the sample is always subjected to an undesired intermediate shear process which may have a crucial influence on the test result.

&⌒   For "Mr. and Ms. Cleverly"

8.9.2 Superposition of oscillation and rotation

With some rheometers it is possible to perform oscillatory and rotational motion simultaneously. This test mode is called "superposition of oscillation and rotation" (or "oscillation with superimposed rotation"). The aim of this method is to measure viscoelastic properties using an oscillatory test while the sample is simultaneously stressed at a constant, steady-state shear load.

Example 1: Testing leveling behavior of an emulsion paint by superposition of O & R

Directly after the coating process, a paints showed continuously surface leveling which did not stop in a desired period of time. Indeed after application, a coating is still stressed by its own weight, and therefore, it may not be immediately motionless but may continue to flow very slowly. However, when performing a sole oscillatory test at the shear conditions of the LVE range it is merely the behavior in the state at rest which will be characterized.

Description of the test
1) Pre-test
Determination of the yield stress τ_y as the point at which the LVE range is exceeded, for example, by performing an **amplitude sweep** which is presented in a diagram with τ on the x-axis (see also Chapter 8.3.4 with Figure 8.12: yield point τ_y and flow point τ_f, showing $\tau_y < \tau_f$).

2) Sole oscillation
With the result of the pre-test, the upper limit of the amplitude τ_A is determined since it is useful now to select $\tau_A < \tau_y$. Only in this case, the user can be sure that the stress amplitude is not exceeding the LVE range. The first part of the test is carried out without superimposed rotation, e.g. as a frequency sweep. The aim is here to evaluate firstly the behavior in the LVE range of the still non-flowing sample.

3) Superposition of O & R
In subsequent tests, the τ-value of the rotation is continuously increased in steps and superimposed onto the unchanged τ_A-value of the oscillation. At the beginning, when the sum is still $(\tau_A + \tau) < \tau_y$ and therefore the yield point or the limit of the LVE range, respectively, has not yet been exceeded. As a consequence, the results do not differ from those of the sole oscillatory test. But when $(\tau_A + \tau) > \tau_y$ the difference becomes clear: The larger the proportion of the rotational stress, the more is the degree of structural yielding of the sample. Finally when $(\tau_A + \tau) > \tau_f$, thus exceeding also the flow point, the sample will show the character of a liquid: The structure is breaking more and more and the sample begins to flow. The following example will explain step by step the test method "superposition of O & R".

Practical example (to Example 1): Emulsion paint (see Table 8.7)
1) Determination by an amplitude sweep: yield stress $\tau_y = 2.5$Pa and flow stress $\tau_f = 8.0$Pa.
2) The shear stress amplitude of the subsequently performed frequency sweep is selected as $\tau_A = 1$Pa. Without superimposed rotation, the frequency sweep at $\omega = 0.i$rad/s results in $G' > G''$, showing $G' = 50$Pa. **Viscoelastic character:** Here, the paint displays "gel-like character", since $G' > G''$. **Curve slope:** The G'-curve remains on an almost constant level, even at low ω-values.
3) Superimposing rotation with $\tau = 10$Pa, the frequency sweep at $\omega = 0.1$rad/s results in $G'' > G'$, showing $G' = 10$Pa and $\tan\delta = G''/G' = 2$. **Viscoelastic character:** Now, the paint exhibits the character of a viscoelastic liquid, since $G'' > G'$. **Curve slope:** Towards lower ω-values, the G'-curve slopes down slightly.
4) After increasing the shear stress to $\tau = 20$Pa, the frequency sweep at $\omega = 0.1$rad/s results in $G'' \gg G'$, showing $G' = 5$Pa only, and $\tan\delta = 10$. **Viscoelastic character:** Finally, the paint indicates clearly the character of a flowing liquid, since $G'' \gg G'$. **Curve slope:** Towards lower ω-values, the G'-curve clearly falls downwards now.

Example 2: Drilling fluids at pulsing flow conditions
Situation: A drilling fluid used to **exploit oil springs** shows sedimentation during the drilling process, although it contains a sufficient amount of thickener and gellant. Background: The position of a **drill**

Table 8.7: Testing a coating by superposition of oscillation and rotation: When increasing the values of the superimposed rotational shear stress τ, the structural strength is decreasing, here in terms of G'

τ [Pa] (by super-imposed rotation)	0	10	20
G' [Pa] at $\omega = 0.1$ rad/s	50	10	5
viscoelastic character	$G' > G''$	$G'' > G'$	$G'' \gg G'$

pipe usually is not perfectly in the center of a borehole and it is often **moving continuously back and forth**. This means that the drilling fluid in the non-concentric annular gap is additionally subjected to **oscillations**.

In laboratory tests, a model fluid at rest exhibited no sedimentation and showed an appropriate yield point and also flow point, both as well in rotational tests as well as in oscillatory tests. This lead to the idea to simulate the real process by superposing rotation and oscillation (R & O).
Preset in six steps (S_1 to S_6):
S_1: **Amplitude sweep** to determine the yield point τ_y at the limit of the LVE range and the flow point τ_f at the crossover point G' = G''. Result: τ_y = 2.0Pa and τ_f = 2.5Pa
S_2: **Frequency sweep** in the range of ω = 100 to 0.1rad/s without superposed rotation, presetting the shear stress amplitude τ_A = 1.0Pa < τ_y
S_3 to S_6: **Frequency sweeps** with τ_A = 1.0Pa and superposed rotation, first with τ_3 = 0.5Pa/ then with τ_4 = 1.0Pa/then τ_5 = 1.5Pa/and finally with τ_6 = 2.0Pa.
Determination of the value of the superposed rotational shear stress at which the gel structure collapses, the so-called the **"break stress"**.
Result, read off at ω = 0.1rad/s each time, as "almost at rest":
a) In S_2 without rotation, at (τ_A + 0) = 1Pa occurs G' = 40Pa > G'' and tanδ = 0.2, i.e. gel-like
b) In S_3 at (τ_A + τ_3) = 1.5Pa, occurs G' = 35Pa > G'' and tanδ = 0.25, i.e. it is still gel-like
c) In S_4 at (τ_A + τ_4) = 2.0Pa, there is already G'' > G' and tanδ > 1000. Suddenly the fluid shows ideally viscous behavior, although the yield point of τ_f = 2.5Pa, which was measured by pure oscillation, is not reached yet (see above).
Summary: In this case, compared to a sole ocillatory test it is clear to see: The behavior of the drilling fluid in the bore hole can better be simulated by superposition of R & O.

♊   End of the Cleverly section

9 Complex behavior, surfactant systems

In this chapter are explained **structures in the nano- and micro-range** and their sometimes extraordinary rheological behavior, using surfactant systems as an example. The focus here is on aqueous micellar solutions, colloidal dispersions, emulsions, viscoelastic liquids and gels. The aim is to illustrate simple and complex, natural and synthetic material systems and their superstructures to facilitate practical users to better understand the often unexpected effects of the rheological behavior of these systems.

&ℰ   For "Mr. and Ms. Cleverly"

9.1 Surfactant systems

9.1.1 Surfactant structures and micelles

a) Surfactant molecules

A simple surfactant molecule has the shape of a chain showing a head group and a tail group. There are also surfactants with two or more heads and tails. Examples of this are lecithin, which is the most important emulsifier in the food industry, with two tail groups, and biological surfactants with up to seven tails [220, 372].

Usually the head group is **hydrophilic**, i.e. "loves water" or attracts water, and **lipophobic**, i.e. rejects fat, respectively (Greek: hydor means water, and lipos means fat). Usually these molecules are **polar** and therefore electrostatically charged, they are trying to get contact to the likewise polar H_3O^+ ions of the aqueous phase (hydrogen ions, better: hydronium ions). They can interact with the ions in its surroundings, for example in the form of hydrogen bridges (H-bridges), forming together dipoles.

The tail group, usually a fatty acid chain, is **lipophilic**, i.e. "loves fat" or attracts fat, and **hydrophobic**, i.e. rejects water, respectively. This group is usually **non-polar**, and therefore electrically neutral.

Information about head groups [40, 158, 309]: Depending on their charge and charge distribution, surfactants can be classified as ionic, non-ionic and amphiphilic surfactants. Ionic surfactants again can be divided into anionic and cationic surfactants.

Anionic surfactants have a **negatively charged head** which attracts positively charged ions of the solution. The head is therefore anionic-active, e.g. trying to get contact to H_3O^+ ions or Na^+ ions of an aqueous salt solution. **Cationic surfactants** have a **positively charged head** which is therefore cationic-active trying to get contact to negative ions of the solution, e.g. Cl- ions. The term **amphiphilic surfactant** emphasizes that the molecules show a double character, as they consist of one or more anionic groups and one of more cationic groups. Another term for this is **amphoteric** or ampholytic (Greek: amphis means both, two-sided). The term **gemini surfactants** also means dual ionic, showing two heads which are connected by a spacer of variable length and rigidity. The term **bipolar** is used when two parts exist which carry different electrostatic charges.

Order of magnitude: Typical surfactant molecules have a length between 0.5 and 5nm. **Units for dimensions in the nano- and micro-range** are: 1nm **(nanometer)** = $10^{-3}\mu m$ (micrometer)

Thomas G. Mezger: The Rheology Handbook
© Copyright 2011 by Vincentz Network, Hanover, Germany
ISBN 978-3-86630-864-0

= 10^{-6}mm = 10^{-9}m. According to recent standards, the unit Angström: 1Å = 10^{-10}m = 0.1nm should not be used anymore.

Example: A surfactant molecule with a fatty acid chain of 16 C-atoms is around 2.5nm long. Stretched out as much as possible, the length of a tail group would be 2.5nm approximately, and around 2nm in a non-stretched state. The size of a typical head group is around 0.5nm.

In an aqueous solution at a low surfactant concentration, the surfactant molecules are at first individually and evenly distributed in the liquid which can be called a monomeric solution then, see Figure 9.1 (left). If the surfactant concentration is increased, a **single layer** of uniformly oriented **surfactant molecules** appears on the liquid's surface or at the interface to the surrounding air, respectively. This mono-molecular interface in the shape of a thin film is also termed a **monolayer**. Here, the hydrophobic part of the surfactant is pointing outwards then, see Figure 9.1 (center and right). This is also called a **self-assembling monolayer (SAM)**. However, this surface layer is not rigid, as the molecules are constantly changing their position since they continuously compete with each other for space at the interface. It is therefore a dynamic system even when at rest, above all driven by Brownian motion. **Surfactants as surface-active substances** in aqueous solutions are reducing the relatively high value of the surface tension of water.

Examples: Hair shampoos, clothes conditioners, and washing up liquids. This effect is referred to in advertising as " ... makes the water softer".

Note: Surfactants, and measurement of the rheological behavior at interfaces
Rheological behavior of a so-called **"two-dimensional" film of surfactant molecules on a surface**, e.g. of a liquid surrounded by air, or at an interface between two liquids which do not mix, can be investigated using a special measuring system for **interfacial rheology** (see Chapter 10.8.3.1).

b) Superstructures of surfactants (micelles) in aqueous solutions, dispersions and gels
Due to electrostatic and therefore physical-chemical effects, spontaneous self-organization of the surfactant molecules **leads to a formation of micelles in the form of highly ordered molecule clusters**. Other terms used are: self-emulsifying structures or self-assembling systems. The structural development aims at achieving a balance or equilibrium of the active forces or energies, respectively. The structures formed by the aggregated single molecules are called **association colloids**, aggregates or associates.

Micelles can show a variety of different shapes depending on the following factors:
- Chemical structure of the surfactant molecules, molar mass, number of double bonds
- Shape of the surfactant molecules, size of head and tail groups, ratio of their volume and the effective surfaces
- Physical-chemical, electrostatic interaction with the surrounding, on the one hand by the ionic strength of the polar groups of the surfactants and on the other hand by the ions of

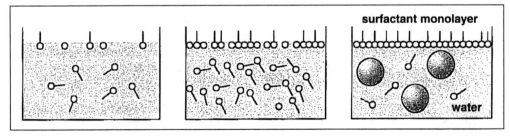

Figure 9.1: Configuration of the molecules in an aqueous solution at an increasing surfactant concentration, right: formation of a monolayer on the surface or at the interface to the surrounding air, respectively [403]

the solution since both together are forming dipoles; network formation via van-der-Waals forces and H-bridges
- Surfactant concentration, volume fraction (volume ratio of the surfactants and the entire solution), spatial obstruction of the micelles
- Salt concentration and pH value of the aqueous solution
- Temperature
- Sample preparation, type and extent of the shear load, time-dependent structural regeneration after a shear process ("thixotropic behavior")

c) Spherical micelles

In aqueous solutions above the **critical micelle concentration (cmc)**, spherical or ellipsoidal micelles might occur spontaneously. The hydrophilic part of the surfactant orients outwards to form the outer boundary of the micelle in the form of a monolayer surface of the spherical shell; see Figure 9.2. The hydrophobic part of the surfactant molecules point to the inside of the sphere then. Many micelles can be used to transport other hydrophobic molecules in their center, which are therefore shielded from the surrounding water by the spherical shell. As an illustrative example you can imagine a balloon with its very thin and flexible rubber skin being sensitive to deformation.

Order of magnitude: Typical **spherical monolayer micelles** have diameters between d = 1 and 10nm, and in extreme cases up to 50nm [259].

d) Rod-like, cylindrical and worm-like micelles

At a higher surfactant concentration, a transition to superstructures with an enlarged surface may take place. This results in a higher ratio of the boundary area to the outside and the volume inside, i.e., there is a larger volume-specific surface. Then, rod-like or cylindrical micelles may occur, see Figure 9.3. "Worm-like micelle chains" may also form; see Figures 9.4 and 9.5. In principle, the shape of these aggregates is comparable to the thread-like macromolecules of unlinked polymers [111, 188, 301].

Each one of these kinds of micelle chains consists of semi-solid components showing the **persistence length** L_p. The term "persistence length" refers to a relatively rigid, inflexible chain link of the micelle. L_p is often between 2 and 20nm, and it is rarely larger than 100nm. The size of L_p depends predominantly on the chemical and physical structure of

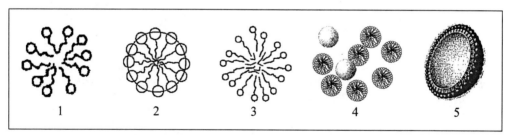

Figure 9.2: Spherical micelles showing a monolayer shell, no. 3 is a micelle which includes an ingredient, and no. 5 is a relatively large example [18, 109, 144]

Figure 9.3: Rod-like and cylindrical micelles [27, 109, 217, 220]

the surfactant molecules, on the interaction forces and on the temperature. The cross-section of these micelles can be circular or ellipsoidal.

As an illustrative example you can imagine a long train with individual wagons. Then, the total length of the train corresponds to the so-called **contour length** of the whole micelle chain, and the length of the individual wagons represents the persistence length of the individual chain links. For worm-like micelles, the contour length is considerably larger than the persistence length.

The external conditions affect the length of the occurring chains, since here, individual surfactant aggregates in single or multiple persistence lengths continuously and very quickly connect and disconnect in a sort of "chain exchange". Depending on the shear load, the resulting dimensions of the micelles are a result of the equilibrium state of the acting forces. The antagonists are the shear force and the structural strength at rest, e.g. as dipole forces. These long micelles often have a slightly curved shape or they are even coiled showing a certain internal mobility. However, compared to the superstructures of the thread-like macromolecules of synthetic polymers, their complex superstructures mostly are considerably less flexible.

Worm-like micelles may show entanglements without building up a solid, permanent and integrated three-dimensional network, see Figure 9.6. In this case, the continuously changing superstructure is comparable to a temporary network which is based on the entanglements of unlinked polymer molecules. This kind of network exists only as long as the surfactant system is forced to carry out relatively fast motion, e.g. when performing oscillatory tests at high frequencies. In both cases however, in the subsequent period of rest there remains no stable and coherent structure, in other words: There exists no permanent gel-like structure. Besides linear worm-like structures, however, also long-branched worm-like micelles may occur, showing transient intermicellar branching points and physical-chemical bonds. In this case, there may be formed a more or less pronounced network-of-forces as a meta-stable superstructure [411].

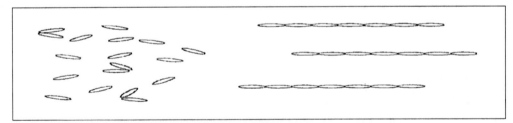

Figure 9.4: Individual small ellipsoidal surfactant aggregates, being micelles in the persistence length, cluster to form long-chained micellar superstructures[109]

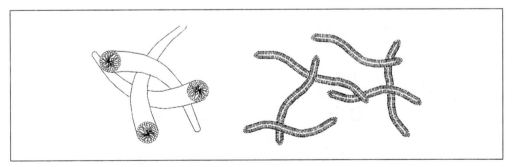

Figure 9.5: Worm-like micelles[18, 109]

Order of magnitude: The typical size of the total length of worm-like micelles is L = 10 to 1000nm. Typical diameters of these kinds of micelles are approximately of the same size as the diameters of spherical micelles which may form at a lower concentration, i.e. between d = 1 and 10nm, and in maximum 50nm. However, even longer worm-like micelles showing a length of several μm and d = 10 to 20nm may occur [411].

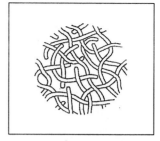

Example: For L = 500nm and d = 5nm the ratio L/d = 100:1. Illustrative comparison: A piece of spaghetti with a diameter of d = 1mm and L/d = 100:1 would therefore be L = 100mm = 10cm long. A further comparison using another material system: For a linear PE polymer molecule with a molar mass of 100kg/mol,

Figure 9.6: Temporary network due to entanglements of unlinked worm-like micelles [220]

with d = 0.5nm and L = 1μm the ratio results in L/d = 2000:1 [119]. This corresponds to a piece of spaghetti which would be 2m long (see also Chapter 3.3.2.1).

This comparison makes it clear that worm-like micelles are typically around ten times thicker compared to the thread-like polymer chain macromolecules, and therefore, they are of course considerably less flexible. However, the flexibility of the worm-like micelles is not so much based on the mobility of the complete micelle but predominantly on the **very fast connection, disconnection and re-connection of the small individual surfactant molecules, molecule groups or micellar clusters** in a single or multiple persistence length. This process is called **fluctuation**.

However, a large difference to polymer molecules which is important from the application point of view is the fact, that polymer molecules remain permanently destroyed after breaking since they cannot regenerate their original structure. Their molar mass remains irreversibly reduced then, due to "degradation". In contrast, **micelle structures** are able to always change and rebuild themselves again and again, and therefore, they may show **reversible behavior**.

e) Planar sheet micelles (lamellae)

At an increased surfactant concentration, superstructures of micelles may occur, showing flat structures in the form of planar sheets. As an illustrative example, you can imagine a thin, flat film or sheeting. Structural formations which may appear are explained below.

Single bilayers of surfactants

A surfactant bilayer consists of two layers of surfactant molecules with the head groups uniformly oriented into the same direction on both sides of the sheet, for example, with all heads pointing outwards. The result is a sheet-like superstructure in the form of a bilayer, see Figure 9.7. This is also called a lamellar structure, **lamellar phase**, or it is referred to as **uni-lamellar**. The following superstructures may form: small **disk-like micelles**, see Figure 9.8; or flat layers showing large planar surfaces, see Figure 9.9; or lamellae showing a curved shape, concave or convex, see Figure 9.10.

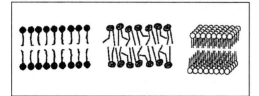

Figure 9.7: Planar bilayer micelles [109]

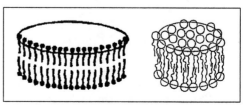

Figure 9.8: Disk-like bilayer micelles [109]

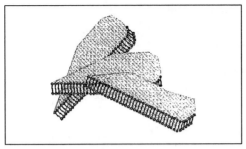

Figure 9.9: Planar sheet-like bilayer micelles showing large surfaces [18]

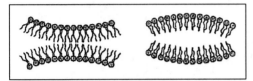

Figure 9.10: Curved bilayer micelles [109]

Example: Bio-membranes

Examples from biology are **phospholipid bilayers** of bio-membranes. Typical **biological cell membranes** show a thickness of around 3 to 5nm [38]. Comparison: The size of human body cells is around 10 to 100μm, for example, typical skin cells are around 10μm. Developers of biological tissues, for example in the field of regenerative medicine, aim to produce **synthetic bio-compatible membranes** (tissue designers, tissue engineered medical products TEMPs) [206].

Note: Direction-dependent properties

Lamellar structures often show **anisotropic behavior**, i.e. their behavior is direction-dependent. Typical examples are **liquid crystals** (LC).

Multiple layer systems of bilayer lamellae

For highly concentrated surfactant systems, the surfactant phase often shows a multiple layer or **multi-lamellar** structure in the form of **multi-bilayers**, see Figure 9.11. There is a possibility to store aqueous solution in between the individual bilayers. Liquid of these **inter-layers** can act as a lubricant making it easier for the lamellae to slide along each other. This means flow behavior can be considerably improved when using these so-called **"fluid lamellae"** or **"fluid systems"**. The flexibility of such a system greatly depends on the dimensions as well of the individual bilayers as well as of the whole group of layers as a cluster, on the number of layers and clusters, on the packing density, and on the distance between the individual layers and clusters.

Order of magnitude: Typical dimensions for the thickness of a bilayer h = 1 to 10nm, and in maximum 50nm approximately. Inter-layers are often in the range of 1 to 100nm.

Examples 1: Binary lamellar system, showing the two layer thicknesses of 4 and 8nm, and an inter-lamellar distance of 1 to 3nm; see Figure 9.11 (right).

Example 2: Multi-lamellar planar bilayers with encased, encapsulated, shielded hydrophobic molecules, the so-called "solute"; see Figure 9.12.

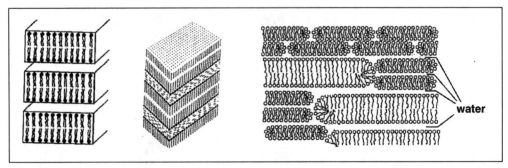

Figure 9.11: Multi-lamellar planar micelles, on the right there is a binary surfactant system showing two different layer thicknesses of the lamellae, containing inter-lamellar water which is stored between the layers [27, 109]

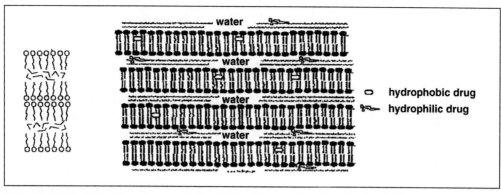

Figure 9.12: Multi-lamellar, planar micelles filled with an encapsulated hydrophobic solute, e.g. a medicine[109, 144]

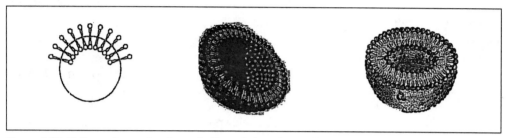

Figure 9.13: Hollow spheres showing a shell consisting of a single surfactant bilayer[109, 144]

f) Vesicles
Vesicles as hollow spheres showing a single surfactant bilayer

Sometimes hollow spheres are occurring with an outer skin or shell consisting of a single surfactant bilayer, e.g. with hydrophilic surfaces both outside and inside, see Figure 9.13. These kinds of spheres are also known as "vesicles" (Latin: vesicula, meaning small bubble), and "core/shell aggregates" or "core/shell particles". The inside of the sphere can be hollow, or it can be filled with an aqueous solution or a solid nano-material. As an illustrative example, you can imagine a football with its thick leather shell which is deformable to a certain degree.

Example 1: Biological vesicles, liposomes

All typical natural **bio-membranes** or synthetic lipid membranes have a common basic structure. They all consist of bilayer phospholipid lamellae build up by "bio-surfactant" molecules. Therefore, they do not show static but dynamic structures since the membrane molecules are able to diffuse laterally, for example, and they might also fuse with other membranes. By the way, the short-term dynamics of these time-dependent structural changes can be investigated today, for example by rheo-optical measurements even in the millisecond range (see also Chapter 10.8.2.5: SAXS tests). An example of bio-membranes are **liposomes in the form of hollow "fat globules"** or little "fat bubbles", as "transporters" for the essential metabolism and information exchange, e.g. between nerve cells. Their shell consists of a lipid matrix with embedded protein molecules which are "swimming" in or on this flexible superstructure, and these proteins are controlling the exchange, e.g. of ions.

Example 2: Vesicle systems as "capsules in the capsule"

Here, stable spherical containers with a total diameter of around 4μm serve as the outer capsule of **"micro- or nano-transporters" for pharmaceutical active agents**. Their shell consists of a biopolymer layer with a thickness of around 500nm. On the inside wall there are liposomes as "inner capsules" which are charged with active agents. Inside of such a "capso-som system" can be transported even thousands of these small vesicles showing diameters between 100 and 200nm.

Similar to active agents which are encapsulated in surfactant structures used in the medical sector, nano- and micro-capsules are also in use in other industrial branches such as food technology, environmental technology. More and more they are incorporated also in resins, varnishes, printing inks, adhesives, sealants and **functional coatings** for special applications, which in creative marketing sometimes are called "smart coatings".

Example 3: Microspheres or microcapsules in food

Aromas and active agents like omega-3-fatty-acids or salts are often encapsulated by natural hydrocolloids such as agar-agar, alginate, gelatin and pectin or by fats and waxes as the shell material. For a **"core/shell encapsulation"** the capsule consists of a solid shell and a liquid or solid core. For a **"matrix encapsulation"** the active agent is evenly distributed throughout the homogeneous matrix inside of the sphere. Here, the range of grain size is often between 50μm and a few millimeters [52].

Example 4: "Self-healing" or "self-repairing" coatings showing the "reflow effect"

"Self-healing" coatings and building materials contain encapsulated liquids as "functional agents" such as oils or other single components of reactive resins, which may flow out into scratches or cracks of the coating if the polymer shell of the capsules is also destroyed **("reflow effect")**. They may fill the cracks and then harden, for example, due to oxidation in air or when getting contact to a second reaction component. These capsules have diameters e.g. between 60 and 250μm [361].

Vesicles showing multiple surfactant bilayers

With single surfactant bilayers a low mechanical solidity must be expected. Vesicles as multi-shell and multiple bilayer systems are showing correspondingly higher stability, see Figure 9.14. They can occur as more or less hollow spheres which can also be filled with liquid or nanoscopic solid material such as crystals. These kinds of complex micelles have a multi-lamellar spherical shell structure consisting of multi-bilayers. As an illustrative example, you can compare it to onions. These vesicle systems are often clearly poly-disperse, i.e. their spheres are showing a wide particle size distribution.

Order of magnitude: The dimension of smaller vesicles is usually between 10 and 100nm, so-called "nano-spheres". More complex systems can show an outer diameter of several μm, for example with ten bilayer shells as "micro-spheres" or "macro-spheres". Sometimes due to shear forces, originally planar structures may become rolled up into onion-like vesicles. Such a closely packed stable system is often called a **"vesicle gel"** then.

Comment: Vesicles and transporters in bioscience

Under **vesicular transport** in living organisms ("in vivo"), people in the medical sector include the process of substances passing through individual cells, and also the exchange of "bio-

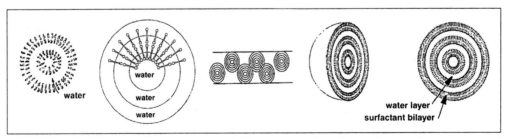

Figure 9.14: Vesicles showing several spherical shells consisting of multiple surfactant bilayers [18, 109, 217, 220]

active transporters" between neighboring cells. These substances are encapsulated in membrane capsules, and are therefore "dissolved" or "solubilized" in water. Creative marketing then talks of "nano-containers", "drug vehicles", "nano-carriers", "drug taxis", "nano-submarines", "Trojan horses", "drug-targeting systems" or simply of "carrier systems".

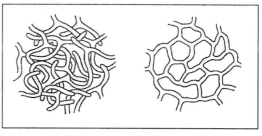

Figure 9.15: Gel structure based on a network of surfactant structures, left: incompletely linked, right: completely cross-linked [220]

Example: SEDDS (self-emulsifying drug delivery systems) used as drug transporters

Using such a vesicle as a "retard system", after reaching the target object such as tumor cells, there can be controlled the delayed dosing process of active agents such as antibodies or antigens.

g) Complex superstructures

There are also other forms, such as hexagonal, cubic, rhombic, helical (spiral) or crystalline and liquid-crystalline superstructures, which can be built up by spherical or cylindrical micelles. These formations often show a phase transition if the concentration is changed, this is called **lyotropic behavior**. When a phase transition occurs if the temperature is changed, this is called **thermotropic behavior**. Typical materials displaying this kind of behavior are liquid crystals.

h) Sol-like and gel-like micelle structures

Gels are superstructures showing a three-dimensional (3D) network which is connected via inter-micellar interaction forces, for example, due to dipoles or H-bridges; see Figure 9.15. As well the degree reached by the micelles when growing together to form superstructures, as well the finally appearing network density, are greatly depending on the surfactant concentration. **Hydrogels** are gel structures occurring in an aqueous solution, and **organogels** are gel structures appearing in organic solvents.

In gels the internal cohesion forces are stronger throughout the whole "gel body" compared to the weight force due to gravity, and therefore, gels are **"coherent systems"** showing a **stable structure-at-rest**. However, it is important to remember that even a solid and rigid gel structure in an aqueous surfactant system is predominantly based on continuously opening and closing dipoles or H-bridges of the hydrate shells. This kind of structure is therefore based on physical-chemical, but reversible secondary bonds only, and not on the considerably stronger chemical primary bonds.

Examples: Hydrogels used for artificial reproduction of tissues which are similar to biological ones [206].

In contrast to gel-like structures, in fluid systems the intermolecular forces between the micelles are only weak or not pronounced at all. This is referred to as the **character of a liquid** or it is called an **incoherent colloid system**, such as a **hydrosol** which is forming in an aqueous solution. **Organosols** are fluid systems based on organic solvents.

Example: Unlinked spherical micelles which indeed obstruct each other when flowing, but they show **no dispersion stability** when at rest; see Figure 9.16.

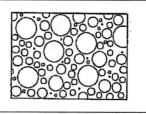

Figure 9.16: Highly concentrated but incoherent system showing properties of a hydrosol, and therefore, instability is to be expected when at rest [109]

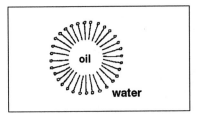

Figure 9.17: Oil-in-water emulsion (o/w) [217]

9.1.2 Emulsions

Emulsions are mixtures of at least two phases, usually water and oil. Without the stabilizing effect of emulsifiers they are unstable, showing phase separation due to the different density values of the phases. **Surfactants are ideal emulsifiers.**

a) Oil-in-water emulsions

In oil-in-water (o/w) emulsions, the hydrophilic components of the surfactant molecules try to get contact to the water phase while the hydrophobic parts try to get contact to the oil phase. The head groups are forming a spherical shell around the oil droplets. This process stabilizes the oil phase in the surrounding water, see Figure 9.17. At room temperature, this kind of emulsion **usually is a liquid.** The most important natural emulsifier is **lecithin.**

Examples: o/w emulsions

Salad dressing made of oil, e.g. olive oil, and water, e.g. from white wine vinegar, can be stabilized with a little amount of mustard or mustard seed oil. The taste of **meat juices** as aqueous solutions from roast pork, for example, can be refined by adding butter or cream. The smaller the fat droplets the creamier is the consistency of the emulsion [372].

b) Water-in-oil emulsions

As described for o/w emulsions, also in water-in-oil (w/o) emulsions the hydrophilic components of the surfactant molecule try to get contact to the water phase and the hydrophobic parts try to get contact to the oil phase. However, since here the amount of oil or fat dominates, the surfactant heads turn inward and form so-called **inverse micelles** showing a spherical or cylindrical shell around the included water phase. This stabilizes the water phase in the surrounding oil or fat. Here, the hydrophobic tail groups point outwards now, see Figure 9.18. This kind of emulsion **usually is a semi-solid formulation** as can be found for **stable dispersions, creams or ointments.** Usually, water-in-fat emulsions are also called w/o emulsions.

Examples: w/o emulsions

Mayonnaises, made of oil and egg yolk which also contains water and the emulsifier lecithin, are semi-solid. **Butter** consists, for example, of around 82% fat and 16% water [372].

c) w/o/w and o/w/o emulsions

Complex systems can be formulated as **"emulsion in the emulsion".** In (w/o)/w emulsions, very small, nanoscopic or microscopic water droplets are emulsified in oil and this w/o phase is distributed again in a surrounding water phase in the form of larger droplets. These kinds

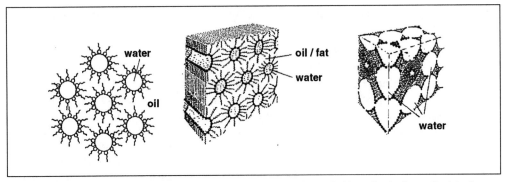

Figure 9.18: Water-in-oil emulsion (w/o): the water phase is surrounded by spherical or cylindrical micelles [27, 109]

of emulsions usually show the character of a liquid. There are also (o/w)/o emulsions. Here, oil-in-water droplets are emulsified again in a surrounding oil or fat phase. These kinds of emulsions mostly exhibit gel-like character.

9.1.3 Mixtures of surfactants and polymers, surfactant-like polymers

Hybrids are **multi-component systems**. In the context of this section, these are, for example: dispersions containing surfactants showing interactions to co-surfactants and to other molecules, micelles, droplets, aggregates, associates, agglomerates, inorganic and organic particles.

For water-based systems, often **associative thickeners** are used. Their molecules can link with one another, they are "associating". This usually occurs **via hydrophobic groups**, i.e. via non-permanent secondary bonds. As a comparison: Other thickeners – if not hydrophobically modified – such as cellulose or polyacrylic acid, can only interact with the water phase, since they are hydrophilic. These kinds of rheological additives cannot build up a physical-chemical network between its molecules. They can only form a mechanical network via the entanglements of the macromolecules. However, this is not a permanent network then (see Chapter 3.3.2.1 and 3.3.7b).

Example 1: Surfactants for stabilizing latex particles in a water-based dispersion
Water-based latex dispersions or polymer dispersions consist of monomeric, pre-polymerized or polymerized molecules which are stabilized by surfactants. The hydrophobic organic polymer molecules can be found in the core of the micelle, and therefore, they are shielded from the water. Compact latex particles which are homogeneously distributed in water appear as spherical micelles; see Figure 9.19 (core/shell particles). Micelles in typical colloids and nanoscopic, microscopic and macroscopic latex dispersions can show diameters between 20nm and 20µm; often they are around 50 to 500nm. Comparison: The size of inorganic particles of pigments and fillers for coatings is around 0.1 to 10µm, and often 0.5 to 5µm, i.e. ten times larger than latex micelles; and organic pigment particles are around 10 to 1000nm, and often 50 to 500nm, i.e., they are showing the same dimension like latex micelles.

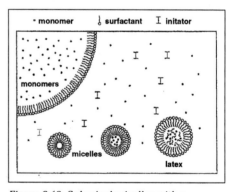

Figure 9.19: Spherical micelles with a surfactant shell for stabilizing a water-based latex dispersion [109]

Example 2: Surfactants building bridges between polymer molecules

Figure 9.20 shows a system containing as well surfactants as well a water-soluble hydrophobically modified (HM) polymer, for example, hydroxyethylcellulose (HM-HEC). A three-dimensional (3D) network may be formed, on the one hand due to direct bonds between the small hydrophobic side groups of the polymer and on the other hand due to bonds between the surfactant micelles and the hydrophobic side groups of the polymer. This method is used to achieve gel formation in cosmetic products, and also to thicken coatings.

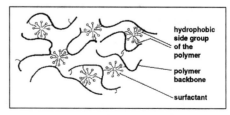

Figure 9.20: Gel structure formed by surfactant micelles building bridges between the polymer molecules [217]

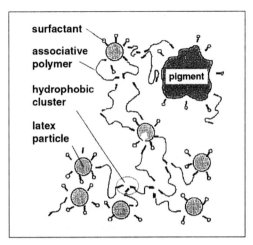

Figure 9.21: Water-based dispersion containing as well latex and pigment particles encapsulated by surfactants as well as a polymer associative thickener showing hydrophobic end groups [302]

An example of this is **HASE**, i.e. hydrophobically modified alkali soluble emulsion. It is an acrylate thickener whose effectiveness is dependent on the pH value. Typically it shows a molar mass M of around 100kg/mol. These kinds of **polymer associative thickeners** consist of a hydrophilic main chain and hydrophobic side chains [176].

Example 3: "Surfactant-like polymers" as associative thickeners of dispersions
A **water-based dispersion** containing latex and pigment particles is thickened by adding a polymer associative thickener, for example, a modified polyurethane (PU); see Figure 9.21 and Figure 3.34 in Chapter 3.3.7. Both **end groups of the polymer molecules are hydrophobic**, and therefore, on the one hand they may associate or **cluster via loose bonds** with one another, and on the other hand they may also try to find contact to the hydrophobic particle surfaces. Sometimes, the polymer chain between the end groups is hydrophilic to a certain degree. In this case, due to their double character, these kinds of molecules are called **amphiphilic organic polymers**. Therefore, it is possible for the thickener to build bridges through the aqueous phase, forming a three-dimensional structure finally. While most hydrocolloids are hydrophilic polymers and therefore are able to thicken even a pure water phase via H-bridges, the associative thickeners with their hydrophobic groups are merely **system thickeners** which are only effective if there are also other components available in the aqueous solution showing hydrophobic contact points, such as the surfaces of dispersed filler particles or pigment particles. Although associative polymers thickeners carry surface-active molecular groups, they are not called surfactants. However, sometimes they are referred to as **"surfactant-like polymers"** [208].

The special advantage of these systems is that the soft gel-like structure at rest is not as strong as it is obtained when using other thickeners, for example hydrocolloids or clays, **since the hydrophobic bonds are not rigid**, because the individual molecules undergo indeed a very fast but **continuous exchange via fluctuation** (see also Chapter 3.3.7).

An example of this is **HEUR**, i.e. hydrophobically modified ethoxylated urethane. This polyurethane (PU) thickener typically shows a molar mass M of around 10kg/mol. These kinds of **polymer associative thickeners** consist of a hydrophilic main chain and hydrophobic end groups, and the effect of the polymer chain may occur therefore in three parts: hydrophobic – hydrophilic – hydrophobic [176].

Example 4: Dispersions containing associative polymers and surfactants
A **water-based dispersion containing also surfactants** is thickened by adding a polymeric associative thickener. These kinds of coating systems may contain surfactant molecules in the formulation, for example as wetting agents (for particles and substrates), dispergents, defoamers, deaeration additives or leveling agents. Additional surfactants can get into the dispersion – welcome or not – via other components of the mixture, for example, together with color pigments or due to contamination. In this case, **the hydrophobic end groups of the associative thickener** and the hydrophobic tail groups of the **surfactant molecules compete** for a space **on the particle surfaces**, see Figure 9.22. The thickening effect is therefore greatly influenced by the surfactant concentration. To form a 3D structure, bridges must be

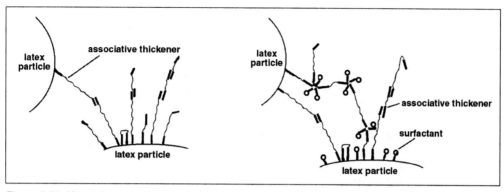

Figure 9.22: Network structure via polymer associative thickeners of different lengths using their hydrophobic end groups; on the left side without surfactants, and on the right side containing additional surfactants in the mixture, draft acc. to [316]

built through the aqueous phase. This can occur as well via hydrophobic clusters between the polymer molecules of the associative thickener (see Figures 9.21 and 9.22, left) as well as by integrating additionally surfactant micelles into the superstructure (see Figure 9.22, right, and Figure 3.34: no. 3b).

9.1.4 Applications of surfactant systems

Surfactants are used today in all industrial branches, and besides other purposes also **as stabilizing agents and thickeners** of solutions, dispersions (i.e. suspensions, emulsions, foams), creams, pastes and gels, and therefore, for all kinds of liquid and semisolid formulations. In the following is listed a number of applications where surfactants can be found.

Consumer and home care products, detergents
Household cleaners, liquid soaps, dish and textile washing liquids, textile conditioners, fat removers; **stabilizing agents** in cleaning dispersions

Personal care, health and beauty care products, cosmetics
Liquid soaps, shower gels, shampoos, hair sprays, shaving creams, bubble bath, toothpaste and tooth-gels, perfume oils (for paper towels, diapers, sanitary towels), washing creams, lipid replenishing creams, cosmetic creams (containing nano-capsules, vesicles, liposomes, showing delayed release of the ingredients on the skin surface), wetting agents for dry substrates (e.g. for contact lenses); viscoelastic surfactants (VES) with "body" (to give the consumer the impression of a "product having volume")

Life sciences, pharmaceuticals, bio-tech
Medical solutions and dispersions (liquid, semi-solid, e.g. emulsions), bio-compatible and self-emulsifying "drug carriers", encapsulated active agents, target-oriented pharmaceutical retard systems with delayed drug release, **hydrogels**, bio-membranes, artificial tissues (tissue engineered medical products TEMPs), organic-inorganic hybrid composite materials

Food technology
Emulsifiers, dispergents, **thickeners, gel formation agents**, wetting agents, stabilizing agents for water-insoluble vitamins, foams and opacifiers, aroma solvents
Examples: Surfactants to coat and **stabilize particles and droplets** in lemonades, fruit juices, mayonnaise, ice cream, margarine, chocolate, herb extracts, instant powders, sausage products, and emulsions to assure stability also after several freezing and melting cycles; "low fat" yogurts [4]

Surface treatment, paints and "smart coatings"
Adhesion improvers, leveling agents, wetting agents (for solid surfaces, substrates, pigments and filler particles), dispergents, substrate detergents, grease removers, defoamers, deaeration additives, heat transfer liquids, cooling agents, for **stabilizing latex dispersions** and abrasive agents for cleaning hard surfaces such as glass and ceramics, wear protection agents; **polymeric associative thickeners** for water-based emulsion paints, polymer dispersions (e.g. latex); **system thickeners**

Petrochemical industry, mining, transportation in pipe lines
Agents for oil extraction, drilling fluids, solvents, fluids used for pipe transport of **fluidizing sludge**, flotation agents, fracture fluids; fluids to break, separate or destabilize emulsions
Example: Extraction of residual oil using a surfactant-polymer flooding process for tertiary oil recovery. First step: An aqueous solution (water, surfactant, oil, alcohol, salt) is injected into the pores of the rock of the oil spring to reduce as well the capillary forces in the pores as well as the surface tension between rock and crude oil mixture (petroleum, water, natural gas) and to coagulate the oil droplets. Second step: A polymer solution containing **drag reduction agents (flow improver for viscosity reduction)** is injected to optimize the rate of oil production.

Agrochemicals, crop science
Stabilization of dispersions used as herbicides, fungicides, insecticides and pesticides

9.2   Rheological behavior of surfactant systems

Rheological experiments describe macroscopic effects, i.e. not the behavior of single atoms, molecules, molecule clusters or quantum dots consisting of max.10,000 atoms, but the behavior of their superstructures in the order of magnitude from at least one micrometer up to several millimeters. Reproducibility of the measurements is only possible if the surfactant system is in an equilibrium state between the shear forces and the forces causing structural regeneration. This regeneration may take a long time also in a state at rest, and sometimes even up to several days. **Rheologically relevant viscoelastic effects** occur in surfactant systems **only if there are relatively large superstructures of micelles showing interaction.**

Note: Surfactants, and measurement of the rheological behavior at interfaces
Interfacial rheology is used to characterize rheological behavior of monomolecular surfactant films which are forming a quasi two-dimensional (2D) interface on a **surface**, for example, between a liquid and air, or at an **interface** between two liquids which do not mix (see also Chapter 10.8.3.1).

9.2.1 Typical shear behavior

Ideally viscous flow behavior
Surfactant systems at a low concentration show ideally viscous or Newtonian flow behavior, as shown in Figures 3.3 and 3.4 (no. 1 in both). In this case, there may exist already some pre-micellar structures, but not yet superstructures. Perhaps some single spherical micelles can be found, however, with a relatively large distance from one another.

Shear-thinning flow behavior
Shear-thinning flow behavior often occurs at an increased surfactant concentration, for example at a volume fraction of $\Phi > 30\%$, as shown in Figures 3.3 and 3.4 (no. 2 in both). Sometimes in the low-shear range, i.e. below the shear rate $\dot{\gamma} = 1\,\mathrm{s}^{-1}$, the **plateau of the zero-shear viscosity** is indicated – similar to the behavior of unlinked polymer molecules; as presented by the logarithmic viscosity function in Figure 3.10 (range 1).

Dispersions at a higher concentration are showing correspondingly higher viscosity values. The reason for this is that during the flow process there is of course more friction between the now increased number of micelles. Above a critical shear rate value – when exceeding the limiting value of the zero-shear viscosity range – new superstructures may be formed continuously in the flowing liquid. Then there is often a **shear-induced and direction-dependent flow** which is influenced by the degree of the shear rate, leading to orientation and formation of new micellar superstructures. These structures are adapting more and more to the changing shear conditions which facilitates to glide along one another more easily. This is supported by the **very rapid exchange or "fluctuation" of single molecules or molecule groups** in the persistence length. In this case, the shear process may often result in shorter chain structures and this usually leads to a more pronounced shear-thinning effect; as shown in Figures 3.3 and 3.4 (no. 2 in both) and Figure 3.10 (range 2).

Above a certain value of the surfactant concentration, **self-organizing worm-like micelles** may be formed as a superstructure. These long micelles may entangle to build a **temporary and transient network**, being in a transition state where parts of the superstructure are continuously exchanged with time, similar to the behavior of long-chained polymer molecules. However, this network does not show a permanent and coherent structure as is the case for a stable gel.

Here, the individual micelles exhibit a more or less structured shape which under shear load continuously may break into parts which are showing a single or several persistence lengths. However, these flexible structures may also reform immediately again. Therefore, a continuous process takes place of dissolving or breaking up and reforming or fusing of parts of the superstructure, dependent on the shear forces and interaction forces between the micelles. Related to the viscosity **thickening effect** however, **surfactant systems show the following advantage compared to polymers:** If polymer chains are broken under a high shear load, then the molecular "degradation" is irreversible and therefore associated with a permanently remaining reduction of the molar mass. In contrast, after **breaking** even relatively large **micellar structures**, these may **reform again and again** afterwards in a period of rest, or when applying a lower shear load then.

Strongly pronounced shear-thinning behavior can also occur if **sheet-like lamellae are gliding along one another** which may be facilitated due to their **planar surfaces**. This process is of course supported if aqueous layers act like a lubricant in the inter-lamellar regions. In this case, there exists a "fluid lamellar system". **Phase separation** may occur then. This may result in **the existence of areas with a higher amount of micelles and areas containing only few micelles** which again may lead to different flow velocities in different parts of the sample: The result would be an inhomogeneous flow field, a non-constant shear gradient and therefore a complex, often uncontrollable deformation behavior (see Chapter 9.2.2: shear-banding).

No plateau value of the viscosity function occurs in the low-shear range if a coherent network of interaction forces exists showing a stable **three-dimensional gel-like structure** in the linear-viscoelastic (LVE) deformation range. In this case, behavior is similar to those of dispersions showing a yield point.

Viscoelastic behavior
Elastic effects usually become more pronounced both at increasing surfactant concentration and decreasing temperature. With worm-like micelles, stronger viscoelastic effects may occur with increasing strain rate if the formation of a temporary network of entangled structures takes place in the shear field. In this case, the behavior is similar to those of unlinked polymers. Here, further **oscillatory tests** would be useful, comparable to tests on gels used in the

food, cosmetics and pharmaceutical industry and in biotechnology. Also here, testing begins with an **amplitude sweep** to investigate the behavior in the linear-viscoelastic (LVE) deformation range, see Figures 8.9 to 8.12. When performing **frequency sweeps**, the following information may be achieved about the occurring structures.

Behavior of viscoelastic liquids

A sample behaves like a viscoelastic fluid, if at low frequencies $G'' > G'$ and at high frequencies $G' > G''$, i.e., if the curves display a crossover point, as shown in Figure 8.21 (left). Such a system is liquid at rest, often showing a plateau of the zero-shear viscosity. In this case when at rest, even small air bubbles would rise up through the fluid system, even when this takes a long time. At high frequencies, there is a **temporary network** consisting of micelles, and additionally of macromolecular entanglements if there are also dissolved polymers available. Behavior is similar to those of polymer solutions containing unlinked polymer molecules and can, if certain criteria are met, be described by the Maxwell model (see also Chapter 8.4.2.1a) [162]. Some surfactant systems may deviate from this model at high frequencies (see also Example 1 in Chapter 8.4.2.1b) [220].

Example: Hair shampoo containing viscoelastic surfactants (VES) which are forming a **worm-like micelle system**. See also Example 3 to Note 2 in Chapter 3.3.3: Dilatancy peak in viscosity functions.

Behavior of viscoelastic solids, gel-like structures

A viscoelastic gel structure exists, if there is $G' > G''$ across the whole frequency range, i.e., if the curves of G' and G'' do not cross, see Figure 8.21 (right). At rest, these kinds of systems are stable showing a **permanent network** of interaction forces. Therefore, they are not liquids, and even over a long period of time at rest, not all included air bubbles would rise up here. Behavior is similar to those of custards containing physical-chemically linked components or to soft elastomers consisting of polymer molecules showing a low degree of cross-linking.

Example: Oil-in-water micro-emulsions with a stable physical-chemical **network**, whose **cross-linking additives**, e.g. **associative polymers**, on the one hand are in contact with the water phase via its hydrophilic polymer chains, and on the other hand whose hydrophobic groups cluster with one another via loose secondary bonds or dock onto the surface of the oil droplets.

Thixotropic behavior, and time-dependent structural regeneration

After an interval of shear load which was leading to a partial destruction and subsequently being now in a period of rest, micellar structures are mostly regenerating until a steady-state is reached at the equilibrium of forces. Until then a time-dependent reversible process of "thixotropic behavior" takes place which finally leads to a more or less complete regeneration of the structure.

As with all viscoelastic samples, surfactant systems require a certain period of **time to regenerate** their structure. If the periods of decomposition and recombination of the micelles take on average around the same amount of time, some users talk of a **relaxation time**, like when analyzing the behavior of polymers.

For viscoelastic surfactant systems it is useful to carry out step tests in the form of **rotational or oscillatory tests**, comparable to measurements of polymer solutions; as presented in Figures 3.40 and 3.43, as well as in Figures 8.28 and 8.31. With oscillatory step tests, and measuring at low-shear conditions after an interval of high-shear, particularly the time point of the crossover of the curves of G' and G'' provides useful information, since this is the **sol/gel transition**, or briefly, the "gel point".

9.2.2 Shear-induced effects, shear-banding and "rheo chaos"

Particle size distribution or polydispersity of the micelles or the length of worm-like micelles, all depend on several conditions, such as on the surfactant concentration, pH value, temperature and on the shear conditions. These kinds of materials are mostly **dynamic systems with alternating**, and therefore often **unpredictable, complex rheological behavior**. Due to **shear-induced structures (SIS)**, **shear-thickening** (dilatant or rheopectic) **behavior** may occur, as shown in Figures 3.3 and 3.4 (no. 3 in both), as well as in Figures 3.41 and 8.29. When testing low-viscosity fluids, it should be taken care of turbulent flow behavior such as Taylor vortices, which may appear when cylinder measuring systems are used (see also Chapters 3.3.3 and 10.2.2.4b) [411].

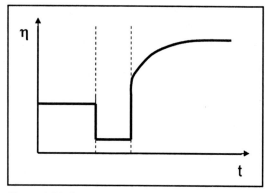

Figure 9.23: Rotational step test, first at a constantly low shear rate, then at a constantly high shear rate, and finally again at a constantly low shear rate, resulting in a remaining shear-induced structural change: Here showing a higher flow resistance finally in terms of the viscosity function.

However, often rheological behavior of complex micelles is neither dilatant nor rheopectic, since just anew occurring SIS which are differing from the original structures may remain stable for a certain period of time. This may be a short-term or a long-term process, and it may be reversible or may lead to a permanent and irreversible state, see Figures 9.23, 9.24 and also 3.18. SIS may show **anisotropic** and therefore **direction-dependent properties**, which can be observed when performing rheo-optical investigations

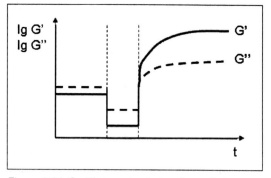

Figure 9.24: Oscillatory step test performed at a constant frequency, first at a constantly low shear strain, then at a constantly high shear strain, and finally again at a constantly low shear strain, resulting in a remaining shear-induced structural change: Here showing a higher structural strength finally in terms of the functions of G' and G".

(see Chapter 10.8.2: using e.g. microscopy and SALS, and analyzing e.g. effects like flow birefringence or dichroism). In contrast, a surfactant solution at rest may be absolutely isotropic, therefore showing no direction-dependent behavior at all.

For the step tests presented in Figures 9.23 and 9.24, the same kind of test conditions was preset as for the tests shown in Figures 3.43 and 8.31. Figure 9.23 displays the time-dependent viscosity function obtained by a rotational test, and Figure 9.24 illustrates the functions of G' and G" achieved by an oscillatory test. For both figures, in the measuring interval shown on the left-hand side, there was firstly applied a constant low shear load, then in the medium interval a constant high shear load, and finally in the right-hand interval again the same constant low shear load like in the first interval. Result: In contrast to Figures 3.43 and 8.31, a **stronger structure** is forming in the third interval **compared to the initially existing structural strength** from the first interval. The oscillatory test makes it clear that even the kind of the viscoelastic character may be changed here. In this case it changes **from an originally fluid structure when at rest**, showing G" > G', **to a gel-like structure** with G' > G" **finally**.

a) Reasons for shear-induced structures

The term **shear induction** includes the **production, orientation, enlarging or reduction of structures**. In complex surfactant systems or surfactant-like materials, the originally homogeneous arrangement of the structure may possibly be changed so effectively **by the shear process** that different **micro-structural superstructures with locally separated phases** may occur in the sample [196, 243, 411]. The process of **phase separation** can develop increasingly during a shear process between a phase showing a high amount and a phase showing a low amount of surfactants. **Separation of** flowing layers with different **concentrations of surfactants** can occur during the transition from a low to a high degree of orientation of the micelles. Separation can also take place due to the formation of layers containing broken and intact vesicles, or because short aggregate structures are partially transforming into an **entangled network of long micelles** or even into **gel-like solid regions**. As a result, locally inhomogeneous shear gradients may appear, as well in the direction of the forced deformation or flow as well as in all directions of vortices occurring at turbulent flow conditions.

b) Effects on the flow type

Due to the shear-dependent formation of superstructures, spatial separation into two, three or even more simultaneously existing, macroscopic layer structures can occur in a fluid. These layers may move along one another at different speeds showing different viscosities, either without a clear separated borderline or possibly even with an oscillating interface between the appearing layers. This occurrence is called **shear-banding**, see Figure 2.9 (no. 6 and 7). This behavior results in **inhomogeneous flow** which depends on the shear gap dimension and is often associated with **wall-slip** for all types of measuring systems. Particularly with cone-and-plate and parallel-plate measuring systems, **inhomogeneous behavior causing edge effects** may occur.

c) Results when measuring deformation and flow behavior

Layers showing different flow behavior may differ in all types of rheological behavior: Already beginning in the dependence on the pre-treatment of the sample; in the deformation and flow behavior such as shear-thinning or shear-thickening, thixotropic or rheopectic, or viscoelastic; or in varying stable or meta-stable structures depending on time and on the location in the shear gap. Effects such as **inhomogeneous blocking and yielding, or dissolving and breaking** of on the one hand just disintegrating and on the other hand just developing **superstructures** is to be expected when, with concentration-dependent behavior, for example planar lamellar phases are rolling up to form onion-like multi-lamellar vesicles. These effects may occur immediately and diminish again after a certain time. The duration of the effects resulting in orientation of shear-induced superstructures and their more or less complete reformation and stress relaxation in the subsequent period of rest, may take only seconds and minutes, or even several hours. Sometimes the behavior is even irreversible and the shear-induced structure remains permanently in a gel-like state after all. Unfortunately, sometimes the rheological response is chaos. Despite the greatest care taken with setting the test parameters, the measuring results might be not reproducible then, and in this case, they are often indescribable and hardly to explain, since that is **"rheo chaos"** [240, 248].

Shear-banding is sometimes indicated in diagrams in the form of **plateau-like steps occurring in flow curves**. In contrast to the "common" shear-thinning flow behavior as shown in Figure 3.10, there **may appear several viscosity plateaus**, as well in the low-shear range as well as in an intermediate shear rate range. Therefore, there are perhaps two shear rate values which can be analyzed as the limiting values of these plateaus, as shown in Figure 9.25: $\dot{\gamma}_1$ and $\dot{\gamma}_2$. This can also occur with amplitude sweeps. Then, in contrast to the "common" viscoelastic deformation behavior as presented in Figure 8.12, there **may appear two yield points** τ_{y1} and τ_{y2}, as shown in Figure 9.26.

Examples of materials which may show shear-banding

Aqueous "semi-diluted" or highly concentrated surfactant solutions including worm-like micelles, semi-flexible aggregates, lamellar phases, multi-lamellar vesicles; liquid crystals; concentrated colloidal dispersions (suspensions, emulsions, foams) such as pastes; lubricants; polymers with functional groups and copolymers; viscoelastic surfactants (VES)

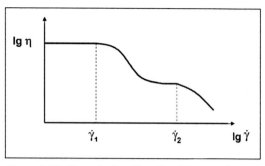

Figure 9.25: Viscosity function showing several plateau-like steps due to shear-banding

Note 1: Rheo-optical observation of micelle structures in a shear field

Special rheo-optical devices such as **microscopes**, and measuring cells for **SALS** (light scattering), **SAXS/WAXS** (X-ray scattering), **flow birefringence** and **dichroism** enable the user to observe the **occurrence and change of micellar superstructures** under defined shear conditions (see also Chapter 10.8.2: rheo-optics).

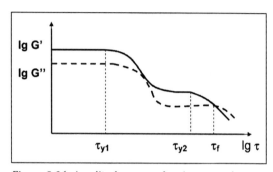

Figure 9.26: Amplitude sweep showing several plateau-like steps due to shear-banding

Note 2: Rheo-optical observation of flow velocity profiles

Velocity profiles of flow fields can be investigated using a device called particle imaging velocimetry (PIV) or particle tracking velocimetry (PTV), respectively. With these kinds of measuring cells, **inhomogeneous flow fields showing shear-banding effects or turbulent flow behavior** can be observed and presented, e.g. in the form of diagrams, and they can also be recorded on video if desired (see also Chapter 10.8.2.7).

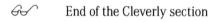 End of the Cleverly section

10 Measuring systems

10.1 Introduction

The previous chapters have covered the basics of rheology and test methods to measure deformation and flow behavior by rotational, creep, relaxation and oscillatory tests. In the chapter at hand information is given on typical measuring systems used to perform these kinds of tests. In this Chapter as a simplification, the term **"measuring system"** is abbreviated to **"MS"**.

10.2 Concentric cylinder measuring systems (CC MS)

10.2.1 Cylinder measuring systems in general

10.2.1.1 Geometry of cylinder measuring systems showing a large gap

A cylinder MS consists of an inner cylinder (bob) and an outer cylinder (cup), see Figure 10.1. Concentric cylinders mean that both cylinder-shaped components are showing the same symmetry axis or rotation axis, respectively, if mounted in the working position, briefly, the CC MS. A synonym for concentric is **coaxial**. CC systems are described by the standards ISO 3219 (1993) and DIN 53019-1 (firstly in 1976 by DIN 53018).

The definitions of the rheological parameters as specified in Chapters 2.2, 4.2 and 8.2 can also be used here despite the rounded areas of the cylinder walls, as long as the MS is showing a relatively narrow gap. In this case, the areas can be considered to be relatively even, corresponding to the Two-Plates-Model (see Figures 2.1, 4.1 and 8.1). However, the larger the gap, the less favorable are the preconditions for this assumption and for the definition of the mentioned rheological parameters.

A large gap may lead to secondary flow effects (see Chapters 3.3.3 and 10.2.2.4: Flow instabilities and turbulent flow of low-viscosity liquids), time-dependent behavior (see Chapter 3.3.1b: Transient effects, and Figure 2.9: no. 5), or inhomogenous deformation behavior (Chapter 3.3.4.3d: Plastic behavior, and Figure 2.9: no. 4).

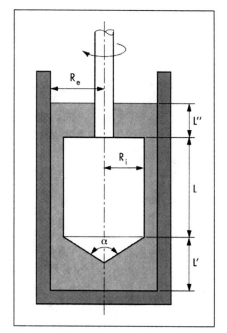

Figure 10.1: Concentric or coaxial cylinder measuring system (CC MS)

10.2.1.2 Operating methods

For rotational tests there are two modes of operating:

a) Searle method
Here, the bob is set in motion as the "rotor", and the cup is stationary as the "stator". In industrial laboratories, almost all rheometers are working on

Thomas G. Mezger: The Rheology Handbook
© Copyright 2011 by Vincentz Network, Hanover, Germany
ISBN 978-3-86630-864-0

this principle which is named after *G.F.C. Searle* (in 1912, [330]). However, after first experiments by *C.A. Coulomb* (in 1784, [80]) probably *A. Mallock* (in 1896, [238]) was the first who used this method systematically; see also [18, 107, 289].

The disadvantage of this method is that turbulent flow conditions may occur in the form of so-called "Taylor vortices" when measuring low-viscosity liquids at high rotational speeds. This is due to centrifugal forces and inertial effects of the liquid, respectively (see Chapter 10.2.2.4a).

b) Couette method
Here, the cup is set in motion and the bob is stationary, or it is deflected to a negligible degree only. Only few rheometers are designed for this method, which is named after *M. Couette* (in 1888/1890, [79]). However, probably *J. Perry* (in 1882, [285]) was the first who worked with this method (using a "rotated vessel and a hollow cylindric body"), after this method was already proposed by *M. Margules* (in 1881, [241]); see also [107, 289].
The advantage of this method is that Taylor vortices are not occurring like when working with the Searle method. However, one disadvantage is decisive if temperature is controlled by a liquid bath circulator: the rotating cup has to be sealed against the circulator liquid, and this often causes problems due to leakage. In this case, friction of the sealing may additionaly influence the measuring result significantly and this is no longer acceptable for modern test requirements. In order to achieve accurate results, therefore corresponding tests should not be carried out using a circulator but preferably in a temperature-controlled room, or using a closed test chamber for heating and cooling which is surrounding the whole test equipment.

The following text and the calculations are related to MS which are operated according to the Searle method, thus, using a rotating or oscillating measuring bob and a stationary cup which is mounted onto the frame of the measuring device.

&ℓ   For "Mr. and Ms. Cleverly"

10.2.1.3 Calculations

The following formulas are developed for large-gap cylinder MS which do not meet the requirements as stated in ISO 3219 (for these kinds of narrow-gap cylinder MS: see Chapter 10.2.2). In the section at hand, i.e., **when using large-gap cylinder MS, the shear stress and shear rate are related to the bob surface**.

a) Shear stress in a large cylinder gap
In the general form: $\tau(r) = M/(2\pi \cdot L \cdot r^2)$

with the torque M [Nm], and the length L [m] of the cylindrical part of the bob (see Figure 10.1).

$\tau(r)$ is the distribution function of the shear stress in the shear gap, i.e. τ changes with the radius r. Here, r is the distance between the rotation axis and any layer of the liquid, with $R_i \leq r \leq R_e$, where R_i is the bob radius (or "inner" or "internal radius") and R_e is the cup radius (or "outer" or "external radius"). For cylinder MS showing a large gap, the shear stress is usually related to the bob surface (i.e. to R_i), then:

Equation 10.1 $\tau_i = M/(2\pi \cdot L \cdot R_i^2)$

b) Shear rate in a large cylinder gap
In the general form:

Equation 10.2 $\dot{\gamma}(r) = \dfrac{1}{r^2} \cdot \dfrac{(2 \cdot R_i^2 \cdot R_e^2)}{(R_e^2 - R_i^2)} \cdot \omega$

$\dot{\gamma}(r)$ is the distribution function of the shear rate in the shear gap (see Figure 10.2). The angular velocity is, with the rotational speed n [min^{-1}]:

Equation 10.3 ω [rad/s] = $(2\pi \cdot n)/60$

Sometimes, ω is specified using the unit [s^{-1}], see also the Note in Chapter 8.4; and n in [rpm], i.e. in "revolutions per minute" which is the same like [min^{-1}]. The following applies:

Equation 10.4 $v = \omega \cdot r$

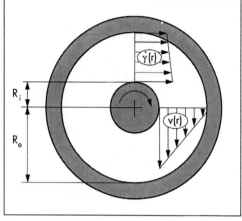

Figure 10.2: Cross-section of a concentric cylinder measuring system showing a large shear gap

v(r) is the distribution function of the circumferential velocity in the shear gap (see Figure 10.2). For cylinder MS showing a large gap, the shear rate is usually related to the bob surface (i.e. to R_i), then:

Equation 10.5 $\dot{\gamma}_i = [(2 \cdot R_e^2)/(R_e^2 - R_i^2)] \cdot \omega$

c) The viscosity in a large cylinder gap
Related to the bob surface (with r = R_i) holds:

Equation 10.6 $\eta = \tau(r)/\dot{\gamma}(r) = \tau(R_i)/\dot{\gamma}(R_i) = \dfrac{(R_e^2 - R_i^2)}{4\pi \cdot L \cdot R_e^2 \cdot R_i^2} \cdot \dfrac{M}{\omega}$

✍   End of the Cleverly section

10.2.2 Narrow-gap concentric cylinder measuring systems according to ISO 3219

Concentric cylinder MS showing a narrow gap are described by the standards ISO 3219 (1993, firstly in 1991) and DIN 53019-1 (firstly in 1980). The basic idea of these standards is to postulate a limited dimension of the annular gap in order to achieve shear rate values as constant as possible within the entire shear gap.

10.2.2.1 Geometry of ISO cylinder systems

For the values of the circumferential velocity counts, depending on the shear gap dimension (see Figure 10.3):
1) **Narrow gap:** The v(r)-curve shows almost the shape of a straight line, which is desired.
2) **Large gap:** The v(r)-function displays a curvature, which is not desired. This leads to inaccurate results when liquids are tested which are not exhibiting ideally viscous flow behavior. If the sample shows clearly cohesion forces, inhomogenous deformation behavior should be taken into account. In other words: Then, deformation and shear rate are not constant throughout the entire shear gap (see also Chapter 2.2.2 and Figure 2.3: Two-Plates-Model, as well as Chapter 3.3.4.3d and Figure 2.9: Plastic behavior).

a) The standard geometry: Ratio of the radii of cup and bob
In order to limit the shear gap dimension, in the ISO standard the permissible maximum value of the ratio δ_{cc} (pronounced "delta-cc") is specified as follows:

Equation 10.7 δ_{cc} = R_e/R_i = 1.0847   or $(R_i/R_e)^2$ = 0.85,   respectively.

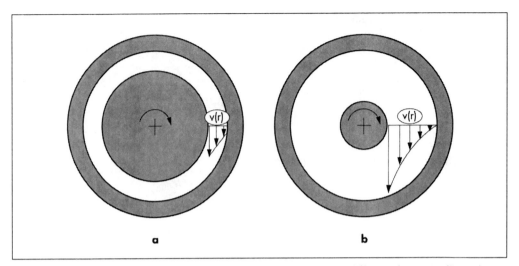

Figure 10.3: Cross-section of two cylinder measuring systems showing different shear gap dimensions

This is the elementary requirement of ISO 3219. Some scientists prefer even narrower gaps. Table 10.1 demonstrates the considerable effect of the ratio of the radii on the resulting absolute gap dimension (here for R_i = 12.5mm).

Note 1: ISO specifies only the ratio of the radii
It is only the ratio of the radii of cup and bob which is specified in the standard, and not the absolute values of the dimensions of the radii.

Note 2: Denomination of the ratio of the radii as δ_{cc} (and not as δ)
According to the standards (ISO and DIN), the ratio of the radii is called δ. However, since the sign δ is already used for the phase shift angle of oscillatory tests (see Chapter 8.2.3), the term δ_{cc} has been chosen here to avoid confusion (the index "cc" for "concentric cylinders").

b) The standard geometry: Cylinder dimensions according to ISO and DIN
The following values are stated for the standard geometry (see Figure 10.1):
L/R_i = 3, with the length L of the cylindrical part of the bob,
L'/R_i = 1, with the distance L' from the lower edge of the cylindrical part of the bob to the base of the cup,
L''/R_i = 1, with L'' as the immersed part of the bob shaft,
R_s/R_i = 0.3, with the radius R_s of the bob shaft (this is only defined by DIN, but not by ISO),
α = 120°, with the (inner) angle α of the apex of the bob.
Thus, the dimensions of L, L', L'' and R_s all are related to R_i.

Only when meeting the above specifications, a concentric cylinder MS can be called an **ISO cylinder measuring system** or a DIN-MS, respectively.

For "other geometries" may be chosen (according to ISO):
$\delta_{cc} \leq 1.2$, and $L/R_i \geq 3$, and $L'/R_i \geq 1$, and $90° \leq \alpha \leq 150°$
Please note, that cylinder MS showing the dimensions of the "other geometries" no longer should be called "standard systems".

Table 10.1: Relation between the ratio of the radii of cup and bob, and the resulting gap dimensions

R_i [mm]	δ_{cc}	R_e [mm]	Gap ($R_e - R_i$) [mm]
12.5	1.2000	15.000	2.5
12.5	1.0847 (ISO)	13.559	1.06 ≈ 1.0
12.5	1.0100	12.625	0.125 ≈ 0.1

&ᴄ   For "Mr. and Ms. Cleverly"

10.2.2.2 Calculations

In order to analyse tests performed with cylinder MS showing a narrow gap, **the so-called "representative" rheological parameters** are defined, and this means, they **are related to the middle of the shear gap**.

a) The representative shear stress in a narrow cylinder gap
Definition:

Equation 10.8 $\tau_{rep} = (\tau_i + \tau_e)/2$

with the shear stresses τ_i on the bob surface and τ_e on the cup surface. Then:

Equation 10.9 $\tau = \tau_{rep} = \dfrac{(1 + \delta_{cc}^{2})}{2 \cdot \delta_{cc}^{2}} \cdot \dfrac{M}{2\,\pi \cdot L \cdot R_i^{2} \cdot c_L} = C_{ss} \cdot M$

with the "end-effect correction factor" c_L that accounts for that part of the total torque which is occurring due to the conical area of the apex of the bob (usually is taken: $c_L = 1.10$), and the MS constant C_{ss} [Pa/Nm or m^{-3}] as the conversion factor between M and τ. C_{ss} depends only on the geometrical dimensions of the MS (i.e. L, R_i and δ_{cc}). Increased sensitivity for low torque or low shear stress values, respectively, is achieved when using a bob with a larger radius.

The following applies to all standard ISO cylinder MS (i.e., when $\delta_{cc} = 1.0847$):

Equation 10.10 $\tau = (0.0446 \cdot M)/R_i^{3} = C_{ss} \cdot M$

Summary for practical users in order to select the optimal MS geometry
For low-viscosity liquids it is preferable to use a large MS, i.e. a MS with a large diameter, and therefore, showing a large **shear area**. Correspondingly, for highly viscous samples better a smaller MS should be selected.

Note: The conversion factor for viscometers displaying relative torque values
When using simple viscometers, often the values of the relative torque M_{rel} are displayed only, thus, in percent or in per thousand. In this case, the MS constant C_{ss} is related to M_{rel}. Here, the maximum torque $M_{rel} = 100\%$, and the lower torque values are specified in terms of a percentage of this, for example, as $M_{rel} = 10\%$.

b) The representative shear rate in a narrow cylinder gap
Definition:

Equation 10.11 $\dot{\gamma}_{rep} = (\dot{\gamma}_i + \dot{\gamma}_e)/2$

with the shear rates $\dot{\gamma}_i$ on the bob surface and $\dot{\gamma}_e$ on the cup surface. Then:

Equation 10.12 $\dot{\gamma} = \dot{\gamma}_{rep} = \dfrac{(1 + \delta_{cc}^{2})}{(\delta_{cc}^{2} - 1)} \cdot \omega = C_{sr} \cdot n$

with the MS constant C_{sr} [min/s] as the conversion factor between n and $\dot{\gamma}$. C_{sr} depends only on the ratio of the radii of cup and bob. The following applies to all standard ISO cylinder MS (i.e., when $\delta_{cc} = 1.0847$): $\dot{\gamma}_{rep} = 12.33 \cdot \omega$, and with $\omega = (2\pi \cdot n)/60 = 0.1047 \cdot n$,
then: $\dot{\gamma}_{rep} = (12.33 \cdot 0.1047) \cdot n$ or

Equation 10.13 $\dot{\gamma}_{rep} = 1.291 \cdot n$

Summary for practical users in order to select the optimal MS geometry
Independent of the size, for all ISO cylinder MS holds: When presetting the same rotational speed or deflection angle, respectively, there is also resulting the same shear rate or strain, respectively. A higher shear rate would be achieved at the same preset rotational speed when

using a cylinder system with a smaller ratio of the radii (i.e., when $\delta_{cc} < 1.0847$), and therefore, with an even narrower gap as stated by ISO 3219 (see also the Example in Chapter 10.2.4: high-shear cylinder MS).

Note: The shear rate range in the gap of an ISO cylinder MS
However, when setting a constant rotational speed, the shear rate in an ISO cylinder MS is not really constant throughout the entire shear gap. Even in an ISO standard geometry and even for fluids showing ideally viscous flow behavior, there is still a difference in the shear rates of approximately 8% between the middle of the gap, showing $\dot{\gamma}_{rep}$ there, and the bob surface. The same relative value occurs between the middle of the gap and the cup surface. Therefore, the relative total difference in the shear rates between the two surfaces of the MS results in approximately 16%. For scientific investigations, possible effects due to these non-constant shear conditions should be taken into account. When performing experiments on liquids showing clearly shear-thinning, shear-thickening or viscoelastic behavior, these kinds of effects are occurring of course even more pronounced then.

c) The representative shear viscosity in a narrow cylinder gap
Related to the middle of the shear gap holds:

$$\eta_{rep} = (\tau_{rep}/\dot{\gamma}_{rep}) = (0.0446 \cdot M)/(R_i^3 \cdot 1.291 \cdot n) \qquad \text{or}$$

Equation 10.14 $\eta_{rep} = (0.0345 \cdot M)/(R_i^3 \cdot n)$

10.2.2.3 Conversion between raw data and rheological parameters

a) Torque M and shear stress τ
$\tau = C_{ss} \cdot M = (0.0446 \cdot M)/R_i^3$

Example: When $R_i = 12.5mm$, i.e., with a diameter of 25mm, then M = 10mNm corresponds to $\tau = 228Pa$.

b) Rotational speed n and shear rate $\dot{\gamma}$
$\dot{\gamma} = C_{sr} \cdot n = 1.291 \ [min/s] \cdot n$

The C_{sr}-factor is the same for all ISO cylinder MS, independent of the radii of bob and cup since it always shows the value $C_{sr} = 1.291min/s$.
Example: $n = 77.5min^{-1}$ corresponds to $\dot{\gamma} = 100s^{-1}$.

c) Deflection angle φ and shear deformation γ
$\gamma = (60 \ [s/min] \cdot C_{sr} \cdot \varphi)/2\pi = 9.55 \ [s/min] \cdot C_{sr} \cdot \varphi$

for $C_{sr} = 1.291min/s$, then $\gamma = 12.3 \cdot \varphi$ or $\varphi = 0.0811 \cdot \gamma = 81.1 \cdot 10^{-3} \cdot \gamma$

Example: $\varphi = 0.0081rad = 8.1mrad$ corresponds to $\gamma = 0.1$ (= 10%).

10.2.2.4 Flow instabilities and secondary flow effects in cylinder measuring systems

Secondary flow effects may occur when measuring **low-viscosity liquids at high shear rates**. This may lead to **turbulent flow behavior**, and as a consequence, to an increasing flow resistance. The following refers to ideally viscous fluids only, showing no significant elastic behavior.

a) Taylor vortices and the Ta number in the annular gap of a Searle cylinder MS [354]
If the bob is rotating and the cup is stationary, for liquids flowing in the annular gap of a concentric cylinder MS there is a critical upper limit between laminar and turbulent flow behavior at which flow instabilities are occurring (i.e. when using the Searle method, see Chapter 10.2.1.2a). This is **due to centrifugal forces and inertial effects caused by the mass of the fluid**. It

is possible to determine the corresponding limiting value using the Taylor number (Ta) which depends on the rotational speed, here in terms of the angular velocity ω in [rad/s] or in [s⁻¹], and on the radius R_i [m] of the bob, on the density ρ [kg/m³] and viscosity η [Pas] of the liquid, as well as on the shear gap dimension of the cylinder MS, here in terms of the ratio of the radii δ_{cc}.

Note: Taylor vortices do not appear in cylinder MS showing a rotating cup and a stationary bob (i.e. when using the Couette method, see Chapter 10.2.1.2b).

Calculation of the Ta number (according to DIN 53019-3):

Equation 10.15 $Ta = [\omega_c \cdot \rho \cdot R_i^2 \cdot (\delta_{cc} - 1)^{3/2}]/\eta \geq 41.2$

Therefore, the **stability criterion** for the critical angular velocity ω_c when Ta vortices are occurring is met if

Equation 10.16 $\omega_c = (41.2 \cdot \eta)/[\rho \cdot R_i^2 \cdot (\delta_{cc} - 1)^{3/2}]$

Example: Measurement of water using an ISO cylinder MS (at T = +20°C)
Presets: η = 1mPas = 10^{-3}Pas = 10^{-3}kg/(s · m), ρ = 1g/cm³ = 1000kg/m³,
R_i = 12.5mm = $1.25 \cdot 10^{-2}$m, and δ_{cc} = 1.0847

Calculation
$\omega_c = (41.2 \cdot 10^{-3})$kg · m³/[1000 · (1.25)² · 10^{-4} · (0.0847)³/² s · m · kg · m²] = 10.7rad/s (or s⁻¹)
With $\omega_c = (2\pi \cdot n_c)/60$, the following holds:
The critical rotational speed: $n_c = (60 \cdot \omega_c)/2\pi = (60 \cdot 10.7s^{-1})/2\pi$ = 102min⁻¹
The critical shear rate: $\dot{\gamma}_c = 1.291 \cdot n_c$ = 132s⁻¹
For η = 10mPas, then $\dot{\gamma}_c$ = 1320s⁻¹; and for η = 100mPas, then $\dot{\gamma}_c$ = 13,200s⁻¹.

Note: According to DIN 53019-3, the above formula used to calculate the Ta number is only valid if $1 < \delta_{cc} < 1.04$. When δ_{cc} = 1.0847 (as stated in ISO 3219), therefore a more complex formula should be taken. However, the difference is not more than 5% when comparing the critical velocity values achieved by use of the two different formulas.

More information on this topic can be found in [246, 261, 276, 352], and images of **axi-symmetric "Taylor vortices"** or **ring-like "Taylor cells"** are illustrated in [45].

b) Reynolds number in the circular gap of a cylinder MS [305]
Turbulent flow behavior also occurs if the critical Reynolds number (Re) is reached. The Re number characterizes the **ratio of the forces due to mass inertia of the fluid and flow resistance of the fluid**.

Note: Flow instabilities according to Reynolds are occurring for both kinds of cylinder MS, and therefore, as well when working with the Searle method as well as with the Couette method.

Calculation of the Re number in the general form:

Equation 10.17 $Re = (v_m \cdot L \cdot \rho)/\eta$

with the density ρ [kg/m³], the viscosity η [Pas], and the mean velocity v_m [m/s] of the fluid (sometimes referred to as the "characteristic" or "typical speed"), the geometrical dimension L [m], (or the "characteristic" or "typical length").

For an annular gap: $L = (R_e - R_i)$

Related to the middle of the gap holds, using the "representative values" according to ISO 3219:

$v_m = \omega \cdot R_{rep}$, with $R_{rep} = (R_e + R_i)/2$

with the angular velocity ω [rad/s]. Therefore:

$Re = [\omega \cdot (R_e + R_i) \cdot (R_e - R_i) \cdot \rho]/(2 \cdot \eta) = [\omega \cdot \rho \cdot (R_e^2 - R_i^2)]/(2 \cdot \eta)$ or

Equation 10.18 $Re = [\omega \cdot \rho \cdot R_i^2 \cdot (\delta_{cc}^2 - 1)]/(2 \cdot \eta)$

There are **two critical Re numbers** existing:

1) $Re_{c1} \geq 1$: occurrence of "end effects", causing flow instabilities around the edges of the upper and lower end of the cylindrical part of the bob
2) $Re_{c2} \geq 1000$ (to 10,000): occurrence of turbulence in the circular gap itself
Therefore, the critical angular velocity ω_{c2} [rad/s] is reached if
$\omega_{c2} = (2 \cdot \eta \cdot Re_{c2})/[\rho \cdot R_i^2 \cdot (\delta_{cc}^2 - 1)]$, and $\omega_{c1} = 0.001 \cdot \omega_{c2}$
For more information on the Re number: see Chapter 11.3.1.3, and DIN 53019-3, [276, 352].

Example: Measurement of water using an ISO cylinder MS (at T = +20°C)
Presets: $\eta = 1mPas = 10^{-3}Pas$, $\rho = 1g/cm^3 = 1000kg/m^3$, $R_i = 12.5mm$, and $\delta_{cc} = 1.0847$

Calculation
$\omega_{c2} = (2 \cdot 10^{-3} \cdot 1000\ kg \cdot m^3)/[1000 \cdot (1.25)^2 \cdot 10^{-4} ([1.0847]^2 - 1)\ s \cdot m \cdot kg \cdot m^2] = 72.5rad/s$
The critical rotational speed: $n_{c2} = (60 \cdot \omega_{c2})/2\pi = (60 \cdot 72.5s^{-1})/2\pi = 692min^{-1}$
The critical shear rate: $\dot{\gamma}_{c2} = 1.291 \cdot n_{c2} = 893s^{-1}$ (and $\dot{\gamma}_{c1} = 0.9s^{-1}$)
For $\eta = 10mPas$, then $\dot{\gamma}_{c2} = 8930s^{-1}$; and for $\eta = 100mPas$, then $\dot{\gamma}_{c2} = 89,300s^{-1}$.

Note: Critical values of the angular velocity
It should be taken into account that the first flow instability effects are occurring already at the critical values of ω_{c1} (or n_{c1} or $\dot{\gamma}_{c1}$, respectively). These values indeed are only one per thousand compared to the values with the index "c2", at which clearly visible turbulent effects are appearing finally. Of course, these flow irregularities are also influenced by the degree of the smoothness or unevenness of the walls of the MS, respectively.

 ᘓ⁓   End of the Cleverly section

10.2.2.5 Advantages and disadvantages of cylinder measuring systems

a) Advantages
1) **Low-viscosity fluids, and liquids showing a low surface tension cannot flow off the shear gap**, and that counts **even for high shear rates**. This is a great advantage compared to cone-and-plate (CP) MS and parallel-plate (PP) MS.
2) Good temperature control of the sample is garanteed due to the relatively large contact area at the wall of the cup.
3) Even at high rotational speeds, the annular gap still remains filled with the sample, and this counts even when measuring viscoelastic materials which are showing rod-climbing or the "Weissenberg effect", respectively (see Figure 5.4). These effects are occurring due to normal forces (see Chapter 5.3). Precondition is that the sample has been filled into the cup up to the level L" (see Figure 10.1).
4) In order to counter wall slip effects, measuring bobs and/or cups with sandblasted or profiled cylinder surfaces might be used (see Chapter 10.6.1).
5) In order to prevent solvents evaporating, a cover might be used, and if desired, also a solvent trap.
6) Disposable cups, and if required also disposable bobs, might be used when cleaning is difficult after the test, or if it is impossible, for example, when testing curing materials.

b) Disadvantages
1) A relatively large amount of sample is required.
2) Cleaning after the test is often time-consuming.
3) When testing paste-like samples, unnoticed air bubbles might be included.
4) When measuring low-viscosity liquids, flow instabilities and turbulent flow effects might occur at high rotational speeds (e.g. Taylor vortices, see Chapter 10.2.2.4).

5) When using a small bob with a correspondingly thin shaft, radial eccentricity may occur after a certain time of usage, and as a consequence, turbulent flow may already appear at relatively low shear rates.

10.2.3 Double-gap measuring systems (DG MS)

Double-gap MS are concentric cylinder MS **designed for tests on low-viscosity liquids**, briefly, DG MS. Probably *F. Moore* and *L.J. Davies* (in 1956, [256]) were the first who used this method (called a "double viscometer"), [384]. Before rheometers equipped with air-bearings were available, it was not easy to obtain accurate measuring results on low-viscosity liquids at shear rates lower than $\dot\gamma = 100s^{-1}$. Therefore, special MS were designed showing a large shear area to achieve torque values as high as possible also in this shear rate range.

Here, an additional inner cylinder is mounted in the center of the cup, and therefore, the cross-section of the cup shows an annular gap (see Figure 10.4; the corresponding standard DIN 54453 is meanwhile

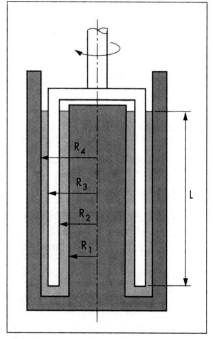

Figure 10.4: Double-gap measuring system (DG MS)

withdrawn, however, for reasons which are not referring to the geometry of the MS). The bob has the shape of a hollow cylinder, thus, showing both an inner and an outer surface, and as a consequence, there is an increased shear area available. Since this area corresponds to the contact area with the sample, this kind of MS is better suited to detect lower torques compared to the usual cylinder MS according to ISO 3219. The aim is to achieve uniform shear conditions as well in the inner as well as in the outer gap, and therefore, the same shear rate values in both gaps. The following is stated for the ratio of the radii (according to DIN 54453):
$\delta_{cc} = R_4/R_3 = R_2/R_1 \leq 1.15$ (the four radii are illustrated in Figure 10.4).
The length of the immersed part of the bob should be $L \geq 3 \cdot R_3$.

When using double-gap MS, there is the option to control the temperature of the sample as well via the wall as well as via the inner cylinder of the cup, if the hollow cylinder in the center of the cup is also connected e.g. to a water bath circulator. A disadvantage of the DG MS is the rather time-consuming cleaning procedure.

10.2.4 High-shear cylinder measuring systems (HS MS)

Concentric cylinder MS **for testing low-viscosity fluids at high shear rates** are designed with **a very narrow annular gap**. In extreme cases, they can be compared to a "loose fit", for example, showing a shear gap of 100μm only. Therefore, even a very slight eccentricity of the bob would cause a significant influence on the test results, since in this case, there is the risk of exceeding the range of laminar flow and to get turbulent flow conditions then. For tests on dispersions should be taken into account, that in any case the gap dimension should be in minimum five (or even better ten) times larger than the maximum particle size.

Example: High-shear cylinder measuring systems
With the radii of the cup R_e = 29.3mm and the bob R_i = 28.7mm, therefore showing the shear gap dimension of $(R_e - R_i)$ = 0,6mm = 600μm, and the ratio of the radii $\delta_{cc} = (R_e/R_i)$ = 1.0209,

there is a conversion factor between the rotational speed and the shear rate of $C_{sr} = 5.34$. Compared to the factor of the standard ISO cylinder MS ($C_{sr} = 1.291$), at the same rotational speed there are achieved more than four times higher shear rates (see Chapter 10.2.2.2b).

Note 1: Viscous heating at high shear rates
In principle, when performing rotational tests, viscous shear-heating of the sample is occurring, and this effect has to be taken into account especially for high shear rates. In general, flow, i.e. viscous behavior, is connected with relative motion between the molecules or particles, and therefore, with frictional heating which of course is more pronounced at higher shear rates. Thermal effects always have a considerable influence on the rheological properties. **It is essential to consider this when applying shear rates higher than $\dot\gamma = 1000s^{-1}$**, and it is crucial if $\dot\gamma \geq 10,000s^{-1}$ (see also the Note in Chapter 2.3.1b: viscous shear heating).

Note 2: High-pressure capillary viscometers for testing at very high shear rates
High-pressure capillary viscometers are used to perform tests at extremely high shear rates up to $\dot\gamma = 100,000$ or 1 million, and even up to 10 mio.s^{-1} (see also Chapter 11.4.2).

10.3 Cone-and-plate measuring systems (CP MS)

Cone-and-plate MS are described by the standards ISO 3219 (1993) and DIN 53019-1 (firstly in 1976 by DIN 53018).

10.3.1 Geometry of cone-and-plate systems

A cone-and-plate MS consists of a relatively flat circular cone and a plate, briefly, CP MS. Usually, the cone as the rotor is the upper part of the measuring geometry, and the bottom plate is the stationary part which is mounted onto the rheometer stand. The dimensions of the slightly conical area are defined by the **cone radius R** and the **cone angle** α (see Figure 10.5). In the ISO standard it is recommended to use the angle $\alpha = 1°$. Larger angles may be used, but ISO states here $\alpha < 4°$. Therefore, this MS is sometimes also called "small-angle CP geometry". According to DIN to the dimension of the radius applies: $10mm \leq R \leq 100mm$, i.e., the diameter should occur between 2 and 20cm.

Note: Optional specifications for angle units, degrees or rad
Angles can be specified as well in the radian measure [rad] as well as in the angular dimension [°], i.e., in degrees. The following applies:
2π rad ≈ 6.283rad corresponds to 360° (using the circle contstant $\pi = 3.1416$),
1 rad corresponds to about 57.3°, and 1° corresponds to approximately 0.0175rad = 17.5mrad.

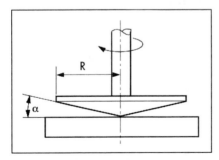

Figure 10.5: Cone-and-plate measuring system (CP MS)

᪥   For "Mr. and Ms. Cleverly"

10.3.2 Calculations

a) Shear stress in a CP gap
Equation 10.19 $\tau = (3 \cdot M)/(2\pi \cdot R^3) = C_{ss} \cdot M$

with the MS constant C_{ss} [Pa/Nm or m^{-3}] as the conversion factor between M and τ. C_{ss} depends only on the cone radius R. An increased sensitivity for low torque or low shear stress values, respectively, is achieved when using a cone with a larger radius.

Summary for practical users in order to select the optimal MS geometry

For low-viscosity liquids it is preferable to use a large MS, i.e. a MS with a large diameter and therefore showing a large shear area. Correspondingly, for highly viscous samples better a smaller MS should be selected.

Table 10.2: Dependence of the factor C_{sr} on the size of the cone angle α

α	0.5°	1°	2°	3°	4°	6°
C_{sr}	12	6	3	2	1.5	1

b) Shear rate in a CP gap

The shear gap dimension h between cone and plate is increasing with the distance r from the rotation axis. For r counts: $0 \leq r \leq R$, with r in the center, and $r = R$ at the edge of the cone. Therefore, the maximum gap size is reached at the edge of the cone. Here counts:

$\tan\alpha = h_{max}/R$ or $h_{max} = R \cdot \tan\alpha$

The circumferential velocity v is also increasing from the inside to the outside reaching the maximum value: $v_{max} = v(R) = \omega \cdot R$. Therefore:

$\dot{\gamma}(R) = v_{max}/h_{max} = (\omega \cdot R)/(R \cdot \tan\alpha)$ or

Equation 10.20 $\dot{\gamma}(R) = \omega/\tan\alpha \approx \omega/\alpha = C_{sr} \cdot n$

with the MS constant C_{sr} [min/s] as the conversion factor between n and $\dot{\gamma}$. C_{sr} depends only on the cone angle α. For cones showing small angles, there can be assumed that $\tan\alpha$ and α are showing approximately the same value if the angle is specified in [rad]; i.e., $\tan\alpha \approx \alpha$. According to ISO and DIN, this counts for $\alpha < 0.05$rad or $\alpha < 3°$, respectively. As mentioned above, it is recommended to use cone angles not greater than $\alpha = 1°$. Then, the simplified formula can be taken, using α instead of $\tan\alpha$.

For CP MS, the shear rate is independent of R, and that counts for each value of the radius in the whole range of the conical gap, thus, for $0 \leq r \leq R$. Therefore, when presetting a constant rotational speed, also the shear rate is showing a constant value within the entire shear gap. Correspondingly, when presetting a constant deflection angle, a homogeneous shear deformation occurs within the whole sample. The following holds:

Equation 10.21 $C_{sr} = 2\pi/(60 \cdot \alpha)$, with α in [rad],   or $C_{sr} = 6/\alpha$, with α in degrees

Some examples are listed in Table 10.2 showing the inverse proportionality of the C_{sr}-factor and the cone angle α. Therefore, at the same rotational speed a higher shear rate is achieved when using a smaller cone angle.

Example

With a cone angle of 1°, at the rotational speed of n = 100min⁻¹, the resulting shear rate $\dot{\gamma} = 600s^{-1}$. However, with the cone angle of 2° at the same speed only $\dot{\gamma} = 300s^{-1}$ will be achieved. Therefore, when taking a doubled cone angle and operating at the same speed, only the half shear rate value will be obtained.

Summary for practical users in order to select the optimal MS geometry

In ISO 3219 is recommended to use preferably the cone angle $\alpha = 1°$. When using a larger cone angle, independent of the cone radius, at the same rotational speed a lower shear rate will be obtained, and correspondingly, when using a smaller cone angle a higher shear rate will be achieved.

c) Viscosity in a CP gap

Equation 10.22 $\eta = \tau/\dot{\gamma} = (3 \cdot M \cdot \alpha)/(2\pi \cdot R^3 \cdot \omega)$

10.3.3 Conversion between raw data and rheological parameters

a) Torque M and shear stress τ

$\tau = C_{ss} \cdot M = (3 \cdot M)/(2\pi \cdot R^3)$

Example: When R = 25mm, i.e. with a diameter of 50mm, then M = 10mNm corresponds to τ = 306Pa.

b) Rotational speed n and shear rate $\dot{\gamma}$

$\dot{\gamma} = C_{sr} \cdot n = (6 \cdot n)/\alpha$,   if the cone angle α is specified in degrees [°]

Example: When α = 1°, then n = 16.7min^{-1} corresponds to $\dot{\gamma}$ = 100s^{-1}.

c) Deflection angle φ and the deformation γ

$\gamma = (60 \ [s/min] \cdot C_{sr} \cdot \varphi)/2\pi = 9.55 \ [s/min] \cdot C_{sr} \cdot \varphi$

for $C_{sr} = 6/\alpha$ and α in degrees [°],   then $\gamma = (57.3 \cdot \varphi)/\alpha$

or $\varphi = 0.0175 \cdot \alpha \cdot \gamma = 17.5 \cdot 10^{-3} \cdot \alpha \cdot \gamma$

Example: When α = 1°, then φ = 0.0018rad = 1.8mrad corresponds to γ = 0.1 = 10%.

Please note: Here, the cone angle α is specified in degrees, however φ is given in rad or mrad.

10.3.4 Flow instabilities and secondary flow effects in CP systems

Secondary flow effects may occur when measuring **low-viscosity liquids at high shear rates**. This can lead to **turbulent flow behavior**, and as a consequence, to an increasing flow resistance. Turbulent flow behavior appears if the critical Reynolds number (Re) is reached due to inertial effects of the liquid (see also Chapter 10.2.2.4b).

However, for a conical gap – which is open to the outside – it is not possible to specify a useful value for the "characteristic length" L, since there is no constant value for the geometry of the "flow channel". When turbulent flow appears, a low-viscosity liquid is uncontrollably flowing off the open conical gap due to centrifugal forces or inertia effects of the fluid, respectively. In this case it is better to use a cylinder MS.

When testing viscoelastic materials using CP MS, edge effects in the form of **flow instabilities** and **surface fracture**, streaks and migration off the gap may occur even at low rotational speeds due to the elastic behavior of the sample. These effects can be compared to "melt fracture" when extruding polymers, or to rod-climbing or the "Weissenberg effect" when stirring. Both behaviors are caused by normal forces (see also Chapters 5.2.1.2b and 5.3).

$\mathcal{GF}$ End of the Cleverly section

10.3.5 Cone truncation and gap setting

In order to perform useful rheological tests, it is recommended to use a CP MS cone showing a **truncated tip**. Here, **exact gap setting** is required between the cone and the bottom plate; see the distance "a" in Figure 10.6. This measure "a" is the distance required by the imaginary cone tip if it still was present, to touch the lower plate exactly in the contact point. This requirement demands accurate operation by the user when setting the cone distance manually. Therefore, most modern rheometers are equipped with features like **"automatic gap setting" AGS** and **"automatic gap control" AGC** in order to simplify the gap setting procedure and to improve the reproducibility of the test conditions. For more information about AGS and AGC: see Chapter 10.4.6.

There are different concepts concerning the **degree of cone truncation**.
Concept 1: Some manufacturers specify the distance between cone and bottom plate to be **always at a defined distance** of a = 50µm for all common CP MS showing the cone

angles of $\alpha = 1°$ or $2°$. Illustrative comparison: This is the thickness of a human hair. Cones showing smaller angles, for example $\alpha = 0.5°$, are often designed for a = 25μm; and for larger angles there is a correspondingly larger size of the measure "a".

Concept 2: Here, the design is specified by the manufacturer **with the aim of generating a similar flow field** which is independent of the geometric dimensions. The ratio of the flow field coefficient should always be the same for a CP system with cone truncation compared to a system without (e.g. showing the ratio 1.004).

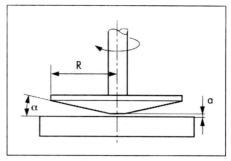

Figure 10.6: CP measuring system showing a truncated cone

According to DIN 53019-1, the flow field coefficient is defined as the ratio of the MS-constants C_{sr} and C_{ss} (see Chapter 10.3.2a/b). This results in different values for the measure "a", depending as well on the radius R as well as on the cone angle α. Examples:

CP 25-1 (showing the diameter 2R = 25mm and $\alpha = 1°$): a = 50μm
CP 25-2 (with 2R = 25mm and $\alpha = 2°$): a = 105μm
CP 50-1 (with 2R = 50mm and $\alpha = 1°$): a = 100μm
CP 50-2 (with 2R = 50mm and $\alpha = 2°$): a = 210μm

Note: Exactly gap setting is essential and the required gap dimension has to be kept strictly on this value during the whole test.

Usually for each individual CP MS, the gap dimension is specified by the manufacturer in the corresponding data sheets and in the analysis software, respectively. Gap setting at any other dimension would lead to wrong measuring results.

Cones are truncated for the following reasons:
1) There is no wear by abrasion on the surfaces of both the rotating cone and the stationary bottom plate. Otherwise the dimension of the shear gap, and therefore the geometrical conditions, would be changing with progressive use.
2) There is no direct friction between the rotating cone and the stationary bottom plate. Without cone truncation, the measurement of the torque value would consist not only of the torque which is generated by the sheared viscous sample, but also of the torque resulting from the static and kinetic friction between the contact partners cone and plate. Of course, these kinds of forces would influence the test result significantly.
3) A positive side effect is that even materials containing particles up to a certain size can be measured usually without any problems. For example, when gap setting at a = 50μm, particles may be tolerated showing dimensions up to 5 (or even 10) μm.

10.3.6 Maximum particle size

When using CP MS, they should be taken for samples only which contain solid components such as particles up to a limited maximum particle size. Otherwise there would be not enough free space available between the particles when in motion, which would of course influence the deformation and flow behavior. In this case additionally, a great number of particles would directly touch the surfaces of the MS, and as a consequence, the occurring friction forces would falsify the test result. This also applies to samples containing **aggregates, associates, agglomerates, flocculates, gel particles, crystals, fibers, semi-solid and other rigid superstructures** which therefore should not be larger than the dimension d_{max} of the maximal permissible particle diameter.

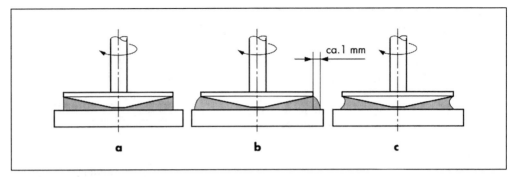

Figure 10.7: Optimal filling of a CP measuring system:
a) good, but not useful for practical tests, b) recommended and correct, and c) underfilled

As a rule of thumb for practical users is recommended:
$d_{max} \leq a/5$, or d_{max} should be not greater than 20% of the gap setting.
Example: When $a = 50\mu m$, then d_{max} should not be larger than $10\mu m$.
In many R&D laboratories people are using $d_{max} \leq a/10$ to be on the safe side.
Example: When $a = 50\mu m$, then the permitted size is merely $d_{max} \leq 5\mu m$.

10.3.7 Filling of the cone-and-plate measuring system

In order to achieve correct test results, the shear gap should be filled completely. When performing rotational tests on low-viscosity liquids, it should be taken into account that the sample may migrate off the gap due to inertia effects or centrifugal forces, respectively. With viscoelastic samples there may appear streaks, surface effects and "melt fracture" due to elastic effects or normal forces, respectively. In order to prevent this, or at least to delay its occurrence, it is recommended to overfill the cone at approximately 1mm beyond the edge, and to strip off the rest using a soft spatula (see Figure 10.7).

10.3.8 Advantages and disadvantages of cone-and-plate measuring systems

a) Advantages
1) Using CP MS, **homogenous shear conditions are achieved since the shear rate or the shear deformation**, respectively, **is constant in the entire conical gap**. Therefore, most scientists prefer the use of CP MS to all other MS.
2) Only a small amount of sample is required.
3) Cleaning after the test is a very simple and therefore a fast action.
4) Already when setting the gap, due to the conical shape, most of the air bubbles which may possibly be contained in a liquid sample are pressed out of the gap.

b) Disadvantages
1) **When testing dispersions, the maximum particle size is limited**. It is not useful to test samples showing a three-dimensional internal structure such as gels or solids.
2) When setting the gap on highly viscous and viscoelastic samples such as polymer melts, a long period of rest might be required until the equilibrium state is reached between the internal structural strength during the relaxation process and the external forces. Often half an hour or even more might be required for this equilibration process (see also Chapter 7.3.3.3b).
Note 1: Therefore, for tests on polymer melts is recommended to use a larger cone angle, e.g. $\alpha = 2°$ instead of $1°$. However, experience showed: For these kinds of samples it is even better to use a parallel-plate MS, setting a gap of e.g. 1mm.

Note 2: In order to guarantee "soft" gap setting, the option "normal force control" NFC, might be used (see Chapter 10.4.6).

3) The following might occur at the edge of a rotating cone: flow inhomogeneities, turbulent flow, inertia effects of the sample due to centrifugal forces, discharge off the gap by creep or flow of the sample, skin formation, streaks, surface effects, melt fracture due to elasticity or due to normal forces, respectively, and evaporation of solvents. All of this may be summed up as the so-called "edge failure" or "edge defects". These appearances might significantly influence the test result (see also Chapter 10.3.7 and Figure 10.7: optimal filling of a MS).

Note: The degree of evaporation might greatly be reduced when using a cover, and if desired, also a solvent trap.

4) The temperature gradient in the sample is increasing with the radius of the cone if the temperature is directly controlled via the bottom plate only. This applies particularly when using a larger cone angle.

Note: As a counter-measure, a hood might be used enabling temperature control of the inner space of the test chamber (see also Chapter 11.6.6e: "active hood").

5) CP tests should be performed at a single constant temperature only, due to the strong influence of the thermal expansion of sample, cone, plate and measuring device on the relatively small shear gap dimension of the CP system.

Note 1: When testing over a temperature range wider than $\Delta T = 20K$, it is better to use a parallel-plate MS, setting a gap of e.g. 1mm.

Note 2: In order to counteract undesired, uncontrolled changes in the gap dimension, the options "automatic gap control" AGC and "TruGap" function may be used (see Chapter 10.4.6).

6) Sandblasted or profiled surfaces are not useful for cones, since after this rough procedure, the exact geometrical conditions such as conical surface, cone angle and measure of the cone truncation cannot be guaranteed any longer.

Summary: When performing rotational tests, despite of the many possible disadvantages mentioned, the first advantage outweighs them all, i.e., the constant shear rate in the conical gap. Therefore here, CP MS should be prefered to all other measuring systems, unless conditions dictate otherwise, such as the particle size. When performing oscillatory tests, as long as measuring in the LVE range, the advantage of the constant shear strain in the conical gap counts no longer, since within this shear range also for all other MS counts: The results obtained are independent of the strain.

Note: "High-shear cone-and-plate viscometer"

Sometimes, for simple QC tests on **coatings** still the **ICI cone-and-plate viscometer** is used, and the measured results often are called **"high-shear viscosity"** values, or **HSV** "according to ICI". This simple CP instrument was designed for high shear rates, aiming to measure in the shear rate range of about 10,000s^{-1} (acc. to ISO 2884-1, ASTM D4287, DIN 53229 and BS 3900, for coatings, paints, varnishes) [118]. Therefore, this device is also termed the "High-Shear Viscometer".

By calculation, for example, at the constant rotational speed of n = 750min^{-1} and using a 0.45° cone, the shear rate of $\dot{\gamma}$ = 10,000s^{-1} will be reached. In ISO 2884-1, a cone diameter of 24mm is mentioned as typically [254]. With some HS-CP viscometers available, it is possible to preset the following rotational speeds: n = 5/50/500/750/900min^{-1}, and the following cone angles are typical: α = 0.45°/1.8°/3.0° (see e.g. [55]).

However, since **cones without truncation** are being used, **the test results are only of limited use** for today's requirements. As a matter of fact, they are actually useful just **for very simple QC tests**. The reasons for this evaluation are:

1) The rotating cone is directly contacting the stationary bottom plate which of course causes friction between the two solid surfaces of the measuring system.

2) When testing dispersions in this very narrow conical shear gap, the particles are in a direct frictional contact to the surfaces of the rotating cone and the stationary bottom plate. Therefore, the gap is "infinitely small" in the center of the cone.

Comment: This is not a scientific test method. Both effects, the direct solid friction on the bottom plate as well of the cone as well as of the particles, are considerably influencing the measuring results since they have a significant direct influence on the measured torque. Therefore, this apparatus should merely be used for very simple QC tests and the results should not be overrated.

10.4 Parallel-plate measuring systems (PP MS)

Parallel-plate MS are described by the following standards ISO 6721-10 (1999) and DIN 53019-1 (firstly in 1976 by DIN 53018).

10.4.1 Geometry of parallel-plate systems

A parallel-plate MS consists of two plates, both showing even surfaces, briefly, the PP MS. Usually, the upper plate as the rotor is the upper part of the measuring geometry, and the bottom plate is the stationary part which is mounted onto the rheometer stand. The dimension of the upper plate is defined by the **plate radius R** (see Figure 10.8). According to DIN for the **distance H between the two plates** is stated: $H \ll R$.

Note: Specifications of PP MS dimensions for testing polymer melts (ISO 6721-10)
For testing polymer melts (e.g. polyethylene PE at T = 190°C, or polypropylene PP at T = 210°C) is recommended: Plate diameter (i.e., 2R) between 20 and 50mm, gap H between 0.5 and 3mm, and the ratio (2R/H) in the range of 10 to 50. For example if 2R = 25mm, gap setting should be between H = 0.5 and 2.5mm. The maximum roughness of the plates is specified as 0.25μm.

The larger the selected gap dimension the larger is the risk of the following effects:
- **Low-viscosity liquids** might show secondary flow effects with vortices (see Chapters 3.3.3, 10.2.2.4 and 10.4.4: Flow instabilities and **turbulent flow** behavior).
- **Polymer samples** might show transient behavior (see Chapter 3.3.1b and Figure 2.9, no. 5: **Time-dependent effects**).
- **Viscoelastic materials** might show **edge failure** (migration and creep off the gap, see Chapter 5.3: Normal forces, and Chapter 10.4.4).
- **Gel-like and paste-like samples, and samples containing surfactants** might show **inhomogenous deformation** behavior (Chapter 3.3.4.3d and Figure 2.9, no. 3/4: Plastic behavior) and shear-banding effects (Chapter 9.2.2 and Figure 2.9, no. 6/7).

This is especially important when performing rotational tests. With oscillatory tests, a larger gap dimension H might be not so critical when working still in the linear viscoelastic (LVE) range, since at these shear conditions, the samples usually are deformed homogeneously throughout the entire shear gap (see also Chapter 10.4.5: Recommended gap setting).

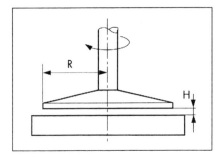

Figure 10.8: Parallel-plate measuring system (PP MS)

᠌᠎   For "Mr. and Ms. Cleverly"

10.4.2 Calculations

a) Shear stress in a PP gap

Equation 10.23 $\tau(R) = (2 \cdot M)/(\pi \cdot R^3) = C_{ss} \cdot M$

with the MS constant C_{ss} [Pa/Nm or m^{-3}] as the conversion factor between M and τ. C_{ss} depends only on the plate radius R. An increased sensitivity for low torque or low shear stress values, respectively, is achieved when using a plate showing a larger radius. In order to calculate the "mean shear stress" see below Chapter 10.4.2b and Equation 10.26.

Summary for practical users in order to select the optimal MS geometry
For low-viscosity liquids it is preferable to use a large MS, i.e., a MS with a large diameter, therefore showing a large shear area. Correspondingly, for highly viscous and rigid samples, better a smaller MS should be selected.

b) Shear rate in a PP gap
The distance between the plates is constant in the whole shear gap. The circumferential velocity v is increasing with the distance r from the center to the edge of the plate. For r counts: $0 \leq r \leq R$, with r = 0 in the center and r = R at the edge of the plate. For the shear rate counts
$\dot{\gamma}(r) = v/H = (\omega \cdot r)/H = (2\pi \cdot n \cdot r)/(60 \cdot H)$
Even at a constant rotational speed, **the shear rate value $\dot{\gamma}$ is not constant** throughout the entire shear gap of the PP MS since it is dependent on the distance r from the rotational axis. This is the great and negative difference compared to the cone-and-plate MS. The shear rate is increasing linearly from the center of the plate: in the center $\dot{\gamma} = 0$, and it is maximal at the edge of the plate $\dot{\gamma}(R) = \dot{\gamma}_{max}$. Therefore, the shear rate distribution is dependent on the radius. This inhomogeneous, not constant shear conditions by scientists are of course considered a disadvantage of the PP MS in principle.

The circumferential speed reaches the maximum value at the edge of the plate, showing:
$v_{max} = v(R) = \omega \cdot R$. Then applies:

Equation 10.24 $\dot{\gamma}(R) = v/H = (\omega \cdot R)/H = (2\pi \cdot n \cdot R)/(60 \cdot H) = C_{sr} \cdot n$

with the MS constant C_{sr} [min/s] as the conversion factor between n and $\dot{\gamma}$. C_{sr} depends on R and H, and therefore with R, it is related to the edge of the plate.

For the **calculation of the shear rates** and analysis of the measuring results, different concepts can be found.
Concept 1 ("maximum shear rate"): The specification of $\dot{\gamma}$ is related to the edge of the measuring plate. Calculation and analysis of the test results are based on the maximum shear rate in the gap (see Equation 10.24).
Concept 2 ("mean shear rate"): Here, the value of the shear rate refers to the location $r_m = (2/3) \cdot R$, in order to work with the "effective mean value of the shear rate". The "mean shear rate" $\dot{\gamma}_m$ [s^{-1}] is the sum of the radius-dependent local, volume-weighted shear rate contributions. The following applies:

Equation 10.25 $\dot{\gamma}_m = (2/3) \cdot \dot{\gamma}(R)$

Similarly, for the "mean shear stress" τ_m [Pa] the following applies:

Equation 10.26 $\tau_m = (2/3) \cdot \tau(R)$

The shear rate is also dependent on the distance H between the plates. Setting a larger distance at the same rotational speed, a lower shear rate is achieved. Correspondingly holds:

In this case, when using the same deflection angle, a lower value for the strain or deformation is obtained. It should be taken into account that a larger shear gap causes the rheological conditions to change for the worse since there is a risk of an increasing degree of inhomogeneous shear behavior. This counts especially for gap dimensions larger than H = 1mm.

Note: Comparing the measuring systems PP and CP in terms of the shear rates
Both disadvantages, the dependence of the shear rate on the radius and on the gap dimension, are not occurring when using a cone-and-plate MS.

Summary for practical users in order to select the optimal MS geometry
When presetting the same rotational speed, a higher shear rate will be obtained, on the one hand when using the same plate diameter and a smaller plate distance, or on the other hand when taking the same plate distance and a larger plate diameter.

c) Viscosity in a PP gap
The following applies for the viscosity, independent of which concept is used for the calculation of the shear rate and the shear stress:

Equation 10.27 $\eta = \tau/\dot{\gamma} = \tau_m/\dot{\gamma}_m = (2 \cdot M \cdot H)/(\pi \cdot R^4 \cdot \omega)$

The only difference is: When using Concept 1 with $\eta = \tau/\dot{\gamma}$, viscosity values are related to the maximum value of the shear rate occurring at the edge of the plate. When using Concept 2 with $\eta = \tau_m/\dot{\gamma}_m$, however, viscosity values are related to the lower shear rate values of $\dot{\gamma}_m = (2/3) \cdot \dot{\gamma}$ which of course makes a difference for samples which are not showing ideally viscous flow behavior.

Many practical users have made the experience that test results which are calculated by Concept 2 are better to compare to results obtained when using other geometries such as cone-and-plate or cylinder MS. This is particularly true for samples showing deviations from the ideally viscous flow behavior.

10.4.3 Conversion between raw data and rheological parameters

a) Torque M and mean shear stress τ_m
$\tau = (2/3) \cdot C_{ss} \cdot M = (4 \cdot M)/(3\pi \cdot R^3)$
Example: When R = 25mm, i.e., with a diameter of 50mm, then M = 10mNm corresponds to $\tau_m = 271$Pa.

b) Rotational speed n and mean shear rate $\dot{\gamma}_m$
$\dot{\gamma} = (2/3) \cdot C_{sr} \cdot n = (\pi \cdot R \cdot n)/(45 \cdot H)$
Example 1: When H = 0.5mm and R = 25mm, i.e., with a diameter of 50mm,
then n = 28.7min^{-1} corresponds to $\dot{\gamma}_m = 100$s^{-1}.
Example 2: When H = 1mm and R = 12.5mm, i.e., with a diameter of 25mm,
then n = 115min^{-1} corresponds to $\dot{\gamma}_m = 100$s^{-1}.

c) Deflection angle φ and mean shear deformation $\dot{\gamma}_m$
$\gamma_m = (2/3) \cdot (60\,[\text{s/min}] \cdot C_{sr} \cdot \varphi)/2\pi = 6.37[\text{s/min}] \cdot C_{sr} \cdot \varphi$
for $C_{sr} = (2\pi \cdot R)/(60 \cdot H)$, then $\gamma_m = (2R \cdot \varphi)/3H$ or $\varphi = (3H \cdot \gamma)/2R$
Example 1: When H = 0.5mm and R = 25mm,
then $\varphi = 0.003$rad = 3mrad corresponds to $\gamma_m = 0.1 = 10\%$.
Example 2: When H = 1mm and R = 12.5mm,
then $\varphi = 0.012$rad = 12mrad corresponds to $\gamma_m = 0.1 = 10\%$.

10.4.4 Flow instabilities and secondary flow effects in a PP system

When measuring **low-viscosity liquids at high shear rates**, this may lead to **turbulent flow conditions** (see also Chapter 10.3.4). When testing **viscoelastic materials**, even at low rotational speeds flow instabilities, **streaks, creep and migration off the gap** might occur **due to elastic effects** of the sample, and rod-climbing or the "Weissenberg effect" and **"melt fracture"** with polymers (see also Chapters 5.2.1.2b and 5.3).

&⌐   End of the Cleverly section

10.4.5 Recommendations for gap setting

Note 1: Testing resins, polymer melts, and silicones (unlinked PDMS)
1) **When testing polymer melts**, usually good results are achieved when using a plate with the diameter of 2R = 25mm if a shear gap is selected in the range of 0.5mm ≤ H ≤ 1mm.
2) According to ASTM D4440, gap setting at H = 1 to 3mm is useful as a good operating range **when testing thermoplastic resins and polymer melts**; and **H = 0.5mm** is recommended **as a minimum**. Even when large plates are used for testing low-viscosity materials, the suggested minimum gap H = 0.25mm.
3) According to ASTM D4473, gap setting for testing neat resins should be approximately at H = 0.5mm, and for self-supporting compositions showing several plies, H = 1 to 2mm is proposed.

Note 2: Maximum particle size, and required gap setting
Gap setting should be at least 5 times larger – and 10 times larger is even better – than the largest dimension of the particles, agglomerates, gel particles, fibers, semi-solid and other rigid components or superstructures of the sample. However, also **for dispersions and gels** counts: Well approved for daily measuring practice is **H = 1mm (or 0.5mm ≤ H ≤ 1mm)**.

Note 3: Special analysis is required when gap setting is less than 0.3mm
When performing rotational tests, for gap dimensions smaller than H = 0.3mm, pre-tests are required using corresponding standard oils. The reason for this measure is to get first of all information on the degree of the internal or **viscous shear-heating of the sample** which is occurring more pronounced in a narrow gap geometry. Experience showed that in this case, lower viscosity values may be obtained then. Therefore, it is recommended to determine a corresponding correction factor by previously performed tests on standard oil showing a comparable viscosity in order to get a better base for the calculation of the viscosity values finally. Please note: The smaller the gap the worse will be the test results if plane parallelism of the plates cannot be taken for granted.

&⌐   For "Mr. and Ms. Cleverly"

10.4.6 Automatic gap setting and automatic gap control using the normal force control option

a) Target: Gap setting and gap control
1) **Speed-controlled gap setting** without normal force control, as **"automatic gap setting"** AGS, in the form of a "force-insensitive" gap setting option.
Example: It is aimed to go to the target gap dimension of H = 500μm = 0.5mm, at an axial speed of v = 0.1mm/s downwards, without regard to the sample's structure.
2) **Speed-controlled and "normal force-controlled"** NFC gap setting, thus, both simultaneously, as a **"soft" gap setting option** in order not to destroy the sample's structure.

Example: It is aimed to go to the target gap dimension of H = 0.5mm at v = 0.1mm/s, but the normal force F_N = 1N should never be exceeded.

3) "Automatic gap control" AGC and "TruGap" function

This option controls the plate distance, always returning to the desired value if it should have been changed due to any reasons [10].

b) Target: Forced contact of sample and plates at a constant compression force

This option may be used to guarantee always contact between the sample and the surfaces of the plates, or to perform compression tests or shear tests in the form of rotational, oscillatory, creep or relaxation tests while the sample is stressed by a defined compression force or positive normal force, respectively. Here, the gap dimension may change during testing, which is detected as resulting raw data then, to be recorded in the test protocol if desired.

Example: It is aimed to move the measuring plate downwards until reaching the target value of F_N = 1N, subsequently keeping this value constantly as a compression force onto the sample.

c) Target: Compression-less but guaranteed contact between the sample and the plates

This option may be used to compensate thermal expansion or shrinkage of sample and test device when heating or cooling, respectively. Here, the gap dimension may change during testing which is detected as raw data then, to be recorded in the test protocol if desired.

Example: It is aimed to move the measuring plate downwards just until contact with the sample is detected, i.e. until shortly F_N > 0, keeping afterwards the constant target value at F_N = 0.

Note: Bottom measuring plate made of wood to measure wood adhesives [395, 410]

A parallel-plate measuring system with a bottom plate made of wood, e.g. plywood or beech veneer, is used to test the curing behavior of wood adhesives. Here, oscillatory tests are carried out, selecting the small distance between the plates of H = 0.2 mm only, to simulate the application conditions. **The sample volume is shrinking with time since a part of the solvent migrating off the adhesive is absorbed by the porous wood.** Therefore, these kinds of testings are performed using the option "normal force-controlled gap setting" maintaining compression-less contact at the target value of F_N = 0.

10.4.7 Determination of the temperature gradient in the sample

Special **temperature calibration sensors** are available in order to measure the actual temperature values and the temperature gradient in the shear gap between the plates. For example, the following calibration plates are available (with the diameter d and thickness H):

1) d = 25mm and H = 2mm, equipped with two temperature sensors in its center, the one close to the top, and the other one close to the bottom.

2) d = 50mm and H = 2mm, equipped with four temperature sensors at the following positions: top inside, top outside, bottom inside, and bottom outside.

$\mathcal{GS}$ End of the Cleverly section

10.4.8 Advantages and disadvantages of parallel-plate measuring systems

a) Advantages

1) It is possible to measure dispersions containing relatively large particles, samples showing three-dimensional structures, hardening and curing materials, and even soft solids in the form of pre-formed disks. Examples: gels, pastes, melts of filled polymers, pellets, elastomers, rubbers and hard cheese

Note: In order to guarantee always contact between the sample and the surfaces of the plates, the option "normal force control" NFC might be used (see Chapter 10.4.6.)

2) When setting a large gap (e.g. H = 1mm), **highly viscous and viscoelastic samples such as polymer melts and silicones** (unlinked PDMS) are requiring a shorter preparation time which facilitates testing. Here, **less pre-stress effects**, and therefore, shorter relaxation times are occurring to reach the sample's equilibrium state between the internal and external stresses.

Note: In order to guarantee "soft gap setting", the option "normal force control" NFC might be used (see Chapter 10.4.6.)

3) The shear rate range might be varied easily by changing the gap dimension H.

4) When setting a larger gap the following measuring error might be reduced, compared to cone-and-plate MS: When performing tests over a wide temperature range, the gap dimension might change in an uncontrollable way during the test due to thermal expansion or contraction of both sample and measuring device when heating or cooling, respectively.

Example: When setting a narrow gap of only H = 50μm, a change of $\Delta H = 10\mu m$ might have a relatively strong influence on the test result (since here, $\Delta H/H = 20\%$). With H = 1mm, however, this kind of error would be relatively small, showing a clearly smaller relative value of only $\Delta H/H = 1\%$ then.

Note: In order to counteract uncontrolled gap failures, the option "automatic gap control" AGC or the "TruGap" function might be used (see Chapter 10.4.6)

5) Cleaning after the test is a very simple and therefore a fast action.

6) **Sandblasted, serrated or profiled surfaces** might be used to counteract wall slip effects, either for the upper plate as the rotor only, or for both plates (see Chapter 10.6.1).

7) **Disposable plates** might be used for the upper plate only, or for both plates, if cleaning is difficult or if it is impossible, e.g. when testing curing materials.

b) Disadvantages

1) **The shear conditions are not constant** in the gap between the plates since the value of the shear rate is increasing from zero in the center of the plate to the maximum at the edge, and the same applies to the shear deformation. For most samples the test results are influenced significantly by this effect. **However, this effect is negligible when performing oscillatory tests at very small deformation values in the LVE-range.**

2) The following might occur at the edge of the plate: flow inhomogeneities, turbulent flow, inertia effects of the sample due to centrifugal forces, discharge off the gap by migration of the sample, skin formation, streaks, surface effects, melt fracture due to elasticity or due to normal forces, respectively, and evaporation of solvents. All of this may be summed up as "edge failure" effects. These appearances may significantly influence the test result (see also Chapter 10.3.7 and Figure 10.7: Optimal filling of a MS).

Note: The degree of evaporation might greatly be reduced when using a cover, and if desired, also a solvent trap.

3) The larger the gap, the larger is the temperature gradient in the sample if the temperature is directly controlled via the bottom plate only.

Note: As a counter-measure, a hood might be used enabling temperature control of the inner space of the test chamber (see also Chapter 11.6.6e: "Active hood").

4) When performing tests over a wide temperature range, equipment-dependent thermal expansion or contraction might influence the test result. Of course, this effect is more pronounced when setting a narrower gap.

Note: In order to counteract undesired, uncontrolled changes in the gap dimension, the option "automatic gap control" AGC and the "TruGap" function may be used (see Chapter 10.4.6).

Summary: When performing rotational tests, despite of the many advantages mentioned, the first disadvantage outweighs them all by far, i.e. the not constant shear rate in the shear gap.

Therefore here, PP MS should only be used when conditions make it necessary, for example, when testing dispersions containing large particles. With oscillatory tests, this disadvantage has no effect as long as measuring in the LVE range. When testing viscoelastic samples, for practical users the shorter time required for gap setting may even count as an advantage, e.g. when testing silicones or polymer melts.

10.5 Mooney/Ewart measuring systems (ME MS)

Measuring systems according to *M. Mooney* and *R.H. Ewart*, briefly ME MS, are described in DIN 53019-3, [46, 155, 255]. They consist of a **combined concentric cylinder (CC) and cone-and-plate (CP) geometry**. Sometimes, an ME MS is also termed a "cone-cylinder" [18], a "conicylinder" [233] or a "Mooney/Couette viscometer" [152]. Like a concentric cylinder MS, it consists of an inner and an outer cylinder in the form of a bob and a cup. The bob shows a conical lower end which forms a cone-and-plate geometry together with the bottom of the cup, when installed in the working position (see Figure 10.9).

The aim of this design is to attain the same value for the mean shear rate $\dot{\gamma}$ as well in the annular gap between the cylinders as well as in the conical gap at the bottom. Therefore:
$$\dot{\gamma} = \dot{\gamma}_{cc} = \dot{\gamma}_{cp}$$
with the shear rates $\dot{\gamma}_{cc}$ in the annular cylinder gap and $\dot{\gamma}_{cp}$ in the conical gap.

The surface of the conical part is considerably smaller than those of the cylindrical part, and therefore, the proportion of the torque from the conical shear area is relatively small compared to that one from the cylindrical area. The geometrical conditions are given by the following two limitations: the maximum of the permitted ratio of the radii δ_{cc} of the two cylinders of cup and bob (according to ISO 3219, see Chapter 10.2.2.1a), and the maximum of the permitted cone angle $\alpha \leq 0.1\,\mathrm{rad}$ or $\alpha \leq 6°$. Therefore holds as the condition for the design:

Equation 10.28 $\qquad \alpha = (\delta_{cc}^2 - 1)/(1 + \delta_{cc}^2) \qquad$ or $\qquad \delta_{cc}^2 = (\alpha + 1)/(1 - \alpha)$

with the cone angle α in [rad]

Example: Dimensions of a Mooney/Ewart measuring system
With a bob radius $R_i = 22.7\,\mathrm{mm}$ and a cup radius $R_e = 24.0\,\mathrm{mm}$, thus, the ratio of the radii $\delta_{cc} = R_e/R_i = 1.057$. Then, the following applies for the cone angle: $\alpha = 0.055\,\mathrm{rad}$ or $3.15°$.

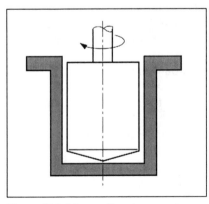

Figure 10.9: Mooney/Ewart measuring system (ME MS)

An advantage of ME systems compared to ISO cylinder systems is the smaller amount of sample required. Gap setting has to be performed very carefully as is the case for all cone-and-plate systems. A disadvantage is the time-consuming cleaning procedure, as is the case with all cylinder systems. As a consequence, ME systems are rarely used in industry labs, for example, they might be used when testing suspensions which are settling rapidly. The space between the lower end of the bob and the bottom of the cup is showing a relatively small volume when compared to the ISO cylinder system, and therefore, settling of particles is a little delayed since they are disturbing each others' motion during the sedimentation process.

10.6 Relative measuring systems

The calculation of rheological parameters in the form of **absolute units** which are independent of the individual measuring system used, from raw data which are indeed measured by the instrument, is only possible if standardized geometries of **absolute** measuring systems are used (see Chapters 10.2 to 10.5: concentric cylinder, cone-and-plate, parallel-plates, Mooney/Ewart, according to ISO and DIN).

Example: A liquid is tested using two ISO cylinder measuring systems of different size. With the larger MS, the rheometer measures of course a higher torque value although the same rotational speed was preset. Nevertheless, finally the same viscosity value is calculated from the two different raw data, since there are different conversion factors of the two measuring systems to calculate the same shear stress value from the different torque values.

When using relative measuring systems however, the shear conditions in most cases are not clearly defined as required for an accurate rheological analysis.

When performing rotational tests using relative measuring systems, fluids are often flowing inhomogeneously, **showing secondary flow effects like turbulent flow** with vortices instead of the desired laminar flow, or **time-dependent effects** are occurring **such as transient flow effects**. In these cases, it is impossible to calculate shear rate values from the raw data, since here with relative systems, neither the geometry nor the dimension of the shear gap are meeting the required conditions according to the ISO or DIN standards. Therefore then in general, viscosity values cannot be calculated, measuring point by measuring point, according to the law of Newton, since here as mentioned above, defined shear rate values are not available at all. In other words: **When using relative measuring systems, the shear conditions required to determine the viscosity values of liquids** as stated in Chapter 2.2 **are not met**, e.g. in the form of laminar flow.

Also when performing creep tests, relaxation tests and oscillatory tests using relative measuring systems to determine the viscoelastic behavior of gels, pastes and solids, the required shear conditions are not exactly defined. Here is obtained an **inhomogeneous shear process** if the sample shows **"plastic behavior"** (see also Chapter 3.3.4.3d and Figure 2.9). In this case in general, the values of the viscosities or shear moduli as well based on the laws of Newton and Hooke or on the relations of Maxwell, Kelvin/Voigt and Burgers, as well as according to the ISO and DIN standards, cannot be calculated exactly. This is true since here, measuring point by measuring point, defined deformation values are required in order to calculate the corresponding values. However with relative systems, neither the geometry nor the dimensions of the shear gap are meeting the required conditions for an accurate calculation, and therefore, these values are not available at all. In other words: **When using relative measuring systems, the shear conditions required to determine the rheological properties of viscoelastic samples** as stated in Chapters 4.2 and 8.2 **are not met**, e.g. in the form of homogeneous deformation behavior in the entire shear gap.

Therefore, test results attained with relative measuring systems should be specified in terms of **relative values** and not as absolute units, i.e., not as viscosities in Pas and not as shear moduli in Pa; and the same holds for the storage modulus G' and the loss modulus G" obtained from oscillatory tests. Then, the results should be presented in terms of **instrument units or measuring system units**, for example, in terms of **relative torque values**, in percent of the total torque.

10.6.1 Measuring systems with sandblasted, profiled or serrated surfaces

When effects like **wall slip** are occurring on the surfaces of a measuring system, it is useful to take a system with a rough surface instead of the usually smooth and polished surface, in

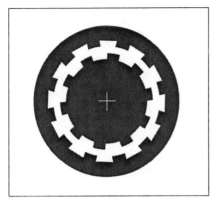

Figure 10.10: Cross-section of a cylinder measuring system, here both bob and cup are profiled, [77]

order to guarantee adhesion between the sample and the wall of the measuring system. For the corresponding samples, mostly parallel-plate systems with sandblasted surfaces are used. As a comparison: In ISO 6721-10 for smooth plate surfaces a maximum roughness of $0.25\mu m$ is recommended. With modified surfaces, either only the surface of the rotating plate is roughened, or both surfaces are treated correspondingly, the upper and the stationary bottom plate.

For all samples containing oil and fat it is recommended to use sandblasted surfaces for at least that part of the measuring system which is set in motion. This applies therefore for many samples from pharmaceutical, cosmetics, medical and food industry; and it is also valid for many petrochemicals such as lube greases, waxes and vaselines.

For very slippery samples and for materials exhibiting interfacial slippage or gliding effects along external surfaces, and for sliding rigid solids, it is sometimes necessary to use a measuring system with a **profiled or serrated surface**, for example, when testing **gels, waxes, elastomers, rubbers, or hard cheese**. Also here, often the rotating plate is profiled only. However, when measuring materials with a strong tendency to slip, the test result will be improved significantly, when also a profiled stationary bottom plate is used.

Sometimes even bobs of cylinder systems with sandblasted surfaces are taken, and rarely also corresponding cups, or they are milled lengthways (see Figure 10.10) or in the form of spiral lines.

Scientifically working users should be aware of the following: All surfaces which are not smooth are disturbing the laminar flow conditions, and therefore, turbulent flow conditions may be generated causing vortices in the boundary layer between the liquid sample and the surface of the measuring system. However, since laminar flow is a pre-condition for the validity of the viscosity law of Newton (see Chapter 2.2), results which are attained with rough surfaces should be considered relative values in principle. This holds even though the obtained measuring data are sometimes quite similar to those attained when using smooth surfaces. The difference between the results is sometimes hardly noticeable, especially at low-shear conditions, i.e., below the shear rate of $1s^{-1}$. But the effects may be significant when determining the yield stress and the flow stress, when measuring in the yield zone and when performing LAOS tests (see Chapters 3.3.4.3 and 8.3.4.3: Yield zone, e.g. with amplitude sweeps between yield and flow point, and Chapter 8.3.6: LAOS).

10.6.2 Spindles in the form of disks, pins, and spheres

Here, the measuring systems mostly consist of a single part only which is usually called a "spindle", and sometimes it is termed a "rotor". These kinds of measuring systems are showing a variety of different designs and can be found in many QC laboratories, since they are frequently used for simple tests, e.g. according to the following standards:

ISO 1652 (rubber latices), ISO 2123 (silicates), ISO 2555 (resins);

ASTM C965 (glass melts), C1276 (mold powders), D115 (varnishes), D789 (polyamides, PA), D803 (tall oils), D1076 (latices and rubbers), D1084 (adhesives), D1417 (synthetic rubber latices), D1439 (carboxy-methyl-cellulose, CMC), D1824 (plastisols), D1986 (polyethylene waxes), D2196 (non-Newtonian materials), D2336 (wood coatings), D2364 (hydroxy-ethyl-celluloses,

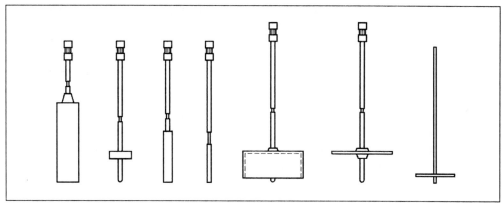

Figure 10.11: Various spindles geometries, from the left-hand side: type LV in the form of cylinders, thick disks and pins, type RV as a hollow cylinder and thin disks, and on the right: a T-bar

HEC), D2396 (PVC resins), D2556 (adhesives), D2669 (petroleum waxes, hot melts), D2983 (automotive fluid lubricants and gear oils at low temperatures), D3236 (hot melts adhesives and coatings), D3468 (liquid Neoprene and modified PE sealants), D3716 (emulsion polymers in floor polishes), D3791 (asphalts), D4016 (chemical grouts), D4300 (adhesives), D4402 (unfilled asphalts), D4878 (polyols for PU), D4889 (isocyanates for PU), D5018 (coal-tars, petroleum pitches), D5133 (lubricating oils at low temperatures), D6080 (hydraulic fluids), and D6267 (hydrocarbon resins at elevated temperatures up to 300°C), and D7110 (gelation of engine oils at low temperatures)

Typical spindles take the shape of **disks** exhibiting a thickness h and a diameter d comparable to coins, usually showing d = approx. 13 to 50mm and h = approx. 2 or 7mm; or of **pins** showing d = approx. 3 or 6mm and a cylindrical length L = approx. 15 to 50mm; or of **cylinders** showing d = approx. 10 or 20mm and L = approx. 55 or 65mm.

Mostly the following **three series of spindle sets** are used (see Figure 10.11):
1) The spindles LV-1 to LV-4, sometimes also including an LV-5: LV-1 as a cylinder, LV-2 and LV-3 as disks showing different diameters and thickness, LV-4 and LV-5 as pins showing different lengths, looking like a "spindle shaft without a disk"
2) The spindles RV-1 to RV-7 (as HA and HB): RV-1 as a **hollow cylinder**, RV-2 to RV-6 as thin disks showing different diameters, and RV-7 as a pin
3) The **cross-bar** or so-called **"T-bar"** for the "Heli-path" drive. Here, a pin is attached right-angled to the rotation axis. There are six bars available showing different lengths. This system is used to measure non-flowing, paste-like samples. When operating the "Heli-path" System, the T-bar is rotating while it is slowly lifted upwards through the sample which is remaining in its stationary container. Thus, the T-bar is moving on a spiral path upwards through the sample. The name "Heli-path" was chosen because of the Greek words "helikos" or "helix" for "winding" or "screw-like".

Here, the test results are often specified in the form of a relative torque value M_{rel} with the unit [%], related to the maximum torque of the viscometer used. Some users just mention the values of the dial reading DR then. For most of the spindles, the shear stress might be calculated from the measured torque since the geometry of the shear area is known.

The shear rate, however, is not defined since there is no narrow shear gap existing like for the ISO or DIN measuring systems (see e.g. Chapter 10.2.2.1a, and Figure 10.3). For analysis, the shear gap is therefore termed "infinitely wide", i.e. the cup radius is assumed to be $R_e = \infty$. The geometrical conditions are for example called a "dip-in spindle being in an infinite sea of fluid" [18], e.g. when measuring in a typical 600ml glass beaker showing an inner diameter of

85mm. For this reason, homogeneous shear conditions throughout the entire sample cannot be guaranteed then. Also, when using a disk-like spindle, the cylindrical length of the shear gap corresponds in fact just to the very small thickness of the disk which is often 2mm only. In this case of course, a useful shear gap dimension cannot be specified. Due to this, these kinds of measuring geometries are producing more incalculable edge effects, and therefore, more turbulent flow conditions showing vortices, than laminar flow conditions.

Summary: Using spindles, viscosity values are not absolute values but relative values
According to Newton's viscosity law, it is not possible to calculate viscosity values without defined shear rate values (see Chapter 2.3.1 and Equation 2.11). Therefore, viscosity specifications which are based on tests performed with these kinds of spindles are **relative viscosity values** and should be termed as such, or as values which are dependent on the instrument or measuring system used. For example, many users are speaking of **"Brookfield units"**, or briefly **BU**, after the name of a widespread viscometer type, [55]. Comparison of "Brookfield viscosities" which are specified in mPas or Pas, to absolute viscosity values which are measured when using absolute measuring systems, in almost all cases will result in deviations due to the above reasons. Remark: By the way, besides these spindles, company Brookfield is also offering absolute measuring systems according to ISO and DIN.

Indeed, after many comparative tests, it might be possible to find a certain approximative correlation between test results obtained with an individual spindle and an individual ISO measuring system. However in this case, the correlation only applies for these two measuring geometries and only for that individual liquid sample used, and even then only, if this fluid shows ideally viscous flow behavior. Comparability cannot be expected for non-Newtonian materials, and that counts particularly for different rotational speeds. If conversion formulas are given in literature, they are based on empirical tests which are performed in a limited range of rotational speeds which was individually chosen for a certain group of samples from a certain application field. These kinds of formulas should not be overrated since in most cases they are only useful for a rough estimate of the rheological behavior of a sample.

Note 1: In the coatings industry, when performing simple QC tests with spindles, sometimes the term **low-shear viscosity** is used, or **LSV** "according to Brookfield". Here is assumed, that testing is carried out in the shear rate range of about 1 to $10s^{-1}$, [118].

Note 2: Disk-like and spherical rotors according to ISO 2884-2
In ISO 2884-2 are described simple viscometers for testing coatings, which are operating at the constant rotational speed of $n = 562min^{-1}$, using the following spindle types (rotors):
1) Type 1: Disk showing the diameter d = 58mm and the thickness h = 7mm; the disk exhibits four ellipsoidal holes which are about 9mm long and 3mm wide.
2) Type 2: Sphere showing d = 31.75mm, and Type 3: Sphere with d = 19.05mm.
When testing, the spindles are dipped into a paint container showing the volume V = 250ml, inner d = 74mm, height 61mm and the filling level of "about 15 to 20mm from the top", [254]. Also here it is best to specify the test results in the form of the relative torque M_{rel} using the unit [%], related to the maximum torque of the viscometer used. This counts, since also here, **relative viscosity values** are achieved, even when the following three average shear rate values are mentioned in the ISO standard: 200, and 44, and $20s^{-1}$. These estimated values are assumed to be obtained for the different three disk types 1 to 3, when setting the rotational speed of $n = 562min^{-1}$. This viscometer is also called "Rotothinner", to be used for example to continuously monitor the sample's viscosity whilst a solvent or thinning agent is added, [336].

Note 3: Testing the solidification of liquid films using a T-bar spindle and a trough
In order to analyze film formation and curing of coatings, there is a patented **measuring system** consisting of a trough and a T-bar, [334]. Troughs showing a diameter between $d_1 = 2$ and

50mm (preferably d_1 = 15 or 25mm) and a depth between h_1 = 5μm and 5mm (preferably h_1 = 100 to 200μm) are mentioned. A thin, **T-shaped spindle** made of a rigid wire with a diameter between d_2 = 0.01 and 2mm (preferably d_2 = 50 to 100μm) is immersed into the liquid sample. The immersion depth is chosen in a way that the distance to the bottom of the trough is between 10 and 100μm. The spindle is driven by a rheometer to perform **oscillatory motion**.

Preset: Oscillation at a constant angular frequency of ω = 1 or 25rad/s and at a constant amplitude of the deflection angle $φ_A$. The deformation amplitude is specified as $γ_A$ = 100%, which of course should merely be considered a relative value here.

Measurement of two results: Firstly, the torque value is increasing when the sample is solidifying more and more. Higher torque signals can be achieved when using modified immersion spindles consisting of several straight or sloped bars. Secondly, there is a decreasing value of the phase shift angle δ between the preset sinusoidal oscillations and the measured response, since the time delay between these two signals is continuously becoming smaller during the solidification of the sample. The reason for this is the transition from a viscous flow behavior of the initially liquid coating via a viscoelastic and gel-like behavior, finally to an almost purely elastic deformation behavior of the solid film.

Presentation: In the patent specification, the resulting curves are shown in terms of the functions of the storage modulus G' and the loss modulus G" as well as of the complex viscosity η*. Of course, also these specifications have to be regarded as relative values, since the measuring geometries used are not meeting the required geometrical conditions for the determination of the deformation values which are needed to calculate the values of G' and G".

10.6.3 Krebs spindles or paddles

The "Krebs-Stormer viscometer" is used for simple routine tests in the **coatings** industry, for example, according to ASTM D562 for **paints**, ASTM D856 and ASTM D1084 for free-flowing adhesives [246, 281, 309]. It is named after the Krebs Pigment and Color Corporation where the spindle system was developed, and after the inventor of these devices *E.J. Stormer* (see also Chapter 13.3: 1909, [349]).

The Krebs-spindle consists of a stirrer exhibiting two blades, thus, a shaft on which are attached two **paddle-shaped agitator blades**, each one showing the size of 24 mm · 8 mm approximately (see Figure 10.12). Therefore, some users call it just a "paddle".

For the first generation of these kinds of instruments, the spindle was set in motion by a weight-and-pulley system, e.g. using weight pieces with a mass between 25 and 500g. The result was read off in terms of the weight which was required to move downwards along a defined path in a defined time period. With later viscometer types the weight was determined to reach the rotational speed of n = 100 or 200min⁻¹, which was checked by comparison with the signals of a stroboscope generating light flashes of a known frequency.

Nowadays, this viscometer type is equipped with an electric drive, and the desired speed of usually n = 200min⁻¹ can be set directly. The torque is detected directly by a spiral spring or it is determined indirectly via the electric power required to maintain the desired rotational speed.

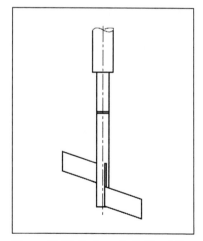

Figure 10.12: Krebs spindle ("paddle measuring system")

The measurement is carried out after the "paddle" is immersed into a can containing the liquid sample. Typical can sizes which are commonly used in the paint industry are e.g. $1/2$ liter (or 1 pint), $1/4$ liter (or $1/2$ pint), and 1 liter (or 1 quart). The **measured torque** values are displayed **in Krebs units**, or briefly, **KU**. Conversion: 40 to 141KU correspond to the weight of m = 32 to 1099g.

Viscosity values cannot be determined in absolute units when using Krebs spindles since the rotating paddle due to its geometrical shape, is always performing a stirring process which is generally generating **turbulent flow** with the formation of vortices. Since laminar flow conditions are not available, shear rate values cannot be calculated in principle. Therefore here, all test results are to be considered relative values. Viscosity values specified in mPas or Pas by the manufacturer of these instruments are empirical determinations which are based on previously performed comparative tests. These **relative viscosity values** cannot be compared to absolute viscosity values which are measured when using absolute measuring systems according to ISO or DIN.

Note: In the **coatings industry**, when performing simple QC tests with Krebs-spindles, sometimes the term **medium-shear viscosity** is used, or **MSV** "according to Krebs-Stormer". Here is assumed, that testing is carried out in the shear rate range of about 100s⁻¹, [118].

10.6.4 Paste spindles and rotors showing pins and vanes

"Spindles" and "rotors" in the form of pins and vanes are used when testing pasty materials which are not flowing homogeneously, or if they are containing large particles. Therefore in principle, for these kinds of systems holds that shear rate values cannot be calculated. Thus also here, the test results generally should be specified in terms of raw data, e.g. as rotational speed n in [min⁻¹] and as relative torque M_{rel} in [%].

Pin rotors consist of several pins screwed at a right angle onto the rotational axis. Examples: spindles showing 6 pins, each one with the length of 20 or 50mm and the thickness of 1 or 2mm (see Figure 10.13). These kinds of rotors are also known under the name "paste spindles" or "RS", [55, 171].

With **vane spindles**, several rectangular vanes are attached radial to the shaft (see Figure 10.13), [233, 324]. Examples are spindles showing 4 or 6 vanes with the thickness 1.5mm, the vane length L and the spindle diameter d (all in mm) showing L/d = 25/13, 43/22, 69/34, or 9/10, 16/22, 60/40. These kinds of spindles and similar ones are also known under the name "FL" or "flag impeller", [55, 171].

However, when using a vane rotor at high rotational speeds, there is the risk that this part of the sample being in between the vane areas might not be sheared at all, and that there might occur inhomogeneous, "plastic" deformation behavior. In this case, the whole sample material enclosed between the vane areas might merely rotate in the form of an undeformed cylinder.

Note: Vane rotors for testing gel-like samples
Foodstuffs like **yogurt** and other **dairy products**, **desserts** and **sauces** are often showing an inflexible gel structure. This three-dimensional structure might be destroyed already when immersing a bob of a common standard cylinder measuring system, or when setting the gap when using a parallel-plate system. For these kinds of samples it might therefore be preferable to select a vane system for the following reasons:

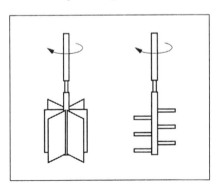

Figure 10.13: Vane rotor (left), and pin rotor

1) **Vane spindles can be immersed into shear-sensitive samples without changing their structural strength significantly.** For example, a measurement might be performed directly in a glass container or in a can ("in situ"), immediately after the production of the sample and the filling process into containers, [165, 277].

2) Since vane rotors do not show smooth cylindrical walls like standard measuring bobs, **wall slip might be prevented**, for example, **when determining the yield point and the flow point**. Usually here, the controlled torque method is selected, either performing a rotational test and analyzing by use of a fitting straight line according to Chapter 3.3.4.2, or performing an oscillatory amplitude sweep according to Chapter 8.3.4. However, it should be taken into account that the measured values generally are relative values, and therefore should merely be presented in terms of raw data. This means: For rotational tests should be displayed the deflection angles φ [rad] or the rotational speeds n [min^{-1}] versus the torque M [mNm] or the relative torque values M_{rel} in [%], respectively; and for oscillatory tests should be shown φ and the phase shift angles δ [°] versus M.

10.6.5 Ball measuring systems, performing rotation on a circular line

Ball measuring systems, briefly ball MS, are developed for tests on **semi-solid dispersions** containing **particles up to 5mm in diameter** [260, 368]. These kinds of systems consist of especially designed rotors and a cup showing an inner diameter of approximately 115mm for a sample volume of approximately 500ml (see Figure 10.14). The rotors are designed as follows: An arm with the length L_1 is mounted radially to the rotor shaft, projecting at a right-angle. At the outer end of this arm a pin with the length L_2 is mounted right-angled pointing downwards, and the end of this pin is carrying a sphere of the diameter d (called "ball"). Balls with the following diameters are usually in use (all dimensions in mm): d = 15/12/8; the arm length L_1 is between 35 and 40, and the pin length L_2 = 30 approximately.

With the rotor in motion, the ball is drawn through the sample on a circular path showing the radius L_1. Only during the first rotation, the ball penetrates unsheared material which is not yet cleared of particles. However, even when performing only a single complete rotation, it is possible to obtain a flow curve over several decades of the rotational speed, if a rheometer is used which is able to control the speed very fast at each individual measuring point, e.g in the speed range of n = 0.001 to 10min^{-1}.

For the evaluation are taken either raw data, i.e. rotational speed n and torque M, or it is based on a so-called "dimensional analysis", and empirically obtained comparative values are taken by assuming conditions of a so-called "displacement-flow". Hence, analysis is not based on laminar flow conditions which are usually taken as a basis for the rheological analysis when measuring with absolute measuring systems. Thus, the viscosity values which are attained when testing with the ball MS, cannot be compared to the absolute viscosity values determined when using the measuring systems according to ISO and DIN.

After elaborated investigations on the flow behavior of **granular debris and muds** containing **particles up to 10mm in size**, for yield stress and flow stress

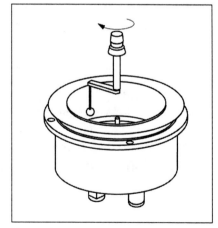

Figure 10.14: Ball measuring system: the ball performs a rotational motion on a circular path

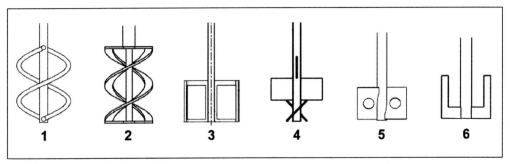

Figure 10.15: Relative measuring systems as stirrers for dispersions; from left (1) and (2) showing a helical shape, (3) for construction materials, (4) for starches, (5) a blade stirrer, and (6) showing the shape of an anchor

fluids an improved approach was found which is based on the Metzner/Otto theory [319]. In the following, some samples are listed which can be measured using a ball MS.

Examples 1: Multi-phase materials containing large particles or fibers such as **construction materials, mortars, plasters, ceramic tile adhesives**, and coarse-grained dispersions which are showing separation or wall slip effects when testing; e.g. to investigate creep and flow of **sludges, muds and soil**, for **geological investigations** to simulate the behavior of landslides and avalanches or of "solifluction" (see also Examples 2 to 4 in Chapter 3.3.4.3d).
Examples 2: Foodstuffs such as **sauce** "alla Bolognese" containing small meat chunks, or **jam** and **marmalade** containing fruit or other pieces.

10.6.6 Further relative measuring systems

In Figure 10.15 are presented some relative measuring systems which are used in industrial laboratories as stirrers for dispersions, for example, in the construction and food industry [123].

10.7 Measuring systems for solid torsion bars

Some rheometers are equipped with a measuring system to perform torsion tests on bar-shaped solid specimens. A torsion bar fixture consists of two clamps to hold the specimen. For example, the stationary bottom clamp is mounted onto the frame of the rheometer, and the upper clamp is connected to the shaft of the rheometer drive, usually to be set in oscillatory motion.

In most cases, these kinds of experiments are performed to investigate the temperature-dependent behavior of solid specimens at constant dynamic-mechanical conditions (see also Chapter 8.6). This kind of testing is often termed **"dynamic-mechanical thermoanalysis" DMTA** (or "dynamic thermomechanical analysis" DTMA) as in ISO 6721 and ASTM D4065. Information on shear modulus values in the form of G and G* of various solid materials is presented in Chapter 4.2.2, Table 4.1.

a) Determination of the temperature gradient in a specimen
Temperature control should be performed carefully to keep the temperature gradient in the specimens as small as possible. Of course, the design of the test chamber and the selection of an appropriate temperature control system are crucial for this purpose (see also Chapter 11.6.6). Before testing, the user might estimate which actual temperature and which temperature gradient within a specimen has to be expected by use of a specially designed **temperature calibration sensor**. For example, the following tool is available in the form of a

rectangular calibration bar showing the length L = 54mm, width b = 12.7mm, and thickness a = 2mm, equipped with four temperature sensors at the following positions: top, middle top, middle bottom, and bottom.

      For "Mr. and Ms. Cleverly"

b) Automatic setting and control of the clamp distance

The specimen should always be fully stretched between the clamps. In order **to prevent** uncontrolled deformation by compression stresses, **contraction, bending and buckling** of the clamped specimen due to **thermal expansion** of specimen and test device when heating, or to prevent tensile stresses and **stretching** due to thermal **shrinkage** when cooling, respectively, the rheometer should be equipped with an **option for setting and control of the distance between the clamps, e.g. by "normal force control" NFC or by "automatic gap control" AGC** (see also Chapter 10.4.6).

Note: Positive normal forces F_N are **compression forces** acting on the specimen, and negative F_N-values are tensile forces. Please note: These kinds of testings performed on solid specimens as it is decribed in this section are not the classic tensile or compression tests, since there, the forces applied are several dimensions higher.

Examples

1) **Target: Guaranteed stretching of the specimen bar by a constant tensile force**

The option "normal force control" NFC may be used to guarantee always a stretched specimen, e.g. to perform oscillatory tests while continuously a defined tensile force is acting onto the sample. Here, the clamp distance may change during testing, which is detected as resulting raw data then, to be recorded in the test protocol if desired.

Example: In order to keep the specimen bar always stretched, a constant tensile force is preset in the form of a negative normal force of $F_N = -1N = const.$

2) **Target: Stress-free compensation of changes in the length of a specimen bar**

"Normal force control" NFC might be used to compensate thermal expansion or shrinkage of specimen and measuring device when heating or cooling, which may lead to contraction or stretching of the specimen, respectively. Here, the clamp distance may change during testing, which is detected as resulting raw data then, to be recorded in the test protocol if desired.

Example: It is aimed to move the upper clamp downwards as long as neither a tensile nor a compression force is detected deriving from the specimen, i.e., until there occurs shortly $F_N = 0$, and to maintain afterwards the constant target value at $F_N = 0$.

3) **Adjustment of the clamp distance**

The option "automatic gap control" AGC controls the clamp distance, always returning to the desired value if it should have been changed due to any reasons.

10.7.1 Bars showing a rectangular cross section

Specimens for torsion tests are usually showing a rectangular cross-section (according to ASTM D4065). The following **dimensions** are specified, with **the free length L, the width b, and the thickness a** (all in mm):

1) According to ISO 6721-2:

L = 50 (40 to 120), **b = 10** (5 to 11), **a = 1** (0.15 to 2)

2) According to ASTM D4065, the following is found to be usable in many cases:

L = 50, b = 9.4, a = 0.75

3) In ASTM D5279, the following is stated:

L = 64, b = 13, a = 3

4) It is reported about good results when using specimens showing the following ratios [114]:

b:a = 3:1 for thermoplastic polymers and elastomers

b:a = 10:1 for reinforced laminates

L is the free specimen length between the clamps and therefore the length which is **outside the grips**. Therefore, the total length of the specimen is larger than L by the measure of the additional clamping length which is enclosed by the clamps when testing (e.g. $2 \cdot 6$, or $2 \cdot 7$mm).

Note: Preparation of the specimens
It is recommended to tighten a specimen in the fixture once more after the lowest test temperature is reached.

a) Conversion between raw data and rheological parameters for rectangular bars
The dimensions of the specimen are: free length L [m], width b [m], and thickness a [m]. In the following section, the parameters shear stress τ, torque M, deflection angle (torsional angle) φ and shear strain (deformation) γ are specified in terms of the amplitude values of the time-dependent sine functions, i.e., as τ_A, M_A, φ_A and γ_A (see Chapter 8.2.5: Presentation of presetting and resulting sinusoidal parameters).

1) Torque M and shear stress τ

Equation 10.29 $\tau_A = M_A / W_t$

The following applies to the **torsional section modulus W_t [m³]**:

Equation 10.30 $W_t = c_2 \cdot b \cdot a^2$

with the factor c_2 from Table 10.3. Then:

Equation 10.31 $\tau_A = M_A / (c_2 \cdot b \cdot a^2)$

Example
reset: $M_A = 10$mNm $= 0.01$Nm; b $= 10$mm $= 10^{-2}$m; a $= 1$mm $= 10^{-3}$m
Thus, with the ratio b/a $= 10$, according to Table 10.3: $c_2 = 0.312$
Calculation: $\tau_A = 0.01$Nm$/[0.312 \cdot 10^{-2}$m $\cdot (10^{-3}$m$)^2] = 3.2 \cdot 10^6$N/m²
Result: $\tau_A = 3.2$MPa

2) Deflection angle φ and deformation γ

Equation 10.32 $\gamma_A = (I_t \cdot \varphi_A)/(W_t \cdot L)$

The following applies to the **torsional geometrical moment of inertia I_t [m⁴]**:

Equation 10.33 $I_t = c_1 \cdot b \cdot a^3$

with the factor c_1 from Table 10.3. Then:

$\gamma_A = (c_1 \cdot b \cdot a^3 \cdot \varphi_A)/(c_2 \cdot b \cdot a^2 \cdot L)$ or

Equation 10.34 $\gamma_A = (c_1 \cdot a \cdot \varphi_A)/(c_2 \cdot L)$

Example
Preset: $\varphi_A = 100$mrad $= 0.1$rad; L $= 50$mm $= 5 \cdot 10^{-2}$m; b $= 10$mm $= 10^{-2}$m;
a $= 1$mm $= 10^{-3}$m
Thus, with the ratio b/a $= 10$, according to Table 10.3: $c_1 = 0.312$ and $c_2 = 0.312$
Calculation: $\gamma_A = (0.312 \cdot 10^{-3}$ m $\cdot 0.1)/(0.312 \cdot 5 \cdot 10^{-2}$ m$) = 2 \cdot 10^{-3}$
Result: $\gamma_A = 0.2\%$

Table 10.3: For rectangular torsion bars: Factors c_1 and c_2 depending on the ratio b/a, used to determine the torsional section modulus W_t and the torsional geometrical moment of inertia I_t[112]

b / a	1	1.5	2	3	4	6	8	10	∞
c_1	0.141	0.196	0.229	0.263	0.281	0.298	0.307	0.312	0.333
c_2	0.208	0.231	0.246	0.267	0.282	0.299	0.307	0.312	0.333

b) Calculation of the complex shear modulus |G*|

According to Hooke's law applies: $|G^*| = \tau_A/\gamma_A$

Example

with $\tau_A = 3.2 \text{MPa}$, and $\gamma_A = 0.2\%$, then:

$|G^*| = (3.2 \cdot 10^6 \text{Pa})/(2 \cdot 10^{-3}) = 1.6 \cdot 10^9 \text{Pa} = 1.6 \text{GPa}$

Note: Other formulas to calculate the G-value are presented in ASTM D1043 (and in ISO 537, which is meanwhile withdrawn).

Example: Testing a solid polymer specimen in the form of a torsion bar (at T = +20°C)

An example for the interpretation of the curve functions of G' and G" which are attained by performing an oscillatory test on a specimen in the form of a torsion bar is presented in Note 3 in Chapter 8.3.3.1: amplitude sweep on a solid polymer, determination of the limiting value of the LVE range, increase in the G"-curve and **occurrence of micro-cracks**.

10.7.2 Bars showing a circular cross section

a) Conversion between raw data and rheological parameters for circular bars

The dimensions of the specimen are: free length L [m] and diameter d [m].

1) Torque M and shear stress τ

Equation 10.35 $\tau_A = M_A/W_p$

The following applies to the **polar section modulus W_p [m³]**:

Equation 10.36 $W_p = (\pi \cdot d^3)/16$

Then:

Equation 10.37 $\tau_A = 16 \cdot M_A/\pi \cdot d^3$

Example

Preset: $M_A = 100 \text{mNm} = 0.1 \text{Nm}; d = 5 \text{mm} = 5 \cdot 10^{-3} \text{m}$
Calculation: $\tau_A = (16 \cdot 0.1 \text{Nm})/[\pi \cdot (5 \cdot 10^{-3} \text{m})^3] = 4.1 \cdot 10^6 \text{N/m}^2$
Result: $\tau_A = 4.1 \text{MPa}$

2) Deflection angle φ and deformation γ

Equation 10.38 $\gamma_A = (I_p \cdot \varphi_A)/(W_p \cdot L)$

The following applies to the **polar geometrical moment of inertia** I_p [m⁴]:

Equation 10.39 $I_p = (\pi \cdot d^4)/32$

Then: $\gamma_A = (16 \cdot \pi \cdot d^4 \cdot \varphi_A)/(32 \cdot \pi \cdot d^3 \cdot L)$ or

Equation 10.40 $\gamma_A = (d \cdot \varphi_A)/(2 \cdot L)$

Example

Preset: $\varphi_A = 10 \text{mrad} = 10^{-2} \text{rad}; L = 50 \text{mm} = 5 \cdot 10^{-2} \text{m}; d = 5 \text{mm} = 5 \cdot 10^{-3} \text{m}$
Calculation: $\gamma_A = (5 \cdot 10^{-3} \text{m} \cdot 10^{-2})/(2 \cdot 5 \cdot 10^{-2} \text{m}) = 5 \cdot 10^{-4}$
Result: $\gamma_A = 0.05\%$

b) Calculation of the complex shear modulus |G*|

According to Hooke's law applies: $|G^*| = \tau_A/\gamma_A$

Example

with $\tau_A = 4.1 \text{MPa}$, and $\gamma_A = 0.05\%$, then:
$|G^*| = (4.1 \cdot 10^6 \text{Pa})/(5 \cdot 10^{-4}) = 8.2 \cdot 10^9 \text{Pa} = 8.2 \text{GPa}$

10.7.3 Composite materials

As increasing numbers of **composites** (composite materials) are being used, this section gives a short overview of these kinds of materials and explains some terms used in the corresponding branches. Lightweight materials are selected, for example, in the transportation branch for cars, trucks, railroad cars, airplanes and boats, in the architecture and building branch, and for equipment used for leisure and sportive activities. **Hybrid materials** consist of several different types of materials such as e.g. inorganic and organic components. Examples are glass fibers embedded in a polymer matrix as GFRP, alternating layers of metal and polymer as "Glare", organically modified silicates as chemically cross-linked compounds such as organo-silanes (see also Note 2 in Chapter 8.6.3b), steel girders embedded in a concrete matrix as ferroconcrete, special concrete containing textile fibers embedded in a resin, ceramic fibers embedded in metals such as aluminum alloys, and wood flour embedded in a polymer matrix as wood plastic composites WPC.

Fiber-reinforced plastics (FRP) are used to resist high mechanical loads. **Examples: Glass-fiber reinforced plastics GFRP, CFRP (carbon-fiber** reinforced plastics), SFRP (**synthetic-fiber** reinforced plastics) such as AFRP (**aramid-fiber** reinforced plastic; aramid is an aromatic polyamide, e.g. available since around 1970 under the brand name "Kevlar"), and **fibers of natural materials**. For technical applications, glass fibers usually show diameters of d = 9 to 24μm and a density of around ρ = 2.5g/cm^3, carbon fibers display for example d = 5 to 8μm and ρ = 1.6 to 2.0g/cm^3, and the lightweight aramid fibers exhibit e.g. d = 12μm and ρ = 1.45g/cm^3. As a comparison: A human hair has d = 50 to 60μm.

Prepregs are dry fabrics or tapes made of rovings which are impregnated with resin. A **roving** is a strand made of many spun fibers (e.g. of 1000, or even up to 24,000). It is therefore a strong **thread, band or tape**. Prepregs are rovings which are already preimpregnated, wetted with resin and **embedded in a polymer matrix**. This guarantees a liquid, sticky and finally viscoelastic solid compound between the single threads and layers of tissues, before, during and after the curing process.

The following criteria are important for the **matrix resin: Viscosity** should not be too high to ensure that the fibers are wetted sufficiently, **pot life and open time (or gelation time)** should be balanced to ensure the desired processing time (see also Chapter 8.5.3b), the maximum **temperature** which occurs when curing should be not too high **and** the period of **time for the curing process** should be not too long (see also Chapters 8.5.3 and 8.6.3).

Examples of applications: In airplane construction as well thermosetting resins (such as UP, EP, PU, PF) are used, as well as thermoplastic polymers (such as PE, PP, PA, PVC, PET, PEEK, PES; see also Chapter 8.6.2.1).

The following applies to FRP: If the fibers are all oriented into a single direction (unidirectional), the materials are able to withstand a high tensile force in this direction, but they may fail relatively fast under a torsional load. In spite of this, by use of these kinds of materials many requirements can be fulfilled since there are a lot of different ways of combining fibers and matrix, and optimizing the fiber orientation. If all fibers are arranged in the direction of the extension (i.e., showing the orientation angle 0°), the compound will be very resistant to tension and compression, where the fibers should absorb the tensile force, and the **matrix** as the **embedding and formable polymer mass** should absorb the compression force. However here, the stability is considerably lower in the transverse direction. In this case, **anisotropic properties** are occurring which are **direction-dependent**.

If the fiber orientation is 45° to the direction of the load, then there is also a certain degree of torsional and shear resistance. Designers therefore have many possibilities to develop **mul-**

tilayer compounds containing differently oriented fabric components. For example, a multi-directional laminate showing the fiber orientations of 0°/ 45°/90°/-45° in a radial arrangement can be produced which is resilient in all directions. This concerns **isotropic properties** which are **independent of the direction of loading**.

Note: Testing fibers and rovings without the polymer matrix

Of course, fibers and rovings can also be tested by tensile tests in the dry state, hence, without being embedded in a resin matrix. To prevent sagging, this kind of testing is normally carried out **under a certain degree of pretension**. See also Chapter 4.2.2 (tensile modulus E), and Chapter 10.8.4.1 (tensile tests and complex tensile modulus E^*).

A **sandwich construction** is a combination of solid and stiff top layers and a core which is a space holder filled with a lightweight material. This is, for example, light wood or hard foam made of PVC or PU. These **lightweight materials** are often used in a hexagonal shape for stabilizing, as the so-called "honeycomb" structures.

Glare means glass-fiber reinforced aluminum. This composite material was developed in 1985; it is a multilayered **fiber-metal laminate (FML)**. Thin layers of around 0.3 to 0.5mm of on the one hand aluminum, and on the other hand a resin reinforced by glass fibers in the form of GFRP-prepregs, are stuck together in alternate layers by heating. There may vary the layer thicknesses, the number of layers and the orientation of the fibers. **Application:** Glare is used in the **construction of airplanes** since 1996, because compared to pure aluminum, on the one hand its density is 10 to 20% lower and on the other hand its resistance to corrosion is higher.

🙰   End of the Cleverly section

10.8 Special measuring devices

The enormous expansion of measuring and analysis methods since the beginning of the 1990s has lead to increased interest in, and to new insights into the internal structure of materials. Using rheometers equipped with special measuring devices, it is nowadays possible to determine the rheological properties under separately controlled external influences, and also to observe micro- and nanostructures under defined shear conditions (see also Chapter 9). The following test methods are mentioned here only briefly, since the aim is to give the user just an overview of the numerous specialist terms and measuring methods used in this context.

🙰   For "Mr. and Ms. Cleverly"

10.8.1 Special measuring conditions which influence rheology

The rheological behavior of a sample is influenced by many external measuring conditions. A sample's dependence on time (Chapters 3.4 and 8.5), on temperature (Chapters 3.5, 8.6 and 11.6.6) and on pressure (Chapter 3.6) are described in the given chapters. The following sections present further experiments which can be performed to control the deformation and flow behavior of materials.

10.8.1.1 Magnetic fields for magneto-rheological fluids

Magneto-rheological fluids MRF consist of magnetically polarizable particles suspended in a carrier fluid. When a magnetic field is applied, the particles are oriented in the direction of the field and form a **chain-like superstructure**. For this reason the structural strength increases with increasing magnetic field strength until the maximum orientation of the particles is achieved. After overcoming the flow point, MRFs are flowing in an activated state.

Here, the chains are either breaking, or they are pushed along as a whole. Also **ferrofluids**, which contain much smaller particles than MRFs, can be investigated in a magnetic field. All rheological tests are possible at a constant or variable **magnetic field strength H** [A/m] or **magnetic flux density B** [T], respectively (the unit of B is Tesla). Preset is the electrical current I [A], with the unit Ampere, [222, 223, 396].

Examples of applications: adaptive shock absorbers; vibration dampers; loudspeaker systems which are self-adjusting to the environmental conditions; torque or force transmission in valves, braking and clutch systems; medical prosthetises

Measuring examples
1) Rotational tests
1a) Viscosity η at a constant shear rate $\dot{\gamma}$, first at H_1 = 0, then a step to H_2 = const
1b) η = f(H) at $\dot{\gamma}$ = const, and at ramp-like increasing or decreasing H-values, as **"magneto-sweep in rotation"**
1c) Flow curve $\tau(\dot{\gamma})$ and $\eta(\dot{\gamma})$ at H = const, at various field strengths H
1d) Function of $\eta(T)$ in a certain temperature range at $\dot{\gamma}$ = const and H = const

2) Oscillatory tests
2a) Storage and loss modulus G' and G'' at a constant strain γ and a constant angular frequency ω, first at H_1 = 0, then a step to H_2 = const
2b) G' and G'' = f(H) at γ = const and ω = const, and at ramp-like increasing or decreasing H-values, as **"magneto-sweep in oscillation"**
2c) Amplitude sweeps and frequency sweeps at H = const, at various field strengths H
2d) Functions of G'(T) and G''(T) in a certain temperature range at γ = const and ω = const, and H = const

10.8.1.2 Electrical fields for electro-rheological fluids

Electro-rheological fluids ERF contain electrically polarizable particles in a carrier fluid. When an electrical field is applied, the particles are oriented in the direction of the field, forming **chains**. ERFs are therefore very similar to MRFs, also showing a similar deformation and flow behavior. All kinds of rheological tests are possible at a constant or variable **electrical field strength E** [V/m]. Preset is the voltage U [V], with the unit Volt, [185].

Measuring examples of rotational and oscillatory tests: All kinds of tests can be carried out with ERFs as described above for magneto-rheological fluids MRF, when replacing the magnetic field strength H by the electrical field strength E.

Examples of applications: Liquids used in hydraulic bearings and brakes

10.8.1.3 Immobilization of suspensions by extraction of fluid

Here, dispersions are measured using a parallel-plate measuring system with a perforated stationary bottom plate, which is therefore showing many small holes at equal distances. On this plate a porous and absorbent paper, a membrane or a filter material is placed as the substrate for the sample. Typical samples are **paper coatings**. After setting the measuring gap and starting the measuring program, an under-pressure (vacuum) is applied under the perforated plate. As a consequence, an increasing amount of suspension liquid is extracted from the sample to be sucked through the substrate. The particles of the suspension therefore are approaching one another until they are getting direct contact, becoming immobilized finally. This is called the **immobilization point**. The decrease in sample volume is compensated by adjusting the distance of the upper plate using the option "normal force control" NFC by presetting "compression-less contact" in terms of the normal force F_N = 0 = const (see also Chapter 10.4.6). The gap width h(t) between the measuring plates which decreases with time can also be analyzed [388, 396].

Measuring examples
1) Rotational test as a time-dependent viscosity function η(t) at a constant shear rate $\dot{\gamma}$ or at a constant shear stress τ, consisting of two measuring intervals: Firstly without and then under vacuum.
2) Oscillatory test as time-dependent functions of the storage modulus G'(t) and the loss modulus G''(t) at a constant strain γ and a constant (angular) frequency ω; consisiting of two measuring intervals: First without and then under vacuum. With oscillatory tests there are the following advantages:
a) If the sample is liquid at the start of the test, showing G'' > G': The **immobilization point** can be analyzed accurately **at the crossover point of G' = G''** or via the loss factor when tanδ = G''/G' = 1, respectively.
b) If the sample is not liquid at the start of the test, therefore showing already here G' > G'': In this case the inflection points of the curve of G' (and if desired, also of G'') can be analyzed.
c) The measurement can also be continued beyond the immobilization point, i.e., when the sample is in a solid state.

10.8.1.4 UV light for UV-curing materials

This measuring method is suitable for samples containing molecules with functional groups which may trigger a chemical cross-linking process using the energy from ultraviolet (UV) light. A parallel-plate measuring system is used consisting of a transparent bottom plate, e.g. made of quartz glass. The UV light source is installed under the glass plate. After setting the measuring gap and starting the measurement, UV light can be radiated into the sample from below with the desired beam intensity I_{UV}. The unit of the light intensity I_{UV} is [W/cm²], or [%] if the values are related to the maximum intensity [304].

For **samples showing shrinkage when curing**, also the gap width h(t) between the measuring plates should be evaluated when becoming smaller with time. The decrease in sample volume is compensated by adjusting the distance of the upper plate using the option "normal force control" NFC by presetting "compression-less contact" in terms of the normal force $F_N = 0$ = const (see also Chapter 10.4.6). Shrinkage is a crucial parameter, for example, for the quality of tooth fillings.
Examples of applications: UV curing coatings, printing inks, adhesives and dental materials

Measuring example
Two-step **oscillatory test**, firstly without and then with UV light at I_{UV} = const, as time-dependent functions of the storage modulus G'(t) and the loss modulus G''(t) at a constant strain γ and a constant angular frequency ω. Analysis of the **sol/gel transition at the crossover point G' = G''** or when the loss factor tanδ = G''/G' = 1, respectively (see also Chapter 8.5.3 and Figure 8.35: curing).

Tips for users
1) **For very fast curing processes**, the measurement should be carried out **at a higher (angular) frequency**, e.g. at ω = 50 or even 100 rad/s to achieve a higher frequency of measuring points.
2) **Dark and** therefore **strongly absorbing samples** should be measured using a very small gap of around 0.1mm, otherwise the UV rays cannot completely penetrate the sample, and correspondingly, it may cure only partially.

10.8.2 Rheo-optical measuring devices

The aim of rheo-optical methods is to visualize the behavior when shearing **microstructures and nanostructures**. In particular, there is much interest in generating shear-induced superstructures of the following systems: Solutions, colloids, dispersions (i.e. suspensions, emulsions, foams), synthetic polymers and biopolymers such as proteins, surfactant systems showing associates such as micelles, vesicles, liposomes (see also Chapter 9), systems with self-organizing and combined superstructures, gels such as hydrogels, liquid crystals and "soft matter" such as biological tissues in contrast to crystallizing and therefore stiff solids. Recent development of **visualization methods** has lead to breathtaking results.

For example, researchers ask the following questions: What is the shape and size of the observed polymer molecules, what about their spatial distribution and arrangement (amorphous, liquid-crystal, partially crystalline), orientation or anisotropy? What about droplets, particles, aggregates, agglomerates, associates, domains or partial structures? Where is the position of the atomic nuclei and the electron density distribution at rest and do they change in a deformed state? Are there transient structures occurring with phase transitions? Can we observe even chemical reactions? Which kind of bonding occurs: primary/covalent or secondary, polar, ionogenic, or multiple bonds? Are there weak interactions? Are there any hydrogen bridges, or even hydration shells? Which distances are between the intra-molecular and inter-molecular bonds? What is the driving force behind these processes?

10.8.2.1 Terms from optics

Unfortunately, terms from the field of optics are often used in a different sense. Therefore, some terms are explained in the following section.

a) Light, visible and invisible
Physicists refer to each kind of electromagnetic radiation in the whole spectrum as "light", this therefore includes both visible and invisible light.

b) Radiation and wavelengths, light quanta or photons
Radiation can be interpreted as being a flow of particles or as a wave ("wave/particle duality"). A beam is transporting energy in the form of "mass-less" energy quanta which represent the smallest amount of energy, referred to as light quanta or photons. Different wavelengths λ_w can be distinguished: "hard" gamma-rays with $\lambda_w < 0.01$nm, X-rays with $\lambda_w = 0.01$ to 1nm, UV radiation (ultra-violet) with $\lambda_w = 1$ to 400nm, **VIS radiation (visible) with λ_w = 400 to 750nm**, IR radiation (infra-red) with $\lambda_w = 750$nm to 1mm, microwaves with $\lambda_w = 1$mm to 1m, and radio waves with $\lambda_w = 1$m to several km.

c) Lasers and monochromatic light showing only a single wavelength
Natural light consists of a broad radiation band showing many colors which we perceive in the form of a mixture as the color "white". "Laser" means "light amplification by stimulated emission of radiation". Ideally, this light source produces a light beam showing a single wavelength λ_w, and therefore, a monochromatic (one-color) light. Examples: Blue light laser with $\lambda_w = 420$nm, yellow sodium light with 589nm, red light with 658nm.

d) Polarization as a linear orientation in a single plane of oscillation, and polarimeters
Natural white light is non-polarized. The light waves show no uniform orientation right-angled to the direction of propagation. A light beam can be oriented linearly or polarized using a **polarizer**. Such a light wave is oscillating in a single plane only, the **plane of polarization**. The **angle of polarization** can occur between 0° and 90°, i.e. between the horizontal and the vertical direction. When penetrating optically active, impure or knowingly deformed materials the beam's direction of polarization may be changed. For an evaluation, the analyzer may be installed in parallel to the polarizer or it may be set as a **depolarizer** right-angled to it. Of course, it can also be positioned at any other angle, e.g at 45°. The degree of optical rotation of a polarization angle is material-specific and, for example, dependent on the concentration of a solution. It is determined by use of **polarimeters**.

e) Transmission, and a few coincidental interactions of light beam and sample

In measuring technology is usually used a polarized light beam. The light beam is split after entering a homogeneous sample. The following applies for the intensity (in %): Primary beam (100%) = **transmission (usually > 99.99%)** + reflection + refraction + absorption + scattering + diffraction.

Transmission means that the light beam passes directly through the material. Therefore, most of the primary beam is shot through the sample and usually only loses less than 0.01% of its intensity due to interactions with the sample, i.e., it loses less than a 10,000[th] part of its intensity! By **reflection**, some light quanta are reflected back, e.g. from phase interfaces or particle surfaces. Only the remaining minimal rest is "scattered light". The reason for this is that the photons are so tiny that they rush through each kind of material unless they happen to meet something by chance. To illustrate the dimensions: An atomic nucleus diameter is around (1 to 2) $\cdot$ 10^{-14}m (= 10^{-5}nm) in size and the entire atom diameter is around 0.1 to 0.2nm including the outer electron shells. This means that the ratio of the cross-section areas of the whole atom and the nucleus is approximately 10^8: 1. In other words: Here, a photon is surrounded by scarcely anything but empty space! The result is that only when applying an intensive primary beam, it is possible to achieve a sufficient number of interactions with the atomic particles, which can be measured and analyzed in a practice-relevant period of time.

f) Refraction of light and change of its direction, refractive index and refractometers

When moving into denser material, i.e. at surfaces or internal interfaces such as on gas bubbles of foams or on droplets of emulsions, a light beam is slowed down and therefore changes its direction. The **propagation speed is dependent on the "optical density"** and the propagation time varies accordingly. In air the light speed is around c = 300,000km/s, and in water it is around 225,000km/s, but in the crystal lattice of a solid diamond it is still 124,000km/s. The **refractive index** is defined as the ratio of light speed in vacuum and sample material. For air this index is 1.00 and for a common glass 1.513; the values for other materials can be found in literature. Refractive indices are material-specific and, for example, dependent on the concentration of a solution. They are determined by use of **refractometers**.

g) Birefringence in optically anisotropic materials and propagation delay

In isotropic materials, the speed of the light propagation is the same in all directions, this counts for gases at rest, liquids and unstressed glasses. Therefore, also the **refractive index** of these kinds of materials is independent of the direction. However, **in optically anisotropic materials**, these parameters are **dependent on the direction**. When passing through birefringent material, in different directions may occur a difference in the **delay of propagation**, for example, of two linearly polarized light waves oscillating right-angled to one another. This results in different refractive indexes in both directions of propagation.

Optical anisotropy can be generated under **defined shear conditions**. There are two types of birefringence: **intrinsic and form birefringence**. For intrinsic birefringence, the polarizability of the molecule is decisive. Polystyrenes, for example, consist besides of main chains also of polarizable side chains. This on its own is not enough to show birefringence, since there is a mean refractive index as long as the molecules are disordered. Intrinsic birefringence occurs if the molecules are all oriented uniaxially, i.e., into a single direction. Form birefringence appears however, if relatively large structures, such as droplets or aggregates, are deformed.

h) Absorption and conversion of energy

The intensity of a primary beam decreases in the sample. It appears as if some photons have been swallowed up. However, the principles of physics tell us that no energy can ever be lost, since it is only changed into another form. Therefore, when the electrons of involved atoms are rearranging, energy may be exchanged into heat or into fluorescent radiation, showing another wavelength afterwards. These effects can also be measured and analyzed.

i) Dichroism, optically anisotropic and bi-colored, using a selective absorber

Dichroitic means bi-colored or two-colored (Greek: dichroos). This describes the property of a material to display **diverse colors into different directions** when a visible, non-polarized light is beamed into it. Examples are "light-absorbing" minerals, jewels, liquid crystals, special mirrors and color filters. These kinds of materials absorb only a part of the light spectrum. In these optically anisotropic materials, a light beam is split into different wavelengths, and therefore, diverse colors can be observed.

In science, dichroism is commonly understood as follows: Linearly polarized light beams oscillating right-angled to one another are absorbed at a different degree. Dichroism in this context is **selective or partial absorption of light depending on the direction of polarization**. Examples are polarization filters and films, beam splitters, color separators or special mirrors. With some materials, this can be achieved also via external magnetic or electric fields (voltage), or via mechanical forces.

Optical anisotropy can be generated **under defined shear conditions**. As for birefringence, there are two types of dichroism: **intrinsic and form dichroism**. With form dichroism, absorption takes place on large structures such as droplets and aggregates, whereas with intrinsic dichroism, the absorption occurs on the side chains of the molecules.

k) Scattering on small particles, coherent bouncing, and incoherent secondary beams

Scattering is often used as the general term for all phenomena which make the difference between the intensity of the primary beam and the transmission beam. Therefore, it is used for all appearances such as reflection, refraction and diffraction. In a very simplified and unscientific manner, here, all effects of the secondary beam, such as change in frequency, spectral composition, direction of polarization, and therefore, also birefringence and dichroism, they all are called "scattering".

Scattering is secondary radiation, whose direction in relation to the primary beam is changed after a contact with **particles which are small compared to the wavelength** λ_w. An object with the size d is referred to as a "punctual emitter" if $d < (\lambda_w/20)$. Then, the result is spherically symmetric scattering. Illustrative example, showing a similar effect: A bright sun beam falls through a narrow gap into a dark room. Dust particles in the air, which usually are invisibly small, sparkle briefly due to the scattered secondary radiation.

There are two types, **coherent and incoherent scattering**. The former shows **the same wavelength** like the primary beam. It is either "in phase", i.e. indicating the maxima and minima of the oscillation amplitudes simultaneously with the primary beam. In this case, it enlarges the amplitude of the intensity, this is called "constructive scattering". Or it is phase-shifted by half an oscillation period compared to the primary beam, in which case this reduces the intensity value, and this is termed "destructive scattering" then. No energy is lost here, for example by absorption, this is therefore also called **"elastic scattering"**. Illustrative example: Photons bounce off the hitten objects without noteworthy time delay. This changes their direction of motion, like a football hitting the goal post. Here, the same photon is forwarded then.

The secondary beam can also be delayed if a beamed-in photon raises an electron of the involved atom onto a higher energy level due to a transfer of energy. These quantum leaps occur on discrete energy levels only. Such valence electrons are therefore shot into another orbit or "energy shell". The atom stores the received energy only for a short time, since the electron is transfered just temporarily into an unstable or "meta-stable" intermedium state. Afterwards it usually returns in discrete steps to the previous orbit, radiating energy in the form of light quanta on its way back to the original orbit. In this case, it is not the same photon which is forwarded. Subsequently, the **scattered beam** is usually coherent.

Or it is **incoherent** afterwards. Then it may be **phase-shifted** to the primary beam, despite having the same wavelength, or it shows a **changed wavelength**. Since the energy content is reduced due to an energy transfer or absorption, this is therefore also called **"inelastic scattering"**. However, often the incoherent, inelastic part of the secondary beam is small compared to the coherent, elastic part. The former does not contribute to structural information, but merely results in disturbances of the measuring signal, for example, in the form of background noise.

l) Diffraction on large particles, and lighting the shaded area

If light spreads out behind an obstacle as it were "bent around the corner" at the edge of an observed object into the shaded area, this is called **diffraction**. Diffraction is secondary radiation, whose direction is changed due to the contact with **particles which are not small compared to the wavelength**. Diffraction is strongest if the dimensions of the object are in the same order of magnitude like the wavelength, i.e., if $d \cong \lambda_w$.

However, if an object is much too small or even as "tiny as a dot", i.e., when there is scattering, then will hardly occur any diffraction (see above). Illustrative example: At night, a thin nail lit up in the beam of a flashlight hardly shows any shaded area. The light almost completely "flows around" the nail. However, if the object is relatively large compared to the wavelength, you can indeed see a sharp edge of the shadow, but relatively little diffraction. Diffraction depends on the wavelength of the primary beam, the size of the object, its shape and orientation, and the distances between the particles. The distance, on the one hand between the object and the light source, and on the other hand between the object and the detector also has a great influence. Usually, the result is interference of the secondary radiation as well showing in-phase and phase-shifted waves as well as different wavelengths. Then, interaction occurs of scattered and diffracted rays with one another as "multiple scattering". Diffraction effects can be shown clearly using narrow light slits and diffraction grating, [153].

The intensity of the diffracted scattered light is detected by a detector via electrical impulses. The effects can be presented as **scattering angle-dependent intensity** in the form of two-dimensional interference patterns. These diffraction patterns or "diffraction figures" occur due to the interference of the waves. Typical **diffraction images** of samples at rest display the following patterns dependent on the "structural parameters" and the degree of orientation of the short-range and long-range order of the superstructure [320]:

1) Concentric circles as "diffraction circles": **isotropic, completely irregular, and disorganized (amorphous), non-oriented configuration**, random distribution of the structure; example: dispersed particles in a liquid

2) Blurred, indistinct rings or "halos", showing graduations or "modulations" on the circumferences, these are dependent on the direction of shearing or stretching: **partially ordered structures, partially crystalline but non-oriented configuration**; example: polymer solutions and polymer melts such as sheared spinning fibers

3) Sickle-shaped pictures: liquid crystalline structures, **partially crystalline and oriented configuration**; example: liquid crystals

4) Regular grid of spots: **completely regular and ordered structure, crystalline configuration**; example: most solids such as stone and steel, as "single crystals"

m) Luminesence, fluorescence and color change, phosphorescence and after-glow

Luminescence is a material's ability to glow when it is hit by a light beam (Latin: lumen, meaning light). Firstly, the exitating energy is absorbed and stimulates electrons to perform quantum leaps onto higher energy levels. After a short time, usually only nano-seconds, the electrons are returning to their basic state. Light is emitted during this process. Two types of luminescence are described here, [66].

1) **Fluorescence** is the ability of a material to absorb light energy, radiating it afterwards as a light showing a longer wavelength. Example: UV light is shifted within the color spectrum towards larger wavelengths, e.g. into the visible range. Compared to the initial state, the elec-

trons return to a lower energy level finally. Part of the energy is lost in the form of heat. When samples are showing multiple fluorescence, they are emitting bright signals with diverse wavelengths.

2) **Phosphorescence** is after-glow, still appearing even when light is no longer radiated onto the sample. This is due to delayed quantum leaps of the electrons occurring during the diffraction process, as an "after-effect".

Since in many cases only these kinds of light effects make it possible to observe partial structures and details, corresponding functional components are often integrated into a sample by scientists on purpose as **"optical markers"**. Examples: Compounds containing photoactivable, **fluorescent chemical colors; fluorescent proteins** as "living colors"; **flourescent surfactants** which occupy the interface between the dispersed and the continuous phase in emulsions; and **fluorescent quantum dots** as nanoscopically small but luminous crystals.

10.8.2.2 Microscopy

Microscopic images display details from a locally focussed observation window, but not an average value of the sample. In contrast, scattering methods such as SALS, SAXS and SANS, generate average values of the whole irradiated sample, but do not provide local details. Thus, the two methods complement each other.

This section only describes microscopy using visible light (VIS), in a wavelength range from λ_w = 400 to 750nm, approximately. Precondition is that the sample is sufficiently transparent.

In order to improve the contrast and therefore the observation possibilities, there are a number of options in microscopy: using **transmitted** or **reflected light, inclined illumination, bright-field or dark-field, phase contrast** or **differential interference contrast** (DIC; which results in relief-like images), and combined methods. For a pretreatment of the samples there are also a number of options: Dying, using **flourescent components** as optical markers, or **phase-selective contrast agents** to achieve differing optical density.

Often in microscopy, the magnification is specified only (e.g. x5 or x50), more decisive however, is an information about the **resolution**, i.e. the selectivity between two details of the object. Up until recently, Abbe's limiting law was valid. This law states that an image loses sharpness when two points are closer than half the wavelength (λ_w/2) of the used light (1873 by *Ernst Abbe*, 1840 to 1905). Therefore in theory, with VIS light the resolution limit is reached at around 200nm. With conventional microscopes, however, usually a resolution of at best 1μm is possible.

Microscopy with visible light has the advantage over electron microscopes that the samples do not have to be investigated in a vacuum as with a scanning electron microscope or briefly SEM, or with a transmission EM or briefly TEM. That is a great advantage because in conventional diffraction experiments using SEM and TEM, material structures in the crystalline form, i.e. in a "frozen state", of course show a different deformation behavior as in a solution or dispersion. Therefore, when using light microscopes, samples can also include radiation-sensitive fluids, **hydrogels, soft matter** and even living "biological systems". Recent developments in observation techniques are attracting great interest at the moment, particularly amongst basic researchers who want to characterize complex structures under defined rheological shear conditions.

Examples of applications: biological tissue cultures (tissue engineered medical products, TEMPs), structures and transport processes in motion within single **living cells and in complex 3D cell systems,** for studies in cell biology or in other life sciences, in the form of high-resolution image series in real time using video microscopy at a very high imaging frequency.

Note 1: STED microscopy, higher resolution due to encircled light points

The resolution limit according to Abbe states that light points cannot be more sharply focussed than on a light point diameter of $(\lambda_w/2)$. By STED microscopy, the limits of the conventional miscroscopy using VIS light can be pushed considerably back (stimulated emission depletion, in 2006 by *Stefan Hell*, *1963) [177].

With STED, two laser beams are pulsed quickly one after the other. The first beam stimulates fluorescence of the sample, and the second beam showing an annular shape which surrounds the stimulating beam, deletes this effect again within its area of influence. The deleting beam cuts out an annular area around the emitting center, that a light point area showing the diameter of **around 20nm** only remains visible **as the resolution limit**. Therefore, the STED method considerably expands the previous observation range, in particular for biological samples, at the moment by the factor ten, compared to the conventional light microscopy. Comparison: Using electron microscopy, a resolution of 0.1nm is possible – the size of atoms – however, this is only possible in a vacuum and this destroys of course any life in biological cells.

Note 2: Confocal 3D microscopy, dot by dot in focus

A problem of high-resolution images achieved by conventional microscopy is that there are disturbances due to multiple scattering effects. Using CLSM or LSCM, however, very sharp **three-dimensional (3D) images** can be generated (confocal laser scanning microscopy). Here, with a laser beam, the sample is radiated and scanned dot by dot. The resulting reflection passes through a very narrow "pinhole" blind, which is installed conjugate focal, or briefly, "confocal". This pinhole blocks all the scattered light effects not having its origin in the corresponding dot to be displayed [66].

In a first step, this process is used to make at each time point an image of only a single dot of the sample, which is stored then. Afterwards, in a second step, all individual dots of an imaging plane are composed to the corresponding two-dimensional (2D) plane. Finally, in a third step, the program combines the individual 2D planes which were scanned in varying depths, and produces a 3D image pile which becomes visible as the resulting 3D image finally. This therefore leads to spatially well-focussed images dot by dot throughout the whole object. The resolution of this microscopy method is around 200nm in the horizontal, i.e. in the x- and y-direction; and it is around 500nm in the vertical, i.e. in the z-direction [296].

10.8.2.3 Devices for measuring anisotropy in terms of optical rotation and birefringence

There are various measuring cells which can be used to detect simultaneously several different parameters of the two fields rheology and optics. While the sample is usually radiated using polarized monochromatic laser light, a rheological rotational or oscillatory test is carried out simultaneously. For example, by applying a controlled shear strain or shear rate, a desired structural orientation can be achieved which can be presented and analyzed then as well via the mechanical data as well as via the optically anisotropic behavior. Anisotropy is direction dependency. Example: **Flow birefringence** and **flow dichroism** of a flowing fluid is investigated simultaneously using a polarized laser beam producing two planes of polarization which are in a right-angled position to one another. In literature, corresponding measuring equipment may be termed, for example, **DORA (dynamic optical rheo-analyzer)**, or similar.

10.8.2.4 SALS for diffracted light quanta

SALS means **small-angle light scattering**. Sometimes people speak of LALLS (low-angle laser light scattering). Here, a monochromatic laser beam is sent into the sample which does not need to be completely transparent. **VIS light** shows **wavelengths in the range of** λ_w = **400 to 750nm**.

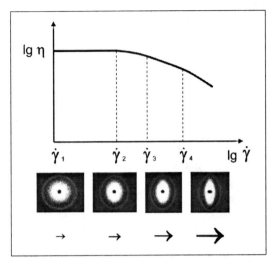

Figure 10.16: Viscosity function of an emulsion and the corresponding scattering patterns at different shear rates, generated by use of a SALS measuring cell

The **resolution** for SALS tests is often specified in the **range of around 0.5 to 100μm**. SALS measurements usually display the **total size of superstructures**, agglomerates, complexes, droplets, and all this also **under defined rheological deformation and flow conditions**. However, SALS tests do not exhibit partial components or details of nano- and micro-structures.

The result of these kinds of experiments is presented in the form of scattering or diffraction patterns showing the intensity of the secondary radiation as a function of the scattering angle. Using these characteristic images, there can be distinguished as well between the following structures of particles and macromolecules as well as their order of magnitude and distribution: Spheres or rods, monodisperse or polydisperse.

Figure 10.16 shows in principle the viscosity function and the corresponding scattering patterns of an emulsion. The scattering images are recorded at the four shear rates $\dot{\gamma} = 1/10/30/100s^{-1}$. The droplets of the internal dispersed phase still are hardly deformed in the zero-shear viscosity range which reaches here up to $\dot{\gamma} = 10s^{-1}$, whereas the ellipsoidal shape of the scattering pattern at higher shear rates indicates a significant deformation and orientation of the droplets.

10.8.2.5 SAXS for diffracted X-rays

SAXS means **small-angle X-ray scattering**. In 1895, *Wilhelm Conrad Röntgen* (1845 to 1923) discovered X-rays by accident. Here, an X-ray beam is sent into the sample. X-ray photons or "X-ray light quanta" can be generated by **vacuum tubes** or cathode ray tubes (or *Braun* tubes). However, if there is a **synchroton** available, the generated X-rays have an intensity of around 100,000 times higher compared to the X-rays produced by X-ray tubes. This leads of course also to a correspondingly higher intensity of the scattered secondary radiation. As a consequence, this allows again considerably shorter registration times, and therefore, a clearly higher resolution of the kinetics of the physical-chemical processes with respect to time; today this is already possible in the millisecond range. **X-rays** show **wavelengths in the range of $\lambda_w = 0.01$ to 1nm**. Examples: using a Mo-K$_\alpha$ tube anode $\lambda_w = 0.0711$nm, with Cu 0.154nm, with Cr 0.229nm, and with Al 0.834nm; or, filtered from a synchroton beam showing a broad spectrum of wavelengts, for example, with 0.1nm.

The **resolution** is in the **range of around 1 to 100nm for SAXS tests** and of **around 0.1 to 10nm for WAXS** tests. **WAXS** means **wide-angle X-ray scattering**. SAXS/WAXS tests (or "SWAXS") are suitable for detecting the electron density and their distribution. Examples: investigation of hollow and filled superstructures such as vesicles, core/shell particles and lamellar systems via the **electron density profile** of their surfactant superstructures (see also Chapter 9); **interactions** of molecules and particles between one another or to the surrounding matrix; **hydrodynamic radii of polymers coils including their hydration shells**. This allows insight into the **type of chemical or physical bonds, also under defined rheological deformation and flow conditions**: Are there primary bonds, covalent, secondary, van-der-Waals bonds or hydrogen bridges?

As a result of a SAXS test, a scattering or diffraction diagram is obtained showing the intensity of the secondary radiation as a function of the scattering angle. Small scattering angles indicate relatively large material structures. With WAXS tests, large scattering angles are analyzed representing small local structures.

On the **development of measuring techniques for X-ray scattering**:
In the 1920s *Karl Weissenberg* presented his "X-ray goniometer". In the 1950s *Otto Kratky* and *Günther Porod* presented the "Kratky camera" for recording small-angle X-ray scattering patterns. At the beginning, only the surfaces of hard materials consisting of solid crystals could be investigated. Today due to the considerably higher intensity of the primary beam, and therefore, also the higher intensity of the secondary radiation, meaningful static and dynamic images are obtainable already after clearly shorter measuring times.

However, the observation time is still relatively long compared to the **reaction and motion time of molecules and structures**. Today in the best cases, observation times of $t = 1\mu s$ can be achieved, and therefore, only a temporal average of the dynamics of the interactions between the molecules can be presented. With tests at low temperature, the molecular motion can indeed be slowed down more or less when reaching an almost frozen state, but then again only the behavior of the crystalline solid form can be investigated as above.

In future, scientists aim to perform measurements in the range of nano-seconds or even pico-seconds (ns or ps), i.e., in the time frame of the **"characteristic motion times"** – and this also under exactly controlled rheological conditions. This means, the required time for a structure to move along a distance which corresponds to the dimension of its individual size:
a) For particles showing a size of $d = 1\mu m = 1000nm$, there is $t = 1s$.
b) For micelles and latex particles with $d = 100nm$, then $t = 1ms = 10^{-3}s$.
c) For polymer molecules showing a molar mass of $M = 10$ to $1000kg/mol$ and $d = 10nm$, then $t = 1\mu s = 10^{-6}s$.
d) For smaller molecules with $M < 10kg/mol$ and $d = 1nm$, then $t = 1ns = 10^{-9}s$.
e) For atoms with $d = 0.1nm$, then $t = 1ps = 10^{-12}s$.

10.8.2.6 SANS for scattered neutrons

SANS means **small-angle neutron scattering**. Here, a neutron "beam" is sent into the sample. Neutrons are particles having a mass, in contrast to light or X-ray quanta. Therefore, here is occurring a particle flow or a "corpuscular beam", and not a wave without any mass. These kinds of atomic particles are either generated in a nuclear reactor by nuclear fission or by use of a synchroton and a spallation source.

Even when a particle flow is not a wave, a range of wavelengths can be simulated and adjusted via the energy value. A typical beam of a spallation source showing a continuous spectrum of electrons, neutrons and protons would correspond to a wide wavelength spectrum of $\lambda_w = 0.001nm$ to $100\mu m$. Via retardation in so-called moderators, the neutrons can be brought to a defined "thermal energy value" which corresponds to the range of around $\lambda_w = 0.1$ to $1nm$, for example, $\lambda_w = 0.17nm$. Comparison: Atom diameters show dimensions of $d = 0.1$ to $0.2nm$. The beam intensity of a neutron beam is controlled on generation via the acceleration energy, and it is around 100,000 times higher than the intensity of common X-rays.

With SANS tests, the **resolution** is in the range **of around 0.5 to 50nm**, hence similar to those of SAXS tests. However, there is a difference due to the considerably higher penetrating power of the neutrons. Whereas the tiny, mass-less X-ray quanta interact with the electrons of the atom shell, the much heavier neutrons with their clearly higher radiation intensity are interacting with the atom nuclei. Around 99.9% of the total atomic mass is located in the nucleus of an atom.

SANS tests are particularly well suited for **detecting the position of atomic nuclei**, the distribution and configuration of the nuclei in relation to neighboring atoms, and the inter-atomic bond lengths under external, mechanical or thermal, excitation. This gives insight into the type of the short-range order of structural complexes or agglomerates; and this holds of course also when applying **defined rheological deformation and flow conditions**. The results of diffraction experiments using SANS and SAXS/WAXS measuring devices complement each other with insights, on the one hand into the positions of the atomic nuclei (via SANS), and on the other hand into the type of bonding and electron density distribution in the atom shell (via SAXS/WAXS).

10.8.2.7 Velocity profile of flow fields

Velocity fields in a flowing material can be characterized using particle-observing devices, usually by performing rotational tests. This kind of system consists of a transparent measuring geometry with concentric cylinders or parallel plates, a laser, special lens optics, a digital camera, an evaluation unit, and the corresponding software program. If required, special **reflecting particles** can be added to the investigated fluid, e.g. silver-coated hollow glass spheres showing a diameter of around $10\mu m$.

The **laser light** reflecting from these **"marker particles"** is recorded by the camera, and therefore, the time-dependent positions of the particles are detected at **defined shear conditions**. The analysis program uses displacement vectors to generate the velocity profile occurring in the shear gap. In this way, even inhomogeneous flow fields can be illustrated, for example, when shear-banding appears (see Chapter 9.2.2) or when turbulent flow occurs showing Taylor vortices (see Chapter 10.2.2.4). In literature, corresponding measuring equipment may be termed, for example, **particle imaging velocimetry PIV** or **particle tracking velocimetry PTV**, or similar [196, 243].

10.8.3 Other special measuring devices

10.8.3.1 Interfacial rheology on two-dimensional liquid films

Interfacial rheology is used for rheological characterization of liquid films on surfaces or between interfaces. In dispersions, the phase boundary between two fluids which do not mix is called an **interface**. Typical measuring samples are emulsions and foams, and especially the interfaces between the liquid phase and the gas phase. A **surface** is in between a liquid and the outer surroundings being usually air (see also Chapter 9.1.1a and Figure 9.1). All types of rheological tests are also possible in the form of interfacial rheological tests. In order to measure the viscous and viscoelastic properties of monomolecular, **two-dimensional (2D) liquid films**, special measuring geometries such as the **bi-cone** is used (or "bi-conus"). In specialized literature, a water/air interface is also called a "Langmuir monolayer".

Analysis is carried out via the parameters **interfacial viscosity** η_i with the unit $Pas \cdot m$, **interfacial storage modulus G_i' and interfacial loss modulus G_i''**, both with the unit $Pa \cdot m$. Fields of application are above all found in basic research on surface-active molecules such as surfactants (see Chapter 9), on emulsions in the food, pharmaceutical and cosmetics industries and in the biotechnology sector, or on aqueous dispersions containing natural and synthetic functional polymers showing hydrophilic and hydrophobic groups. Comparison: In contrast to interfacial rheology, conventional rheological measurements are used to characterize the behavior of the entire mass of the sample, and therefore, the latter is the three-dimensional (3D) "bulk rheology" [124, 222, 252].

10.8.3.2 Dielectric analysis, and DE conductivity of materials showing electric dipoles

Being in an electric field, materials are polarized. As a consequence, in electrically conducting materials an electric current is flowing. In non-conducting and partially conducting materials, so-called **dielectric (DE) materials**, such as **electrically insulating dispersions**, however, local electrical dipoles are occurring which are in sum electrically neutral but may show a certain degree of charge separation on the surface of the particles or within the molecules. The dipoles are showing interaction with the electric field, and orientation, the positive dipoles towards the external negative pole and the negative dipoles towards the external positive pole. In other words: The dipole orientation is polarizing the material. All kinds of rheological rotational and oscillatory tests are possible in combination with **dielectric analysis (DEA)**. In literature, corresponding measuring equipment may be termed, for example, **dielectro-rheological device DRD**, or similar.

DEA measurement, preset: Voltage U [V] as an alternating electric field, sinusoidally oscillating at the frequency f_{DE} [Hz], for example in the range of U = 0.01 to 20V and f_{DE} = 20Hz to 20MHz

DEA measurement, result: Current I [A], oscillating at the same frequency like the preset voltage, and the phase shift δ_{DE} between the amplitudes U_A and I_A of the two sine curves.

At low frequencies of the alternating voltage, reversal of polarization in the DE material occurs immediately, without delay, i.e., in phase with the preset electric field. At higher voltage frequencies, however, the reversal of polarization of the dipoles can no longer follow the frequency of the electric field. This results in a delayed change of the orientation of the dipoles. As a consequence, between the time-dependent sine curve values of the preset U(t) and the measuring result I(t), a further parameter can be observed now: the dielectric phase shift δ_{DE}. The results can be analyzed after a conversion in terms of the complex **permittivity**:
$\varepsilon^{\star}(\omega_{DE}) = \varepsilon'(\omega_{DE}) - i \cdot \varepsilon''(\omega_{DE})$
with $\varepsilon^{\star}$ [F/m], F/m is the unit farads per meter, and $1 F/m = 1\ (A \cdot s)/(V \cdot m) = 1\ (A^2 \cdot s^4)/(kg \cdot m^3)$; and the angular frequency ω_{DE} [rad/s] $= 2\pi \cdot f_{DE}$ and the frequency f_{DE} in Hz; and ε' as the real part of $\varepsilon^{\star}$ and ε'' as the imaginary part of $\varepsilon^{\star}$. The DE loss factor $\tan\delta_{DE} = \varepsilon''/\varepsilon'$.

Permittivity is the ability of a material to polarize in response to an electric field and it characterizes a **material's ability to transmit an electric field** (Latin permittere means to allow to pass, to permit).

The **parameter ε'** is a measure for the **ability of a DE medium to store electromagnetic energy**. When showing a high ε'-value, the sample acts **like a capacitor**. Therefore, it is able to store and to return energy of an electric field without disturbing it, and as a consequence, no or only a very limited amount of conductivity can be observed between the electric poles. The **parameter ε''** on the other hand mirrors the **energy which is lost** in the DE medium, and this occurs mostly in the form of heat. Therefore, ε'' is a measure for the **ability of a DE medium to transfer electromagnetic energy into heat**. The higher the dielectric loss (dissipation) the better are the **conductive properties** of the sample, resulting in a weaker electric field. Application example: Heating of water containing foods in a microwave oven.

Examples of simultaneous **combination of DEA and rheology tests**:
a) Variable preset for rheological tests at constant DEA measuring conditions:
On the one hand, all kinds of rheological tests are possible, and on the other hand, DEA is performed at a constant frequency (e.g. f_{DE} = 1kHz) and a constant amplitude of the voltage (e.g. U_A = 1V); presentation in the form of the usual rheology diagrams.
b) Constant preset for rheological tests at variable DEA conditions: On the one hand, as a rotational test at a constant shear rate or as an oscillatory test at constant strain amplitude and constant angular frequency of the mechanical oscillation. On the other hand, DEA at a variable frequency of the electric field (frequency sweep, e.g. from f_{DE} = 20Hz to 20MHz); presentation in diagrams as $\varepsilon'(\omega_{DE})$ and $\varepsilon''(\omega_{DE})$.
c) Constant preset as well for the rheological tests at a constant shear rate (rotation) or at a constant strain and a constant angular frequency (oscillation), as well as a constant preset for DEA as a test at variable temperature, e.g. in the form of a temperature ramp.

For these kinds of combined tests, during the exactly controlled mechanical deformation of the sample, additionally useful information may be achieved if also the electrical conductivity is changing. Performing a curing process, it is sometimes possible to recognize changes in the DEA data of a material already, even if no significant mechanical (rheological) change can be observed yet.

Applications: multi-component systems, DE-active materials are e.g. water, carbon black powder (soot), and carbon nanotubes (CNTs).

Polymers: conductivity induction in a polymer matrix due to the filler, dispersion quality, interfacial behavior between the soot particles and the polymer matrix when deforming **rubber filled with soot**; orientation of CNTs when deforming correspondingly filled resins; curing process of filled reactive resins, and phase transitions such as in the range of the glass transition temperature T_g (see also Chapter 8.6.2.1a); **water amount**; degree of polymerization; polymer blends

Pharmaceuticals, cosmetics, foods: temperature-dependent phase transitions, ageing processes of **hydrogels**, microstructure of dispersions such as semi-solid **emulsions**, interfacial behavior, droplet size, volume ratio of the phases, **moisture content**, stability in the behavior of frozen food after reheating (freezing and defrosting behavior; see also Chapter 8.6.2.2b: freeze/thaw cycle tests), sensation in the mouth, feel on the skin (haptic behavior), proportions of the ingredients, effect of emulsifiers

Remark: This test should not be mixed up with the tests performed on electro-rheological fluids (ERF), as explained in Chapter 10.8.1.2.

10.8.3.3 NMR, and resonance of magnetically active atomic nuclei

NMR means **nuclear-magnetic resonance**. This kind of testing, performed in a high-frequency magnetic field is used to investigate the structure of a sample. Atomic nuclei consist of protons and neutrons. Depending on its composition, each nucleus shows a **mechanical angular momentum or nuclear spin** and a **magnetic moment**. Without an external magnetic field, the magnetic moments are not oriented. Being in a magnetic field they show orientation. However, there are only limited possibilities for an orientation or for the distribution of orientation. In different directions, there are occurring material-dependent energy differences. Energetic transitions between the states are possible at a certain resonance frequency, dependent on the type of the nucleus and the strength of the magnetic field. Corresponding amounts of energy have to be applied to trigger these transitions. When returning to the low-energy level, the nuclei are emitting a material-specific impulse. When presetting a variable frequency of the magnetic field (frequency sweep), an NMR spectrum is obtained which may show an absorption maximum. This kind of analysis is called **NMR spectroscopy** [178].

Remark: This test should not be mixed up with the tests carried out on magneto-rheological fluids (MRF), as described in Chapter 10.8.1.1.

10.8.4 Other kinds of testings besides shear tests

With many modern rheometers besides shear tests, samples can also be investigated by performing further experiments which are more or less related to rheological shear tests. This section describes a small selection of these kinds of tests.

10.8.4.1 Tensile tests, extensional viscosity, and extensional rheology

a) Extensional viscosity of ideally viscous fluids, and Trouton relation
For ideally viscous fluids at uniaxial extension the following applies, if the values of the **tensile strain rate** $\dot{\epsilon}[s^{-1}]$ and shear rate $\dot{\gamma}[s^{-1}]$ are the same size:

Equation 10.41 $\eta_E(\dot{\varepsilon}) = 3 \cdot \eta(\dot{\gamma})$

with the **extensional viscosity** η_E [Pas] and the shear viscosity η [Pas]. Besides η_E sometimes in literature the symbol η_u is used, for "uniaxial" extensional viscosity [233]. This correlation is called Trouton relation after *Frederick T. Trouton*, who published it in 1904, [365].

b) Extensional viscosity of viscoelastic liquids

In some production processes and **applications**, higher tensile or stretching loads are occurring compared to shear loads. **Application examples:** flow of spray coatings through nozzles, die casting, spinning process of polymer fibers, blow process of films or plastic bottles, flow through filters and porous bodies. In order to investigate the extensional behavior, here, besides rotational tests also tensile tests may be useful. This kind of testing is also possible with rotational rheometers when using a special measuring device, for example, consisting of one or two rollers ("drums") performing a rotational motion [333].

However, for practical users, these kinds of tests only provide meaningful insights for viscoelastic polymer solutions or melts. In this case for a uniaxial elongation the following counts, if the values of the tensile strain rate $\dot{\varepsilon}$[s⁻¹] and shear rate $\dot{\gamma}$[s⁻¹] are the same size:

Equation 10.42 $\eta_E(\dot{\varepsilon}) = 3 \cdot \eta_0(\dot{\gamma})$

with the zero-shear viscosity η_0 [Pas], (see also Chapter 3.3.2.1a)

The Trouton relation is therefore valid for viscoelastic fluids in the low-shear range only. Outside this range there is usually $\eta_E > 3 \cdot \eta$. This is expressed by the dimensionless **Trouton ratio**:

Equation 10.43 $TR = \eta_E(\dot{\varepsilon})/\eta(\dot{\gamma})$

Typical sample dimensions are: free length between the clamps L = 12.7mm, width b = 1 to 10mm, thickness a = 10μm to 1mm. To achieve useful results with this measuring method, a minimum value for the viscosity of η = 10kPas is recommended for the samples used.

The following has to be taken into consideration for tensile tests: **Due to necking of the sample** which is fixed on both ends, **the cross-section area** A(t) [m²] **changes exponentially with time when stretching the sample at a constant strain rate** (see also ISO/DIS 20965; however there, the signs λ and λ_0 are taken instead of L and L_0). Therefore in 1924, *Heinrich Hencky* (1885 to1951) defined the **Hencky strain** ε_H [1 = 100%] as follows, [180]:

Equation 10.44 $\varepsilon_H = \ln(L/L_0)$

with the (free) length L [m] of the sample when testing, and the (free) length L_0 [m] at the start; and ln is the "natural logarithm" to the basis e, using Euler's number (which is e = 2.718...).
Example: When ε_H = 1, then with exp $\varepsilon_H = (L/L_0)$, counts
$L = L_0 \cdot \exp \varepsilon_H = L_0 \cdot e^1 = 2.72 \cdot L_0$

For the **Hencky strain rate** $\dot{\varepsilon}_H$ (sometimes also called the "true strain rate") holds:

Equation 10.44a $\dot{\varepsilon}_H = d\varepsilon_H/dt$

or simplyfied: $\varepsilon_H = \dot{\varepsilon}_H \cdot t$

By the way, since $\varepsilon_H = \ln(L/L_0) = \ln L - \ln L_0$, therefore:
$\dot{\varepsilon}_H = d\varepsilon_H/dt = d[\ln(L/L_0)]/dt = d(\ln L)/dt - d(\ln L_0)/dt = (1/L) \cdot (dL/dt) - 0$,
since L = L(t); but L_0 = const., and therefore, the latter is irrelevant for the time derivative.

Change of the cross-section area of the sample, precondition is the volume conservation:
$V_0 = V(t)$, with the volume V_0 [m³] = $A_0 \cdot L_0$ of the sample at the start, and
V(t) [m³] = A(t) · L(t) when testing, and with the cross-section area of the unstretched sample at the start A_0 [m²], then: $A_0 \cdot L_0 = A(t) \cdot L(t)$,   or $A(t) = A_0 \cdot (L_0/L)$
Thus: $A(t) = A_0 \cdot (1/\exp \varepsilon_H) = A_0 \cdot [1/\exp(\dot{\varepsilon}_H \cdot t)]$, and therefore

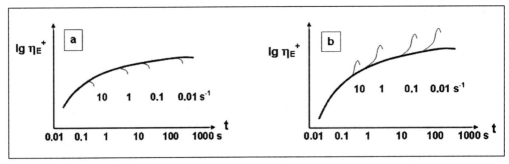

Figure 10.17: Growth curve of the transient extensional viscosity function $\eta_E^+(t)$, on the left-hand side of a linear polymer, and on the right-hand side of a polymer showing long-chain branching

Equation 10.45 $A(t) = A_0 \cdot \exp(-\dot{\varepsilon}_H \cdot t)$

Typical presets of tensile strain rates are in the range of $\dot{\varepsilon}_H$ = 0.003 to 30s^{-1}. A constant tensile strain rate is preset via a constant rotational speed n of the rheometer. However, when working in the tensile stress mode, the preset torque M must be reduced exponentially, since the cross-section area of the sample changes correspondingly, as explained above.

Usually, finally the function curve of the transient (i.e. time-dependent) extensional viscosity $\eta_E^+(t)$ [Pas] is presented as follows:

Equation 10.46 $\eta_E^+(t) = \sigma(t)/\dot{\varepsilon}_H$

with the transient tensile stress $\sigma(t)$ in [Pa]; then: $\sigma(t) = F/A(t)$ with the tensile force F [N], which can be detected via the torque M [Nm] when using a rotational rheometer with a suitable measuring device. In typical result diagrams η_E^+ [Pas] is presented on the y-axis and time t [s] on the x-axis, both on a logarithmic scale.

Example of presetting to obtain **a growth curve of the transient extensional viscosity function $\eta_E^+(t)$**: The measurement is carried out at different Hencky strain rates which are constant for each single test, e.g. at $\dot{\varepsilon}_H$ = 0.005/0.01/0.05/0.1/0.5/1.0/5.0/10s^{-1}. Thus in this case, there are performed eight single tests. Each one of these tests is finished when layers of the sample are overlapping on the drums of the measuring device or if the sample breaks previously. Usually the following is true: The lower the extensional strain rate, the further the η_E^+ value is travelling with a decreasing curve slope asymptotically towards its high final value (see Figure 10.17). As long as the effects of tensile strain softening or tensile strain stiffening are not occurring, the shape of the $\eta_E^+(t)$ curve equals the shape of the growth curve of the transient shear viscosity $\eta^+(t)$, which is used as an envelope curve. The latter curve is achieved by a shear test which was previously performed at appropriate low to medium shear rates. Here, the adapted Trouton relation $\eta_E^+(\dot{\varepsilon}, t) = 3 \cdot \eta^+(\dot{\gamma}, t)$ is taken as a basis of this kind of presentation [167, 276].

Further possible diagrams are, for example:
1) using raw data: force F versus length L, or F vs. extensional velocity v, or F vs. time t
2) using rheological parameters: $\sigma(t)$ as the time-dependent **growth curve of the tensile stress**, showing the same shape like the $\eta_E^+(t)$ curve, since the preset $\dot{\varepsilon}_H$ is constant.

Typical **application examples** are polymer melts, unlinked rubbers and elastomers, thermoplastic elastomers (TPE), films, sheets and laminates, foods such as dough and other highly viscous, viscoelastic and semi-solid preparations, adhesives and sealants.

With tensile tests, often a stronger orientation of molecules is achieved as is the case when flowing under shear conditions. Therefore the former tests are very useful, particularly to describe **polymers showing long-chain branching** (LCB). These kinds of materials often show tensile strain-stiffening (also called "tensile strain-thickening" or "tensile strain-hardening"). The reason for this behavior may be also **"flow-induced crystallization"** (FIC). Therefore, with tensile tests, in comparison to shear tests, it is usually easier to distinguish whether a sample is a LCB polymer or not (see Figure 10.17, the diagram on the right-hand side shows the behavior of a LCB polymer).

Sometimes there is occurring firstly stiffening behavior up to a certain extensional strain rate, which is followed by softening behavior at a higher strain rate. A possible explanation is that at the beginning, both the main chains and the long-chain branches of the side chains are deformed to the same degree in their network of entanglements. The latter however, show increasingly resistance at higher strain values. Finally, the main chains are stretched to the maximum, subsequently beginning to glide in the direction of the extension.

Note 1: Tensile stress relaxation tests, and tensile creep and creep recovery tests
Further possible tests are:
1) Measurement of the time-dependent tensile stress relaxation $\sigma(t)$ after presetting a tensile strain step at ε = const
2) Measurement of the time-dependent creep and creep recovery curve $\varepsilon(t)$ after presetting firstly a tensile stress step at σ = const, followed by a reverse step returning to σ = 0 = const

Note 2: Stretch tests, and the critical point of extension
Presetting a stretch test at a constant tensile strain rate of $\dot{\varepsilon}$ = const for a limited period of time, immediately followed by a tensile stress relaxation $\sigma(t)$. This preset is repeated several times at step-wise increasing, but constant tensile strain rates in order to observe when the critical point of extension is exceeded. This is the case if an increasingly large delay of the relaxation behavior can be observed in the second measuring interval. Then, the inner structure of the sample possibly may show the first signs of **flow-induced crystallization** (FIC), and therefore, **anisotropic behavior**. Perhaps it is also possible to see the effects of a certain degree of constriction of the sample due to viscoelastic yield necking. Microcracks may occur then, and brittle samples may even show the first signs of the final break.

Note 3: Tensile tests on solid films and laminates at a constant strain rate
For details on the tensile stress/strain diagram according to ASTM D882 and the corresponding analysis parameters such as the elasticity modulus, yield strength, tensile strength, break strength, and the corresponding tensile strain values: see Chapter 11.2.14b4.

c) Extensional rheology, and oscillatory tensile tests
Besides shear tests, it may be useful to carry out also tensile tests in the form of oscillatory tests. This is also possible with an oscillatory rheometer when using a special measuring device, for example, consisting of one or two rollers ("drums") performing a torsional oscillatory motion.

Application examples: films, sheets, laminates and fibers; elastomers and thermoplastic elastomers (TPE); flexible multi-layer composite materials; co-extrudates; polymers showing shape memory effects; medical materials such as natural and synthetic membranes, biological tissues and blood vessels

For oscillatory tests, Hooke's law is valid in the following form:

Equation 10.47 $E^* = \sigma(t)/\varepsilon(t)$

with the **complex tensile modulus E^*** [Pa], the sinusoidal time-dependent tensile stress $\sigma(t)$ [Pa] and the sinusoidal tensile strain $\varepsilon(t)$ with the unit 1 = 100%. See also Chapter 4.2.2 showing $E = \sigma/\varepsilon$, here for uniaxial tensile strain (i.e., into a single direction only), by applying constant and very low values; and Chapter 8.2.4 showing $G^* = \tau(t)/\gamma(t)$, here as an oscillatory shear test. Correspondingly, the sinusoidal functions and the calculated values are as follows:

Equation 10.48 $E' = (\sigma_A/\varepsilon_A) \cdot \cos\delta$

Equation 10.49 $E'' = (\sigma_A/\varepsilon_A) \cdot \sin\delta$

Equation 10.50 $\tan\delta = E''/E'$

Equation 10.51 $E^{*2} = E'^2 + E''^2$ (according to Pythagoras, see also Figure 8.6)

with the **storage modulus E'** [Pa], the **loss modulus E''** [Pa], the **loss factor or damping factor tanδ** [1], and the amplitudes of the sinusoidal tensile stress σ_A [Pa] and tensile strain ε_A in [1] or in [%]

In principle, all kinds of **oscillatory tests** can also be carried out as **tensile tests. Amplitude sweeps** as $E'(\varepsilon)$ and $E''(\varepsilon)$, or as $E'(\sigma)$ and $E''(\sigma)$; **frequency sweeps** as $E'(\omega)$ and $E''(\omega)$, with the angular frequency ω [rad/s]; **time-dependent and temperature-dependent tests at constant dynamic-mechanical conditions** with ε_A = const or σ_A = const, and ω = const.

In order to compare the values of E^*, E' and E'' on the one hand, and G^*, G' and G'' on the other hand, the following applies:

Equation 10.52 $E^* = 2 \cdot G^* (1 + \mu)$

with the dimensionless Poisson's ratio μ. For the definition of μ see Chapter 4.2.2. Table 4.1 lists a number of values of E^*, E and μ of various materials. This relation is also true for E' and G', as it is for E'' and G''.

For rigid solids showing μ = 0, the following applies: $\mathbf{E^* = 2 \cdot G^*}$

For the complex tensile modulus E^* this results in values which are twice as high as those obtained of the complex shear modulus G^*.

Concerning the effects of load onto the internal structure, tensile tests may sometimes be clearly more sensitive compared to shear tests. As mentioned in the previous section, effects may occur for **polymer samples** like **tensile hardening or stiffening by strain-induced crystallization**. In particular, a strong point of tensile tests is the characterisation of type and distribution of **long-chain branching** of polymer molecules. Further information is provided about possible effects such as **melt fracture during extrusion** or effects like **orange skin or shark-skin when blowing hollow containers**. When performing tensile tests, the mentioned effects can be characterized scientifically in terms of viscoelastic effects, for example, in order to prevent their occurrence during a production process.

10.8.4.2 Tack test, stickiness and tackiness

The stickiness or tackiness of adhesives is the property of a material to form a connection with a measurable solidity to a substrate after applying a slight pressure for a short contact time (e.g. for t = 1s). There are two types of tack: Solid-tack tests with rigid samples and wet-tack tests with liquid samples. However, tack values are not exactly defined in a scientific sense. Besides adhesives, these kinds of tests are also used to characterize **food, coatings** and **printing inks**.

Typical measurements consist of three intervals:
1) Gap setting or positioning: The measuring system moves downwards to get contact with the sample.
2) Pretreatment:
a) For solid samples: Applying a defined compression force, a typical measuring range for the normal force is F_N = 0.01 to 50N.

b) For liquid samples: Waiting for a certain time at rest or performing rotational motion at a defined shear rate

3) Removal: The upper part of the measuring system is lifted at a removal speed of typically between v = -0.01 and -10mm/s.

For different applications there are specially designed measuring geometries. Often is used a parallel-plate system. Then the upper, movable plate shows a diameter of for example d = 50 or 25 or 15mm for liquid samples, or 15 or 12 or 8mm for solid samples. While the bottom plate is stationary, the upper plate firstly

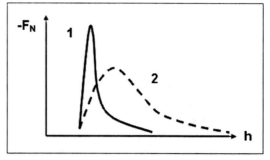

Figure 10.18: Results of tack tests in terms of a force/ path diagram: (1) showing little tack and a low level of stringiness, and (2) with a high level of tack and strong stringiness

moves vertically downwards and then upwards. In this latter interval, the plate may be rotating or not.

Possible test programs are explained by the following **measuring examples**:

a) Testing of a solid adhesive or a tape as a substrate carrying an adhesive layer; preset:
a1) Gap setting: The sample with a thickness of a = approximately 0.5mm is placed on the bottom plate. To set the gap between the plates at the start, the upper plate is moving downwards, controlled by a setting program which is defined by the user, either speed-controlled or force-controlled, until the plate makes contact with the sample. The motion is stopped when F_N is reaching 10N or more. When testing solids, it is important to set **the gap h_0 at the start** not to a fixed value, since here, **self-adjusting to the real sample thickness** should occur.
a2) Compression: For a contact time of t = 5s the upper plate is pressing onto the sample at F_N = 10N = const.
a3) Removal: Finally, the upper plate is moving upwards at a constant removal speed of v = -5mm/s. Negative values mean upwards and positive values mean downwards.

b) Testing of a liquid adhesive; preset:
b1) Gap setting: The sample is placed onto the bottom plate. To set the desired gap of h_0 = 0.250mm between the plates at the start, the upper measuring plate moves downwards, controlled by a setting program defined by the user, either speed-controlled or force-controlled. The upper plate does not rotate in this interval.
b2) Shear by rotation: For t = 1s, the upper plate is rotating at a constant shear rate of $\dot{\gamma}$ = 1000s^{-1}.
b3) Removal: Finally the upper plate moves upwards at a constant removal speed of v = -1mm/s (negative value, meaning upwards). The upper plate does not rotate here.

c) Analysis:
The following **diagrams** are typical, mostly the third measuring interval is analyzed only. Usually here, a linear scale is selected for both axis.
c1) Time-dependent force as a $F_N(t)$-diagram; speed-dependent force as a $F_N(v)$-diagram; path-dependent force as a $F_N(h)$-diagram; it is useful to show the removal force as $(-F_N)$ on the y-axis; see Figure 10.18.
c2) Force maximum $(-F_{Nmax})$ [N], the **time point** t [s] and the **removal path** h [m] at $(-F_{Nmax})$
c3) Calculation: area-related energy of separation (or adhesive failure energy or fracture energy, per unit of the geometric interface) E_{ad} [J/m^2 = N/m]

Equation 10.53 $E_{ad} = (F_N \cdot v \cdot t)/A = (F_N \cdot h)/A$

In the force/path diagram $F_N(h)$, the value E_{ad} corresponds to the value of the **area under the resulting curve**, related to the area $A = \pi \cdot R^2$ of the measuring plate with the plate radius R [m].

Particularly the **shape of the curve** provides information which is relevant for practical work; sometimes the **average curve slope**, e.g. of the declining part of the curve, is determined.

1) If there is a steep and rapid curve incline and a high **spiked force maximum** followed by a steep and rapid curve decline, then the **sample** is stiff, hardly deformable **with a tendency to show cracks**. These samples exhibit **hardly any stringiness, no stickiness and no tack**. Here, the area under the $F_N(h)$ curve is relatively small, and therefore, the value of the **energy of separation is relatively low**.

2) Or there is a **slight curve incline** with an often relatively low value of the force maximum, followed by a **slight curve decline**, which often shows a more or less pronounced "secondary maximum". The latter often looks more like a plateau, or like a so-called "shoulder". Then the sample usually displays more **tack with pronounced stringiness**. Here, the area under the $F_N(h)$-curve is relatively large, and therefore, the value of the **energy of separation is relatively high**.

The following factors influence the curve shape: temperature; degree of the compression force F_N or the pressure $p = F_N/A$, respectively, the latter can be varied over a wide range by the size of the measuring plate used; duration of the contact time; removal speed; average molar mass and molar mass distribution (MMD) of the sample, molar mass of the effective intermolecular entanglements which are forming a temporary network

The following types of breaking behavior can be distinguished:

1) **Adhesive fracture:** The sample detaches from one or both measuring plates. Usually, these kinds of materials show strong cohesion, with little stringiness. Often, these types of samples can be characterized well by shear tests via oscillation in the linear viscoelastic (LVE) range, e.g. using the values of the loss factor tanδ near the glass transition temperature T_g (see also Chapter 8.6.2.1a). At the application temperature they usually show G' > G'' or tanδ < 1, respectively.

2) **Cohesive fracture:** The sample breaks in the middle but continues to be well attached to both measuring plates. Usually there is a stronger stringiness. Often in shear tests there is G'' > G' or tanδ >1, respectively, also in the LVE range; for these kinds of liquid adhesives, however, the tanδ-value should not be too high. Otherwise their elastic part is too low to show enough tackiness.

If the measurement is carried out during a **curing process**, the **tack maximum can be found close to the gel point**, from a rheological point of view this is in the range around the crossover point G' = G'' (see also Chapter 8.5.3). Then, there already exists a loose but still unfinished network structure. The macromolecules show a wide molar mass distribution and a **high degree of long-chain branching**. Possibly there are also some chemical-physical secondary bonds and many **long-chain but purely mechanical entanglements** can be found between the molecules.

Pressure-sensitive adhesives (PSA) are usually **multi-component systems** consisting of elastomers, resins as tackifiers, plasticizers and filler particles. The cohesion force represents the internal strength. An adhesive should show the right balance between cohesion and flow behavior. With the words of a physicist: The ratio of energy storage and energy dissipation should show the optimal value. For users, an adhesive should provide enough resistance to slipping or removal while still being able to wet the substrate sufficiently. PSAs show the highest tack values in the application temperature range if the internal structures consist of the right mixture of liquid and solid phases. In other words: They are **optimal when** they are **balanced viscoelastically**, i.e. if the tanδ-value occurs in the right range. This is often achieved by mixing several components showing different glass transition temperatures.

Note: The Dahlquist criterion for adhesives in order to show sufficient tack
In 1983 *C.A. Dahlquist* published his opinion, that tack only occurs if the stiffness of the sample is not too high. Only in this case PSAs are effective and useful. The following is deduced from his statements:
1) The value of the storage modulus should not be higher than G' = 330 kPa = 0.33 MPa.
2) The molar mass M of the entangled polymer molecules should be higher than
M = 10 kg/mol; [84, 269, 407].
Comment: The suggested G'-value is a good reference value for practical users. Today, however, adhesives should be evaluated as viscoelastic materials. Besides the G'-value, also the G''-value or the ratio tanδ = G''/G', respectively, should be considered in order to characterize the **viscoelastic properties** and the stickiness in particular. As an alternative or additionally to shear tests, of course tack tests may be performed in this case.

10.8.4.3 Tribology

Tribology is the science of friction, sliding and wear behavior of materials. This science is concerned with the effects of friction between non-lubricated, dry or lubricated material partners. There are different designs for **tribometers** or **"rheo-tribometers"** and also for the measuring systems used. As is the case in rheology, tribological measuring parameters are highly dependent on the temperature. Therefore, careful temperature control is necessary, e.g. in the range of T = -40 to +200°C.

Two examples of different measuring geometries:
1) **The ball/pyramid geometry:** The upper part of the measuring system consists of a **ball** (e.g. of a diameter d = 6.35mm) and the bottom part consists of **three small platelets** which are fixed in a holder at an **angle of 45° to the vertical direction**. Since there is a spring-loaded bearing in the bottom part, the holder of the plates always adjusts flexibly to the prevailing geometrical conditions when the ball is pressed onto the platelets. Therefore, the measuring system always centers itself. The **axial force F_A** is preset **vertically from above by the rheometer**, and is evenly distributed via the ball onto the three contact points with the three platelets. Since the ball meets the platelets at an angle of 45°, the following applies: The **normal force F_N** = $F_A \cdot \sqrt{2}$ = 1.41 · F_A is acting right-angled onto the platelets.
2) **The ball/plane geometry:** The upper part of the measuring system consists of a **ball** which is pressed vertically onto a single or several **platelets** which are fixed on a planar bottom plate as a holder. The drive shaft with the ball adjusts flexibly to the prevailing geometrical conditions using a spring-loaded bearing. In this case, the following counts: The normal force F_N corresponds to the axial force F_A.

For dry friction conditions, in contrast to shear tests, there is no defined gap since both of the friction partners are in a direct contact with one another due to the permanently acting compression. Even when the friction is lubricated, there is occurring only a very thin fluid film in the micrometer range between the two materials [185, 345].

Measuring examples: Friction partners are a **ball made of steel and platelets made of plastic** (e.g. polyethylene PE or "Teflon" PTFE), and as a lubricant is used a mineral oil or a lubricating grease.

a) Static friction (friction-at-rest): **Preset** of the torque M as an upwards ramp with a logarithmic measuring point distribution (e.g. from M = 0.01 to 100mNm) and the axial force F_A as the compression force (e.g. F_A = 10N). **Measurement** of the sliding speed v_s in [m/s] or in [mm/s]. **Analysis** of the M-value as the **"start torque"** at which v_s shows a significant increase, or alternatively, the friction force F_f [N]. **Presentation in a diagram** showing M on the y-axis and v_s on the x-axis; or with M versus time t.

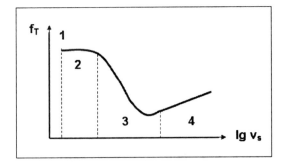

Figure 10.19: Intervals of the Stribeck curve showing the friction coefficient f_T versus the sliding speed v_s: (1) Static friction or friction-at-rest (without any motion), (2) solid friction or boundary friction, (3) mixed friction, (4) hydrodynamic friction or lubrication

Note: With conventional, simple tribometers it is merely possible to preset and to measure the speed and torque in linear steps. Therefore, it is not possible to determine them at any low values.

b) Kinetic friction: Preset of the rotational speed n [min⁻¹] as an upward ramp with a logarithmic measuring point distribution (e.g. from n = 0.1 to 3000 min⁻¹) and the axial force F_A [N] as the compression force. **Measurement** of the torque M as the "operation-torque", or alternatively, the friction force F_f [N]. **Analysis** of the friction coefficient f_T as a function of the rotational speed n or the sliding speed v_s or the sliding path s_s. **Presentation in a diagram** showing f_T on the y-axis and v_s on the x-axis; or with M versus time t.

The dimensionless **friction coefficient** is defined as follows:

Equation 10.54 $f_T = F_f / F_L$

with the friction force F_f [N] and the compression force (load) F_L [N]

F_f is determined via the torque. The load F_L depends on the axial force F_A, on the ball radius and on the contact angle between the ball and the platelets. Please note:

b1) If the platelets are fixed on a ball/pyramid system showing a conical geometry at an indentation angle of 45° to the vertical, the following applies: $F_L = F_N = F_A \cdot \sqrt{2}$

b2) If the platelets are fixed on a ball/plane system, then: $F_L = F_N = F_A$.

c) Static friction, rolling friction and roll-out time of a ball bearing

A special option is to use a ball bearing instead of a ball-and-platelets measuring system. The drive shaft of the rheometer is connected to the inner ring of the ball bearing. Here, besides the static friction and rolling friction, also the roll-out time can be determined. **Preset and measurement:** After a first measuring interval at a high rotational speed n (e.g. at n = 1000 min⁻¹ = const for t = 10s), for the second interval, however, there are made no presettings. Subsequently, the motion of the rheometer axis is detected to determine how long the bearing continues to rotate before coming completely to a standstill finally. This is **presented in a diagram** with the rotational speed n on the y-axis and time t on the x-axis.

Richard Stribeck (1861 to 1950) developed the **"Stribeck curve"** to scientifically evaluate lubricating properties. In the corresponding diagram, the friction coefficient f_T is shown on the y-axis and the sliding speed v_s is displayed on the x-axis. When **presenting on a semi-logarithmic scale** as in Figure 10.19, each one of the curve intervals are clearly visible (here as $\lin f_T$ versus $\log v_s$). The lubricating behavior at an increasing sliding speed can be divided into the following intervals:

1) **Static friction** (friction-at-rest): without any motion, i.e., $v_s = 0$; there is direct contact between the friction partners, no abrasion; f_T shows the maximum value

2) **Solid friction** (boundary friction): v_s is very low; there is still contact between the solid surfaces, strong **abrasion**, irregular **"slip stick"** effects are leading to vibration and noise. Comparison: dry automobile windscreen washers on a dry screen; f_T remains high

3) **Mixed friction:** v_s is low; there is still partially contact between the solid surfaces, less abrasion now; a lubricating film forms at the contact points; f_T decreases. Sometimes for analysis, a further interval is added showing the range in which the Stribeck curve goes through its

minimum, as the range of the "elasto-hydrodynamic lubrication" at a medium value of v_s; now a thin lubricating film occurs between the friction partners; f_T shows the minimum value

4) **Hydrodynamic friction (lubrication):** v_s is high; there is a fully developed lubricating film, the distance between the two friction partners is now larger than the surface roughness; f_T increases slightly but steadily with increasing v_s. These are the optimal lubrication conditions. However, it should be taken into account that the lubricating film can break again at higher speeds.

Conventionally, test programs were preset with a linear measuring point distribution and also the diagrams were shown on a linear scale. The disadvantage of this kind of presentation was that the results in the first two intervals were hardly visible. There are now measuring instruments available which enable the user to both preset and detect torques and speeds with a logarithmic measuring point distribution over a very wide range (e.g. from $M = 0.1\mu Nm$ to 200mNm, and from $n = 10^{-6}$ to 3000min^{-1}). In this case, all the four ranges of the Stribeck curve can be measured and analyzed. This is particularly useful for the practically relevant stip-slick effects which can cause great acceptance problems due to **vibration** and **noise, wear of materials** and therefore **destruction, unnecessary energy consumption** and, last but not least, the resulting **costs**.

Application examples for investigating **friction**:
a) **Behavior between** materials such as **metals, plastics** and **sealants in mechanical engineering**, with and without lubricants such as oils and greases
b) **Behavior of foodstuffs and the "feel": Sensation in the mouth** between the tongue as a movable, rough "elastomer" and the palate. Example: mayonnaises containing different amounts of oil, to simulate **perceptions like creamy**, soft, smooth, yielding, or watery, greasy, sticky and tacky behavior
c) **Behavior of creams** in the **cosmetics, pharmaceuticals and medicine** branch: **haptic sensation, feel when rubbing** into the skin

&∽   End of the Cleverly section

11   Instruments

11.1   Introduction

We know that around 80,000 ago, Stone Age people already used resin from birch trees or tar and pitch from birch bark as sealants, adhesives and also as chewing gums. However, we do not know whether our predecessors ever thought knowingly about the deformation behavior of these kinds of materials. Supposedly there will be also no answer to the question whether the Egyptians, more than 3500 years ago, ever examined the flow resistance of water systematically when they designed water clocks in the form of ceramic jugs showing a defined outlet in order to measure time. It is also unknown whether the Maya when performing their cultic games in the 11th century in Central America, knowingly took advantage of the viscoelastic behavior of their balls made of coagulated caoutchouc. What is certain that there were already many empirical, "pseudo-rheological" methods to characterize flow and deformation behavior before rheometrical tests were carried out on a scientific basis, [98, 146, 152].

11.2   Short overview: methods for testing viscosity and elasticity

In the past, special test methods and instruments were used in each branch of industry, and many of these kinds of tests are still being performed today because usually, the devices are cheap and easy to handle. The corresponding experimental results are often determined in terms of times, forces or temperatures, or they are specified in the form of **apparatus dependent relative values**. Some more or less simple, but also useful test methods are listed in the following.

11.2.1   Very simple determinations

a) Determination whether a material is a liquid or a solid, for official purposes (ASTM D4359). The material to be tested is held at T = +38°C (= 100°F) in a tightly closed can showing a volume of 1 liter or 1 quart, a diameter of 108mm and a heigth of 120mm. The lid is removed and the can is set inverted on a tripod or ring stand. Test result: A material that flows due to gravity a distance of 50mm (or 2 inches) or less within t = 3min is considered a solid, otherwise a liquid. Alternative criterion: It is a solid if the mass flowing out is less than m = 1g.

b) Testing with a spatula for a very simple visual evaluation, e.g. of **printing inks**; **"thick"** ink remains sticking, hardly running off the spatula, while **"thin"** ink is clearly dropping down [274].

c) "Daniel wetpoint" WP and "Daniel flowpoint" FP, as a hand-mixing method
The Daniel WP and FP technique, used **for millbase pigment pastes** of paints and coatings showing a high pigment concentration, is a simple hand-mixing method for characterizing two consistency stages in the take-up of vehicle (solvent/binder mixture) by a bed of pigment particles.

The wetpoint WP is defined as the stage in the titration of a vehicle given to a specified amount of pigment mass (e.g. of 20g), where just sufficient vehicle, as incorporated by vigor-

Thomas G. Mezger: The Rheology Handbook
© Copyright 2011 by Vincentz Network, Hanover, Germany
ISBN 978-3-86630-864-0

ous kneading with a glass rod or a spatula, is present to form a soft and **coherent paste-like mass**. The WP has the unit cm^3/g, as the volume of vehicle per weight of pigment. In this state, the mass shows the following properties: a significant resistance to strong or sudden pressure as the **"incipient dilatancy"**, a poor dispersibility, an adhesive and dull appearing, and a putty-like consistency.

The flowpoint FP is determined by noting what further vehicle is required and incorporated to produce a mixture that **just drops**, flows or falls off under its own weight from a vertically held spatula. Then, the mixture is a mobile, flowing and dispersable dispersion showing a softer consistency now. The FP mixture must allow rapid motion of the spatula across a layer of the millbase without strong drag, without permanent strokes being left behind, and with a fast regain of a glossy appearance. Between the WP and the FP, the mass remains sticking to the spatula without any sign of flow. At the FP, the mass is well-balanced in the ratio of solvent, binder and pigment, showing the **optimum mill base premix and pigment concentration** for charging e.g. a ball mill in order to provide the maximum volume output of the **dispersed pigment blend. The "Daniel dilatancy index"** DDI (in %) = [(FP – WP)/WP] · 100% (see also Chapter 3.3.3) [86, 281].

d) The gel time test, hot-plate method: Evaluation of the **chemical reactivity of resins and powder coatings** on a heating plate (e.g. at T = 200°C). The "gel time" is determined as the period after which a sample starts to show **tacky filaments when stirred manually with a glass rod** (e.g. after t = 30 to 40s). Compare PCI procedure #6: gel time reactivity [282].

e) The gelation time test (DIN 16945): Determination of the time period required to change from the sol state (liquid) to the gel state (no flow), to evaluate the **curing behavior of reactive resins**. After resin and hardener have been mixed for 5 minutes, the mixture is filled into a laboratory glass tube or into a pasteboard cup. Then a **glass rod** is put into the mixture. Determination: After time intervals of ever Δt = 15s the rod is lifted gently **(manual method)**. The "gelation time" is reached when the sample has become so sticky and solid that **when one tries to lift the rod out of a sample, the whole container is lifted as well** (see also Chapter 11.2.11b: gelation time, DIN 16945, using an electric stirrer).

f) The gelation time of unsaturated polyester resins (DIN 53184) is the time after mixing until the temperature has increased by ΔT = 10K due to the exothermic curing reaction. The **curing time** is the time required until the maximum temperature is reached.

g) The finger test to simply determine the **tack and compression behavior** e.g. of **pastes, printing inks, adhesives, bitumen or dough**. The sample is pressed between thumb and index finger and these are then pulled apart (see also Chapter 5.2.1.2c, Figure 5.5). The behavior of a sticky sample is called "long", and of a sample without tack "short" [274]. As an example, evaluation of **lubricating grease**: 1) **Brittle**: It ruptures and crumbles. 2) **Buttery**: It separates in short peaks on the edges with no visible fibers. 3) **Long fiber**: It stretches or strings out into a single bundle of fibers. 4) **Resilient**: It can withstand moderate compression without permanent deformation and rupture. 5) **Short fiber**: It shows short break-off with evidence of fibers. 6) **Stringy**: It stretches or strings into long and fine threads with no visible evidence of a fiber structure.

11.2.2 Flow on a horizontal plane

a) Spreading on a horizontal plate: A defined volume V of the sample is placed on a horizontal plate and is subsequently spreading due to gravity. Alternative determinations:
1) The period of time until a defined diameter d of spreading is reached
2) The actual diameter of spreading after a defined period of time
Example: Diverse **cosmetic emulsions** of V = 50µl on a horizontal glass plate after 15 (or 60) min showed spreading of between d = 10 and 14mm.

b) Drying or curing time of coatings during film formation on a glass plate (ASTM D5895). Evaluated is the **tack by** touching or rubbing gently with **the thumb or the flat of the hand** over the film. Alternative: The pin of a recorder moves in straight lines or in circles across the surface of the film. Classification by **visual evaluation** of the occurring tracks: **set-to-touch time, dry-hard time, dry-through time** [334]. See also Chapter 11.2.8c.

11.2.3 Spreading or slump on a horizontal plane after lifting a container

Usually here, a form is used which is open at the top and the bottom.

a) The "Adams consistometer": diameter of spreading of semi-fluid foods, using a cone-shaped container. The cone, better: a truncated circular cone, containing a defined sample volume is placed in the center of a square and then lifted gently. The square is made e.g. of transparent plastic showing concentric circles on the underside. Determined is the diameter of spreading (in inches) in a defined period of time, called the "Adams consistency". For creamy corn starch, t = 10 to 30s is typical [19, 344].

b) Spreading of self-leveling cast floors, using a hollow metal cylinder of h = 50mm and an inner d = 30mm, i.e. the ratio h/d = 1.67 (method CEN/TC 193). The short tube is placed on a horizontal glass plate, and the dispersion is filled into. Then it is lifted rapidly and the sample is spreading due to gravity. Measured is the **diameter** occurring after 4min as "initial flow". Other cylinders which were filled at the same time were lifted after 5min or even later to determine the "time-dependent flow behavior" or the "thixotropy".

c) Spreading of concrete, using a truncated circular cone of h = 200mm, d = 130mm on top and d = 200mm on bottom. The material is filled into the form made of tin, and then the form is lifted. Determined is the **diameter of spreading**, sometimes **after a defined shaking procedure**. Optional classification, with increasing diameter: stiff (no flow), plastic, soft, flowing [318].

d) The slump test for pastes, thickened slurries, mineral suspensions, tailings in the mining industry, clayey soils, and concretes. A defined volume V of the sample is filled into the form, which is a cylinder or a truncated cone, and placed onto a horizontal plane. Afterwards the form is lifted vertically. Alternative determinations [318]:

1) **Slump s** at rest (in mm), or after a defined shaking procedure. Optional classification of concretes, with increasing slump: stiff, plastic, soft, flowing (if s is too large).

2) **The slump time** until an initially truncated cone shows the form of a cylinder when shaking by a defined procedure, e.g. of concrete.

3) **"Yield stress" estimation via the slump of slurries, e.g. in the mining industry**, using the "Fifty Cent Rheometer" which is a hollow tin cylinder, e.g. of h = 73mm and an inner d = 73mm, i.e. the ratio h/d = 1. Diagrams are imprinted on the cylinder, with the yield stress values on the x-axis (two ranges: from 10 to 200Pa, and 100 to 900Pa), and the slump values on the y-axis (range: s = 2 to 72mm) related to densitiy values 1500, 2000 and 2500kg/m³ of the slurry. This method is used as a simple index test for **on-site measurements** to be made hourly in the field, e.g. to estimate maximum slopes of earth walls to be raised [279].

4) **"Yield stress" estimation via the slump of debris and muds** containing particles up to 30mm grain size, to evaluate the behavior of avalanches in the mountains. A cylinder was used of h = 89mm and d = 103.5mm, containing V = 0.75 l [319].

11.2.4 Flow on an inclined plane

a) Flow on an inclined plate: Testing the "free flow" of **coatings and inks** at room temperature. The samples are filled into **little bowls on a "flow plate"** while the plate is still kept
in a horizontal position, remaining there for a defined time. Afterwards the **plate is raised
to an inclination angle of 45°**, and the flow path of the sample is measured. By selecting
different periods of the time-at-rest before raising the plate (up to 1h), evaluations are made
such as "thixotropy" and "yield point" [186].

b) The flow distance on a heated, inclined glass plate (e.g. ISO 8619 **for resins**). Example: The plate flow test on **powder coatings** which are previously pressed to tablets. After
melting on a pre-heated plate for t = 30s, the plate with the sample is transferred into an oven,
which is usually heated **at T = +180°C**, and placed at an inclined position at the **angle of 65°**
to the bottom. After further t = 15min the flow distance of the melt is determined (e.g. 100 to
120mm). Compare PCI procedures #7: **inclined plate flow**, and #5: melt viscosity of powder
coatings [282].

c) The "inclined plane test" IPT is used to estimate the **"yield point" of debris and muds**
containing particles up to 10mm grain size, to evaluate the behavior of avalanches in the
mountains. The dispersion is poured onto a plane which is lifted slightly on one side e.g. at an
inclination angle of 1.26°, and driven by gravity, the fluid is spreading downslope. When
the flow stops, e.g. after t = 5min, the layer thickness of the suspension is detected. The finally
calculated **"yield stress value" depends on** the inclination angle, the fluid's density, and **the
final layer thickness** [319, 370].

11.2.5 Flow on a vertical plane or over a special tool

a) Flow distance on a vertical plane: A defined volume V of the sample is placed on a
plate which is afterwards put at a vertical position. Determination of the flow distance s due
to gravity after a defined period of time. Example: Diverse **cosmetic emulsions** of V = 50μl
on a vertically positioned glass plate after t = 15 (or 60) min showed a **flow path** of between
s = 40 and 80mm.

b) Sagging of coatings, paints and varnishes: The sample is applied by a **"sag index
applicator"** on a horizontal panel which is put afterwards in a vertical position. Alternatively,
a spray gun may be used to apply the sample on a substrate which is already installed in a
vertical position (ISO 16862).

c) Sag resistance of adhesives. Using a special device, e.g. a **"flow tester"**, a defined amount
of the sample is applied on a horizontal or vertical wall, and the **flow distance** by leveling
or sagging due to gravity is measured in mm. ISO 14678 describes seven methods, also for
curing materials.

d) The Visco-spatula according to Rossmann to roughly estimate the consisteny of **coatings**
used in workshops or construction sites, is a flat spatula showing two slits of different size. It is
dipped into the liquid sample, then rapidly drawn out and held vertically. The dimension of the
slits is designed in a way to achieve for a flow time of t = 5s, a) in the narrow gap a good **"spray
coating viscosity"**, or b) in the wider gap a good **"brush viscosity"** [406]. There are spatulas
available showing different designs and sometimes also a flow path scaled in mm.

e) Test blades to evaluate **leveling and sagging of paints and coatings**: An especially
designed blade is used to produce a film or a **wet coating layer** of a defined thickness on
a substrate. Example: Using a **stepped blade**, several layers are produced simultaneously
showing different thicknesses, e.g. each one of 25μm difference to the next neighbored

layers. The result is a wet **layer** showing **thicknesses** in discrete steps, e.g. of between 75 and 300μm. Afterwards, the coated plate is put in a vertical position with the thicker layers on bottom, and the user oberserves which one layer thickness is just remaining on the spot, while the thicker layers are sagging downwards [54].

11.2.6 *Flow in a channel, trough or bowl*

a) The **Bostwick consistometer**: Determination of the flow distance of **semi-solid foods**, such as purees of fruits and vegetables, along a trough in a defined time under the influence of gravity; e.g. for **tomato purees** t = 30s is typical, and 10s for **tomato sauces**. A sample of V = 100ml is filled into a reservoir which is closed off by a spring-loaded gate, and therefore, can be opened almost instantaneously. **The trough is inclined to a negligible degree** only, dimensions: 5cm wide or narrower, 2.5cm high, 24 or 50cm long which depends on the consistency of the samples, and the bottom is graduated in increments of 0.1 or 0.5cm. The result is reported in terms of cm of flow distance, e.g. for a ketchup at T = +20°C as the **"Bostwick consistency"** B_{30} = 5cm, i.e. in t = 30s [19, 284, 299].

A special type is the "flow distance measuring device" for electrically conducting fluids: The **measurement is carried out capacitively via electric contact points in the base of the trough**. Preset is the measuring time, e.g. t = 7.5 or 30s; specification of the result in mm flow distance. Afterwards, the **average flow velocity** can be calculated in mm/s [165, 207].

b) Matthis fluidometer: The sample, e.g. a **coating**, is poured into a **semi-spherical cavity**, being in a horizontal position. Then it is **put in a vertical position** allowing the liquid to flow under gravity into a groove which is graduated in mm. The distance it flows in t = 10s indicates the sample's **"fluidity"**, and therefore, a sand timer is attached on the device [117].

c) Daniel flow gauge: The sample, e.g. a **thick paint** or a **printing** ink, is poured into a **semi-cylindrical reservoir**, which then is **put in a vertical position**, and the sample flows onto a graduated base plate which is mounted right-angled, i.e. in a horizontal position to the reservoir. The distance it flows in a defined time indicates the sample's **"fluidity"** [117].

d) The **"inclined channel test" ICT** is used to estimate the "yield point" of **debris and muds** containing particles up to 10mm grain size, to evaluate the behavior of avalanches in the mountains. The dispersion is filled into a tank of V = 30 or 100 liters and, after opening a gate, it is flowing downwards through a **rectangular channel** of, for example, 2.1m length and 0.2m width, installed at an **inclination angle of 5.7°**. When the flow stops, e.g. after 5 minutes, the layer thickness of the suspension is detected. The finally calculated **"yield stress value" depends on** the inclination angle, the fluid's density, the channel width, and **the final layer thickness** [81, 319].

e) The **Casagrande apparatus** (DIN 18122) is used to classify **soils and muds** in terms of the **water content** w (in %). The sample is filled into a bowl and then **a 40mm long scratch is drawn through the sample**. Then, using a device equipped with an extender-wheel, **the filled bowl is hit** onto a hard rubber base by a defined procedure **while the scratch is dissapearing more and more**. The water content (w_L in %) for which 25 hits are needed to reduce the scratch length to 10mm is called the **"flow limit"** (in German: "Fliessgrenze"). The water content (w_P in %) at which the sample begins to crumble when it is formed to a roll of a diameter of about 3mm, is called the **"roll limit"** (in German: "Ausrollgrenze"). Classification of the consistency: "liquid" (if $w > w_L$), "plastic" (if $w_L > w > w_P$; subdivision: "pulpy", "soft", "stiff"), and "rigid" (or "hard"; if $w < w_P$).

11.2.7 Flow cups and other pressureless capillary viscometers

Flow cups and simple capillary viscometers are used to measure the flow time due to gravity, but they should only be used for ideally viscous liquids. Finally the kinematic viscosity, i.e., the density-dependent viscosity may be determined.

a) **Flow cups** are used for testing **coatings** such as lacquers, varnishes, paints, liquid inks, **and ceramic suspensions, drilling fluids, and petrochemicals** such as mineral oils and hot bitumen.

Examples: **ISO cup** (ISO 2431), AFNOR, BS, DIN (DIN 53211), Engler (ASTM D1665), Ford (ASTM D1200), Lehmann, Marsh funnel (API 13B), Shell (ASTM D4212), Zahn (ASTM D1084), dip-type viscosity cups (ASTM D3794, D4212), Redwood (IP 70/57; EN 12846), Saybolt Universal/Saybolt Furol (ASTM D88, D244, D2161, D2162, E102), road tar viscometer (DIN 52023, EN 13357, IP 72), dropping point (ISO 2176, ASTM D566, D2256). For detailed information see Chapter 11.3: Flow cups.

b) **Glass capillary viscometers** are used for testing **diluted polymer solutions,** for example, to determine the molar mass via the **"limiting viscosity number"** LVN (or the "Staudinger index", resp.) or the "intrinsic viscosity"; this method is also called "polymer viscometry". Further samples are **petroleum products** such as **mineral oils**, engine oils, gear oils, liquid lubricants, bitumen, asphalt emulsions, for example, to determine the **"viscosity index" VI**, or to use the **SAE classification of mineral oils**. Examples: Ubbelohde, Cannon/Fenske, Ostwald, BS/IP-U-Tube, Poiseuille, Vogel/Ossag (standards: ISO 1628, 2909, 3104, 3105, 3448; ASTM D341, D445, D446, D1243, D1601, D1795, D2170, D2171, D2270, D2493, D2857, D3591, D4603, D4957, D5225; DIN 51366, 51511, 51512, 51519, 51562, 53177, 53727; EN 12595, 12596; IP 71, 222, 319). For detailed information see Chapter 11.4.1: Glass capillary viscometers.

c) The **Kasumeter** is used to estimate the "yield stress" of **clay-dispersions on construction sites**, consisting of a vertical cylinder with the standard dimensions of d = 100mm and h = 300mm for V = 2.3 l (and a large scale type of d = 290mm and H = 800mm for V = 50 l) showing a horizontal **outlet capillary** with the standard dimensions of d = 9mm and L = 200mm for a maximum particle size of 0.25mm (and a large scale type of d = 16 or 25 or 38.5mm and L = 300mm for a maximum particle size of 5 or 10mm). The suspension is filled into the cylinder while the outlet is kept closed by a valve. After opening the valve, the fluid is flowing out and stops to flow when the corresponding final "stagnation level" in the cylinder is reached by the fluid; this is typically after 0.5 to 60min. The finally calculated **"yield stress value" depends on** the suspension's density, the diameter and length of the outlet capillary, and the **final stagnation level** [319, 326].

11.2.8 Devices showing rising, sinking, falling and rolling elements

a) The **bubble viscometer** is used for transparent and liquid samples such as **oils, resins and solutions** which are poured e.g. into a laboratory **glass tube. An air bubble is rising** when the tube is inverted. Alternative methods: Measurement of the rising time in seconds (ASTM D803 and D1545: "Bubble time method", specification of viscosity in terms of **"bubble seconds"**, and the meanwhile withdrawn ASTM D1131 and D1725); or evaluation according to Gardner/Holdt measuring the "GH viscosity"; or comparison to calibration tubes, each one containing a liquid of known viscosity such as "standard oils A to Z" of kinematic viscosities from $v = 0.5$ to $106,000 mm^2/s$ [62, 158].

b) Falling ball, falling rod, falling bar, falling cylinder or falling needle devices
Here, viscosity values are determined via the time of an object to travel a defined path **through a liquid or semi-solid sample**; i.e., to fall, sink, slip, slide, roll, or tumble downwards. Alternative methods are: The motion of the object occurs solely due to gravity or by an additional force, e.g. by a weight of a defined mass m in g or kg, or by an electromagnetic force. Examples:

1) **Falling-ball viscometers for oils and solutions** (for detailed information see Chapter 11.5): The fluid to be tested is filled into a measuring tube made of glass, and a ball made of steel or glass is sinking downwards through the sample, driven by gravity.

1a) **The inclined-tube method** (ISO 12058-1, DIN 53015), at the standard inclination angle of 80° to the horizontal (or 10° to the vertical, respectively), with an inner tube of d = 15.94mm, diverse balls are specified with d = 11, 14, 15.2, 15.6 and 15.81mm, and the measuring path s = 100mm.

1b) The **free-falling ball or vertical-tube method** (ISO 12058-2), with an inner tube of d = 16mm, and diverse balls of d = 1.6, 2, 2.4, 3.2 and 4mm, the measuring path is usually s = 50mm (optional up to 104mm). Example: The **Gibson/Jacobs falling-ball viscometer** with s = 150mm.

1c) **Micro falling-ball viscometers** are available for testing also non-transparent liquids such as **inks**. Here, the detection of the ball's position takes place via induction using electromagnetic detectors. Example: Balls are available of d = 1.5, 2.5 or 3mm, and glass tubes or capillaries of d = 1.6, 1.8, 3 or 4mm for a sample volume of less than V = 1cm^3, the inclination angle of the tube can be varied between 15 and 90° to the horizontal [10].

2) Falling-rod viscometers

2a) The **"Laray viscometer"** is used **for offset printing inks and varnishes** (ISO 12644, ASTM D4040, DIN 53222, [233, 274, 384]). A steel rod (e.g. of r = 5.98mm, a length of 300mm and m = 133g) travels a distance of s = 100mm through an aperture ("ring", e.g. of an inner R = 6.00mm, and the bore length L = 27mm). Thus here, the width of the annular gap is 20μm. For low-viscosity inks instead of a steel rod, a glass rod may be used; e.g. of only m = 32.5g. Weights to be loaded additionally on top of the rod are specified between m = 50g and 5kg. Examples: For all samples the constant load of m = 1kg is preset to measure the resulting falling time, e.g. between t = 40 to 80s. Alternative: To achieve high shear rates, it is aimed to reach a very short travel time between t = 1.0 to 5.0s, therefore corresponding loads are selected as the "high-weight method".
Estimation of the "shear rate": SR = L/[r · ln(R/r) · t]. Several tests are performed (e.g. using five different loads), to generate a "flow curve diagram". For analysis, methods may be used e.g. such as the fitting functions **Power Law or Casson**. Further determinations are the two viscosity values at SR = 2.5s^{-1} (as V2.5) and at SR = 2500s^{-1} (as V2500), and the following parameters:

2a1) the **"pseudo yield value"** PYV = 2.5 (V2.5 − V2500)
2a2) the **"shortness factor"** SF = (PYV/V2500)
2a3) the **"viscosity ratio"** VR = (V2.5/V2500)
2a4) the **"yield point"**, in terms of the shear stress value at SR = 2.5s^{-1}
2a5) the **"high-shear viscosity"** which is V2500.

2b) The **"Pochettino viscometer" for polymer melts, rubber compounds, asphalts, bitumens**. Shear rates of about 100,000s^{-1} are assumed to be achieved for a test time interval of only t = 0.01s, when using a cylinder of r = 25mm at a shear gap of 0.25mm and applying for example gas pressure. The speed is measured e.g. via induction [155, 237, 290].

3) **Falling needle viscometers for coatings and paints** (ASTM D5478), with a diameter of the needle d_1 = 4mm, and different tubes of d_2 = 19, 8 and 5mm; visual analysis or via induction by an electromagnetic sensor.

c) The **rolling-ball test** according to Wolff and Zeidler: A test plate is painted with a **coating** at a desired wet layer thickness (e.g. of 50μm), and then put at an inclination angle of 60° towards the horizontal. A steel or glass ball of d = 2.5mm is rolling down over the coating being still in a wet state. Measured is the rolling time over a defined distance. Alternative, according to Rossmann: **A ball or a wheel is drawn over a horizontal coating film**, e.g. at a defined speed of v = 300mm in t = 24h. This test is also used to determine the drying time of coatings. Alternative methods: A ball is drawn at a defined force, [54, 158]. See also Chapter 11.2.2b.

d) The **softening point (SP) analyzer**, e.g. using the **ring-and-ball (R&B) method**: When heating, the softening point or melting point of a sample is reached when a steel ball, laid on the surface of an initially cold and therefore rigid sample, is immersing a defined distance into the sample. The sample and a ball of d = 9.5mm are placed on a steel ring of an inner d = 15.9mm. The **temperature** is determined at which the ball is sinking downwards due to gravity over a defined path of s = 25.4mm. For bitumen holds: If SP = +40°C or lower it is "soft"; if SP = +63°C or higher it is "hard". According to ISO 4625-1 (for **paints, varnishes, resins**), ASTM D36 (for **asphalt, tar, pitch and bitumen**, from T = +30 to +157°C), ASTM E28 (for resins), DIN 1995 (for bitumen), EN 1427, EN 1871, EN 13179-1, IP 58, [95, 318].
Comparable is the **cup-and-ball method**, ISO 4625-2 (for paints, resins, varnishes), and ASTM D3461 (for asphalt, pitch) to determine the SP in the range of T = +50 to +180°C.

e) The **gelation time** using the **falling ball principle** for **unsaturated polyesters** (DIN 53184). Around 3min before the expected gelation, small balls of d = 3mm are put on the surface of the resin mixture in time intervals of ever 15s until the last ball indicates no more motion downwards. The gelation time is the time between mixing of the sample's components and the time point at which the last ball is showing no more sign of penetration.

f) The **oscillating piston viscometer** for testing **polymer solutions, low-viscosity inks and medicals** (ASTM D7483): This device measures **the cycle time of a travelling magnetic cylinder** (e.g. of d = 2.5mm and L = 20mm), which is drawn back and forth through the sample driven between two coils by a constant **electromagnetic force**. The piston is completely surrounded by the sample. The higher the viscosity of the fluid the longer is the cycle time for the piston to travel through the test chamber. The path is about s = 5mm, the sample's volume V = 6 to 8ml and the typical viscosity range is 5 to 25mPas. Example: For a fluid of 15mPas, the two-way cycle time is about t = 4s [63].

11.2.9 Penetrometers, consistometers and texture analyzers

Here, the structural strength is determined by pressing an object such as a **needle, cone, ball, pin, cylinder**, piston, ram, plate, blade or cutting wire into the sample. This test method is used e.g. **for semi-solid foods, petrochemicals, bitumens, asphalts, waxes, greases, sealants, gels** and **pastes**. Alternatives: Preset of the force and determination of the depth or velocity of penetration, or vice versa. Practical examples are (with pen = penetration):
ISO 2137 (cone pen, petroleum products), ISO 6873 (needle pen, dental gypsum products),
ISO 13737 (low-temperature cone pen, lubricating greases),
ASTM D5 (needle pen, bituminous materials; for pen-values up to 500),
ASTM D217 (cone pen, lubricating greases),
ASTM D937 (cone pen, petrolatum such as vaselines, paraffins, waxes),
ASTM D946 (pen-graded asphalt cements for use in pavement construction, classification),
ASTM D1321 (needle pen, petroleum waxes),
ASTM D1403 (cone pen, lubricating greases using one-quarter and one-half scale cone),
ASTM D1831 (cone pen, roll stability of lubricating grease using one-quarter and one-half scale cone; the difference in pen before and after a defined rolling procedure is an indicator of the shear stability),

ASTM D2884 (yield stress of heterogeneous propellants, by cone pen),
ASTM D5329 (sealants and fillers, hot-applied, for joints and cracks in asphaltic and Portland cement concrete pavements, by cone pen),
DIN EN 1426 (needle pen, bitumen), DIN 1995 (bitumen), DIN 10331 (butter hardness), DIN 51818 (lubricating greases), DIN 51579 (needle pen, paraffines), DIN 51580 (cone pen, paraffines), EN 1426, DIN EN 13880 (joints), IP 49 (bitumen), and IP 50 (greases)

Methods: Determination of the penetration path of a **needle** of d = 1mm with a mass of m = 100g due to gravity, read off after t = 5s, specified in tenths of a millimeter, i.e., **in steps of 0.1mm**, in terms of the **"pen-number"**. Examples:

a) The **"bitumen number"**, at T = +25°C, [95, 318], classification (DIN 1995 and EN 13179-2):
B 25 ("hard") if pen = (20 to 30) · 0.1mm,   B 45 if pen = (35 to 50) · 0.1mm,
B 65 if pen = (50 to 70) · 0.1mm,   B 80 if pen = (70 to 100) · 0.1mm,
B 200 ("soft") if pen = (160 to 210) · 0.1mm

b) "Penetration grades" of asphalt used for pavements, classification (ASTM D946):
40-50, 60-70, 85-100, 120-150, and 200-30

c) "NLGI consistency numbers" of lubricating greases, specified by the NLGI (National Lubrification Grease Institute, USA; ASTM D217 and D1403, and DIN 51818) at T = +25°C as pen-values in steps of 0.1mm read off after t = 5s, however here, using a **cone** of m = 150g; classification:
class 000 for pen = (445 to 475) · 0.1mm (fluid, for enclosed gears; "like thick cream")
class 00 for pen = (400 to 430) · 0.1mm (semi-fluid, for enclosed gears; "like tomato sauce")
class 0 for pen = (355 to 385) · 0.1mm (very soft, for centralized lube systems; "like mustard")
Thus, classes 000, 00 and 0 are used for "fluid" to "very soft" greases.
class 1 for pen = (310 to 340) · 0.1mm (soft, for low temperatures)
class 2 for pen = (265 to 295) · 0.1mm (semi-soft, standard for rolling element bearings)
class 3 for pen = (220 to 250) · 0.1mm (semi-hard, for high-speed ball bearings)
class 4 for pen = (175 to 205) · 0.1mm (hard, stiff, for very high speed and low load; "like soft cheese")
class 5 for pen = (130 to 160) · 0.1mm (very hard, very stiff, for low-speed journal bearings, "like semi-hard cheese")
class 6 for pen = (85 to 115) · 0.1mm (solid, for slow moving journal bearings; "like a block of wax")
Thus, classes 4 to 6 are used for "hard" to "solid" grease pastes, and also for thick layers and large gaps. For a very simple evaluation of lube greases by the "finger test", see Chapter 11.2.1g.

d) Consistometers have been widely spread for testing rheological behavior before rotational viscometers came into use. Probes still are in use with a variety of **geometries like cones** (being spiked or flat, e.g. of an inner angle of 30°, 60°, 90° or 120°, or of 20°, 44° or 53°), **needles, pins, flat-rods, plates, rams** (e.g. called "elastometer" for testing viscoelastic solids), **balls, tubes, hollow and open cylinders with through-bored holes** (for testing liquids). The sample is stressed **under the load of a defined mass** or by its own weight due to gravity, respectively. Determination of the **"elasticity"** as the quotient **force/penetration path** (F/Δs), or the **"viscosity"** as the quotient **force/penetration velocity** (F/v) which is calculated as the path travelled in a certain time interval (v = Δs/Δt). Other parameters used are termed as elastic, "plastic", "plasto-elastic", temperature-dependent "thermo-plastic", time-dependent "visco-elastic" **deformation**; or the "thermo-reformation" after applying a deformation and subsequently a heating procedure. Various types and names are used such as "**drawing-sphere viscometer**" (DIN 52007 **for bitumen**) or "**compressing-sphere vis-cometer**", "**viscobalance**", "**rheoviscometer**" [171, 194, 276, 324].

e) Modern "texture analysers" are equipped with an electric drive to control the com-pression or tensile force F [N], or the penetration path s [mm] or velocity v [mm/s]. In most cases is used a conical probe at a controlled compression force to determine the "**consistency**" or "**firmness**" **of semi-solid foodstuffs** such as butter, margarine, creams, bread; "**stiffness**" or "**gel strength**" e.g. of pectin, gelatin, fruit jellies; "**tenderness**" e.g. of cereals, fruit, peas, beans, for example, using the "Kramer shear cell"; **cutting** behavior e.g. of cheese, sausages, for example, using the "Warner-Bratzler blade" or the "Volodkevich bite jaws" or the "wire cut-ter"; "**hardness**" e.g. of vegetable fats, cookies, crackers; **puncture, penetration, fracture** and **rupture** behavior e.g. of fruits, vegetables, dough, cheese, gels; **bending** behavior e.g. of biscuits, chocolate, bread, vegetables using the "**three point bend rig**"; "**toughness**" e.g. of meat and fish; **extrusion** behavior e.g. of highly viscous liquids, gels, pastes, creams; **tension** e.g. of spaghetti, chewing gum sticks; and **adhesion, stickiness, stringiness** and **tackiness** e.g. of rice, pasta and starch pastes. A special device is the **acoustic analyzer** to measure the sound when breaking crackers, biscuits, cookies or solid chocolate [72, 342, 344, 382].

f) The **Bloom gelometer for testing gelatins**: The weight of lead pellets is determined which is required to press a cylindrical plunger of d = 12.7mm a distance of s = 4mm into a gel-like sample. The gel strength is given in Bloom degrees, which correspond to the weight in grams. A more modern alternative is to preset an immersion velocity of e.g. v = 0.5mm/s and to mea-sure the required force, presented in a diagram versus time [151, 378].

11.2.10 Pressurized cylinder and capillary devices

a) Dispenser or cartridge tests, e.g. to characterize the flow behavior of **soft solder pastes through a defined dispenser nozzle** in a defined time (DIN 32513). Alternative determi-nations:
1) The extruded volume or mass (weight) at a defined pressure p (force): "**flowability**"
2) The force (pressure) required to extrude a defined amount (weight or volume) of the sample.

b) The flow pressure: Evaluation of **lubrication greases at low temperatures** (e.g. the **Kesternich method**, DIN 51805). Determination of the pressure on a piston, which is driven mechanically by a motor or by gas pressure, required to **press a sample through a standard nozzle**, e.g. in the range of T = -60 to +30°C and p = 0 to 200kPa (= 2bar), [20].

c) Pressurized capillary viscometers, "melt testers", "extrusion plastometers", "capillary tubes"
1) **MFR and MVR testers driven by a weight for testing polymer melts**, to determine the **melt mass flow rate (MFR)** or the **melt volume flow rate (MVR)**, specification in g/10min or cm^3/10min (ISO 1133, ASTM D1238, DIN 53735). The polymer is filled into a vertically installed cylinder (of an inner d = 9.55mm and L = 115 to 180mm) and then molten. **Due to gravity**, under the load of an additional weight (of a mass between m = 0.325 and 21.6kg), a piston travels downwards within the cylinder, pressing the melt through a die (of d = 2.095mm and L = 8mm). Typical measuring range T = +125 to 300°C. This test method is explained in detail in Chapter 11.4.2.1.
2) **High-pressure capillary viscometers driven by an electric drive for testing poly-mer melts, PVC plastisols, greases and sealants** (ISO 3146, ISO 4575, ISO 11443, ASTM D1092, ASTM D1823, ASTM D3364, ASTM D3835, DIN 54811). A piston travels in a cylinder (e.g. of an inner d = 15mm, or between d = 9.5 and 24mm, and L = 280mm) pressing the sam-ple through a die (of d = 1mm and L = 16mm) or a short blind (of L = 0.25mm only). Typical

measuring range T = +20 to 400°C (or -40 to +500°C). Alternative presets: a) Drive speed v (e.g. in the range of v = 1μm/min to 1m/min, b) Drive force F (e.g. in the range of F_{max} =10 to 100kN). This test method is explaned in detail in Chapter 11.4.2.2.

3) **High-pressure capillary viscometers driven by gas pressure for testing paper coatings, mineral oils and polymer solutions** (ASTM D4624, ASTM D5481, DIN 53014). The liquid sample is pressed by gas pressure (up to p = 35MPa = 350bar) through a capillary made of glass or steel (e.g. of an inner d = 0.5mm, or between d = 0.2 and 1mm, and L = 50mm, or between L = 30 and 90mm). Here, shear rates up to 2 million s^{-1} are reached, and when using slit dies even up to 10 million s^{-1}. Typical measuring range T = +20 to 80°C, and enhanced -20 to +150°C. This test method is explaned in detail in Chapter 11.4.2.3.

4) **The "flow tester"** or **"extrusion test nozzle"** is used for testing the flow behavior of **PVC plastisols, adhesives and sealants** used e.g. in automotive industry. The sample is forced to flow through an orifice of defined size and shape (e.g. of d = 2 or 3 or 5mm and a minimum length of L = 3mm) or it is **dispensed directly from a cartridge**. The recommended through-put quantity is e.g. around m = 100g. Options: Automatic method using a software-controlled **pneumatic stand**, or manual method using a **pressurized-air gun** and a stop watch. Alternative determinations:

4a) **The rate of extrusion** (RoE) is the extruded mass m in a defined period of time **at a defined constant air pressure** (e.g. RoE = Δm/Δt = 2 to 80g/min at p = 0.3 or 0.4MPa = 3 or 4bar).

4b) The **extrusion force** [N] required to extrude the sample **at a defined travel speed** (e.g. at v = 60mm/min) of the piston (e.g. of d = 41mm), for example, using a compression machine.

d) The plasticity tester equipped with a mouthpiece

The plasticity tester is used to investigate the behavior of "ductile" and "rigid plastic" **clay and ceramic masses** (e.g. containing 55% parts per volume of ceramic powder, plasticizers such as cellulose ether and wax, and approx. 15 to 20% water), for example, when extrusion molding. The mass is filled into a **closed cylinder** (e.g. of an inner d = 25mm), in which there is installed **a "mouthpiece"** of a defined geometry (e.g. showing a funnel-shaped tapering towards its central, circular opening of d = 2 to 5mm). The flow resistance of the sample results in a counterpressure which is used to determine the **"flow pressure"** p_F (e.g. in the range of p_F = 0.1 to 58MPa). Alternative measuring methods:

1) **Moving outer cylinder, stationary inner mouthpiece**: The cylinder is drawn at a preset constant feed rate or by a constant pressure over the stationary mouthpiece on which the resulting counterforce or pressure is measured (e.g. with a travel path of s = 125mm) [393].

2) **Moving inner mouthpiece, stationary outer cylinder**. Using e.g. a hydraulic drive, the mouthpiece is pushed at a preset constant speed or by a constant pressure through the stationary cylinder on which the resulting counterforce or pressure is measured [204].

11.2.11 Simple rotational viscometer tests

a) For the **"20-minutes gelation test"** on **ceramic suspensions a rotational viscometer** is used, e.g. with spindles. At a constant temperature are preset two test intervals, each one for 10min, the first one at a constantly high rotational speed (e.g. of n = 100min^{-1}), and the second one at a constantly low speed (e.g. of n = 10min^{-1}). It is aimed to characterize the **physical structure recovery after shearing** in terms of the **"Bingham buildup"** BBU. The BBU-value indicates the change of the relative viscosity values between the end of the second, low-shear interval and the end of the first, high-shear interval as BBU = Δη = $η_L$ - $η_H$. The **"rate of build-up"** RBU is the initial change in the first two minutes of the second interval related to the total change in the whole second interval (see also the Note of Chapter 3.4.2.2c). Further determinations: The **"yield stress"** YS and **"plasticity index"** PI (see also Note of Chapter 3.3.6.4a), and **"pseudoplastic index"** PPI (see also Note of Chapter 3.3.2) [99].

b) The gelation time test (DIN 16945) using a rotational stirrer: Evaluation of the **curing behavior of reaction resins** at a constant temperature (i.e., at isothermal conditions), in terms of the period of time required to change from the sol state (liquid) to a gel-like state (no flow). After the resin and the hardener are mixed for 5 minutes, the mixture is filled into a laboratory glass tube or into a pasteboard cup containing a glass rod. Method: **Rotation until the forced standstill of the rod** is reached. The glass rod is connected via a magnetic coupling to the rotational axis of an electrically driven rotational device providing a constant torque. The "gelation time" is reached when the motion of the rod finally comes to a stop due to the increased flow resistance after the onset and development of the curing process. See also Chapter 11.2.1e: Gelation time, DIN 16945, manual method.

c) Viscometers equipped with torsion wires

1) **The "Gallenkamp torsional viscometer"** is used for testing **ceramic suspensions, sludges and slurries**, and for gelation tests [149]. Determinations:

1a) **"Fluidity",** i.e. the reciprocal value of viscosity: **A cylindrical bob** is immersed into the sample and then **deflected manually by a complete revolution**, showing afterwards a certain pretension against the **torsion wire** of the viscometer corresponding to the torsion angle of 360°. After the bob is released, it travels the 360° back and usually even over the initial zero-position. The return is caused by the torsion wire, and the angle of the "overshoot" is determined in terms of the number of angle degrees.

1b) Time-dependent flow behavior or **"thixotropy"** is evaluated by use of the above procedure, however, here the bob is released after waiting for one (or five) minutes as a regeneration time of the sample.

2) A similar type is the **"Payne torsion wire viscometer"**, with a standard bob diameter of $d = 17.5mm$, but there are also bob versions available of $d = 6.5, 13, 29$ and $41mm$.

3) **The gel tester according to Säverborn** is a double cylinder system to determine **elasticity of gels** at small deflection angles [264]. The cylinders C1 and C2 are profiled, C1 as the cup of $d_e = 36mm$, and C2 as the measuring bob in three versions of $d_i = 5/15/20mm$, the bobs all being $L = 43mm$ long. The sample is filled into the cup as a hot and liquid sol, and when cooling, C2 will be completely "gelled in". The bottom of C1 is **covered with mercury to prevent the gel sticking to it** (of course, working conditions like this are no longer acceptable nowadays). The measurement is performed 24 hours after the solidification of the gel. **Preset:** The approximately 0.75m long torsion wire, on which C2 is hanging on bottom, is deflected manually at the top by the preset angle φ_1. **Measurement:** Force equilibrium occurs when the reset torque of the gel and the actual torque M of the wire are of the same size. Detected is the occurring deflection angle φ_2 of C2, under the condition of $\varphi_2 \leq 2°$. Depending on φ_1 and φ_2 and the torsion constant of the wire, the value of M can be determined. In order to obtain a good angle resolution of φ_2 there is a small **mirror** mounted on the **torsion wire** reflecting and deviating a **light beam. Analysis:** Finally, the G-modulus of the gel is calculated, using the values of d_e, d_i and L, φ_1 and φ_2, and M.

d) Determination of the **gelation point** when cooling **mineral oils** using a rotational viscometer (ASTM D 5133 and D7110: viscosity at **low temperature and low shear rate, LTLS viscosity**). The oil is cooled at a constant cooling rate of $\Delta T/\Delta t = 1K/h$ (or 1°C/h, resp.) into the range of $T = -10$ to $-40°C$. The "gelation point" is reached at the temperature when the oil viscosity has increased to $\eta = 40Pas$. For further determinations such as the **"gelation index"** GI and the **"GI temperature" GIT** see Note 1 in Chapter 3.5.3.

e) The "mini-rotary viscometer" MRV is taken to measure the **"yield point"** and the **viscosity of motor oils at low temperatures and low shear rates (LTLS viscosity**, e.g. at $T = -40°C$). Using a rotating inner cylinder of $d_i = 17mm$ and $L = 20mm$ and a stationary external cylinder of $d_e = 19mm$, there is a gap width of 1mm. **Preset is the torque** by a weight-and-pulley system taking a mass which is suspended on a cord, moving downwards due to gravity. Determination of the yield point: Beginning with m = 10g, increasing the load in steps of ever 10g. Determination of the viscosity with m = 150g, assuming a shear stress of τ = 525Pa at this load; measurement of the resulting rotational speed and calculation of the corresponding "shear rate", it is aimed to obtain a value between 1 and 15 and in maximum $50s^{-1}$, [20]. Examples:

1) **The low-temperature "borderline pumping temperature" BPT** is determined during a 16-hour cooling procedure in the range of T = 0 to -40°C (ASTM D3829).

2) **The "pumping viscosity"** is measured to evaluate the **low-temperature pumpability** and the **"yield stress" of unused engine oils** during a long-term cooling procedure of t = 45h into the range of T = -15 to -40°C (ASTM D4684), or of **used engine oils** during a cooling procedure to T = -20 or -25°C (ASTM D6896), or of **gear oils and automatic transmission fluids** after a controlled preheat and cooling procedure to the final temperature in the range of T = -40 to 10°C (ASTM 6821).

f) The "tapered bearing simulator" TBS or **"tapered plug viscometer" TPV** are electrically driven single or multi-speed rotational devices to determine viscosity values of **engine oils at high temperatures and high shear rates (HTHS viscosity**; ASTM D4683: TBS, and ASTM D4741: TPV), [20]. Typical operating conditions are T = +150°C, at an assumed shear rate of $10^6 s^{-1}$, in a typical viscosity range of η = 1.5 to 5.6mPas. The measuring system is a slightly tapered inner cylinder ("plug") rotating in a correspondingly shaped outer stator which is therefore also a slightly tapered cylinder. Thus, a very narrow shear gap occurs which can be compared to a loose fit. Preset of a rotational speed, e.g. n = $3000min^{-1}$, and measurement of the torque. Calibration is made using viscosity standard fluids, assuming an extremely narrow **shear gap of only 3 to 4μm**. See also IP-370: Test method for measuring the viscosity of lubricants under conditions of high shear, using the **Ravenfield viscometer**, which is a TPV.

g) The "cold cranking simulator" CCS (ASTM D5293 and DIN 51377: both for T = -5 to -35°C) is used to determine the **viscosity of motor oils at low temperatures and high shear rates (LTHS viscosity)**. With this electrically driven instrument, a constant torque is preset via a short cylinder; measurement of the resulting rotational speed, assuming shear rates in a range of 10^4 to $10^5 s^{-1}$. As an enhanced temperatures range is mentioned T = -1 to -40°C.

h) The "trident measuring system" (ASTM D3232) is used to measure the "consistency" and "heat resistance" of **lubricating greases at high temperatures** and at a heating rate of $\Delta T/\Delta t$ = 5 K/min. Using a rotational viscometer and presetting a rotational speed of n = $20min^{-1}$, as a result is measured the torque.

i) The "torque tester" for lubricating greases (ASTM D1478, and IP 186) is used to determine the **"start torque"** M_S and the **"operation torque"** M_O **at low temperatures** (T < -20°C), using a standardized **ball bearing system**. After a defined start-up procedure, the two torque values are measured at a constant rotational speed of n = $1s^{-1}$. See also Note in example of Chapter 8.3.4.3.

k) The "Malcolm spiral pump viscometer" is used to test **solder pastes and other pastes**, as a double-cylinder system with a rotating outer cylinder showing an inlet on top and an outlet on bottom. Caused by the rotation, the paste is "pumped" into the measuring system. The stationary inner cylinder shows spiral-shaped grooves and it is connected to a torque sensor. Typical sample volumes for solder pastes are V = 100 or $300cm^3$ or m = 0.5 or 1.5kg. The preset is performed in steps, each at a constant rotational speed in the range of n = 1 to $50min^{-1}$. Typical measuring temperatures are between T = +15 and +30°C.

Example of a testing procedure: JIS Standard Operation, a Japanese standard.
Preset in eight steps (S1 to S8), at T = +25°C:
S1: Mixing, preshearing (target: homogeneous consistency), 3 (to 5) min long at n = $10min^{-1}$
S2: Measurement, at first for t = 6min at n = $3min^{-1}$
S3 and S4: then for t = 3min at n = $4min^{-1}$/then for t = 3min at n = $5min^{-1}$

S5 and S6: then for t = 3min at n = 10min^{-1}/then for t = 1min at n = 20min^{-1}
S7 and S8: then for t = 1min at n = 30min^{-1}/and finally for t = 1min at n = 10min^{-1}
Result: The "Malcolm viscosity" values (these are, of course, relative values) over time t at each one of the individual speeds n or shear rates $\dot\gamma$, assuming $\dot\gamma = 0.6 \cdot n$.

l) The "Stabinger viscometer" SVM (ASTM D 7042) is used to determine viscosity η and density ρ, and to calculate the kinematic viscosity v, of **crude oils and liquid petroleum products such as mineral oils, lubricants and fuels**. This is a **rotational viscometer** consisting of a pair of rotating concentric cylinders. The outer cylinder is a steel tube, with an outer d = 6mm and an inner d_e = 4mm, which is driven by an electric motor at a constant rotational speed. The inner cylinder called rotor, which is a lightweight hollow cylinder made of titanium of d_i = 3.1mm and L = 30mm, is completely surrounded by the liquid sample. Measured is the resulting constant rotational speed of the rotor which mirrors the equilibrium of two torques: 1) The driving torque due to the viscosity-dependent flow resistance force or shear stress, respectively, of the test fluid which is set in rotation by the constant motion of the outer tube, and 2) the retarding torque due to eddy currents which are induced in the surrounding copper casing due to the motion of a permanent magnet which is mounted inside of the rotor. Calibration is performed via standard viscosity fluids. The sample volume V = 3ml. Typical measuring ranges: viscosity η = 0.2 to 20.000mPas, density ρ = 0.65 to 3.0g/cm^3, and temperature T = -40 to +100°C [10].

11.2.12 Devices with vibrating or oscillating elements

a) Sensors, oscillating or vibrating at a constant deflection amplitude or frequency
After immersing into the liquid sample, the damping behavior is measured in terms of the change of the preset motion, either as a frequency decrease Δf or as a decrease of the path amplitude Δs, assuming these changes are proportional to the viscosity. Calibration is performed via standard viscosity fluids. Examples:
1) The sensor, e.g. **a sphere** of d = 32mm, **or a rod** of d = 11mm and L = 127mm, is exited on the surface to a **constant torsional path amplitude** of s_A = ±1μm. Here, as a **result, frequencies** are measured **between f = 650 and 750Hz** [139, 262].
2) The sensor, e.g. **a rod** of d = 3mm and L = 40mm, or a tool showing the shape of a **"tuning fork"**, is exited to a **constant flexural path amplitude** of s_A = ±2mm. Here, as a **result, frequencies** are measured **between f = 200 and 400Hz** [340].
3) **The "sine-wave vibro-viscometer"**: The sensor, e.g. a plate, is stimulated to maintain a **constant frequency of f = 30Hz**, and measured is the electric current required to maintain this preset. Here, as a result occurs a **flexural path amplitude** of around s_A = ±1mm, [3].
4) **The "vibrating needle cure meter" VNC**, was designed to evaluate **gel formation** and **curing behavior**. Presetting an **up-and-down motion of a needle** within the sample **at a constant frequency** between f = 35 and 300Hz (e.g. with 40Hz), as a result was **measured a deflection path** of s_A = ±0.1mm when unloaded (i.e. when vibrating in air), becoming increasingly smaller with the increasing resistance of a curing sample [287].
5) **The "Strathclyde curemeter"** was used to determine the **gel formation** and **curing behavior**. Preset was an **up-and-down motion** of a **vibrating element** showing the shape of a **"paddle" or plate,** at a **constant deflection path amplitude** between s_A = ±0.1 and 3mm (e.g. ±0.5mm) **and at a constant frequency** (e.g. f = 2Hz). As a result, **measured** were the **phase shift angle δ between the sine curves of the preset and the resulting oscillation** in order to determine the **gel point** when reaching δ = 45°, [287].
6) **The "Formograph"** is a device to characterize the **gelation process of dairy products**. **Preset of the strain** (deformation): On bottom, a container with the sample volume of V = 10ml is **oscillating** showing a **linear motion** at **a constant frequency** f_1 and **a constant path amplitude** ±s_1. **Measuring result:** The sensor is a wire loop ("pendulum") immersed into the sample to **detect** as well **the amplitude of the deflection path** ±s_2 as well as **the**

resulting force F. At the beginning can be observed the behavior of a low-viscosity fluid showing a small flow resistance, thus, $\pm s_2$ and F are small. During gel formation viscoelastic behavior occurs, $\pm s_2$ increases until $s_1 = s_2$, and also F increases; the **phase shift angle** δ as the delay between the preset and the response is decreasing continuously. Measuring data are recorded via a light signal using a small mirror which is mounted onto the sensor. **Resulting diagram:** Force F(t) and deflection s_2(t) versus time. **Analysis example**, and possible questions: a) When s_2 is beginning to show a significant value? (e.g. after t_1 = 15min); b) When occurs s_2 = 20mm? This **"coagulation point"** t_2 corresponds to the so-called **"curd firmness"** (e.g. after t_2 = 20min); c) Which value is reached by s_2 after t_3 = 30min? (e.g. s_2 = 40mm)

b) Measurements in the range of very high frequencies; examples:

1) **The "quartz viscometer"** for testing liquids such as **oils and paints**. Presetting **torsional oscillation of a quartz crystal** at a **frequency of f = 56kHz** at a minimal deflection, as a result is measured the **damping behavior** of the crystal caused by the viscous sample in terms of a decrease of the frequency values [140].

2) **The "acoustic wave viscometer"**: Presetting a high vibration frequency by a "shear acoustic wave resonator", determined is the change in the damping behavior when in contact with the test liquid. Measured is the power consumption by the energy dissipation of **a quartz crystal element** via an **oscillator circuit operating at a high frequency of f = 160MHz**. The unit of the result is called the "acoustic viscosity" AV. Calibration is performed using mineral oils. The sensor size is (in mm) 33 · 28 · 10, [43].

11.2.13 Rotational and oscillatory curemeters (for rubber testing)

a) The **"Mooney shearing disk viscometer"** or "Mooney consistometer": This **rotational device** is used to determine the **pre-vulcanization behavior of rubbers and elastomers**. **Standards:** ISO 289 (rubbers, unvulcanized), ASTM D1417 (rubbers, synthetic lattices) ASTM D1646 (rubber viscosity, stress relaxation, pre-vulcanization characteristics), ASTM D3346 (rubber property, processability of SBR emulsions), ASTM D4483 (rubbers and carbon-black industries), and DIN 53523 (elastomers) [276, 398].

Standard dimensions: stationary test chamber (10.6mm high, inner d = 50.9mm), rotor showing the shape of a cylindrical profiled disk (5.54mm thick and d = 38.1 or 30.5mm). The specimen is laid in the form of cold layers into the chamber (sample volume V = 25 ± 3cm³).

Preset: rotational speed of n = 2min⁻¹; detection of the torque M in terms of the **"Mooney units" MU**, sometimes called "Mooney viscosity", which may lead to misunderstandings since this is a relative viscosity value. Definition: 100MU correspond to M = 8.3Nm.

Typical test conditions: T = +100°C (or 120, 130, 140, 160°C; acc. to ASTM D1349), pre-heating time t = 1min, total test time t = 4min.

Requirement: A minimum of 40MU should be reached.

Typical diagram: time-dependent function MU(t). Analysis:

1) The "viscosity minimum" in MU (as MU_{min}) after heating and melting

2) The "scorch time t_5" as the time to reach 5MU over MU_{min}

3) The "scorch time t_{35}" as the time to reach 35MU over MU_{min}

Scorch time means: time to reach a certain amount of cure.

4) The "rise time" or "cure index": $\Delta t = t_{35} - t_5$ (as a measure to characterize the mean vulcanization rate)

5) Stress relaxation test according to ASTM D1646, after an abrupt stop of the rotational motion of the measuring disk.

5a) Determined is the time t_x [s] until the torque has decreased by x%. Example: When reaching a decrease of 80% as t_{80} = 16s.

5b) Determined is the decrease of the torque in percent as x_y [%] after y seconds. Example: After t = 30s there is x_{30} = 86%.

b) Oscillating disk curemeters ODC to evaluate the vulcanization behavior of rubbers and elastomers

Standards: ISO 3417: vulcanization characteristics, with the ODC; ISO 6502: rubbers; ASTM D1349: rubber, standard temperatures for testing; ASTM D2084: rubber property, vulcanization, using ODC; ASTM D5289: rubber property, vulcanization, using rotorless cure meters; ASTM D6204/withdrawn: unvulcanized rubbers, rotorless shear rheometers; ASTM D6601: rubber, cure and after-cure dynamic properties, using a rotorless shear rheometer, and DIN 53529, [276, 398].

Two types of measuring systems are described:

1) **Rotorless curemeters:** one part of the test chamber is set in motion, the other part is stationary. In the test chamber there is only the specimen.

2) **Oscillating disk curemeter:** The entire test chamber is stationary. Within the chamber are both the specimen and **a grooved bi-conical disk** which is completely surrounded by the sample. Example (ASTM D2084): Disk of d = 35.6mm; specimen: d = 30mm and 11.5mm thick with a volume of V = approx. 9cm^3; standard T = 160°C, max. torque M_{max} = 2.5 or 5 or 10 or 20Nm (here, the torque M is often specified in deka-newtonmeters: 1dNm = 0.1Nm). Alternatives, when working in an oscillatory mode:

3) **Linear shear** using a **rotorless curemeter**; comparable to the Two-Plates-Model, see Figure 8.1). Typical preset: deflection amplitude $\pm s_A$ = 0.01 to 0.1mm (preferably: 0.05mm), frequency f = 0.167 to 1.67Hz, thus, 10 to 100 cycles per min (preferably: 0.167 or 0.835 or 1.67Hz). Detection of the force F [N], as a time-dependent function F(t)

4) **Torsional shear** as a circular motion, for both rotorless and disk curemeters. Typical preset: amplitude of the deflection angle $\pm\varphi_A$ = 0.1 to 3° (preferably: 0.2, 0.5 or 1°; or as strain $\pm\gamma$ in steps: 2.8%, 3.5%, 7%, 14%, 21%, or 1%, 2%, 5%, 10%, 20%, 50%, 100%, 300%, 500%), at a frequency f = 0.05 to 20Hz (preferably: 0.835 or 1.67Hz, i.e. 50 or 100 cycles per min, or in steps: 0.1, 0.2, 0.5, 1, 2, 10, 20Hz). Detection of the torque M [Nm], as a time-dependent function M(t). Testing at a constant temperature between 100 and 200°C (i.e. isothermal)

5) Analysis: Plot of the time-dependent function M(t) or F(t), respectively. Analysis of the following parameters (here explained for M; but the same applies to F):

5a) **The "start value"** M_{min} at the minimum of the M(t)-function showing the value of the total turnover of x = 0% after pre-heating and melting,

5b) **The "maximum value"** M_{max} of the M(t)-function when curing is finished,

5c) **The "intermediate value"** M_t at a selected time point t,

5d) **The "incubation time"** t_i determined at the intersection of the following two fitting straight lines: The first one through the minimum value M_{min}, and the second one through the inflection point of the rising M(t)-function during the curing reaction.

6) **The "vulcanization or cure time t_x"** to reach a turnover of x%, with:

x = [(M_t – M_{min})/(M_{max} – M_{min})] · 100%, e.g. when x = 50 or 90% is reached as $t_{0.5}$ or $t_{0.9}$; the value of x is assumed to be proportional to the relative degree of cross-linking, vulcanization or curing, respectively). Alternative evaluations (ASTM D2084):

6a) **The "scorch time t_{s1}"** to reach 1dNm (= 0.1Nm) over M_{min},

6b) **The "cure rate index" CRI** = 100/(cure time – scorch time)

When using modern devices, also **parameters such as G', G'', G*, tanδ, η', η'', η*** can be analyzed, at both constant strain and constant frequency; as well by performing strain sweeps at variable deflection or torque amplitudes and at a constant frequency, as well as in the form of frequency sweeps at variable frequencies and at a constant deflection or torque amplitude.

11.2.14 Tension testers

Standards: ISO 37 (vulcanized rubbers, thermoplastics), ISO 527 (moulding and extrusion; plastics, films, sheets, fiber-reinforced composites), ISO 14129 (fiber-reinforced plastic composites, using the ±45° tension test); ASTM D113 (ductility of bituminous materials), ASTM D412 (vulcanized and thermoplastic rubbers and elastomers), ASTM D638 (plastics), ASTM D882 (thin plastic sheetings and films), ASTM D3039 (polymer matrix composite materials), ASTM D6084 (elastic recovery, bituminous materials and waxes, Ductilometer); DIN 52013 (bitumen), EN 13398 (bitumen), IP 32, [276, 384]. See also Chapter 10.8.4.1: Tensile tests.

a) Tensile creep testers, "ductility meters" or "draw apparatus". Preset: tensile force (tensile stress); result: displacement (elongation) or speed (elongation rate)

b) Tensile relaxation testers, "extensometers" or "stretch testers". Preset: displacement (tensile strain) or tensile speed (tensile strain rate); result: force (tensile stress). Examples:

1) **The "direct tension tester" DTT** (AASHTO/SHRP, ASTM D6723, [11]), to determine the tensile failure strain (deformation) of **bitumen used as asphalt binders at low temperatures** between T = -40 and 0°C (or at +25°); specimen dimensions (in mm) $27 \cdot 6 \cdot 6$. Preset: tensile speed $v = \Delta s/\Delta t = 1$mm/min

2) **"Ductility"** of **bitumen** at T = +25°C. Preset: tensile speed v = 50mm/min. Ductility is determined in terms of the distance to which the sample can be stretched before breaking, e.g. to a value between s = 150 and 1000mm (ASTM D113, DIN 1995, [95, 405]).

3) **Ductilometers** for bituminous materials at T = +25°C (ASTM D6084). Preset of the tensile speed v = 50mm/min until reaching the target length $L_1 = 100$mm. Then, the sample is immediately separated into two parts, subsequently remaining unloaded for t = 60min. L_2 is the final length after this period of rest. Calculation of the **percentage recovery or recovered elasticity** (in %) as: $[(L_1 - L_2)/L_1] \cdot 100\%$

4) **Tensile tests** (ASTM D882) at a constant tensile speed in the range of v = 25 to 500mm/min **for films** up to 0.25mm **and sheetings** (laminates) up to 1mm thickness. Sample dimensions: free length L = 50 to 250mm, width b = 5.0 to 25.4mm, thickness a with b : a ≥ 8 : 1; Example (in mm): $60 \cdot 10 \cdot 0.5$. For thicker samples: see ASTM D638 for plastics. For tensile tests, see also Chapter 10.8.4.1.

Analysis using the $\sigma(\varepsilon)$ diagram: with length L_0, cross-section A_0 and volume V_0 of the sample at the beginning; tensile stress σ and tensile strain ε, with $\sigma = F/A_0$ and ε (in %) = $[(L - L_0)/L_0] \cdot 100\%$; force F_y and length L_y "at the yield point"; force F_B and length L_B at break

4a) **The elastic modulus** : E = σ/ε (in MPa; in the linear viscoelastic range)

4b) **The yield strength**: $\sigma_y = F_y/A_0$ (in MPa)

4c) **Percent elongation at yield**: ε_y (in %) = $[(L_y - L_0)/L_0] \cdot 100\%$

4d) **The secant modulus** : $E_{SM} = \sigma_{SM}/\varepsilon_{SM}$ (in MPa). **For samples** showing no proportionality of σ and ε at the beginning, i.e. **without a linear-elastic, Hookean range**, the analysis straight line is drawn from the zero point of the $\sigma(\varepsilon)$-diagram to the measuring point (σ_y/ε_y) at the yield point. Alternative: A straight line is drawn from the zero point to another measuring point (σ/ε) occurring at a previously defined ε-value (as x%), as the **"offset yield strength** at x% offset".

4e) **The nominal tensile strength** at the force maximum: $\sigma_{max} = F_{max}/A_0$ (in MPa)

4f) **The tensile strength at break** : $\sigma_B = F_B/A_0$ (in MPa)

4g) **Percent elongation at break**: ε_B (in %) = $[(L_B - L_0)/L_0] \cdot 100\%$

4h) **The tensile energy to break TEB** as "toughness", energy density or **volume-specific energy** (in MJ/m³), calculated as the area under the $\sigma(\varepsilon)$-curve. The unit of this area or of the parameter $(\sigma \cdot \varepsilon)$, respectively, is $1Pa = 1N/m^2 = 1J/m^3$.

c) Testing **dynamic mechanical properties in tension, i.e. oscillatory tension**: ISO 6721 (plastics, tensile vibration, p4: non-resonance vibration; p9: sonic-pulse propagation method),

ASTM D4065 (dyn. mech. properties of plastics), ASTM D5026 (dyn. mech. properties of plastics in tension), DIN 53513 and DIN 53535 (both for elastomers). See also Chapters 8.6.4e and 10.8.4.1c: DMA and DMTA.

11.2.15 Compression testers

Standards: ISO 3384 (rubber, stress relaxation in compression), ASTM D1074 (compressive strength of bituminous mixtures).

a) Compression creep testers, "press" or "penetrometers". Preset: compression force (compressive stress); result: displacement (compressive strain) or speed (compressive strain rate)

b) Compression relaxation testers. Preset: displacement (compressive strain) or compression speed (compressive strain rate); result: force (compressive stress)
Example 1: The "Marshall apparatus" (ASTM D1559/withdrawn: resistance of plastic flow of bituminous mixtures using the Marshall apparatus; DIN 1996). Preset: compression speed $v = 50$mm/min on a cylindrical **bituminous material** (of $d = 101.6$mm, 63.5mm thick) at $T = +60°C$; determination: **"Stability"** is the maximum of the compression force F_{max} at break, the **"flow value"** is the deflection at this point, specified in steps of 0.1mm, [318, 405].
Example 2: Defo testers for testing the **compression** behavior **of raw rubber and rubber mixtures** at $T = +20$ and $+80°C$ (DIN 53514/withdrawn) [276, 324]. Typical samples are cylinders of $d = 10$mm showing the initial height of $H_1 = 10$mm. Measuring geometries are usually two plates with the sample in between. **Preset:** Compression of the sample by 60%, i.e., to the final height of $H_2 = 0.4 \cdot H_1$. **Measurement:** The **force** F, which is required to compress the sample in $t = 30$s to H_2 is called the **"defo-hardness" DH**. The portion of the total compression which regenerates, e.g. $t = 10$min after the load is removed, is called the **"defo-elasticity" DE**. **A series of compression and regeneration tests** can be used to evaluate the **fatigue behavior** of the sample.

c) Testing **dynamic mechanical properties in compression, i.e. oscillatory compression**: ASTM D4065 (dyn. mech. properties of plastics), ASTM D4473 (cure behavior of thermosetting resins using dyn. mech. procedures), ASTM D5024 (dyn. mech. properties of plastics in compression), and DIN 53535 (elastomers). See also Chapter 8.6.4e: DMA and DMTA.

11.2.16 Linear shear testers

Standards: ISO 14130 (fiber-reinforced plastic composites, apparent interlaminar shear strength, using the short-beam method); ASTM D4065 (dyn. mech. properties of plastics); DIN 53513 and DIN 53535 (both for elastomers)

a) Linear-shear creep testers or **"blade testers"**. Preset: shear force (shear stress); result: displacement (shear deformation) or speed (shear deformation rate)

b) Linear-shear relaxation testers. Preset: displacement (shear strain) or shear speed (shear strain rate); result: shear force (shear stress)

c) Testing **dynamic mechanical properties in shear, i.e. oscillatory shear**: ISO 6721 (plastics, p6: shear vibration, non-resonance method). See also Chapter 8.6.4e: DMA and DMTA.

11.2.17 Bending or flexure testers

Standards: ISO 178 (plastics, flexural properties) and ISO 6721 (plastics, p3: flexural vibration, resonance-curve method), ASTM D747 (apparent bending modulus of plastics by cantilever beam), ASTM D790 (flexural properties of unreinforced and reinforced plastics, by three-point bending), ASTM D2344 (short-beam strength of polymer matrix composite materials and their laminates, by three-point bending), ASTM D5934 (modulus of elasticity of rigid and semi-rigid plastic specimens by controlled rate of loading using three-point bending)

a) Bending or flexure creep testers. Preset: load force (bending stress); result: displacement (bending deformation) or speed (bending deformation rate)
Example: The **"bending beam rheometer" BBR** (AASHTO/SHRP, ASTM D6648,[11]), to determine the time-dependent **flexural creep** stiffness of **asphalt binders at low temperatures** in the range of T = -40 to 0°C (or at +25°). Preset: force F = 0.98N, corresponding to the weight of the mass m = 0.1kg; dimensions of the specimen in the shape of a rod or a bending beam (in mm) 125 · 12.5 · 6.5

b) Bending or flexure relaxation testers. Preset: displacement (bending strain) or bending speed (bending strain rate); result: bending force (bending stress)

c) Testing **dynamic mechanical properties in bending, i.e. oscillatory bending using three-point bending** or a **single or dual cantilever beam**: ISO 6721 (plastics, p5: non-resonance method), ASTM D4065 (dyn. mech. properties of plastics), ASTM D5023 (dyn. mech. properties of plastics using three point bending). See also Chapter 8.6.4e: DMA and DMTA.

11.2.18 Torsion testers

Standards: ISO 4663 (rubbers, dyn. behavior of vulcanizates at low frequencies, torsion pendulum method, i.e. damped oscillation), ISO 4664 (vulcanized and thermoplastic rubber, dyn. properties; p1: general guide, free and forced vibration methods, p2: torsion pendulum methods at low frequencies), ISO 6721 (plastics, p2: torsion-pendulum method); ASTM D1043 (stiffness properties of plastics as a function of temperature by a torsion test), ASTM D2236/ withdrawn (plastics, torsion pendulum test); DIN 53445 (solid polymers),[249, 327, 384, 407]

a) Torsional creep testers. Preset: torque (torsional shear stress); result: torsional angle (shear deformation) or "torsional speed" (rotational speed; shear deformation rate, shear rate)

b) Torsional relaxation testers. Preset: torsional angle (shear strain) or rotational speed (shear rate); result: torque (torsional stress, shear stress). Examples:
1) "Mooney shearing disk viscometer" for rubbers: The rotational speed of n = 2min⁻¹ is preset and then suddenly stopped. Afterwards, the time-dependent torque function M(t) is measured (ASTM D1646: rubber viscosity, stress relaxation, pre-vulcanization). See also Chapter 11.2.13a.
2) Other devices: **"Torsional stiffness testers"** (e.g. at deflection angles between 5 and 100°), the **"torsion pendulum"**, torsional oscillators, **generating "free and damped vibration"** or freely decaying oscillation: the specimen is deflected and then released, the result is a damped oscillatory motion, meaning: the deflection angle is decreasing continuously with time,[242].
Comment: This method corresponds no longer to the requirements state-of-the-art. Usually, nowadays is measured by presetting an undamped oscillatory motion.

c) Testing **dynamic mechanical properties in torsion, i.e. oscillatory tests** as explained in Chapter 8. These kinds of tests are performed for each single measuring point **as a controlled and forced, continuously exited motion, as a periodical oscillatory torsional process under constant shear conditions showing a constant amplitude,** as a non-resonance method. Therefore here, as a measuring result there is occurring no free oscillation, and no damped motion with continuously decreasing deformation amplitudes. Important standards:

ISO 6721 (plastics, p7: torsional vibration, non-resonance vibration),
ASTM D4065 (dyn. mech. properties of plastics),
ASTM D4440 (plastics, melt rheology, dyn. mech. properties),

ASTM D4473 (cure behavior of thermosetting resins using dyn. mech. procedures),
ASTM D5279 (dyn. mech. properties of plastics in torsion);
DIN 50100 (general information), **DIN 53535** (rubber, elastomers), **DIN 65583** (fiber-reinforced composites, glass transition temperature).

Summary of Chapter 11.2, which is a short owerview on a variety of **methods used for testing viscosity and elasticity** for samples showing all kinds of rheological behaviors.
After this journey through the world of more or less exotic but also useful test methods, the practical user should be aware of the following: **Most of the above tests results** obtained are **relative values** which in many cases cannot be compared to those results which are achieved when performing scientific rheometrical experiments. See also Chapter 10.6: Relative measuring systems; and relative values obtained when not working under scientifically defined test conditions.

11.3   Flow cups

Sometimes, flow cups are also termed efflux cups, and each type is showing a specific design. Above all it is the diameter of the outlet capillary and the volume of the cup which is defined accurately. When the outlet is opened, the liquid sample flows due to its own weight driven by gravity through the centric outlet (or orifice) in the bottom of the cup. The **flow time** (or efflux time) is measured and from this is calculated the **kinematic viscosity** ν (see also Chapter 2.2.3b). Several types of flow cups are in use (see Figure 11.1). However today, for useful test results, only the **ISO cup** should be taken (ISO 2431: paints and varnishes; ASTM D5125:

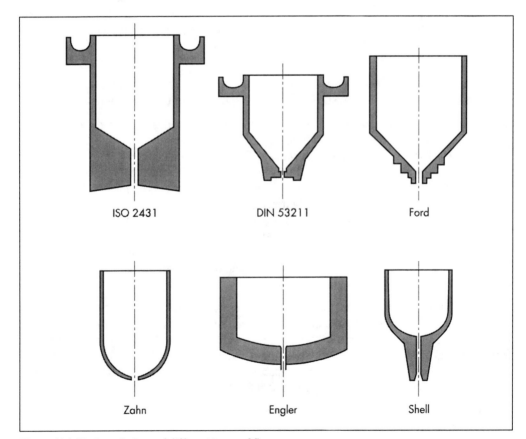

Figure 11.1: Various designs of different types of flow cups

paints by ISO flow cups). Here, the length L of the outlet capillary is specified with L = 20mm, therefore being clearly longer compared to most other flow cups.

The liquid sample is flowing due to gravity, driven by the hydrostatic pressure of the liquid column which occurs vertically over the outlet capillary; and this pressure determines the resulting shear stresses and shear rates. Since there is acting no external pressure, flow cups are considered **"pressureless viscometers"**.

11.3.1 ISO cup

For the different types of flow cups available there are various formulas in use to calculate the kinematic viscosity v [mm²/s] from the measured flow time t [s]. Usually, these formulas are found empirically after calibration tests using viscosity standard oils.

The following applies to ISO cups:
1) for the 3mm orifice: $v = 0.443 \cdot t - (200/t)$
2) for the 4mm orifice: $v = 1.37 \cdot t - (200/t)$
3) for the 5mm orifice: $v = 3.28 \cdot t - (200/t)$
4) for the 6mm orifice: $v = 6.90 \cdot t - (570/t)$
Further ISO cups are available showing orifice diameters of 2 and 8mm. The specified flow time range is limited from t = 30 to 100s. Under this condition the occurring shear rates are depending above all on the capillary diameter used, resulting in values between $\dot{\gamma} = 60$ and $1500s^{-1}$ for the most commonly used cups. When testing **paints and coatings** with flow cups it can be assumed in most cases to measure at mean shear rates between 50 and $500s^{-1}$, [118]. Experience showed that tests with ISO cups are useful merely in the range of the kinematic viscosity of $v = 7$ to 700mm²/s. This counts, of course, only for ideally viscous fluids since other ones should not be tested when using flow cups.

&⁓ For "Mr. and Ms. Cleverly"

11.3.1.1 Capillary length

Longer capillaries are specified for ISO cups compared to most other kinds of flow cups since a certain minimum inflow distance is required for a liquid to generate the desired laminar flow field, showing correspondingly a parabolic flow velocity profile in the capillary. Especially with low-viscosity fluids, in the first part of the capillary length **inlet disturbances** and therefore **turbulent flow conditions** may occur showing vortices. This is not desired since these conditions are unreproducible, and this is of course disadvantageous for accurate measurements. Therefore, the required flow path is usually longer compared to the mostly too short capillaries of most other kinds of flow cups in order to attain the desired laminar flow conditions (see also DIN 53012).

For the following simple calculations there is assumed that the sample shows ideally viscous flow behavior and there are laminar and stationary flow conditions, and no significant influence due to the following effects:
• inlet flow disturbances in the capillary due to the initially increased friction forces when entering the clearly narrower capillary geometry, causing a change of the flow velocity with the risk of appearing turbulent flow conditions; there may occur also outlet flow disturbances
• inclination of the cup, if the stand is not put exactly in a vertical position; therefore, it should be adjusted by use of a water-level
• inaccurate control of the measuring temperature
In order to reduce the influence of the inlet and outlet disturbances, the L/d ratio, i.e. of length and diameter of the capillary, should be not smaller than 10 : 1. However, this condition is not fulfilled

when using ISO flow cups, and therefore, all results which are achieved - especially when testing low-viscosity fluids - should merely be considered relative values, but not absolute values.

11.3.1.2 Calculations

a) The range of the hydrostatic pressure in a capillary

Equation 11.1 $p_l = \rho \cdot g \cdot h$

with the density ρ of the sample, the gravitation constant $g = 9.81 m/s^2$, and the level h of the liquid in the flow cup
Example 1: Using an ISO cup no. 4
Presets: $\rho = 1 g/cm^3 = 1000 kg/m^3$, maximum liquid level $h_{max} = 63mm = 0.063m$
Calculation of the maximum hydrostatic pressure, at $h = h_{max}$:
$p_l = (1000 \cdot 9.81 \cdot 0.063 \, kg \cdot m \cdot m)/(m^3 \cdot s^2) = 618 \, kg/m \cdot s^2 = 618 Pa \, (= 6.18 mbar)$
Please note: The hydrostatic pressure decreases with the level h; for $h \to 0$ holds: also $p_l \to 0$.

b) The shear stress range at the capillary wall according to the Hagen/Poiseuille relation

Equation 11.2 $\tau = (p_l \cdot R)/(2 \cdot L)$

Example 2: Using an ISO cup no. 4
Presets: $p_l = 618 Pa$ as the maximum hydrostatic pressure (see above Example 1),
radius $R = d/2 = 4mm/2 = 0.002m$ and length $L = 20mm = 0.020m$ of the capillary
By the way here, the L/d-ratio is 20mm/4mm = 5 : 1, being clearly less than 10 : 1 which is considered the minimum ratio to neglect inlet and outlet disturbances. Calculation of the maximum shear stress, at $h = h_{max}$:
$\tau = (618 \cdot 0.002 \, Pa \cdot m)/(2 \cdot 0.020 \, m) = 30.9 Pa$
Please note: The shear stress decreases with the liquid level h; for $h \to 0$ holds: also $\tau \to 0$.

c) Shear rates in a capillary
There are two methods to calculate the shear rate values occurring in flow cups.
1) **The mean shear rate value at the capillary wall** acc. to the Hagen/Poiseuille relation

Equation 11.3 $\dot\gamma = (4 \cdot \dot V)/(\pi \cdot R^3)$

Example 3: Using an ISO cup no. 4
Presets: Sample volume $V = 108ml = 0.108 \, l = 1.08 \cdot 10^{-4} \, m^3$,
and $R = 2mm = 2 \cdot 10^{-3} m$, measured flow time $t = 75s$
Calculation of the volume flow rate: $\dot V = V/t = (1.08 \cdot 10^{-4}/75) \, m^3/s = 1.44 \cdot 10^{-6} m^3/s$
Then: $\dot\gamma = (4 \cdot 1.44 \cdot 10^{-6} \, m^3)/[\pi \cdot (2 \cdot 10^{-3} \, m)^3 \cdot s] = 229 s^{-1}$
This single value represents merely just a single shear rate value of the actually wider shear rate range occurring, and this is the value "according to Hagen/Poiseuille"; see the following section.
2) **The shear rate range in a capillary**
The shear rate value calculated above is a mean value, however, the actual shear rates occurring in a capillary flow are depending on the hydrostatic pressure.

Equation 11.4 $p_l = \rho \cdot g \cdot h$

Condition 1: There is stationary flow in a capillary of the radius R and the length L. Then, for the flow resistance force F_R of the liquid and the force F_P due to the hydrostatic pressure acting on the liquid holds: $F_R = F_P$
Using the shear stress τ along the capillary wall area A_l, then:
$\tau = F_R/A_l = F_R/(2\pi \cdot R \cdot L) = \eta \cdot \dot\gamma$ or $F_R = 2\pi \cdot R \cdot L \cdot \eta \cdot \dot\gamma$
with the viscosity η [Pas] of the sample, and the shear rate $\dot\gamma$ which is here $\dot\gamma = dv/dr$, i.e. the change of the flow velocity across the capillary radius.

For the pressure p_2 acting on the circular cross-section area A_2 of the capillary holds:

$p_2 = F_P/A_2 = F_P/(\pi \cdot R^2)$ or $F_P = p_2 \cdot \pi \cdot R^2$

Using condition 1 (i.e., $F_R = F_P$): $2\pi \cdot R \cdot L \cdot \eta \cdot \dot{\gamma} = p_2 \cdot \pi \cdot R^2$

thus: $2 \cdot L \cdot \eta \cdot \dot{\gamma} = p_2 \cdot R$ or $\dot{\gamma} = (p_2 \cdot R)/(2 \cdot L \cdot \eta)$

Using condition 2 (i.e. $p_1 = p_2$): $p_1 = p_2 = \rho \cdot g \cdot h$

Shear rates in a "pressureless" capillary due to the hydrostatic pressure:

Equation 11.5 $\dot{\gamma} = (\rho \cdot g \cdot h \cdot R)/(2 \cdot L \cdot \eta)$

Thus: The shear rate values are depending on the continuously falling level of the liquid h.

Example 4: Using an ISO cup no. 4

Presets: $\rho = 1g/cm^3 = 1000kg/m^3$ and $\eta = 100mPas = 0.1Pas = 0.1kg/(s \cdot m)$;
assumption: ideally viscous flow behavior; radius $R = d/2 = 0.002m$ and $L = 0.020m$, and $h_{max} = 63mm = 0.063m$.

Calculation of the maximum shear rate, at $h = h_{max}$:

$\dot{\gamma}_{max} = (1000 \cdot 9.81 \cdot 0.063 \cdot 0.002 \; kg \cdot m \cdot m \cdot m \cdot s \cdot m)/(2 \cdot 0.020 \cdot 0.1 \; m^3 \cdot s^2 \cdot m \cdot kg)$

Thus: $\dot{\gamma}_{max} = 309s^{-1}$

The maximum shear rate occurs at the beginning of the test, when $h = h_{max}$.

If the cup is still half-full, i.e. at $h = h_{max}/2$, then: $\dot{\gamma} = 155s^{-1}$

The shear rate is reduced to the half value now, and it is decreasing continuously with the level of the liquid; for $h \to 0$ holds: also $\dot{\gamma} \to 0$. Therefore, **for flow cup tests applies: the shear rate values are changing continuously from initially $\dot{\gamma}_{max}$ to finally zero**, i.e., for the shear rate range holds:

$\dot{\gamma}_{max}$ (when $h = h_{max}$) $\geq \dot{\gamma} \geq 0$ (when $h = 0$)

As a comparison: In this case, the shear rate value obtained when using the Hagen/Poiseuille formula was $\dot{\gamma} = 229s^{-1}$, for a fluid of $\eta = 100mPas$ and $\rho = 1g/cm^3$ or $\nu = 100mm^2/s$, using an ISO cup no. 4 of $d = 2R = 4mm$ and $V = 108ml$, and with a flow time of $t = 75s$; see above Example 3. Actually however, there are occurring shear rate values between $\dot{\gamma} = 0$ and $309s^{-1}$. Therefore, even for ideally viscous fluids it should be taken into account that indeed the maximum value of the shear rate in a flow cup capillary is always clearly higher than any mean value which is calculated by the Hagen/Poiseuille relation. This effect will appear even clearer for liquids which are not showing ideally viscous flow, but shear-thinning, shear-thickening or thixotropic behavior, since here, an even wider shear rate range might be obtained, i.e., even higher maximum values.

11.3.1.3 *Flow instabilities, secondary flow effects, turbulent flow conditions in flow cups*

Secondary flow effects may appear in capillary flow when testing low-viscosity liquids at high shear rates. This might lead to turbulent flow conditions, and therefore, the **minimum flow time** is limited to $t = 30s$ for all ISO flow cups.

Turbulent flow behavior may occur when reaching the critical **Reynolds number (Re)**. The Re number expresses the ratio of the force due to the mass inertia of the fluid and the force due to its flow resistance.

The **Re number**, presented in its general form:

Equation 11.6 $Re = (v_m \cdot L \cdot \rho)/\eta = (v_m \cdot L)/\nu$

with the mean velocity v_m [m/s] of the fluid, the geometrical condition L [m], (or "characteristic length"), density ρ [kg/m³], viscosity η [Pas] or the kinematic viscosity ν [m²/s] of the fluid.

For a capillary geometry: $L = 2R$
For capillary flow: $v_m = V/(A \cdot t) = V/(\pi \cdot R^2 \cdot t)$
with the cross section area of the capillary: $A = \pi \cdot R^2$
The Re number for capillary flow: $Re = (V \cdot 2R \cdot \rho)/(\pi \cdot R^2 \cdot t \cdot \eta)$

Equation 11.7 $Re = (2 \cdot V \cdot \rho)/(\pi \cdot R \cdot t \cdot \eta) = (2 \cdot V)/(\pi \cdot R \cdot t \cdot v)$

Re numbers are used to characterize flow conditions, e.g. for flow in capillaries, tubes and pipes as follows [155, 390]:
Re < 1000: laminar flow, 1000 < Re < 2000: transition range, Re > 2000: turbulent flow
Often in text books, **the critical Re number** for the onset of turbulent flow is specified as: Re_c = 2300. Of course, irregularities in a flow field are also influenced strongly by the temperature gradient and by the roughness of the pipe wall. However with extremely smooth walls, Re_c may even reach the value of 100,000. For more information on secondary flow effects: see Chapter 10.2.2.4, DIN 53019-3, [233, 276, 352].

Example 1: Using an ISO cup no. 4
Presets: $V = 108ml = 1.08 \cdot 10^{-4} m^3$, $R = d/2 = 2mm = 0.002m$, flow time $t = 30s$; which is the shortest flow time allowed according to ISO 2431. Calculation of the kinematic viscosity:
$v = 1.37 \cdot t - (200/t) = (1.37 \cdot 30) - (200/30) = 34.4 mm^2/s = 34.4 \cdot 10^{-6} m^2/s$
using Equation 11.7 $Re = (2 \cdot V)/(\pi \cdot R \cdot t \cdot v)$
Then: $Re = (2 \cdot 1.08 \cdot 10^{-4} m^3 \cdot s)/(\pi \cdot 0.002 \cdot 30 \cdot 34.4 m \cdot s \cdot m^2) = 33.3$ (< Re_c = 2300)
Thus, laminar flow conditions are met here; and $35mm^2/s$ is the lowest kinematic viscosity value allowed to be measured when using an ISO cup no. 4.

Example 2: Using an ISO cup no. 3
Presets: $V = 108ml$, $R = d/2 = 1.5mm = 0.0015m$, flow time $t = 30s$; which is the shortest flow time allowed according to ISO 2431. Calculation of the kinematic viscosity:
$v = 0.443 \cdot t - (200/t) = (0.443 \cdot 30) - (200/30) = 6.62mm^2/s$
Then: $Re = (2 \cdot 1.08 \cdot 10^{-4} m^3 \cdot s)/(\pi \cdot 0.0015 \cdot 30 \cdot 6.62 m \cdot s \cdot m^2) = 231$ (< Re_c = 2300)
Thus, laminar flow conditions are met here; and $7mm^2/s$ is the lowest kinematic viscosity value allowed to be measured when using an ISO cup no. 3.

Example 3: Using an ISO cup no. 4 for testing water
Presets: $V = 108ml$, $R = d/2 = 2mm$, flow time $t = 12.5s$; which is too short acc. to ISO 2431. Calculation of the kinematic viscosity (similar to Example 1 using cup no. 4): $v = 1.13mm^2/s$
Then: $Re = 2430$ (> Re_c = 2300). Thus, liquids showing a viscosity as low as water should not be measured with the ISO cup no. 4.

Example 4: Using an ISO cup no. 3 for testing water
Presets: $V = 108ml$, $R = d/2 = 1.5mm$, flow time $t = 22.4s$; which is still too short acc. to ISO 2431. Calculation of the kinematic viscosity (similar to Example 2 using cup no. 3): $v = 0.995mm^2/s$
Then: $Re = 2060$. Thus, the Re number is close to Re_c = 2300. Therefore, liquids showing a viscosity as low as water should not be measured with the ISO cup no. 3.

    ᘯ   End of the Cleverly section

11.3.2 Other types of flow cups

In this section, further types of flow cups are described which previously were frequently used, and some of them are sometimes still in use:
AFNOR cups (French standard; previously: Coupe NFT)
BS cups (British standard)
Consistency viscosity cups (ASTM D1084, for adhesives)

DIN cups (since 1941, DIN 53211/withdrawn; for paints and varnishes)
Engler cups or "Engler viscometer" (since 1884; ASTM D1665, for fluid tar products and starch solutions; DIN 51560/withdrawn), [20]
Ford cups (for lacquers, varnishes, paints, coatings; ASTM D1200: viscosity by Ford viscosity cup; ASTM D333: for clear and pigmented lacquers; ASTM D365: for soluble nitrocellulose base solutions)
Lehmann cups (for ceramic suspensions)
Marsh funnels (API 13B: for drilling fluids, muds, slurries, cement suspensions)
Shell cups (ASTM D4212: dip-type viscosity cups; for paints and inks)
Zahn cups (ASTM D816: for rubber cements; ASTM D1084: for adhesives; ASTM D3794/ withdrawn: for coil coatings; ASTM D4212: dip-type viscosity cups)

For petroleum products such as mineral oils, liquid bitumens, asphalts and tars:
Redwood viscometers (EN 12846, IP 70/57) [20, 276]
Saybolt universal and **Saybolt Furol viscometers** (ASTM D88: Saybolt viscosity; ASTM D244: emulsified asphalts; ASTM D2161: conversion of kinematic viscosity to Saybolt universal viscosity or to Saybolt Furol viscosity; ASTM D2162: basic calibration of master (capillary) viscometers and viscosity oil standards; ASTM E102: Saybolt Furol viscosity of bituminous materials at high temperatures); and [20, 237, 276]

Road tar viscometers (DIN 52023/withdrawn, EN 13357, IP 72), [318]

Different units for the measured **relative viscosity values** or flow times, respectively, are in use for the various cups: e.g. Engler degrees (°E), Redwood (no.1) seconds (RIS), Saybolt universal seconds (SUS), and Saybolt Furol seconds (SFS) [20].

Note: Measuring with flow cups is a single-point testing method
Testing with flow cups is a single-point measuring method, merely generating a single viscosity value. However here, during the time interval of testing, the shear rate values are changing continuously within a certain range depending on the amount of liquid which is still available in the cup, see also Chapter 11.3.1.2c. Therefore, flow cups should only be used for liquids exhibiting ideally viscous flow behavior, since only in this case, viscosity is showing a constant value independent of the continuously decreasing flow velocity or shear rate, respectively.

11.4 Capillary viscometers

11.4.1 Glass capillary viscometers

Capillary viscometers - the **abbreviation CV** is used in this chapter - are termed "pressureless" if the driving force is exclusively deriving from the hydrostatic pressure, i.e. due to the weight of the liquid sample or the gravity force, respectively. These kinds of CVs are usually made of glass, see Figure 11.2. Different types of CVs vary, for example, in the practical use or in the possibility to counteract design-dependent disturbances on the flow conditions [233, 276].

A defined volume of the liquid sample flows under its own weight through **a long capillary of an exactly defined inner diameter d and length L**. Typical dimensions of capillaries used in industrial laboratories are d = 0.25 to 10mm, and L = 70 to 250mm. The flow time is measured which is required for the liquid to flow between two defined level marks. Minimum flow time is specified as t = 200s to be sure that there are laminar flow conditions. This counts when using Ubbelohde CVs, and for Micro-Ubbelohde CVs applies t = 30s.

From the flow time, the kinematic viscosity ν [mm²/s] is calculated. When using typical CVs, according to manufacturers a total viscosity range of ν = 0.3 to 10,000mm²/s

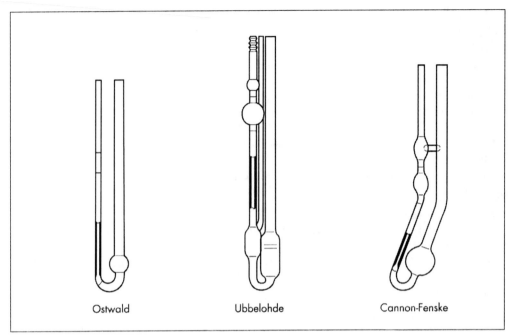

Ostwald   Ubbelohde   Cannon-Fenske

Figure 11.2: Various designs of different types of glass capillary viscometers [323]

may be covered, and with special devices even of 0.2 to 50,000mm²/s. In order to evaluate and calculate the occurring shear conditions which are continuously changing with the level of the liquid in the CV, see Chapter 11.3.1.2c2: shear rate range in a capillary. Usually here, there are shear stresses between 1 and 100Pa, and shear rates between 0.01 and 10,000s⁻¹. Typical range of measuring temperatures T = -40 to +150°C (and enhanced from -60 to +180°C) [221, 237, 323].

Note: Measuring with glass capillary viscometers is a single-point testing method
Concering the flow conditions, the same applies here as explained for flow cups, see the Note at the end of Chapter 11.3.2, even when the shear rate range might be narrower when using a glass CV.

Gotthilf H.L. Hagen (1797 to 1884) designed the first CV in 1839 [172, 233], and almost simultaneously in 1840, also *Jean L.M. Poiseuille* (1799 to 1869) independently built a CV of his own [291, 378]. Until the 1920s, there were almost exclusively CVs to be used when investigating the flow behavior of liquids even for scientific purposes, since in typical industrial laboratories, rotational viscometers did not occur before the 1950s. Nowadays, the following types are frequently in use: Ubbelohde CVs [369], Cannon-Fenske CVs [64], Ostwald CVs (since 1891, [272, 378]), and BS/IP U-Tubes (British Standards, Institute of Petroleum).

For **Ubbelohde CVs** (ISO 3105; DIN 51562), 16 capillaries are available of d = 0.36 to 6.4mm (for ν = 0.3 to 30,000mm²/s) and L = 90mm, and for **Micro-Ubbelohde CVs** there are 5 capillaries of d = 0.40 to 1.26mm (for ν = 0.4 to 800mm²/s).

For **Cannon-Fenske CVs** (ISO 3105; ASTM D445), 12 capillaries are available of d = 0.30 to 4.1mm (for ν = 0.4 to 20,000mm²/s), here, also non-transparent liquids may be investigated when using the **reverse-flow** type (ASTM D446).

For **Ostwald CVs**, 5 capillaries are available of d = 0.3 to 0.7mm (for ν ≥ 0.3mm²/s), and for **Micro-Ostwald CVs** there are 5 capillaries of d = 0.43 to 1.36mm (for ν = 0.4 to 800mm²/s).

For **BS/IP U-Tubes**, d = 5.52 to 9.75mm (for ν = 6000 to 300,000mm^2/s). Further types which are rarely used nowadays are the Poiseuille CV and the Vogel/Ossag CV (DIN 51561/ withdrawn).

See also the following standards:
ISO 307 (viscosity number, polyamides), ISO 1628 (polymers in dilute solution, viscosity, viscosity number, and limiting viscosity number LVN), ISO 2909 (petroleum products, viscosity index VI), ISO 3104 (petroleum products, kinematic viscosity and calculation of dynamic viscosity), ISO 3105 (glass capillary kinematic viscometers, specifications and operating instructions), ISO 3448 (industrial liquid lubrificants, **ISO viscosity classification of liquid lubricants**);

ASTM D341 (viscosity-temperature charts for liquid petroleum products), ASTM D445 (kinematic viscosity of transparent and opaque liquids), ASTM D446 (specifications and operating instructions for glass capillary kinematic viscometers), ASTM D1243 (dilute solution viscosity of vinyle chloride VC-polymers, e.g. inherent viscosity), ASTM D1601 (dilute solution viscosity of ethylene polymers, e.g. inherent, intrinsic, reduced, relative, specific viscosity, and the viscosity ratio), ASTM D1795 (intrinsic viscosity of cellulose), ASTM D2170 (asphalts, e.g. bitumens at T = +60°C, asphalt binders at T = +135°C), ASTM D2171 (asphalts by vacuum cap. viscometer, at T = +60°C), ASTM D2270 (petroleum products, calculating viscosity index VI from kinematic viscosity at 40 and 100°C), ASTM D2422 (**ASTM viscosity classification of industrial fluid lubrificants**), ASTM D2493 (viscosity-temperature charts for asphalts), ASTM D2532 (viscosity and viscosity change after standing at low temperature of aircraft turbine lubricants), ASTM D2857 (dilute solution viscosity of polymers), ASTM D3591 (logarithmic viscosity number of PVC in formulated compounds), ASTM D4603 (inherent viscosity of poly-ethylene terephthalate PET), ASTM D4957 (asphalt emulsion residues and non-Newtonian bitumens by vacuum cap. viscometer), ASTM D5225 (solution viscosity of polymers with a differential (cap.) viscometer);

DIN EN ISO 1157 (using Ubbelohde CVs, e.g. viscosity number), DIN EN ISO 1628 (using Ubb. CVs, e.g. limiting VN), **DIN EN** 12595 (bituminous binders), DIN EN 12596 (bituminous binders); DIN 51366 (mineral oils, hydrocarbons, using Cannon/Fenske CVs), DIN 51562 (using Ubb. CVs), DIN 53177 (resins and oils, using Ubb. CVs), DIN 53728 (using Ubb. CVs, p1: cellulose, p3: PET and PBT, p4: Staudinger Index of PE and PP); **IP** 71, IP 222, IP 319

      For "Mr. and Ms. Cleverly"

11.4.1.1 Calculations

The liquid sample is flowing only due to its weight, i.e. driven by gravity, and the acting pressure is the hydrostatic pressure which is proportional to the resulting shear stress and shear rate. Since there is acting no external pressure, glass CVs are considered **"pressureless viscometers"**.

For the following simple calculations there is assumed that the sample shows ideally viscous flow behavior and there are laminar and stationary flow conditions, and no significant influence due to the following effects:
- Inlet flow disturbances in the capillary, e.g. along a certain inlet distance due to increased friction forces when entering into the now narrower capillary geometry, causing a change of the flow velocity, with the risk of occurring inhomogeneous, turbulent flow conditions until laminar conditions are reached (Hagenbach/Couette correction); there may be also outlet disturbances [79]
- Tube inclination, if the stand is not adjusted exactly in a vertical position
- Inaccurate control of the measuring temperature, thermal expansion of tube and test liquid, heating of the flowing sample due to internal friction or viscous shear heating, respectively, occurring between the molecules

- Air bubbles and foam formation, air buoyancy, evaporation of solvents
- Surface tension, draining off the glass wall, wettability, film formation
- Inaccurate meniscus detection, parallax error

In order to reduce the influence of the inlet and outlet disturbances, the L/d-ratio, i.e. of length and diameter of the capillary, should be larger than 10 : 1. This condition is usually fulfilled when glass CV are used.

a) The range of the hydrostatic pressure in a capillary

Equation 11.8 $p_l = \rho \cdot g \cdot h$

with the density ρ of the sample, the gravitation constant $g = 9.81 m/s^2$, and the level h of the liquid (sometimes termed "level of the hydrostatic pressure"). For most glass CVs counts: h_{min} and h_{max} correspond to two defined level marks on the CV tube, and for the medium level holds:

$h_m = (h_{min} + h_{max})/2$

Example 1: Using an Ubbelohde CV
Presets: $\rho = 0.9 g/cm^3 = 900 kg/m^3$, $h_{max} = 150mm$, and $h_{min} = 110mm$
Therefore: $h_m = (110 + 150)mm/2 = 130mm = 0.130m$
Calculation of the medium hydrostatic pressure, at $h = h_m$:
$p_l = (900 \cdot 9.81 \cdot 0.130 \, kg \cdot m \cdot m)/(m^3 \cdot s^2) = 1150 kg/m \cdot s^2 = 1150 Pa$ (= 11.5mbar)
The hydrostatic pressure is decreasing with the liquid level h, the maximum value is 1330Pa at $h = h_{max}$, and the minimum value is 973Pa at $h = h_{min}$, hence covering the relative range of:
$p_l \pm [(p_{lmax} - p_{lmin})/(2 \cdot p_l)] = p_l \pm 15\%$.

b) The shear stress range at the capillary wall according to the Hagen/Poiseuille relation

Equation 11.9 $\tau = (p_l \cdot R)/(2 \cdot L)$

Example 2: Using an Ubbelohde CV
Presets: $p_l = 1150$ Pa, as the medium hydrostatic pressure (see above Example 1),
radius $R = d/2 = 0.84/2$ mm $= 4.2 \cdot 10^{-4}$ m and L = 90mm = 0.090m of the capillary
By the way here, the L/d-ratio is 90mm/0.84mm = 107:1, therefore being clearly larger than 10 : 1 which is considered the minimum ratio to neglect inlet and outlet disturbances.
Calculation of the medium shear stress, at $h = h_m = 130mm$:
$\tau = (1150 \cdot 4.2 \cdot 10^{-4} \, Pa \cdot m)/(2 \cdot 0.09 \, m) = 2.68 Pa$
The shear stress is decreasing with the liquid level h, the maximum value is 3.09Pa at $h = h_{max}$, and the minimum value is 2.27Pa at $h = h_{min}$.

c) The shear rate range in a capillary

Equation 11.10 $\dot{\gamma} = (\rho \cdot g \cdot h \cdot R)/(2 \cdot L \cdot \eta) = (g \cdot h \cdot R)/(2 \cdot L \cdot v)$

with the (shear) viscosity η and the kinematic viscosity v of the sample (see Chapter 11.3.1.2c)

Example 3: Using an Ubbelohde CV
Presets: $\rho = 0.9 g/cm^3 = 900 kg/m^3$ and $\eta = 10 mPas = 0.01 Pas = 0.01 kg/s \cdot m$; assumption: ideally viscous flow behavior; radius $R = d/2 = 4.2 \cdot 10^{-4}$ m and L = 0.090m, and $h_m = 0.130m$
Calculation of the medium shear rate, at $h = h_m$:
$\dot{\gamma} = (900 \cdot 9.81 \cdot 0.130 \cdot 4.2 \cdot 10^{-4} \, kg \cdot m \cdot m \cdot m \cdot s \cdot m)/(2 \cdot 0.090 \cdot 0.01 \, m^3 \cdot s^2 \cdot m \cdot kg) = 268 s^{-1}$
The maximum shear rate occurs at the beginning of the test, when $h = h_{max} = 150mm$, showing $\dot{\gamma}_{max} = 309 s^{-1}$, and it is decreasing with the level h to $h_{min} = 110mm$, reaching $\dot{\gamma}_{min} = 227 s^{-1}$ then; covering a relative range of $\dot{\gamma} \pm [(\dot{\gamma}_{max} - \dot{\gamma}_{min})/(2 \cdot \dot{\gamma})] = \dot{\gamma} \pm 15\%$.
Summary: Using glass CVs or other "pressureless" capillary viscometers, viscosity values which are obtained when testing liquids which are not showing ideally viscous flow behavior

should not be overrated, since even here, the shear rates values are changing continuously within a certain range.

11.4.1.2 Determination of the molar mass of polymers using diluted polymer solutions

Hermann Staudinger (1881 to 1965) introduced in the 1920s the following method to determine the viscosity-average molar mass M of polymers via viscosity measurements of polymer solutions, [343]. For polymers holds roughly: M > 10,000g/mol (see also Chapter 3.3.2.1a: values of M_{crit} of diverse polymers). When testing polymer solutions, the following terms and symbols should be used (according to DIN 1342-2 or IUPAC, the International Union of Pure and Applied Chemistry [201]):

- η [Pas] is the viscosity of the polymer solution
- η_s [Pas] is the viscosity of the solvent
- c is the mass-specific concentration of the dissolved polymer in the solution

Units: $1g/cm^3 = 1000kg/m^3 = 1g/ml$, or $1kg/m^3 = 1g/l$

Testing is assumed to be performed in the low-shear range, thus, at very low shear rates. For concentrated polymer solutions, this is the range of the zero-shear viscosity η_0 (see also Chapter 3.3.2.1a and Figure 3.10). For more information see [215, 246, 268].

a) The relative viscosity or the "viscosity ratio"
The "relative viscosity" is the viscosity of the solution related to the viscosity of the solvent.

Equation 11.11 $\eta_r = \eta/\eta_s$

with the unit [1]. Two recommendations concerning useful polymer concentrations:
1) Solve 0.2 to 1g of the polymer sample in 100ml of the solvent.
2) Operate in a range of $1.2 \leq \eta_r \leq 1.5$ (according to ISO 1628-1).

b) The logarithmic viscosity number or the "inherent viscosity"
The "logarithmic viscosity number" is the natural logarithm of the ratio of relative viscosity and mass-specific concentration. Nowadays, this parameter is rarely used.

Equation 11.12 $\eta_{ln} = \ln(\eta_r/c)$

with the unit [cm^3/g]

c) The specific viscosity or the "viscosity relative increment"

Equation 11.13 $\eta_{sp} = (\eta - \eta_s)/\eta_s = \eta_r - 1$

with the unit [1]
Note 1: DIN 1342-2 (of 2003) recommends to use no longer the term "specific viscosity" and the sign η_{sp} since the unit is not [Pas] like of a viscosity, and therefore, in DIN it is termed the **"relative change of viscosity"**, and it is expressed only in terms of $(\eta - \eta_s)/\eta_s$ and therefore, without using neither η_{sp} nor η_r.

d) The reduced viscosity or the "viscosity number" VN (or the "Staudinger function")
The "reduced viscosity" or the "reduced specific viscosity" is the concentration-dependent specific viscosity.

Equation 11.14 $\eta_{red} = (1/c) \cdot (\eta - \eta_s)/\eta_s = (\eta_r - 1)/c = \eta_{sp}/c$

with the unit [cm^3/g]
Note 2: DIN 1342-2 recommends to use no longer the term "reduced viscosity" and the sign η_{red} since the unit is not [Pas] like of a viscosity, and therefore, in DIN it is termed the **"Staudinger function"** with the sign J_v and it is merely presented in the following form:
$J_v = (1/c) \cdot (\eta - \eta_s)/\eta_s$

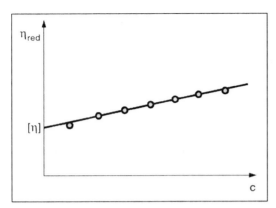

Figure 11.3: Reduced viscosity η_{red} (or Staudinger function J_v resp.) as a function of the polymer concentration c; the value of the intrinsic viscosity [η] (or Staudinger index J_g resp.) is determined at the crossover of the straight fitting line and the y-axis after extrapolating the fitting line to c = 0

e) The intrinsic viscosity IV or the "limiting viscosity number" LVN (or the "Staudinger index")

Equation 11.15
$$[\eta] = \lim_{c \to 0} [(1/c) \cdot (\eta - \eta_s)/\eta_s] = \lim_{c \to 0} \eta_{red}$$

with the unit $[cm^3/g]$
When reducing the polymer concentration more and more coming close to zero, the limiting value of the "reduced viscosity" is approaching the value of the "intrinsic viscosity".

Note 3: DIN 1342-2 recommends to use no longer the terms "intrinsic viscosity" and "limiting viscosity number" and the sign [η], since the unit is not [Pas] like of a viscosity, and therefore, in DIN it is termed the **"Staudinger index"** with the sign J_g and then holds:

Equation 11.16 $$J_g = \lim_{c \to 0} J_v$$

Experience showed that the value of [η] depends on the following parameters: the average molar mass M, the molar mass distribution MMD, the chemical structure (e.g. the degree of branching), the effectivity of the polymer/solvent system, the temperature, and the amount of measuring points determined of solutions showing different polymer concentrations. And of course, there should be no physical-chemical interactions between the polymer molecules.

Steps to determine the value of the intrinsic viscosity [η]
1) Preparation of several solutions showing different concentrations of the polymer. It is recommended to have in minimum seven (better ten) different concentrations available.
2) Measurement of the viscosity of the solvent, and of each solution.
3) Calculation of the reduced viscosity η_{red} of each solution.
4) Plot of the single measuring points in a **diagram with η_{red} on the y-axis, and the concentration c on the x-axis**, both on a linear scale, see Figure 11.3.
5) Fitting of a straight line through the single measuring points; sometimes, this line is termed the "regression line".
6) **Determination of the value of the intrinsic viscosity [η] at the intersection of the straight fitting line and the y-axis after extrapolating the fitting line to c = 0.**

Several scientists developed **mathematical functions of fitting lines** which they found empirically. Examples: **Huggins** (using the Huggins constant K_H), **Schulz-Blaschke** (with K_{SB}), **Kraemer** (with K_K), **Martin** (with K_M), **Billmayer**, and **Solomon-Ciuta** (using a curve). If the fitting constant of a polymer-solvent system is already known from previous tests, the value of [η] can be determined from a single measurement since the slope of the fitting line of the $\eta_{red}(c)$-diagram is already known then.

f) The [η]-M relation
There are various names used for the relationship between the intrinsic viscosity [η] and the molar mass M of the polymer sample: Some call it the [η]-M relation, and others the MH-relation (for Mark-Houwink [244, 195; 233]), or MKH-relation (for M.-Kuhn-H. [214]), or KMHS-relation (for K.-M.-H.-Sakurada [215]), or the SMH-relation (Staudinger-M-H [343; 54]).

Equation 11.17 $[\eta] = K \cdot M^a$

and in the logarithmic form:

Equation 11.18 $\lg [\eta] = \lg K + a \cdot \lg M$

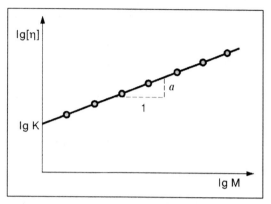

with the molar mass M [g/mol], and the two constants a and K which have to be determined by experiments. The unit of $[\eta]$ is presented in literature in various ways, e.g. as $[cm^3/g] = [ml/g]$, or as $[m^3/kg] = [l/g]$.

The linear relation between lg M and lg $[\eta]$ is used to be illustrated in **a diagram on a logarithmic scale presenting lg M on the x-axis and lg $[\eta]$ on the y-axis**, see Figure 11.4. A straight fitting line through the single points of the $[\eta]$-M relation is used to determine the value of the constant

Figure 11.4: Intrinsic viscosity $[\eta]$ versus the molar mass M; the constant K is determined at the intersection of the straight fitting line and the y-axis, and the constant a is the slope value of the fitting line

K at the crossover of the straight fitting line and the y-axis. The value of the constant a is the slope value of the fitting line, thus, $\Delta \lg [\eta]/\Delta \lg M$ between two points of the fitting line.

The structural parameters a and K of a polymer are assumed to show constant values for a polymer-solution system at a constant temperature, and the corresponding values of all relevant polymers can be found e.g. in the "Polymer Handbook" [53]. However, it should be taken into account the dependence of the constants a and K on the following parameters: the molar mass distribution (MMD), the chemical structure (e.g. the degree of branching), the effectivity of the polymer-solvent system, and therefore, the shape of the coiled macromolecules, the polymer concentration, the temperature, the shear rate, and the amount of measuring points determined at different polymer concentrations. And of course, no physical-chemical interactions should occur between the polymer molecules.

The value of the constant a depends on the shape of the solved polymer molecules:
$a = 0$ for aggregated and compact spheres,
$a = 0.5$ for unperturbed coils (in a poor solvent),
$a = 0.8$ for expanded, perturbed coils (in a good solvent),
$a = 1$ to 2 for semi-flexible rod-like or helical structures,
$a = 2$ for rigid rods or stretched molecules
The constant a usually covers a range between $a = 0.5$ and 1.
Critical note: Since the value of the constant a **depends also on the solvent used**, it cannot be attributed exclusively to the polymer to be investigated.

The value of the constant K mostly covers a range of $1000 < K < 100,000$ if $[\eta]$ is specified in $[cm^3/g]$. However, if $[\eta]$ is expressed in $[m^3/kg]$, then $1 < K < 100$.

Steps to determine the constants a and K if M is known, or after a previous molar mass determination:
1) Preparation of several polymer fractions, each one of a different molar mass M showing an MMD as narrow as possible.
2) Determination of the molar mass M of each fraction (e.g. using methods such as sedimentation, diffusion by ultra-centrifuges; scattering of light, X-rays or neutrons; membrane or vapor pressure osmometry, electron microscopy, gel permeation chromatography GPC, or mass spectroscopy).
3) Preparation of several solutions of different polymer concentrations of each fraction. It is recommended to use in minimum seven different concentrations of each fraction.

4) Measurement of the viscosity of the solvent, and of each solution of each fraction.

5) Plot of the concentration-dependent $\eta_{red}(c)$-function for each fraction (like in Figure 11.3).

6) Plot of the fitting line for each diagram.

7) Determination of the intrinsic viscosity value [η] for each molar mass fraction, read off from the diagram at the crossover of the fitting line and the y-axis.

8) Plot of the molar mass-dependent [η](M)-diagram for each molar mass fraction (like in Figure 11.4).

9) Fitting of the straight analysis line for each diagram.

10) Determination of the constant K at the crossover on the y-axis, and of the constant a which is the slope value of the fitting straight line, read off from the diagram for each molar mass fraction.

Steps to determine the molar mass M if the constants a and K are known

Determination of the intrinsic viscosity value [η]: see the steps 1) to 6) in sector e) of the Chapter at hand. Then, calculation of the molar mass using the Equations 11.17 and 11.18:

$[\eta] = K \cdot M^a$ or $\lg [\eta] = \lg K + a \cdot \lg M$

thus: $M = ([\eta]/K)^{(1/a)}$ or $\lg M = (\lg [\eta] - \lg K)/a$

g) Notes

Note 1: The Fikentscher K-value to characterize VC and PVC resins and polymers

Besides the explained methods used to determine "intrinsic viscosity" and the "limiting viscosity number" LVN (or the "Staudinger index"), there are also other methods in use to characterize the degree of polymerization or the average molar mass, respectively. An example is the determination of the Fikentscher K-value to characterize VC (vinyl-chloride) and PVC (poly-VC) resins and polymers (by *H. Fikentscher* in 1932, [135]). Here, the following steps are performed:

1) Preparation of several solutions of different concentrations of the PVC sample (e.g. with 0.25g PVC in 50ml cyclohexanone).

2) Measurement of the viscosity η_s of the solvent and of each solution.

3) Calculation of the relative viscosity value $\eta_r = \eta/\eta_s$ for each solution.

4) Calculation of the Fikentscher K-value using a complex formula or the corresponding values from a table of DIN 53726: $\lg (\eta_r) = f(K, c)$. This is a relation between η_r, the K-value and the concentration c.

Fikentscher K-values of thermoplastic PVC are usually between 50 and 80, and very high-molecular PVC types may even display $K > 90$, [190, 268]. In literature can be found two different kinds of Fikentscher formulas: either with a k-value (showing a small letter k) or with a K-value (showing a capital letter K), and $K = 1000 \cdot k$ [41], which is leading of course to confusion among practical users. Therefore, and since this method is only valid under special conditions in a limited range as well for the polymer concentration as well as for the molar mass, it should not be used any more [189, 215]. Nevertheless, still today leading polymer producers are presenting their products using this number, albeit sometimes with the statement: "Although showing roughly the same K-value (for example here, K = 35), the four xxx-products differ from one another in their viscosity and in the flow characteristics of their solutions" (e.g. mentioned under "other properties" of a chlorinated binder used as a raw material for coatings, printing inks and road marking paints [21]). Corresponding standards, although meanwhile withdrawn, are ISO 174 (plastics, resins of vinyle chloride, viscosity number in dilute solution), and DIN 53726.

Note 2: Confusion about the variety of constants K, parameters K, and K-values

The parameter K of the [η]-M relation should not be mixed up with the different constants K of the straight fitting (or "regression") lines of the $\eta_{red}(c)$-diagram (such as K_H, K_{SB}, K_K, K_M), or with the various other constants used as parameters to characterize polymer structures (such as the Fikentscher values k or K). Unfortunately, in all these cases there was chosen the same letter.

Note 3: The different molar mass values M_w, M_n and M_v

Only for polymers showing a uniform average molar mass M and a very narrow molar mass distribution MMD applies: the values of the mass average M_w, number average M_n and viscosity average M_v of the molar mass are equal, and then $M = M_w = M_n = M_v$. The latter may be determined by viscometry, for example, using glass capillary viscometers as explained in this Chapter, or micro falling ball viscometers (see Chapter 11.5) or Stabinger viscometers, see Chapter 11.2.11L.

Note 4: Capillary viscometry on polymers solutions is a relative measuring method
1) Capillary "viscometry" is a test method for diluted polymer solutions only

When determining the average molar mass M of polymer solutions by capillary viscometry, it should be taken into account that **this is not an absolute measuring method** since the result depends also on the solvent used and on the assumed shape of the polymer molecules in the solvent, influencing above all the value of the constant a. Beside this, the [η]-M relation is only valid if there are no molecular interactions at all.

2) As a comparison: Rheometry is a test method also for polymer melts

Absolute measuring methods to characterize polymers in terms of the average molar mass M and molar mass distribution MMD are explained in Chapters 3.3.2.1a (rotational tests: shear viscosity function and zero-shear viscosity η_0), 6.3.4.1 (creep tests and η_0), 7.3.5 (relaxation tests and MMD) and 8.4.2.1a (oscillatory tests, frequency sweeps: η_0 and MMD). Here, as well **the shear-rate dependent viscosity function as well as the viscoelastic behavior of polymer melts is measured** by simulating the practical conditions clearly closer to the real conditions occurring in industrial practice. This counts in particular for the concerning **temperatures** since hot melts are tested instead of solutions at room temperature. This applies also to the speed range by simulating low-shear to high-shear conditions, i.e., from the η_0-range or LVE range up to high shear rates or frequencies, respectively. It holds as well for the forces or pressures which can be simulated in a wide shear stress range. And it is valid also for the molecular configuration, since here in the form of a melt, **the polymer molecules are investigated in an entangled state**. A large advantage of this method is the fact that molecular flow and relaxation processes can be simulated closer to practice, because in polymer solutions, these kinds of processes may take place faster compared to the behavior in polymer melts.

Summary: Rheometry is also the better tool to achieve information as well on the average molar mass M as well as on the molar mass distribution MMD. And additionally, only with rheometrical measuring methods there can be simulated process conditions such as thermoforming, extrusion and injection moulding.

11.4.1.3 Determination of the viscosity index VI of petrochemicals

Capillary viscometers are frequently used to determine the viscosity index VI of petrochemicals such as mineral oils, in order to characterize their temperature-dependent behavior. This method was presented by *E.W. Dean* and *G.H.B. Davis* in 1929, [88, 95]. The arbitrary relation is based on the behavior of two typical oils, a paraffinic one was set to show VI = 100 and a naphthen-based one to VI = 0. Nowadays, there may also occur values of VI > 100, achieved by enhanced raffination technologies and the use of synthetic materials. The higher the VI value the smaller is the influence of the temperature on a sample's viscosity.

Standards: ISO 2909 (petroleum products, calculation of the VI from kinematic viscosity); ASTM D2270 (petroleum products, calculating VI from kinematic viscosity at 40 and 100°C).

The analysis consists of the following steps:

1) **Measurement** of the kinematic viscosity ν [mm^2/s] of the sample, **at T = +40°C and at +100°C** (e.g. using a capillary viscometer acc. to ISO 3104, or a Stabinger Viscometer acc. to ASTM D7042), resulting in the two values: ν (40°C) and ν (100°C).

Analysis method A

2) Check whether the following condition is kept, which is required to use method A:
$2\text{mm}^2/\text{s} < \nu\ (100°C) < 70\text{mm}^2/\text{s}$

3) If yes, calculation of the VI value using the formula: $VI = [(L - U) \cdot 100]/(L - H)$
The values of L and H are related to the ν (100°C) value and are listed in Table 1 of ISO 2909 (based on tests with standard oils), and $U = \nu$ (40°C).

Example 1: Of three oils was measured ν (40°C) = 75mm^2/s (= U), and
a) for oil 1: ν (100°C) = 7mm^2/s, from ISO table 1: L = 78.00, and H = 48.57, then: VI = 10
b) for oil 2: ν (100°C) = 8mm^2/s, from ISO: L = 100.0, H = 59.60, then: VI = 62
c) for oil 3: ν (100°C) = 9mm^2/s, from ISO: L = 123.3, H = 71.10, then: VI = 93

4) Final check whether the following second condition is fulfilled, which is required to use analysis method A: VI < 100. Summary: This condition is fulfilled for all the three samples of Example 1.
However, if $2\text{mm}^2/\text{s} < \nu\ (100°C) < 70\text{mm}^2/\text{s}$, but **VI > 100**, the following formula should be used to calculate the VI value:

Analysis method B

$VI = [(10^n - 1)/0.00715] + 100$
with $n = (\lg H - \lg U)/\lg Y$, and H from table 1 of ISO 2909, $U = \nu$ (40°C) and $Y = \nu$ (100°C). If ν (100°C) > 70mm^2/s, then: $H = (1.684 \cdot Y^2) + (11.85 \cdot Y) - 97$

Example 2: Of three oils was measured ν (40°C) = 75mm^2/s (= U), and
a) for oil 4: ν (100°C) = 10mm^2/s, from ISO table 1: H = 82.87; then: n = 0.04, and VI = 113
b) for oil 5: ν (100°C) = 15mm^2/s, from ISO: H = 149.7; then: n = 0.254, and VI = 211
c) for oil 6: ν (100°C) = 20mm^2/s, from ISO: H = 229.5; then: n = 0.369, and VI = 287

৬৴   End of the Cleverly section

11.4.2 Pressurized capillary viscometers

11.4.2.1 MFR and MVR testers driven by a weight ("low-pressure capillary viscometers")

These kinds of testing devices are used for testing **polymer melts at medium shear stresses and medium shear rates**. Two parameters of polymer melts are often checked for simple quality control to get an idea about the flow behavior in injection molding machines: besides density, this is the MFR or MVR value. The measuring cell of such an **"extrusion viscometer"** consists of **a vertical cylinder in which a piston is set in motion** under a defined weight due to gravity, and **an extrusion die (capillary)** at the bottom of the cylinder, see Figure 11.5. For these types of instruments, sometimes also terms such as "extrusion plastometer", "capillary extruder" or "capillary tube" are used. See also "flow testers" in Chapter 11.2.10e.

MFR or MVR testers are widespread in the polymer industry **(ISO 1133)**. The dimensions of the steel cylinder (length L_1 between 115 and 180mm, e.g. L_1 = 162mm, inner diameter d_1 = 9.55mm), steel piston (d_2 = 9.48mm), and die (L_3 = 8mm, d_3 = 2.095mm) are specified in the mentioned standard, and so are the nominal loads to be selected by the user. These loads are weight pieces, and here, also the mass of piston and piston rod are taken into consideration: m_{nom} = 0.325/1.20/2.16/3.80/5.00/10.0 or 21.6kg (sometimes also 1.00 and 1.05kg).

With $F_{nom} = m_{nom} \cdot g$, using the gravitation constant $g = 9.81 m/s^2$, these masses correspond to the weight forces $F_{nom} = 3.19/11.8/21.2/37.3/49.1/98.1$ or $212N$.

a) Test procedure

Testing of polymers is performed as follows:

1) Heat the test device, cylinder and piston, to the selected constant measuring temperature. Specified are T = 125, 150, 190, 200, 220, 230, 250, 280 and 300°C. Some devices can be used for tests between T = +50 and 400°C, and others even from room temperature RT up to 500°C.

2) Wait for temperature equilibration for a period of 15min after reaching the desired temperature.

3) Remove the piston, and charge the cylinder by filling the polymer to be tested into the pre-heated cylinder within 1min to keep any contact to air as short as possible in order to avoid reactions of samples which are sensitive to oxidation. Typical is a mass of m = 3 to 8g, in the form of powder, granules, strips of films or sheetings, or in pieces of ≤ 5mm.

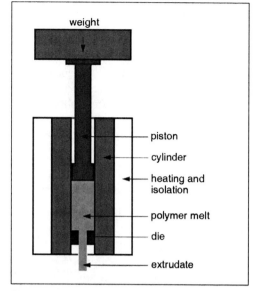

Figure 11.5: Low-pressure capillary viscometer consisting of a vertical cylinder, a piston which is set in motion under a weight due to gravity, and a die or capillary, to determine the MFR value (melt mass flow rate) or the MVR value (melt volume flow rate) of polymer melts

The material is compressed with a packing rod, using manual pressure. Usually, the filling amount should be enough for in minimum three determinations, i.e. extruded cut-offs, of the same sample.

4) Put the piston into the cylinder which is installed in a vertical position.

5) Wait for a pre-heating time of 4min to melt the sample.

6) Place the selected weight piece onto the piston rod.

7) The piston is moving downwards under the load due to gravity, and as a consequence, the melt is pressed out of the die. At first there may occur a filament containing air bubbles.

8) Collect successive cut-offs, using a cutting tool. The single pieces should be each between 10 and 20mm long. The periods between the cut-offs have to be selected depending on the flow rate. For example, low-viscosity melts may be already cut off every t = 10 to 15s, but highly viscous melts only every t = 4min.

9) After cooling, weigh the extruded cut-offs – but only if they are free of bubbles – exactly in mg (milligrams), and determine the value of the average mass of at least three cut-offs for each determination.

b) Determinations

Calculation of the melt (mass) flow rate MFR

Equation 11.19 $\mathbf{MFR\ (T, m_{nom})} = (t_{ref} \cdot m)/t = (600s \cdot m)/t$

unit [g/10min], at the measuring temperature T [°C], under the selected nominal load m_{nom} [kg]; related to the reference time t_{ref} [s], standard is $t_{ref} = 600s = 10min$; the extruded mass m [g], which is the average mass of at least three cut-offs; and the extrusion time t [s], which should be about the same for each cut-off.

Example 1: Using an MFR tester

Test result: Per t = 60s was extruded the average mass of m = 0.5g.

Then: MFR = 5, since MFR = (600s · 0.5g)/60s = 5g/10min
Thus: Related to ten minutes, the extruded mass would have been m = 5g.

Note 1: Classification of poly-ethylene PE via MFR (190/2.16) values, [268]
MFR < 1: high-molecular PE, e.g. for very rigid parts, showing high mechanical stability
MFR = 1.5: for parts showing good mechanical stability
MFR = 5 to 15: for good processiblity, e.g. for shock-proof and form-stable parts
MFR = 15 to 25: good flowing, "low-viscosity types", e.g. for parts showing large areas
MFR > 100: "super-fluid", for high-performance injection moulding over long flow distances

Calculation of the **melt volume flow rate MVR**

Equation 11.20 $\mathbf{MVR\ (T,\ m_{nom})} = (A \cdot t_{ref} \cdot L)/t = (427 \cdot L)/t$

unit [cm³/10min], at the selected measuring temperature T [°C], under the selected nominal load m_{nom} [kg]; with the mean cross-sectional area A [cm²] of cylinder and piston, here: $A = \pi \cdot r^2 = 0.712$cm², with $2r = (d_1 + d_2)/2 = 0.951$cm, or r = 0.476cm; the reference time t_{ref} [s], standard is t_{ref} = 600s = 10min; the distance L traveled by the piston in cm; and the number 427 [cm² · s] which is the product of A [cm²] and the number 600 representing t_{ref} = 600s.

Example 2: Using an MVR tester
Test result: Per t = 60s the average piston travel was L = 0.75cm.
Then: MVR = 5.3, since MVR = (427cm² · s · 0.75cm)/60s = 5.3cm³/10min
Thus: Related to ten minutes, the extruded volume would have been V = 5.3cm³.

Conversion between MFR and MVR values
Since the density ρ = m/V, or the mass m = ρ · V, or the volume V = m/ρ, holds:

Equation 11.21   MFR = ρ · MVR   and   MVR = MFR/ρ

if ρ is specified in g/cm³, and therefore, when ρ = 1g/cm³ holds: MFR = MVR

Example 3: Conversion between MFR and MVR values
MFR and MVR values are available of a polymer melt showing the density ρ = 0.95g/cm³
Then: MFR = 0.95 · MVR
For example, comparing the values of above Examples 1 and 2: (MFR 5) = 0.95 · (MVR 5.3)

The load-dependent flow rate ratio FRR, or the MVR ratio
Since polymer melts usually do not show ideally viscous flow behavior, many users are performing MVR tests in two steps at two different loads, therefore using two weight pieces or masses m_1 and m_2. Afterwards, the FRR or MVR ratio is calculated as follows, here with $m_2 > m_1$:
FRR = MVR (m_2)/MVR (m_1)
The sample shows ideally viscous behavior if the FRR = 1, but usually for polymer melts applies: FRR < 1, therefore indicating shear-thinning flow behavior. If desired, the MFR ratio might be determined correspondingly.

Flow curve determination using MFR and MVR testers
Different loads might be applied step by step to determine a flow curve point by point in the form of the following diagrams: Flow rate versus load, i.e. MVR versus nominal mass m_{nom} or shear rate versus shear stress, respectively.

However, it should be taken into account when using this kind of testing: Due to several effects causing a significant pressure drop or energy loss, respectively, as well in the cylinder as well as in the die, inaccurate test results may be achieved, which therefore are indeed useful for simple quality control, but not for research and development purposes.

𝒢𝒻 For "Mr. and Ms. Cleverly"

c) Calculations

A polymer melt is forced to flow under the load of an installed weight, i.e. due to gravity. This causes a corresponding pressure, which is generating the resulting shear stress and shear rate. Since there is acting an external force or pressure, **MRF and MRV testers are considered "pressurized instruments"**.

For the following simple calculation there is assumed that the melt shows ideally viscous flow behavior, and there are laminar and stationary flow conditions, and no significant influence due to the following effects:

- Inlet flow disturbances in the die due to the initially increased friction forces when entering the clearly narrower die geometry (Couette correction), as well due to the changing flow velocity between cylinder and die which may lead to inhomogeneous, turbulent flow conditions (Hagenbach correction), as well as due to viscoelastic effects which may cause an increased pressure drop or energy loss, respectively (Bagley correction); there may occur also outlet disturbances [17, 79]
- Flow behavior which is not ideally viscous in cylinder and die (Weissenberg/Rabinowitsch correction of the shear rate values) [298, 383]
- Viscous shear-heating of the sample due to friction between the macromolecules of the melt in the cylinder and in the die, and the corresponding energy loss (dissipation)
- Die swell effects and other surface effects of the extruded melt such as melt fracture due to viscoelastic behavior, indicating elasticity and therefore previously absorbed and stored deformation energy
- Friction between piston and cylinder wall
- Inaccurate temperature control, and poor temperature equilibration between device and sample
- Air bubbles and foam formation when melting

All these effects are significantly reducing the real acting pressure, and therefore are clearly influencing the test results. Additionally, in order to reduce the influence of the inlet and outlet disturbances, the L/d ratio, i.e. of length and diameter of the capillary, should be larger than 10 : 1. However, for the dies of typical MFR and MVR testers holds: (8.0mm/2.095mm) < 4 : 1. Therefore, measuring results which are achieved with MFR and MVR testers should be merely considered rough estimates and relative values, but not rheological absolute values.

1) Pressure values in the cylinder and in the die

Using the specified nominal loads or mass pieces (m_{nom} = 0.325/1.20/2.16/3.80/5.00/10.0 and 21.6kg), or the corresponding weight forces (F_{nom} = 3.19/11.8/21.2/37.3/49.1/98.1 and 212N), respectively, the following pressure values are acting on the average cross-sectional area A of cylinder and piston, and the same pressure is also occurring in the die.

Here: $A = \pi \cdot r^2$, with: $r = (d_1 + d_2)/(2 \cdot 2) = (9.55 + 9.48)$ mm/4 = $4.76 \cdot 10^{-3}$m

$p = F_{nom}/A = F_{nom}/(71.2 \cdot 10^{-6}$ m$^2)$

Then: p = (0.0448/0.166/0.298/0.524/0.690/1.38/2.98) · 10^6N/m² or MPa

(= 0.448/1.66/2.98/5.24/6.90/13.8/29.8bar).

Therefore, measurements with MFR or MVR testers are performed in a range of low to medium pressures.

2) Shear stress values at the wall of the die according to the Hagen/Poiseuille relation

$\tau = (p \cdot R)/(2 \cdot L) = (F_{nom} \cdot R)/(2 \cdot L \cdot 71.2 \cdot 10^{-6}$ m$^2)$

with the pressure values p in 10^6Pa (see above, no. 1),

$R = R_3 = d_3/2 = 2.095$mm/2 = $1.048 \cdot 10^{-3}$m, and $L = L_3 = 8$mm = $8 \cdot 10^{-3}$m of the die

Then: $\tau = (p \cdot 1.048 \cdot 10^{-3}m)/(2 \cdot 8 \cdot 10^{-3}m) = 0.0655 \cdot p$

$= (0.0655/71.2 \cdot 10^{-6}m^2) \cdot F_{nom} = (920/m^2) \cdot F_{nom}$

= $(2930/10,900/19,500/34,300/45,200/90,300/195,000)N/m^2$

= $(2.93/10.9/19.5/34.3/45.2/90.3/195)kPa$

Therefore, measurements with MFR or MVR testers are carried out in a medium shear stress range, between 3 and 200kPa.

3) The mean shear rate value at the wall of the die acc. to the Hagen/Poiseuille relation

$\dot{\gamma} = (4 \cdot \dot{V})/(\pi \cdot R^3) = 1.85 \cdot MVR$

with $R = R_3 = 1.048 \cdot 10^{-3}m$, and the volume flow rate

$\dot{V}[m^3/s] = MVR [cm^3/10min] \cdot [(10^{-6}m^3/cm^3)/(600s/10min)]$

= $MVR [cm^3/10min] \cdot 1.67 \cdot 10^{-8} [(10min/cm^3) \cdot (m^3/s)]$

then: $\dot{\gamma} = (4 \cdot MVR \cdot 1.67 \cdot 10^{-8}m^3)/[\pi \cdot (1.048 \cdot 10^{-3}m)^3 \cdot s]$

or: $\dot{\gamma} [s^{-1}] = 1.85 \cdot MVR [cm^3/10min]$

Example 4: Using an MVR tester

Preset: MVR = 5.3; which corresponds to the volume flow rate

$\dot{V} = 5.3cm^3/10min = 5.3 \cdot 10^{-6}m^3/600s = 8.81 \cdot 10^{-9}m^3/s$

then: $\dot{\gamma} = (4 \cdot 8.81 \cdot 10^{-9}m^3)/[\pi \cdot (1.048 \cdot 10^{-3}m)^3 s] = 9.78s^{-1}$ (= approx. $9.8s^{-1}$)

or: $\dot{\gamma} = 1.85 \cdot MVR = (1.85 \cdot 5.3) s^{-1} = 9.81s^{-1}$ (= approx. $9.8s^{-1}$)

The calculated shear rate value is a mean value of the considered flow time interval.

As a comparison: MVR values of 1, or 3, or 5, or 10 correspond to shear rates of 2, or 6, or 9, or $19s^{-1}$ approximately. Therefore, MFR or MVR tests are not carried out in the low-shear range. Even when MVR = 1, there is $\dot{\gamma} > 1s^{-1}$ which has to be considered a medium shear rate value for the mostly highly viscous or viscoelastic polymer melts. Therefore, testing is performed in most cases outside the low-shear range (see also Chapter 3.3.2.1a: Zero-shear viscosity range). Concerning the usually covered shear rate range, for MFR and MVR testers can be stated: MFR or MVR values of 1 to 10 (or 100) correspond to $\dot{\gamma}$ = 2 to 20 (or 200) s^{-1}, approximately.

Note 2: MFR and MVR testers are not simulating industrial operating conditions

For most polymer melts, the values of the **shear rates and shear stresses** achieved with MFR and MVR testers have to be considered **too low to simulate operating conditions**, since extrusion takes place at $\dot{\gamma}$ = 10 to $1000s^{-1}$, and injection moulding at $\dot{\gamma}$ = 100 to $10,000s^{-1}$ (see also Chapter 2.2.2 and Table 2.1).

4) Calculation of viscosity values

Example 5: Using an MVR tester

Presets: Using a load of m = 2.16kg, i.e. applying the shear stress τ = 19,500Pa (see above no. 1 and 2), MVR = 5.3 was measured, i.e. the resulting shear rate $\dot{\gamma}$ = 1.85 · MVR = $9.8s^{-1}$ (see above Example 4).

Calculation: viscosity $\eta = \tau/\dot{\gamma}$ = 19,500Pa/($9.8s^{-1}$) = 1990Pas

Note 3: When using MFR and MVR testers relative values are achieved

Viscosity values should not be overrated if they are calculated from shear stress and shear rate values which are determined by use of MFR and MVR testers. Due to the mentioned effects arising above all from viscoelastic behavior of the samples, the results obtained should be considered **relative values** but not absolute values. Therefore, they are only useful when taken as **comparative values**.

    𝓰𝓸   End of the Cleverly section

Most modern devices are automatically detecting the path traveled by the piston and the corresponding time, and therefore mostly, **MVR values are determined here** (see also Chapter 11.4.2.2: high-pressure capillary viscometers). Often, the following alternative test methods can be selected:

a) Detection of the path traveled by the piston within a defined period of time

b) Measurement of the required time for the piston to travel over a defined distance

Note 4: MFI and MVI are denominations which are out-of-date
In the past, corresponding testers often used to be called "melt indexers", and the test result was termed "melt index" or "melt flow index" MFI instead of MFR, or "melt volume flow index" MVI instead of MVR.

Standards: ISO 1133 (plastics, determination of the MFR and the MVR of thermoplastics); **ASTM** D1238 (flow rates of **thermoplastics** by **extrusion plastometer**; MFR and MVR); (DIN 53735/withdrawn, thermoplastics, MFI, MVI; replaced by ISO 1133). For more information on MFR and MVR testers, see [73, 161, 356, 409].

11.4.2.2 *High-pressure capillary viscometers driven by an electric drive, for testing highly viscous and paste-like materials*

High-pressure capillary viscometers driven by a motor are generating **high shear stresses and medium to high shear rates**. These kinds of devices are used for testing **polymer melts, PVC plastisols, greases, sealants, adhesives, ceramic masses**.
Here, the principle of operation is the same as for MFR and MVR testers.
1) The polymer is filled into a cylinder, mostly showing an inner diameter between d_1 = 9.5 and 24mm, e.g. d_1 = 12.0 or 15.0mm, and a length between L_1 = 250 and 290mm.
2) After melting, the melt is pressed through a **die or capillary showing a circular diameter** with an **L/d ratio of larger than 10** usually, being mostly between 10 and 40. Typical dimensions of dies are d_2 = 1.0mm (or 0.5, 1.2, 1.5, 2.0 or 3.0) and L_2 = 16mm (or 10, 20, 30, 40 or 50).

Example: In order to obtain measuring data for a Bagley correction, three different dies are used of L/d = 10, and 20, and 30 and d = 1mm for each die, i.e., with L = 10, and 20, and 30mm. For special tests, a **blind** is used exhibiting a "very short" die length of only L_2 = 0.25mm which is regarded to be "nearly zero". Also **slit dies** are available showing a **rectangular cross-section** e.g. of the dimensions, in mm: (0.8 to 2.0) · 18
Typical measuring ranges: temperatures from RT to T = +400°C (and enhanced from T = -40 to +500°C), and pressures of up to p = 70MPa = 700bar (and enhanced even up to 200MPa = 2000bar), and a useful shear rate range of $\dot{\gamma}$ = 100 to 10,000s⁻¹. See also e.g. [73, 161, 239, 276].

Standards: ISO 3146 (melting behaviour of semi-crystalline **polymers** by **capillary tube**, etc.), ISO 4575 (plastics, **PVC pastes**, apparent viscosity using the **Severs rheometer** extrusion), ISO 11443 (fluidity of **plastics** using capillary and **slit-die rheometers**) ; **ASTM** D1092 (**lubricating greases** or **adhesives**, using **capillary extruders**), ASTM D1823 (apparent viscosity of **plastisols and organisols** at high shear rates by **extrusion viscometer**), ASTM D3364 (flow rates for **PVC** with molecular structural implications, by extrusion), ASTM D3835 (properties of **polymeric materials** by **capillary rheometer**, by extrusion); **DIN** 54801 (**PVC pastes** at high shear rates using the **Severs capillary viscometer**), and DIN 54811 (**polymer melts**, using a **capillary rheometer**).

a) Test methods
1) **Preset of a volume flow rate** via the drive speed v of a piston (e.g. between v = 0.02µm/s and 40mm/s) using an electric drive. Here, **the resulting melt pressure is measured** by one or several pressure sensors.
2) **Preset of a pressure** via the drive force F acting on a piston (e.g. with F_{max} between 5 and 100kN) by an electric drive or, but rarely used nowadays by gas pressure e.g. using nitrogen to prevent oxidation. Here, the **resulting path traveled by the piston is measured**, and from this **is determined the volume flow rate**.

b) Determinations
Flow curve determination using a high-pressure capillary viscometer
Most modern devices are **speed-controlled, pressure-controlled or force-controlled**, respectively, using an electric drive. Here if desired, usually variable feed speeds or pressures

can be preset making those viscometers useful to determine point by point **flow curves or viscosity functions**, respectively.

Note 1: Determination of elastic behavior using a high-pressure capillary viscometer
Elastic behavior of an extrudate may be **determined indirectly from** the degree of the **die swelling** effect (see Chapter 5.2.1.2a and Figure 5.3), e.g. by a contactless measurement using a laser beam and a light-sensitive sensor. Finally, the calculated result is often expressed in terms of the shear modulus or tensile modulus. **The first normal stress difference** is derived indirectly from die swell, and – when using slit dies – from the pressure drop in the slit die, via two or more pressure sensors. However, **these kinds of specifications should be considered relative values** since die swell is usually measured close to the die when the melt is still in a fully molten state, being therefore still able to swell further afterwards. The period of time for the extrudate to pass from the die to the position of the die swell detector unit may possibly be too short for a full recovery of the elastically stored energy. Nevertheless, this test method may provide meaningful **data of the relative elasticity** at shear rates that may reach up to 5000s^{-1}. But often, **the recommended maximum shear rate is stated as 1000s^{-1} only**, since highly elastic melts at high shear rates are often showing a distorted and even broken-up surface. These effects are called **orange peel, shark skin or melt fracture** due to the inhomogeneous, "elastic-turbulent" flow conditions of the polmer melt. Data measured in this range are useless in a scientific sense, since they are resulting in not reproducible viscosity and elasticity values [324].

Note 2: The "high shear rate Lodge stressmeter"
The "stressmeter" is a pump-operated high-pressure capillary viscometer **showing a slit die** (e.g. of L = 2mm and h = 46µm) which was presented in 1987 by *A.S. Lodge* [229]. It was designed for testing polymer melts up to τ = 130kPa, and for low-viscosity fluids up to $\dot\gamma$ = 5 million s^{-1}. Here, besides the wall shear stress determined via the pressure difference Δp of a stationary capillary flow over a defined distance, additionally the **"hole pressure" p* is measured**. From these values is calculated the **1st normal stress difference N$_1$** (see also Chapter 5.3).

     ✍   For "Mr. and Ms. Cleverly"

c) Calculations
For the following simple calculations there is partially idealized, assuming the same as above for the MFR and MVR testers. Also here the following effects should be taken into account: inlet and outlet flow disturbances causing turbulent flow conditions due to viscous and viscoelastic effects and inertia of the melt (corrections of Hagenbach and Bagley), flow behavior being not ideally viscous (Weissenberg/Rabinowitsch correction of the shear rate values), energy dissipation due to viscous shear-heating, die swell and melt fracture effects due to visco-elastic behavior, friction between piston and cylinder wall, inaccurate temperature control, and air bubbles in the melt. All these effects might significantly reduce the acting pressure, and therefore, they might influence the test results. Hence, **uncorrected test results achieved with high-pressure capillary viscometers should be considered rough estimates only and not rheological absolute values**. Nowadays, the corrections are usually performed using corresponding software programs.

For example, when analyzing the flow behavior of polymer melts, the use of the Hagen/Poiseuille relation without the **corrections according to Bagley and Weissenberg/Rabinowitsch** may lead to errors of the order of 20% and more [324]. Another example: Testing a polystyrene melt of M = 400,000g/mol at T = 190°C and at a shear rate of 200s^{-1}, using a die of L/d = 6:1, a pressure correction according to Bagley was required in a range of more than 50% of the total pressure in order to obtain "true" viscosity data [379].

1) Pressure in the cylinder

With some mostly older instrument types, the pressure p is preset directly, e.g. in the form of gas pressure. The following counts for a force-controlled device:

$p = F/A_1 = F/(\pi \cdot R_1^2)$

with the drive force F acting on the cross-sectional area A_1 of the cylinder with the radius $R_1 = d_1/2$. The same pressure is acting in the die.

Example 1: Using a force-controlled high-pressure capillary viscometer
Presets: $F = 10kN = 10^4N$, and $R_1 = d_1/2 = 15.0mm/2 = 7.5 \cdot 10^{-3}m$
Then: $p = 10^4\ N/[\pi \cdot (7.5 \cdot 10^{-3}\ m)^2] = 56.6 \cdot 10^6 N/m^2 = 56.6MPa\ (= 566bar)$

2) Shear stress value at the wall of the die according to the Hagen/Poiseuille relation

Equation 11.22 $\tau = (p \cdot R)/(2 \cdot L)$

with the radius $R = R_2$ and the length $L = L_2$ of the die

Example 2: Using a force-controlled high-pressure capillary viscometer
Presets: $p = 56.6MPa = 56.6 \cdot 10^6 N/m^2$ (see above Example 1),
$R_2 = d_2/2 = 1mm/2 = 5 \cdot 10^{-4}m$, and $L_2 = 16mm = 0.016m$
By the way, here the L/d ratio of the die is 16mm/1mm = 16 : 1, hence being larger than 10 : 1 which is considered the minimum ratio to neglect the effect of possibly occurring inlet and outlet flow disturbances.
Then: $\tau = (56.6 \cdot 10^6 \cdot 5 \cdot 10^{-4}\ Pa \cdot m)/(2 \cdot 0.016\ m) = 0.884 \cdot 10^6 Pa = 884kPa$

3) Mean shear rate at the wall of the die according to the Hagen/Poiseuille relation

Equation 11.23 $\dot{\gamma} = (4 \cdot \dot{V})/(\pi \cdot R^3)$

with the volume flow rate $\dot{V}$, the die radius $R = R_2 = d_2/2$,
and since $\dot{V}$ is showing the same value in the cylinder and in the die:
$\dot{V} = A_1 \cdot v = (\pi \cdot R_1^2) \cdot v$
with the cross-sectional area A_1 of the cylinder and the cylinder radius $R_1 = d_1/2$,
and the drive speed v
Thus: $\dot{\gamma} = (4 \cdot \pi \cdot R_1^2 \cdot v)/(\pi \cdot R_2^3) = (4 \cdot R_1^2 \cdot v)/R_2^3$

Example 3: Using a speed-controlled high-pressure capillary viscometer
Presets: $R_1 = 7.5 \cdot 10^{-3}m$, $R_2 = 5 \cdot 10^{-4}m$,
and $v = 50mm/min = 50 \cdot 10^{-3}m/60s = 8.33 \cdot 10^{-4}m/s$
Then: $\dot{\gamma} = [4 \cdot (7.5 \cdot 10^{-3}m)^2 \cdot 8.33 \cdot 10^{-4}m]/[(5 \cdot 10^{-4}m)^3s] = 1500s^{-1}$
The calculated shear rate value is constant if the drive speed v is constant in the corresponding period of time.

4) Calculation of viscosity values

Example 4: Using a high-pressure capillary viscometer
Presetting a drive speed of v = 50mm/min, i.e. the shear rate of $\dot{\gamma} = 1500s^{-1}$ (see above Example 3), p = 56.6MPa was measured by the pressure sensor which corresponds to the shear stress $\tau = 884kPa$ (see above Examples 1 and 2).
Calculation: viscosity $\eta = \tau/\dot{\gamma} = 884,000Pa/1500s^{-1} = 589Pas$

Note 3: Data of high-pressure capillary viscometers without correction are relative values

Viscosity values should not be overrated if they are calculated directly from shear stress and shear rate values which are determined by use of high-pressure capillary viscometers, i.e. without regarding corrections according to Bagley and Weissenberg/Rabinowitsch. In this case, due to the mentioned effects arising above all due to viscoelastic behavior of the sample, the test results should be considered **relative values** but not absolute values. Therefore, they are only useful when taken as **comparative values**.

෬෬ End of the Cleverly section

11.4.2.3 High-pressure capillary viscometers driven by gas pressure, for testing liquids

High-pressure capillary viscometers driven by gas pressure, are generating **medium shear stresses at very high shear rates**. These kinds of devices are taken for testing **mineral oils, paper coatings** and similar dispersions containing small particles, and **polymer solutions** used to produce synthetic fibers on spinning machines.

The core of these kinds of high-pressure capillary viscometers is a capillary made of glass or steel showing an exactly defined inner diameter d and length L (e.g. of d = 0.2 to 1mm, often is used d = 0.5mm; and L = 30 to 90mm, often of L = 50mm). The liquid sample is pressed through the capillary by a defined external gas pressure. Typical range of measuring temperatures is T = +20 to 80°C (and enhanced from T = -20 to +150°C).

a) Test methods

Gas pressure is acting directly onto the sample, or indirectly e.g. via a flexible membrane. For example, compressed air or nitrogen is used at a working pressure up to 20 or even 35MPa (= 200 or 350bar). Variable pressures, and therefore shear stresses, can be preset with most devices making them useful to determine **flow curves** point by point and to **calculate** the corresponding **viscosity functions**.

A typical shear rate range covered is $\dot{\gamma}$ = 3000 to 10^6 (= 1 million) s^{-1} when using capillaries showing a circular cross-section. Even higher shear rates up to $\dot{\gamma}$ = $10^7 s^{-1}$ can be achieved when using special slit die geometries. Under certain conditions also lower shear rates can be preset, e.g. at $\dot{\gamma}$ = $500 s^{-1}$.

Standards: ASTM D5481 (**petroleum products**, apparent viscosity at high temperature and high shear rate, **HTHS**, by multicell capillary viscometer), (ASTM D4624/withdrawn, for petroleum products, fluid lubricants, engine oils, apparent viscosity by capillary viscometer at high temperature and high shear rates); and **DIN** 53014 (general use; capillary viscometers showing circular or rectangular cross sections)

b) Determinations
Example 1: Measuring the HTHS viscosity of mineral oils
According to ASTM D5481, single-point tests are performed at T = +150°C and $\dot{\gamma}$ = $10^6 s^{-1}$ to simulate the flow behavior of **lubricating oils** in crankshaft bearings of motors or in turbines, i.e., to determine the HTHS viscosity, hence at high temperature and high shear rate.

Example 2: Paper coating
Presets: Speed v = 1500m/min = 25m/s of the base paper, and a wet layer thickness of h = 9μm, e.g. at the application rate AR [g/m²] = m/A of 8g/m². AR is mass per coating area, with the mass m [g] and the area A [m²] to be coated. Calculation of the shear rate:
$\dot{\gamma}$ = v/h = $25m/(9 \cdot 10^{-6} m \cdot s)$ = $2.8 \cdot 10^6 s^{-1}$
In this case, a **slit die geometry** showing a rectangular cross-section is advantageous since it simulates the blade geometry. For example, the blade occurs in the form of a **"miniblade"** which is coated with a very hard layer of special ceramics, showing a gap dimension of 50μm and a width of 10mm, at a depth of 0.5mm.
Evaluation of **paper coatings**: Low viscosity at all shear rates causes poor water retention. High viscosity at low shear rates causes start-up and pumping problems, and streaks when coating. High viscosity at high shear rates may cause blade bleeding and coating problems, non-uniform application rates, and even web breaks [2].

For "Mr. and Ms. Cleverly"

c) Calculations

For the following simple calculations there is partially idealized, assuming that the sample is showing ideally viscous flow behavior, and that there are laminar and stationary flow conditions, and therefore, no influence due to the following effects: inlet and outlet disturbances, e.g. due to initially increased friction causing turbulent flow along a certain inlet distance until reaching laminar conditions (Hagenbach/Couette correction), flow behavior being not ideally viscous (Weissenberg/Rabinowitsch correction of the shear rate values), viscous shear-heating of the sample due to strong internal friction effects at these high shear rates; inaccurate temperature control; and air bubbles in the sample. All these effects might significantly influence the test results. Hence, **uncorrected test results achieved with high-pressure capillary viscometers should be considered rough estimates only and not rheological absolute values**. Nowadays, the corrections are usually performed using corresponding software programs. In order to reduce the influence of the inlet and outlet disturbances, the L/d ratio, which is the ratio of length and diameter of the capillary used, should be not smaller than 10 : 1.

1) Pressure in the capillary
Here, the pressure p is preset directly by the user or via the instrument.

2) Shear stress value at the capillary wall according to the Hagen/Poiseuille relation

Equation 11.24 $\tau = (p \cdot R)/(2 \cdot L)$

Preset of a constant pressure p for each single measuring point, using a capillary of the radius R = d/2 and the length L.

Example 3: Using a high-pressure capillary viscometer
Preset: p = 10MPa (= 100bar), R = d/2 = 0.5mm/2 = 0.25mm = $0.25 \cdot 10^{-3}$m,
L = 50mm = $50 \cdot 10^{-3}$m. By the way, here the L/d-ratio of the capillary is 50mm/0.5mm = 100 : 1, being therefore clearly larger than 10 : 1 which is considered the minimum ratio in order to be sure that inlet and outlet flow disturbances can be ignored.
Then: $\tau = (10^7 \cdot 0.25 \cdot 10^{-3}$ Pa $\cdot$ m$)/(2 \cdot 50 \cdot 10^{-3}m) = 25{,}000$Pa = 25kPa

3) The mean shear rate value at the capillary wall acc. to the Hagen/Poiseuille relation

Equation 11.25 $\dot{\gamma} = (4 \cdot \dot{V})/(\pi \cdot R^3)$

with the volume flow rate $\dot{V}$, and the capillary radius R
Example 4: Using a high-pressure capillary viscometer
Preset: R = $2.5 \cdot 10^{-4}$m; measurement of the sample volume V = 100ml = 10^{-4}m^3 which is flowing through the capillary in a flow time interval of t = 10s
Calculation of the volume flow rate: $\dot{V} = $ V/t = 10^{-4}m/10s = 10^{-5}m^3/s
Then: $\dot{\gamma} = (4 \cdot 10^{-5}m^3)/[\pi \cdot (2.5 \cdot 10^{-4})^3m^3 \cdot$ s$] = 815{,}000$s^{-1}
The calculated shear rate value is the mean value occurring in the considered flow time interval.

4) Calculation of viscosity values
Example 5: Using a high-pressure capillary viscometer
Presetting a pressure of p = 10MPa, i.e. the shear stress of τ = 25,000Pa (see above Example 3), the volume flow rate $\dot{V} = 10^{-5}$m^3/s was measured, and therefore, here occurs the shear rate of $\dot{\gamma}$ = 815,000s^{-1} (see above Example 4).
Calculation: viscosity $\eta = \tau/\dot{\gamma} = 25{,}000$Pa/815,000s^{-1} = 0.0306Pas = 30.6mPas

Note: Data of high-pressure capillary viscometers without correction are relative values
Viscosity values should not be overrated if they are calculated directly from shear stress and shear rate values which are determined by use of high-pressure capillary viscometers, i.e. without regarding the corrections according to Hagenbach/Couette and Weissenberg/Rabinowitsch. In this case, due to the mentioned effects arising above all due to the turbulent behavior of the sample, the test results should be considered **relative values** but not absolute values. Therefore, they are only useful when taken as **comparative values**.

⤳   End of the Cleverly section

11.5 *Falling-ball viscometers*

Falling-ball tests are one of the oldest tests in rheometry. *Georges G.* **Stokes** (1819 to 1903) began to work scientifically on this field as early as 1851, [348; 233]. A falling-ball viscometer which was designed by *Fritz* **Höppler** and latterly standardized, has been commercially available since 1934 (DIN 53015 and ISO 12058).

The following principle is used: Due to gravity a steel ball is sinking through the liquid sample which is filled into a glass tube. For low-viscosity and low-density samples, a ball made of glass is used, since due to its lower density it is slower moving downwards then. The glass tube shows a defined inner diameter (standard is d = 15.94mm), and balls are available of different diameters (11, 14, 15.2, 15.6 and 15.81mm). **The time is measured for the ball to move downwards over a defined distance** between two marks (usually 100mm). The viscosity value is determined from this, based on calibration tests using viscosity standard fluids. Variable velocities may be simulated if it is possible to set the glass tube at different inclination angles. Standard is here an angle of 80° to the horizontal, or 10° to the vertical, respectively. When setting at different angle positions, also the viscosity function of liquids which are not showing ideally viscous flow behavior may be determined within certain limits.

Advantages of the falling-ball method: ease of use; no evaporation or loss of solvent since the sample is in a closed space; no skin formation on the surface of the liquid.

Disadvantages: samples have to be transparent if the test results are determined visually; **only ideally viscous fluids should be measured** since there are no constant shear conditions, hence no constant shear rate; the flow conditions are clearly inhomogeneous and mostly not laminar, i.e., turbulent.

Micro falling-ball viscometers are available for small sample amounts, e.g. for a sample volume of less than $1cm^3$. Here, glass tubes (called "capillaries") are used of d = 1.6, 1.8, 3 or 4mm, and balls of d = 1.5, 2.5 or 3mm. Transparent and also non-transparent liquids such as inks might be investigated if the motion of the ball is detected via induction using electromagnetic detectors. If the inclination of the measuring tube can be set at different angles, e.g. between 15 and 90° to the horizontal, it is possible to simulate different shear rates due to the resulting variable velocity of the rolling or falling ball. This kind of instrument can also be used to determine the values of the **intrinsic viscosity IV** or the **limiting viscosity number LVN** (or the Staudinger index, resp.) **of diluted polymer solutions**, see also Chapter 11.4.1.2, [10].

⤳   For "Mr. and Ms. Cleverly"

The measuring principle of falling ball viscometers, and calculations
This measuring method is based on the law of Stokes (see also Chapter 2.2.2b), assuming the weight force F_G [N] of the ball and the flow resistance force F_R [N] of the sample are approximately of the same size.

Equation 11.26 $F_G = \Delta m \cdot g = F_R = 3 \cdot \pi \cdot d \cdot \eta \cdot v$

Here with the mass difference Δm [kg] of ball and fluid sample, the gravitation constant $g = 9.81 m/s^2$, the ball diameter d [m], the viscosity of the sample η [Pas], and the velocity [m/s] of the ball.

It holds: $\Delta m = V \cdot \Delta\rho$, with the volume V [m³] of the ball and the density difference $\Delta\rho$ [kg/m³] = $(\rho_k - \rho_{fl})$ of ball and fluid sample; here with the density of the ball as ρ_k [kg/m³] and the density of the fluid as ρ_{fl} [kg/m³].

However, this correlation is only true with the following assumption: the ball is sinking very slowly, therefore showing quasi-stationary motion in a sample being at rest, i.e., at laminar flow conditions showing a Reynolds number of Re ≤ 1 (see also Chapter 10.2.2.4b: Re-number). Then:

$\eta = (\Delta m \cdot g)/(3 \cdot \pi \cdot d \cdot v) = (V \cdot \Delta\rho \cdot g)/[3 \cdot \pi \cdot d \cdot (\Delta s/\Delta t)]$

$\eta = ([(\pi \cdot d^3)/6] \cdot \Delta\rho \cdot g \cdot \Delta t)/(3 \cdot \pi \cdot d \cdot \Delta s) = (d^2 \cdot g \cdot \Delta\rho \cdot \Delta t)/(18 \cdot \Delta s)$

with $v = (\Delta s/\Delta t)$, the traveling path Δs [m] of the ball, in the traveling time interval Δt [s] to be measured, and with $V = [(\pi \cdot d^3)/6]$ of the ball. Known are: d of the ball, the value of g and the path Δs. If all known parameters are combined to the **"instrument constant"** $C_{KF} = (d^2 \cdot g)/(18 \cdot \Delta s)$ in [m²/s²], for the viscosity in [Pas] holds:

$\eta = C_{KF} \cdot \Delta\rho \cdot \Delta t$

The traveling time Δt [s] is measured by the user, and then is calculated $\Delta\rho = (\rho_k - \rho_{fl})$ in [kg/m³], where ρ_k of the ball made of steel or glass is known (see ISO 12058 or DIN 53015), and ρ_{fl} of the sample should also be known. More information – and also to the possibilities for corrections – see [276].

߷   End of the Cleverly section

Note 1: Measuring with falling-ball viscometers is a single-point testing method

Testing with falling-ball viscometers is a single-point measuring method if the shear conditions cannot be varied, e.g. via variable inclination angles of the tube. In this case, merely a single viscosity value is generated. However here, during the time interval of testing, the shear rates may change within a certain range depending on the following conditions: the initial acceleration of the ball; the uncontrolled motion of the ball which may roll, tumble, turn sidewards or even rotate, and therefore, there may occur turbulent flow effects; the distance of the ball to the wall of the tube; and the traveling velocity of the ball. Therefore, this test method should only be used for liquids showing ideally viscous flow behavior since only in this case, viscosity is displaying a constant value independent of the possibly continuously changing velocity of the ball or shear rate, respectively.

Note 2: Measurements which are similar to the falling-ball method

Similar test methods are performed when using **devices** with immersing, descending, sinking, falling and rolling elements such as **rods, cylinders oder needles**, see also Chapter 11.2.8b.

11.6   Rotational and oscillatory rheometers

Rheometer technology advances rapidly, a trend which is set to continue in the future. Some important steps in rheometer development have been, for example:

1784: *C.A. Coulomb* builds a **rotational apparatus** to investigate the torsional elasticity of metal wires and also the "friction of liquids" using **concentric cylinders**. The inner cylinder is suspended to a torsion wire, deflected and released, showing subsequently **free torsional oscillatory motion** which is damped by the resistance of the sample, exhibiting therefore constantly decreasing deflection amplitudes [80, 289].

1863: *J.C. Maxwell* designs a **rotational apparatus** to measure the "inner friction" of gases, detecting the **damping effect** of air and other gases on several **disks** which were indirectly **exited to a torsional oscillatory motion by a variable magnetic field** acting from outside [245, 286].

<u>1863:</u> *F.W.G. Kohlrausch* studies the behavior of glass fibers and rubber threads using a **torsional creep apparatus** [211; 233].

<u>1881:</u> *M. Margules* proposes the use of a **cylinder apparatus** to measure the "coefficient of friction and gliding" (meaning viscosity) [241; 107].

<u>1882/1893:</u> *J. Perry* designs a **cylinder apparatus** and **double-gap systems** (of a mean radius R = 109 or 121mm and a gap of 10mm) [285; 107, 289].

<u>1888:</u> *M.M.A. Couette* measures the "inner friction" of fluids such as water, rape-seed oil, mercury and air using an **electric driven rotational apparatus and a speed-controlled outer cylinder** showing a **concentric narrow gap** (mean R = 114mm, gap of 2.5mm). The **torque is measured** via the inner cylinder using a **torsion wire** on which is mounted a small mirror reflecting a light beam. This is the **first directly driven cylinder viscometer generating continuously ongoing rotational motion** [79; 107, 233, 289].

<u>1888/1896:</u> *A. Mallock* measures viscosity of water and other liquids using cylinders (mean R = 48/82/88mm, gaps of 3.8/11/23mm); first report about an **apparatus with a rotating inner cylinder** [238; 107].

<u>1889/1990:</u> *T. Schwedoff* uses cylinders (mean R = 35mm, gap of 12.6mm), and measures the "cohesion of liquids" (meaning viscosity), and the "modulus of liquids" (meaning elastic effects); and the **shear rate-dependent viscosity** e.g. of castor oils, glycerins and gelatin solutions; first report about **shear-thinning flow behavior** [328; 107].

<u>1909:</u> *E.J. Stormer* builds a **rotational viscometer with a rotating inner cylinder**, driven mechanically by a **weight-and-pulley system**, using diverse geometries such as cylinders, **fork-shaped paddles, flag-like vanes**, etc. Stormer viscometers are sold until the 1960s (see also Chapter 10.6.3) [349; 18].

<u>1912:</u> *G.F.C. Searle* designs a rotational viscometer with a rotating concentric inner cylinder, using a weight-and-pulley system [330; 18].

<u>1913:</u> *E. Hatschek* improves the **"Couette apparatus"** making this cylinder system usable for practical studies, [175; 378].

<u>1915:</u> *F.R. MacMichael* designs an electric driven speed-controlled rotational viscometer operating at a single speed of n = 20min^{-1}, with a rotating cup and a stationary inner disk (of d = 60mm, 5mm thick) suspended to a torsion wire to detect the torque. MacMichael viscometers are sold until the 1940s, however then with a variable speed of n = 10 to 38min^{-1}, [232; 18].

<u>1928:</u> *E. Hatschek* performs tests with the **"Couette/Hatschek rotational viscometer"** using cylinders (cup radius 55.2mm and bobs of R = 35 and 49.6mm); minimum speed n = 1.2min^{-1}, [175].

<u>1931:</u> *M. Mooney* presents the **"shearing disk viscometer"** with a **parallel-disk measuring system**. In 1934, *Mooney* and *R.H. Ewart* use a **cone-and-plate measuring system** (probably for the first time) [255; 233].

<u>1934:</u> *W. Philippoff* builds an **electromagnetic viscometer for continuous oscillatory tests** on liquids such as glycerine and honey, by **immersing a needle** into it; preset of the force via electromagnets, determination as well of the deflection amplitude using a micrometer screw as well as the frequency (f = 30 bis 630Hz) via the tone of a telephon receiver; analysis of the real part η' of the complex viscosity as a function of the frequency (but he did not detect elastic behavior or phase shift) [352].

<u>1938:</u> *H. Roelig* designs an **apparatus generating forces mechanically** in the form of **harmonic oscillation** using a combination of a pre-stressed spring and an eccentrically rotating mass, **determination of the phase shift** between the preset force and the resulting deformation **using an optical system** [308; 384].

<u>1939:</u> *A.P. Aleksandrov* and *Y.S. Lazurkin* build an **apparatus generating forces mechanically** in the form of **harmonic oscillation** using a cam compressing a plate spring (frequency range f = 0.017 to 33Hz), **optical determination of the deformation amplitude** (but no determination of the phase shift) [7; 384].

<u>1951:</u> *K. Weissenberg* presents the **first rheometer with an air bearing, for controlled strain or rotational speed** in the form of **rotational and oscillatory tests**; using a torsion wire or a torsion bar, a **normal force sensor**, and cone-and-plate measuring systems with a rotating bottom plate [383; 174, 233, 237, 307].

<u>1968:</u> *J.J. Deer* presents a **torque-controlled rheometer for low stresses ("CS rheometer")**, with an air bearing, and an **air turbine drive** generating a torque which is depending on the flow-rate of the air. *P. Finlay* adds the **electronics**, later based on microprocessors. Min./max. torque in 1970: 0.1/10mNm (in 1978: 0.01/10mNm). Since 1972, it is equipped with an **electric asynchronous induction motor ("drag-cup")**. Detection of the **deflection angle or rotational speed**, respectively, initially by an **inductive transducer**, and later by an **optical encoder** which is a sectored disk cutting a light beam in countable impulses. They are commercially available since 1969 (air drive) and 1982 (electromotor) [89; 18].

since the 1970s: **Viscometer** control by external programmers with **analog data processing, data output by printers** using light-sensitive paper, **dot matrix printers** or real-time online-recorders; e.g. in 1971 available from company Haake, in 1974 from Contraves, in 1976 from Brookfield and Ferranti.

since the 1980s: **Rheometer** control by **external digital programmers** using **microprocessors** or **computers; software programs** for measurement, **storage** and **analysis of test data**, initially using A/D-D/A (analog/digital) signal converters; e.g. in 1978 from company Haake, and in 1982 from Rheometrics.

1981: Company Contraves launches a speed-controlled viscometer, firstly equipped with a **built-in microprocessor to measure the torque as a function of the power consumption** ("Rheomat 108") [18].

1995: Company Physica presents the first **directly digitally controlled rheometer** using an **electronically commutated (EC) synchronous motor**, showing advantages especially in a fast control of step strain tests such as relaxation tests, **step shear rate tests** (rotation) e.g. to analyze "thixotropic behavior" of coatings and **strain-controlled frequency sweeps** (oscillation) [222].

The names **viscometer** and **rheometer** are not strictly defined. Simple devices, used to measure flow behavior in terms of the viscosity function via speed- or shear rate-controlled rotational tests, are used to be called **rotational viscometers**. Instruments are termed as **rotational rheometers**, when further rheological tests are possible such as torque- or shear stress-controlled rotational tests. With most rheometers, also the viscoelastic behavior of samples can be characterized by performing creep tests, relaxation tests, and oscillatory tests. The following section contains information on the operating modes of viscometers and rheometers which are driven by electromotors.

11.6.1 Rheometer set-ups

In Figure 11.6 are presented two **different set-ups of rheometers**:

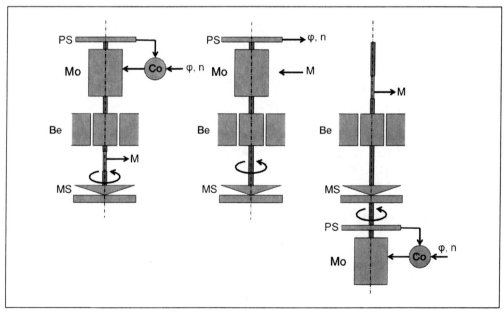

Figure 11.6: Set-up of rheometers: with motor Mo, bearing Be, position sensor PS, measuring system MS – here cone-and-plate, controller Co, deflection angle φ, rotational speed n, torque M
Left-hand side, showing a single head system to be used for controlled strain and controlled shear rate tests: preset of φ or n, and detection of M. Middle, as a single head system to be used for controlled shear stress tests: preset of M, and detection of φ or n. Right-hand side, as a dual heads system showing separated units for drive and detector, to be used for controlled strain and shear rate tests: preset of φ or n (by the motor on bottom), and detection of M (by the transducer on top)

a) Single head system

Both drive (motor) and detector (transducer) are integrated in a single unit, as a single measuring head being on the side of the rotor. This kind of system sometimes is also called a "combined motor transducer" CMT system.

b) Dual heads system

Here, drive and detector are separated, i.e., each one of these two components is installed on different sides of the measuring geometry as separated units. Sometimes this is also called a "separated motor transducer" SMT system. Therefore, **motor and torque sensor are decoupled**. These kinds of instruments are not designed for controlled torque or shear stress tests, and therefore, they are only suitable for controlled speed or shear rate tests and controlled angle or shear strain tests. Deflection angles or rotational speeds are controlled and preset on one side of the measuring geometry, which is usually the side of the bottom plate or cylinder cup, respectively. The torque sensor is installed in the upper part of the instrument and connected with the corresponding part of the measuring system such as cone, upper plate or inner cylinder. This part, the stator, is stationary or can be deflected to a very limited degree only, depending on the torque transducer. A disadvantage of this kind of system is the higher costs since there are two separated units or "two heads".

Usually, for typical applications in industrial laboratories, today's digital servo control techniques make it possible to achieve the same test results regardless of which set-up and control mode is used. However, when performing special modes of testings, there might occur differences in certain details (see also Note 2 in Chapter 8.3.6b: LAOS tests).

&ø&ø&   For "Mr. and Ms. Cleverly"

11.6.2 Control loops

a) Torque control

When performing a controlled shear stress (CSS) test, it is aimed to preset a torque M or shear stress τ at the desired value, see Figure 11.7. The electronic controller provides the motor with an appropriate operating current I. The motor generates a corresponding torque M which is acting via the measuring system upon the sample. Against the motor torque, a torque occurs due to the resistance or reset force of the sample. The resulting deflection angle φ is measured by the position sensor, e.g. in the form of counted impulses generated by an incremental encoder when performing angle steps. The rotational speed n can be calculated from this if also the corresponding time interval is taken into consideration to cover the corresponding angle. Finally as a result, the following measuring data are available: the deflection angle φ or deformation γ and the rotational speed n or shear rate $\dot\gamma$, respectively. Since there is always friction, e.g. of the rheometer bearing, and inertia of the masses in motion, e.g. of the rotating parts of the motor, measuring system and sample, these disturbances have to be corrected by the electronic controller. The following control procedures are occurring if torque-controlled tests are performed:

1) **Controlled shear stress tests in rotation** (see also Chapter 3.2.1b: CSS tests and Table 3.2)
Preset: torque M or shear stress τ, respectively
Result: rotational speed n or shear rate $\dot\gamma$, respectively

2) **Creep and creep recovery tests** (see Chapter 6)
Preset: step to a torque M = const and then to M = 0, or step stress to a constant τ-value and then to τ = 0, respectively
Result: time-dependent deflection and recovery angle $\varphi(t)$, or deformation and reformation function $\gamma(t)$, respectively

3) **Controlled shear stress tests in oscillation** (see Chapter 8.2.5b: CSS tests, and Table 8.2)
Preset: time-dependent sinusoidal oscillation M(t) or $\tau(t)$, resp., with the amplitude M_A or τ_A

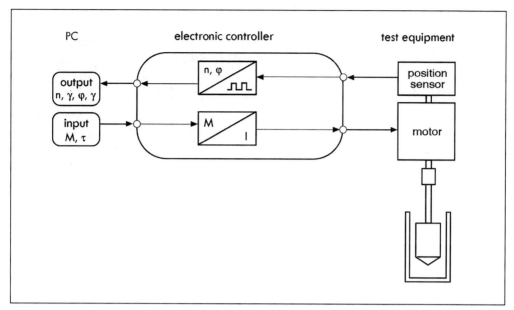

Figure 11.7: Direct control of the torque when using the controlled shear stress mode

Result: time-dependent sinusoidal oscillation $\varphi(t)$ or $\gamma(t)$, resp., with the amplitude φ_A or γ_A, and the phase shift δ between the preset and the resulting sine curve

b) Control of deflection angles and rotational speeds using a closed control loop

When performing a controlled shear strain or controlled shear rate (CSR) test, it is aimed to preset a deflection angle φ or strain γ, or a rotational speed n or shear rate $\dot{\gamma}$, respectively, at the desired value, see Figure 11.8. The electronic controller tries now to provide the motor with an appropriate operating current I to enable the motor to generate the corresponding torque M in order to reach the desired angle φ or the desired speed n. This torque is acting via the masuring system upon the sample, and against this torque a torque occurs due to the resistance or reset force of the sample. As a result, initially an actual deflection angle φ_{act} will be obtained which is determined by the position sensor, and if desired, the initial actual speed n_{act} can be calculated from this if also the time interval is taken into consideration to cover the corresponding angle. In most cases, firstly the actual values of φ_{act} or n_{act} are differing from the desired values of φ or n. After a comparison of the actual and the desired values the controller decides whether the operating current I, and therefore the torque M, have to be adapted or not. If required, a closed control loop follows up until the desired value has been reached. Finally as a result, the following measuring data are available: the torque M or shear stress τ, respectively.

1) **Controlled shear rate tests in rotation** (see also Chapter 3.2.1a: CSR tests, and Table 3.1)
Preset: rotational speed n or shear rate $\dot{\gamma}$, respectively
Result: torque M or shear stress τ, respectively
2) **Step strain or stress relaxation tests** (see Chapter 7)
Preset: step to a deflection angle φ = const or step strain to a constant γ-value, respectively
Result: time-dependent torque M(t) or shear stress function $\tau(t)$, respectively
3) **Controlled shear strain (deformation) tests in oscillation** (see Chapter 8.2.5a: CSD tests, and Table 8.1)
Preset: time-dependent sinusoidal oscillation $\varphi(t)$ or $\gamma(t)$, resp., with the amplitude φ_A or γ_A
Result: time-dependent sinusoidal oscillation M(t) or $\tau(t)$, resp., with the amplitude M_A or τ_A, and the phase shift δ between the preset and the resulting sine curve

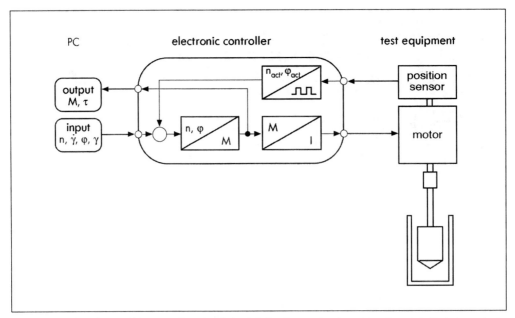

Figure 11.8: Closed control loop to adjust the desired deflection angle when using the controlled shear strain mode, or to adjust the desired rotational speed when using the controlled shear rate mode, respectively

Note: Direct Strain Oscillation (DSO)
This method was developed for testing viscoelastic materials showing very low values of G' and G", such as liquids and other samples with weak gel structures. Traditional strain control of oscillatory tests works as follows: A complete oscillation cycle is carried out, and the resulting strain (deformation) values are measured at different discrete time points. Then, a fast Fourier transform (FFT) is performed to characterize mathematically the actual strain amplitude and phase shift. Using these values, the torque is adjusted for another trial to perform another oscillation cycle. This procedure is repeated until the desired strain value is reached finally. Therefore for each single measuring point, at least a complete oscillation cycle is required. If several adjustment cycles are necessary, it takes of course correspondingly longer to generate each single data point.

The DSO option, however, is working as follows: The strain is adjusted in a **real-time position control mode directly on the sine wave**, i.e., there is no need to wait until a complete oscillation cycle is finished to adjust the actual value [222]. The resulting advantages are obvious for the user:
1) There are required no closed control loops as illustrated in Figure 11.8. This results in a faster data acquisition which is usually finished already within a small part of a single oscillation cycle. Therefore, **faster measurements** are possible.
2) When using a synchronous drive, no strain overshoots are occurring during the adjusting process, even when large amplitude steps are performed. Therefore, **exact strain presetting** is guaranteed without destruction of the sample's structure.
3) There appears no uncontrolled drift of the deflection angle during adjustment and performance of the oscillatory motion, even when measuring in the low-torque range, for example, when testing low-viscosity fluids.
4) **Real strain-controlled oscillatory tests are possible**, i.e., in real-time.
5) Performing DSO with an optimally adjusted instrument, and if all additional disturbances such as internal friction and inertia are taken into consideration for the control process and subsequently compensated by the controller, the following data may be achieved today:

Extremely small deflection angles of φ_{min} = 0.1μrad = 10^{-7}rad can be controlled, **and also extremely low minimum torque values** of M_{min} below 10nNm (i.e. nano-newtonmeters), which is M < 0.01μNm = 10^{-8}Nm. Using DSO therefore allows testing of materials which are very shear-sensitive, e.g. showing a limit of the linear viscoelastic (LVE) deformation range of $\gamma < 10^{-4}$ = 0.01% as may occur when testing solutions of surfactants. Before DSO was available, it was impossible to analyze these kinds of samples under scientifically satisfying test conditions.

In order to carry out the adjustment procedure as fast as possible, **modern rheometers are equipped with digital signal processors (DSP)**, which merely require milliseconds for each closed control loop. Therefore, today's servo control techniques are no longer a limiting factor for the response time of a control system. However, an adjustment to a desired preset value takes a little longer since it is not only dependent on the intelligence of the controller but also on disturbances like friction and inertia contributions of the sample (see also Chapter 11.6.2a). Example: Using **adaptive controllers** and a synchronous motor, the desired value usually can be adjusted within t = 30ms to 1% of the desired value without any significant deviation and without overshoot, independent of the mode of testing, i.e., whether angle/speed-controlled or torque-controlled, and it is solely to a certain degree dependent on the behavior on the sample.

11.6.3 Devices to measure torques

The unit of torque M is newtonmeter [Nm].
It holds: 1Nm = 1000mNm (milli-Nm) = 10^6μNm (micro-Nm) = 10^9nNm (nano-Nm). Previously in use: 1cm · p = 0.0981mNm = approx. 0.1mNm, or 1N · cm = 10mNm.
For exact measurements, the unavoidable internal friction of an instrument, e.g. due to the bearing friction has to be determined by a run at no-load (see also Chapter 11.6.5). Then, it might be compensated either by an internal control procedure or by calculation.

a) Torque measurements using mechanical force sensors
For one of the first rotational viscometers, designed by M.M. Couette in 1888, a torsion wire was taken as the torque sensor. A small mirror was attached to this wire and, using a light beam directed onto the mirror, the radian measure even of relatively small torsion angles was made clearly visible (see also the example in Chapter 11.2.11c3). Still today, mechanical force sensors in the form of spring elements such as **spiral springs, torsion bars or torsion wires** are used in most simple viscometers. Typical for these kinds of **torsion elements** is that their deflection s or deflection angle φ is proportional to the applied force F or torque M, respectively. Mechanical force sensors are measuring relative inaccurately since a test result always includes also the behavior of the force sensor itself. Usually here, a relatively large deflection is required to achieve evaluable results. Therefore nowadays, torsion elements are only used for simple and cheap viscometers.

The **elasticity law of Hooke** holds for all kinds of mechanical force sensors:

Equation 11.27 $F = C_H \cdot s$

with the applied force F [N], the Hookean spring constant C_H [N/m], and the deflection s [m]

Simple viscometers are often displaying values of the relative torque M_{rel} in percent or in per thousand. In this case, the values are related to the maximum value of the measurable torque ($M_{rel, max}$ = 100%), and then, lower M-values are specified in terms of a percentage, for example, as M_{rel} = 10%.

b) Torque measurements via power consumption of an electromotor
For modern rheometers, measuring drives are used to determine the **torque via the consumption of the required operating current** I [A]. Here, for example, the following applies:

$M = c_1 \cdot I$ or $M = c_2 \cdot I^2$ with the instrument-specific constants c_1 and c_2

Preset and detection takes place at a negligible deflection of the detector system. Therefore here, also high torque values can be determined even at very small deflection angles, e.g. when investigating very rigid solids bars at extremely low deformation values.

11.6.4 Devices to measure deflection angles and rotational speeds

When applying a torque, a corresponding deflection angle φ or rotational speed n is occurring at the motor shaft depending on the sample's behavior. **Angles can be specified in radians [rad] or in degrees [°].** In order to convert between the two, a complete revolution corresponds to the angle: 2π rad = 360°   or   1 rad = $360°/2\pi$ = 57.3°

Further applies: 1rad = 10^3mrad (milli-rad) = 10^6μrad (micro-rad) = 10^9nrad (nano-rad)

When measuring the deflection angle and the corresponding time interval which is passing when covering this angle, also the speed of the rotor can be calculated, which alternatively can be specified in terms of the rotational speed n [min^{-1}] or the angular velocity ω [rad/s or s^{-1}]. For the conversion applies:

Equation 11.28 $\omega = (2\pi \cdot n)/60$

For the deflection angle φ [rad] and ω [rad/s] holds:

Equation 11.29 $\omega = \varphi/t$ or $\omega = \Delta\varphi/\Delta t$ or $\omega = d\varphi/dt = \dot{\varphi}$

with the change of the deflection angle $\Delta\varphi$ [rad] in the time interval Δt [s], and the time derivative of the deflection angle, as $d\varphi/dt = \dot{\varphi}$ [rad/s]

Note: The units of ω in terms of [rad/s] or [s^{-1}]
Many users ignore the term "rad" when using the unit of ω, and take [s^{-1}] instead of [rad/s]. This is possible when relating the concerning calculations to the so-called "unity circle" showing the radius r = 1, which is often used in mathematics and physics. According to the standards, the use of both units is correct. The unit "rad" indicates that here, when performing rotational or oscillatory motion in contrast to linear motion: The deflection path at the edge of the measuring system occurs in the form of a circular line when covering a rotational angle.

a) Measurements of rotational speeds using tachogenerators
A tachogenerator is a tool to measure the rotational speed (Greek táchos means speed). These kinds of sensors are operating on the principle of an "inverted electromotor". An electric current is generated via induction in an electrically conducting coil when rotating in a magnetic field. The value of the occurring current depends on the value of the rotational speed. Tachogenerators are measuring relative **inaccurately in the lower speed range**, and they are not suited to determine static deflection angles since a current is only induced if the rotor actually is in motion. For this reason nowadays, tachogenerators are only used for simple and cheap viscometers.

b) Measurements of deflection angles and rotational speeds using incremental encoders
In modern rheometers as **position sensors, opto-electronic** incremental encoders are used. The rotor axis of such an encoder carries a **sectored disk** showing a large number of strips, which are evenly distributed along the circumference at the edge of the disk. The strips on the disk can take the form of slits or small mirror areas. When the rotor is deflected at a certain angle, a laser beam scans the exact position of the rotor. In order to do this, the number of strips which are covered is counted in the form of light pulses. This procedure is called an **incremental**, i.e., a **counting measurement**. Using an incremental encoder, the deflection

angle φ of the measuring system can be determined as the rotation angle of the rotor. Then, the rotational speed might be calculated from this if also the corresponding time interval is taken into account to cover this deflection angle.

11.6.5 Bearings

Bearings are required wherever a rotation axis and a stationary housing have to be connected. In technical applications, for example, ball bearings are taken for this purpose. Practical users should always be aware that the **total torque** M_{tot} [mNm]which is preset or measured by a rheometer may consist of several components:

$M_{tot} = M + M_f + M_i$

Here, M is the torque deriving from the resistance of the sample when it is deformed or when flowing; this term of the formula should be clearly the major part of the total torque. M_f occurs due to **internal friction effects of the rheometer** used. M_i originates from **inertia effects** of the total mass set in motion, i.e., from the rotor of the drive, motor shaft and coupling, measuring bob, but also from the sample. Internal friction effects of a rheometer may have several reasons, e.g. due to bearing friction.

When using a conventional motor type, there is dynamic friction due to the direct contact of stator and rotor between the current supplying and current receiving components of the drive, for example, in the form of brushes of a mechanical commutator system. Rheometer designers aim to keep as low as possible as well the amount of the instrument's internal friction as well as the inertia effects of rheometer and measuring geometries.

a) Mechanical bearings (e.g. ball bearings)

Most viscometers and simple rheometers used for QC tests are equipped with **ball bearings or roller bearings**. Above all in the low-torque range, friction of the balls or cylindric rollers in the bearing cage occurs in the form of **static friction** at the beginning of the motion, **and** as **sliding or dynamic friction** when the rotation is continuously going on. Of course, this leads to limitations or even to errors in the test results. Therefore, at first the instrument manufacturer and later also the user should perform from time to time measures to compensate possible changes of the instrument's friction values, e.g. by performing a **zero-point adjustment** of the instrument.

Directly after an adjustment, according to manufacturers' specifications, values of the minimum torque of ball bearing devices can be expected between M_{min} = 0.05 and 0.5mNm. Usually however, reproducible and useful test results within an acceptable tolerance range of deviation should merely be expected for values which are measured above M_{min} = 0.25mNm. Correspondingly, these values of M_{min} are not sufficient when testing low-viscosity liquids, since here, when presetting low and medium shear rates, of course, only very low shear stress values are occurring. In this case, the torque values resulting from internal friction of the viscometer are often showing the same or even a higher level compared to the torque values caused by the rheological behavior of the sample. This leads to a correspondingly high uncertainty when analyzing these kinds of test results. Therefore, for more and more users there is a trend towards the use of air bearing rheometers, even when performing QC tests.

b) Air bearings

Rheometers equipped with air bearings are showing considerably lower bearing friction compared to ball bearing instruments, since with the former, the rotating parts of the rheometer are floating on an "air bed". The air gap between the rotor and the stationary components (stator) is maintained due to a continuous supply of compressed air.

Example: Operating data of a typical air bearing

The required external **air pressure** is p = 0.5MPa (= 5bar), supplied by a mobile or stationary compressor. This leads to an internal air pressure in the bearing of about p = 0.25MPa

(= 2.5bar). Air consumption is $\dot{V}$ = 10Sl/min = 1.7 · 10^{-4}m³/s (Sl means standard liters, referring to normal conditions). **Cleanliness of the air** has to be guaranteed at a maximum particle size of 0.1μm and a maximum oil concentration of 0.01mg/m³. The dew point should be below T_d = -15°C. For cleaning and drying, an **air filter** and **air dryer** equipped with a dust and oil separator should be installed in the air supply line.

Of course, direct mechanical contact between parts in motion and stationary parts of the air bearing should never occur since this might result in damage. However, many viscoelastic samples are showing significant normal stresses when testing, therefore pushing the rotor upwards which probably may cause the gap in the air bearing to become narrower (see also Chapter 5.3: normal stresses). To be on the safe side, the gap dimension within the air bearing should be checked continuously, e.g. by an electric capacity detector. In order to prevent damage, the overpressure of the air supply has to ensure a certain degree of **stiffness S of the bearing, or the compliance (1/S)**, respectively, has to be kept within certain limits.

Example: In axial direction S ≥ 7.5N/μm or (1/S) ≤ 0.13μm/N, respectively. This example shows that the stiffness of an air bearing usually is clearly higher compared to the stiffness of a total rheometer system including a measuring system.

According to manufacturers' specifications, the limiting value of a relative faultless torque detection of air bearing rheometers is around M_{min} = 0.1μNm. In fact, each user should perform comparative tests, e.g. on water samples or calibration oils, to realistically evaluate these kinds of specifications since the error is increasing when approaching towards the range of lower torques. Here, the error tolerance band shows more and more the shape of a trumpet's bellmouth ("error trumpet"). If M_{min} is taken as 0.25μNm, the sensitivity of air bearing instruments to torque is 1000 times better compared to ball bearing devices (see above, Chapter 11.6.5a). This means for users, that typical air bearing instruments are able to detect torque values which are at least 1000 times lower in comparison. As a consequence, it may be possible to reach a range of correspondingly lower deflection angles or rotational speeds, respectively. This means: When using an air bearing rheometer, rheological parameters can usually be measured in a range of six or more decades even when using only a single measuring system. And when using a rheometer with a synchronous drive and the test mode "direct strain oscillation" DSO, even clearly lower torque values may be achieved (see also the Note in Chapter 11.6.2b).

Air bearings are precision parts, and the following **criteria** should be optimally fulfilled:
1) Minimum internal friction
2) **Constant internal friction:** The internal friction should be proportional to the rotational speed (rotational tests) or to the frequency (oscillatory tests), respectively. This condition must be fulfilled independent of the position of the rotor, i.e., independent of the deflection angle. There should be no "windmill effect", i.e., no position-dependent torque, for example, due to minute scratches on the surface of the bearing, since this would lead to an uncontrolled rotational motion of the rotor caused solely by the flowing air. This effect must be prevented or compensated as much as possible.
3) Good concentricity
4) High stiffness, as well against forces acting from the side (lateral, horizontal) as well as in the axial (vertical) direction.

℘   End of the Cleverly section

11.6.6 Temperature control systems

When performing rheological tests, the **measuring temperature T_M** is one of the most influencing test parameters. It should be always taken into account that a certain difference may occur between the desired T_M and the actual temperature of the sample. For this reason, careful temperature control is essential before the start of each test. This interval should be set to at least 5 to 10min, in order to achieve sufficient temperature equilibration. Using special equipment such as insulating or heating hoods, the temperature gradient can be kept at a minimum. For fast tests, a **heating or cooling rate** might be selected between $\Delta T/\Delta t = 1$ and 4K/min (e.g. 2K/min). For exact measurements, however, the heating rate should not be higher than $\Delta T/\Delta t = 1$K/min (or even better, only 0.5K/min) in order to ensure good temperature equilibration of the complete sample (see also Chapter 8.6: temperature program). The following heating and cooling methods are in use:

a) Liquid bath

The bath container is filled with **water** or **thermo-oil**, to be used **for tests at room temperature** or at another constant medium temperature, and for tests performed at low heating or cooling rates. These devives are often termed by users as **"thermostats"**. According to manufacturers' specifications, there are systems available for -40°C $< T_M <$ +200°C. If only one part of the measuring system is heated (usually the cylinder cup or the bottom plate), an increased temperature gradient in the sample should be taken into account if T_M is higher than +40°C, and the same applies to low temperatures if T_M is lower than +10°C.

b) Electric heating

Electric heating by radiation is used for high heating rates and for tests performed at constantly high measuring temperatures. According to manufacturers' specifications, systems are available in the temperature range of +30°C $< T_M <$ +400°C, and for special applications such as melts of glasses, metals, ceramic composites and basaltes, even up to T = 1600°C.

c) Convection oven for heating and cooling with gas

In convection ovens, a gas takes the function of a transfer medium for the temperature when heating or cooling the sample. Here, usually **nitrogen** is used as an inert gas, if there is the risk of oxidation when using air. For tests performed at low temperatures, the gas is taken from a container, e.g. from a vacuum-insulated Dewar container, in which nitrogen is stored in the liquid form (*James Dewar*, 1842 to 1923). Here, the temperature of the test chamber is controlled indirectly **via convection** to achieve **high heating and cooling rates in a very wide temperature** range of, for example, -150°C $< T_M <$ +600°C. For special applications, systems are available up to T_M = 1000°C.

d) Induction heating

Induction heating is used to obtain high heating rates, and for tests at constantly high test temperatures. The temperature of the whole measuring system is indirectly controlled via electrical coils which surround it. The measuring systems which are usually concentric cylinders, are made of conductive copper which have to be coated with a hard layer (e.g. with nickel) due to its softness. Surrounding non-conductive materials are not heated. For low temperatures, liquid nitrogen is used. According to manufacturers' specifications, systems are available for -100°C $< T_M <$ +500°C. Due to the soft and scratch-sensitive cylinder surfaces, this heating method is not useful for daily practice and therefore rarely used in industrial labs.

e) Peltier element

The Peltier effect is an inverse thermo-electrical effect (*Jean C. A. Peltier*, 1785 to 1845). **The advantage** of this method is the option to achieve as well **high heating rates** as well as **high cooling rates**. Further advantages are **the small size** and **the comparably low purchase and operating costs** of this system.

A Peltier element may be imagined as a thermo-electric "heat pump". It consists of two different materials, e.g. either two metals or two semi-conductor materials. The latter may generate a higher Peltier coefficient. The two components are set up in the form of layers. A contact for an external electric voltage supply is attached to each layer. Ever then, when an electric

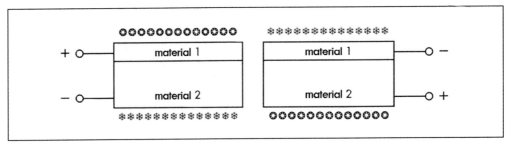

Figure 11.9: The Peltier effect: Reversing the direction of voltage causes reversing of the temperature conditions in the contact area of two corresponding materials

current is flowing, the contact area of the layers is heated or cooled dependent on the direction of the current, see Figure 11.9.

For the generated or dissipated heat energy Q [J] holds:

Equation 11.30 $Q = \Pi \cdot I \cdot t$

with the electric current I [A], time t [s], and the Peltier coefficient Π [J/A $\cdot$ s] which is a material constant; the sign Π is the capital π sign, pronounced: "pi" or "pee".
Typical values are: $\Pi = 4 \cdot 10^{-4}$ to $4 \cdot 10^{-3}$ J/As

The process is completely reversible when reversing the direction of the current. Therefore, if the same quantity of current is flowing in the same period of time, the same degree of heating or cooling capacity will be achieved.

When a certain voltage is applied to the two materials of a Peltier element, and subsequently, a certain amount of current is flowing through it, the element will react immediately, generating a temperature gradient between the two surfaces of a corresponding double layer. Maintaining one area at a constant temperature, the so-called **counter temperature T_C**, the other area might be adjusted to the desired measuring temperature T_M by controlling the current intensity (amperage). The rheometer should be designed in a way to make sure that the temperature-controlled area is mounted very close to the surface of the measuring system and sample, respectively. When performing **tests at higher temperatures, counter-cooling is required** which might be guaranteed by connecting a small and simple water bath, e.g. using just a cheap aquarium pump, or alternatively, a small amount of flowing water.

Materials of Peltier systems are temperature-sensitive, and therefore, they can only be used in a limited temperature range. According to manufacturers' specifications, systems are available for $-40°C < T_M < +200°C$. However, as a typical range for long-term operation in industrial labs is recommended $-30°C < T_M < +150°C$. **A semi-conductor material may show aging effects with time when used continuously at high temperatures** since it then gradually loses the ability to cool. Sometimes in publications, even a wider temperature range is specified for the use of Peltier elements. However, extreme temperatures should only be set for short test intervals, otherwise the effectiveness of the Peltier effect will change and there is a risk of damage. Therefore, the system should be equipped with a safety control continuously checking the temperature on both contact areas, switching the system off when exceeding the minimum or maximum temperature limit.

Note: Information on counter cooling temperatures when using Peltier elements
The difference between the counter temperature T_C and the measuring temperature T_M should not be too large to keep the required power input of the Peltier element at a low level. In order to achieve useful operating conditions, the **conditions** recommended by manufacturers of the Peltier elements should be kept. **Examples:**

1) When controlling low temperatures: $T_C \leq T_M + 35K$
Example 1: When $T_M = -30°C$, then $T_C \leq +5°C$ is required, thus: $T_{Cmax} = +5°C$.

2) When controlling high temperatures: $T_C \geq T_M - 130K$
Example 2: When $T_M = 150°C$, then $T_C \geq +20°C$ is required, thus: $T_{Cmin} = +20°C$.

Peltier elements are available as well for cone-and-plate (CP) and parallel-plate (PP) as well as for concentric cylinder (CC) measuring systems. In order to keep the temperature gradient in the sample at a minimum, **for CP and PP geometries in addition to a Peltier temperature-controlled bottom plate also a Peltier temperature-controlled hood** might be used which is controlled by a second Peltier element. Such a cover encloses the whole measuring area, **providing additional temperature control**. This second temperature control unit might be installed in the upper and side walls of the hood in order **to minimize the temperature gradient** within the whole measuring space, and therefore, also within the sample. Sometimes, this kind of cover is called an **"active hood"** for temperature control, to distinguish it from an open system without any hood, or from a system equipped with a "passive hood" without heating or cooling capacity.

12 Guideline for rheological tests

Situation: You got a sample which should be characterized by a rheological test, but you have no idea how to do this. In this chapter proposals are given for typical test presets, evaluation methods and analysis diagrams in order to help beginners in the field of rheology. The following specifications may be good for many materials, but of course due to the variety of samples to be tested in the different industrial branches, the user has to change here and there some details in order to get useful test results.

Rheological parameters, signs and abbreviations used in this chapter:
viscosity η [Pas], shear stress τ [Pa], shear rate $\dot{\gamma}$ [s^{-1}], strain or deformation γ [%], angular frequency ω [rad/s], storage modulus G' [Pa], loss modulus G" [Pa], complex viscosity $|\eta^*|$ [Pas], loss factor $\tan\delta$ [1] (= G"/G'); scaling of diagrams: lin (linear), log (logarithmic); LVE-range (linear viscoelastic range)

12.1 Selection of the measuring system

a) **Concentric cylinders** for low-viscosity liquids, and for fast drying samples
b) **Parallel-plates** for samples containing particles larger than 5µm, and for highly viscous and viscoelastic materials such as polymer melts
c) **Cone-and-plate** for all other samples

12.2 Rotational tests

12.2.1 Flow and viscosity curves

Preset for a first check of the sample: shear rate ramp (lin or log): $\dot{\gamma}$ = 1 to 500s^{-1} (Figures 3.1 and 3.2)

Measuring results and analysis
Diagrams on a lin scale (Figures 3.26, 3.27, 3.28), or on a log scale (Figures 3.29, 3.30, 3.31), showing the following flow behaviors:
- (1) ideally viscous (Newtonian)
- (2) shear thinning (pseudoplastic)
- (3) shear thickening (dilatant)
- (4) without a yield point
- (5) with a yield point

Determination of the flow curve function, using a flow curve fitting and analysis model on a lin scale (Chapter 3.3.6.1/2) for (1) Newton, for (2) and (3) Ostwald/de Waele, for (5): Calculation of the yield point using the models e.g. according to Bingham, Casson or Herschel/Bulkley. Further tests:

a) If no yield point exists (flow curve types 1 to 4)
Preset: Shear rate ramp (log), in an extended range of $\dot{\gamma}$ = 0.5 to 1000s^{-1} (when using a viscometer with a ball bearing), or $\dot{\gamma}$ = 0.01 to 1000s^{-1} (when using a rheometer with an air bearing).

Thomas G. Mezger: The Rheology Handbook
© Copyright 2011 by Vincentz Network, Hanover, Germany
ISBN 978-3-86630-864-0

Note: To prevent time-dependent (transient) effects, for $\dot\gamma < 1s^{-1}$, i.e. in the low-shear range, holds: The duration for each single measuring point should be t $\geq 1/\dot\gamma$ (see Chapter 3.3.1b, Figure 3.5).

Polymers: Results, and determination of the zero-shear viscosity η_0, for unlinked and unfilled polymers, and polymer solutions showing a sufficiently high concentration
1) The plateau value η_0, e.g. between $\dot\gamma$ = 0.01 and $0.1s^{-1}$ (Figure 3.10),
and, if occurring, also the infinite-shear viscosity η_∞ at high shear rates
2) Calculation of the values of η_0 and η_∞ using a (log) viscosity curve fitting and analysis model, e.g. Carreau/Yasuda, or Cross (see Chapter 3.3.6.3)
Note 1: η_0 is proportional to the average molar mass, thus: $\eta_0 \sim M$ (see Chapter 3.3.2.1a).
Note 2: Using the Cox/Merz relation (see Chapters 8.8 and 12.3.2c)

b) If a yield point exists (flow curve type 5)
Preset: Shear stress ramp (log): M = 0.5 to 5mNm (with ball bearing), or M = 0.5µNm to 5mNm (with air bearing); similar to Figures 3.1 and 3.2, however here, with controlled shear stress

Measuring results, and determination of the yield point τ_y by rotational tests:
1) The yield point is directly read-off from a log/log flow curve diagram as the (almost) constant τ-value in the low shear range, i.e. at $\dot\gamma < 1s^{-1}$, e.g at $\dot\gamma = 0.01s^{-1}$ (Figure 3.22),
2) Using the method of the fitting straight lines, presentation in a diagram lg γ/lg τ on a log scale: The yield point as the limit of the linear-elastic range (Figure 3.23), or using the "tangent crossover method" (Figure 3.24)

12.2.2 Time-dependent flow behavior (rotation)

Preset: Constant shear rate, $\dot\gamma = 1s^{-1}$ (with ball bearing) or $\dot\gamma = 0.1s^{-1}$ (with air bearing); as in Figure 3.36.

Measuring results: Time-dependent viscosity function (Figure 3.39), showing the variants: (1) no viscosity change, (2) decreasing viscosity, e.g. due to time-dependent shear-thinning or viscous shear-heating, (3) increasing viscosity, e.g. due to gelation, hardening, curing, drying

Note: For testing time-dependent behavior, oscillatory tests should be preferred since this kind of testing is resulting in more detailed information (see Chapter 12.3.3).

12.2.3 Step tests (rotation): structural decomposition and regeneration ("thixotropy")

Preset: Three test intervals $\dot\gamma = 1/100/1s^{-1}$ (with ball bearing) or $\dot\gamma = 0.1/100/0.1s^{-1}$ (with air bearing); as in Figure 3.42.
Measuring results, and analysis methods (see Figure 3.43); many users select method M4:
M1) The "thixotropy value": Change of viscosity ($\Delta\eta$) between the time points t_2 and t_3
M2) The "total thixotropy time": The period between t_2 and the time point of complete regeneration, i.e. when reaching 100% of the reference value of the viscosity-at-rest from interval 1
M3) The "thixotropy time": The period required between t_2 and the time point to reach a defined percentage of regeneration, e.g. 75% of the reference value of interval 1
M4) Percentage of regeneration between t_2 and a defined time point of interval 3, e.g. after t = 60s, or after another **"time related to practice"** to be previously defined by the user

Note: For testing time-dependent structural regeneration behavior ("thixotropy"), step tests in oscillatory/rotational/oscillatory mode should be preferred since this kind of testing is resulting in more detailed information (see Chapter 12.3.4).

12.2.4 Temperature-dependent flow behavior (rotation)

Preset: Temperature program, e.g. as a ramp, at a constant shear rate $\dot{\gamma} = 1s^{-1}$ (with ball bearing) or at $\dot{\gamma} = 0.1s^{-1}$ (with air bearing); as in Figure 3.36.

Measuring results and analysis:
1) Typical behavior: When heating, the temperature-dependent viscosity values are decreasing due to softening or even melting (Figure 3.47); and when cooling, the viscosity values are increasing due to hardening, crystallizing or freezing, respectively.
If desired, analysis using the Arrhenius model: Determination of the flow activation energy E_A and of the temperature shift factor, which may be used to calculate viscosity values even for temperatures at which no measuring data are available.
2) Temperature-dependent viscosity functions of gelling, hardening or curing materials (Figure 3.48), showing the minimum viscosity value η_{min} at the onset of the reaction. **Note: For testing temperature-dependent behavior, oscillatory tests should be preferred** since this kind of testing is resulting in more detailed information (see Chapter 12.3.5).

12.3 Oscillatory tests

The following information on oscillatory tests is only valid when using an **air-bearing rheometer**.

12.3.1 Amplitude sweeps

Preset: (log) Ramp of the strain (deformation) amplitude: $\gamma = 0.01$ to 100% and $\omega = 10rad/s$; as in Figure 8.8.

Measuring results and analysis:
1) The limit of the linear-viscoelastic (LVE) range at γ_L as the limiting value of deformation (Figures 8.9 to 8.11: presentation versus γ), or at τ_y as the yield stress (Figure 8.12: presentation versus τ), respectively.
2) Viscoelastic character in the LVE range; question: Is there $G' > G''$, therefore showing gel-like behavior like a viscoelastic solid (Figures 8.9, 8.11 and 8.12); or is there $G'' > G'$, therefore showing the character of a viscoelastic liquid (Figure 8.10)?
3) The G'-value as "gel strength" in the LVE range (if $G' > G''$)
4) The value of the flow stress τ_f at the crossover point $G' = G''$, i.e., when a material begins to flow (Figure 8.12)
5) The yield zone between yield point τ_y and flow point τ_f, when a sample still exhibits gel-like character since $G' > G''$, however, showing merely **partially** reversible deformation behavior (Figure 8.12)

12.3.2 Frequency sweeps

Preset: Frequency ramp (log), e.g. with $\omega = 100$ to $0.1rad/s$ and $\gamma = 1\%$; as in Figure 8.15. Attention: The measurement is performed using a γ-value of the LVE range, i.e. **after a previously performed amplitude sweep** to determine the limit of the LVE range.

a) Polymers: Measuring results and analysis, using the curves of G' and G''
1) Viscoelastic behavior at low frequencies, i.e., long-term behavior (Figure 8.21). Question: Is the sample cross-linked, showing $G' > G''$, or not, showing $G'' > G'$? And if desired: Behavior at high frequencies, i.e., short-term behavior
2) At low frequencies: Maxwellian behavior if the G'-curve shows the slope $2:1$, and the G''-curve the slope $1:1$ (Figure 8.16, i.e., presentation on a log scale)

3) Comparison of unlinked polymers: Value of ω_{co} at the cross-over point G' = G'' which is dependent on the average molar mass M (Figure 8.19)
4) The plateau value G_P at high frequencies (Figure 8.16)
5) Cross-linked polymers at low frequencies: The higher the plateau value of G', the higher is the degree of cross-linking (Figures 8.20 and 8.22)

b) Polymers: Measuring results and analysis, using the curves of η^*

The plateau value of the zero-shear viscosity η_0 at low frequencies, e.g. in the range between ω = 0.01 and 0.1rad/s (Figure 8.17, i.e., presentation on a log scale)
Note: η_0 is proportional to the average molar mass: $\eta_0 \sim$ M (see Chapters 8.4.2.1 and 3.3.2.1a).

c) Further analysis with polymers

1) The Cox/Merz relation for polymer melts and solutions (see Chapter 8.8 and Figure 8.47): The values of $\eta(\dot{\gamma})$ from rotational tests and of $|\eta^*(\omega)|$ from oscillatory tests are identical in the low shear range. However, this conversion is true for Maxwellian fluids only, i.e., for unlinked polymers.
2) Time-temperature superposition (TTS), master curve according to Williams, Landel and Ferry (WLF, see Chapter 8.7 and Figures 8.45 and 8.46)
3) Molar mass distribution (MMD, Figure 7.11) using a special software program

d) Dispersions and gels: Measuring results and analysis

1) Evaluation at low frequencies, e.g. between ω = 0.01 and 0.1rad/s. Question: Is there G' > G'', therefore showing gel-like behavior and long-term stability; or is there G'' > G', therefore showing instability and phase separation, sedimentation or flotation?
2) Structural strength at rest in terms of the G'-value at low frequencies (Figure 8.22)
3) Gel-hardening which may lead to syneresis if G' >> G'', i.e., if the value of tanδ (= G''/G') is too low

12.3.3 Time-dependent viscoelastic behavior (oscillation)

Preset: Constant deformation (as in Figure 8.26), e.g. at ω = 10rad/s and γ = 1%, using a γ-value of the LVE range, i.e. **after a previously performed amplitude sweep** to determine the limit of the LVE range

Measuring results (Figure 8.27)
(1) time-independent behavior, (2) increase and (3) decrease of structural strength

Samples showing gelation, hardening, curing: Measuring results and analysis
1) Curing samples with t_{CR} at the onset of a chemical reaction, as time-dependent G'(t)-curve (Figure 8.35)
2) The sol/gel transition time at t_{SG} as the "gelation time" or the "gel point" (Figure 8.35)
3) The final values of G' and G'' when the reaction is finished

12.3.4 Step tests (oscillation): structural decomposition and regeneration ("thixotropy")

Preset, alternatives:
a) Three test intervals: oscillation/rotation/oscillation (as in Figure 8.32)
1) Interval 1: oscillation, at a constant strain, e.g. at γ = 1% and ω = 10rad/s, using a γ-value of the LVE range, i.e. **after a previously performed amplitude sweep** to determine the limit of the LVE range
2) Interval 2: rotation, e.g. at $\dot{\gamma}$ = 100s⁻¹,
3) Interval 3: oscillation, at the same shear conditions like in interval 1

b) Three intervals: all in oscillation (as in Figure 8.30)
1) Interval 1: oscillation, at a constant strain, e.g. at γ = 1% and ω = 10rad/s, using a γ-value of the LVE range, i.e. **after a previously performed amplitude sweep** to determine the limit of the LVE range
2) Interval 2: oscillation, at a constant strain, e.g. at γ = 100% and ω = 10rad/s, using a γ-value knowingly outside of the LVE range
3) Interval 3: oscillation, at the same shear conditions like in interval 1

Measuring results and analysis methods (Figures 8.31, 8.33 and 8.34); many users select method M4
M1) The "thixotropy value": Change of the G'-value (ΔG') between the time points t_2 and t_3
M2) The "total thixotropy time": The period between t_2 and the time point of complete regeneration, i.e. when reaching 100% of the reference value of G'-at-rest from interval 1
M3) The "thixotropy time" as the period required between t_2 and the time point to reach a defined percentage of regeneration, e.g. 75% of the reference value of interval 1
M4) The "thixotropy time" until the crossover point G' = G" is reached
M5) Percentage of regeneration between t_2 and a defined time point of interval 3, e.g. after t = 60s, or after another **"time related to practice"** to be previously defined by the user

12.3.5 Temperature-dependent viscoelastic behavior (oscillation)

Preset: Temperature program, e.g. as a ramp, at a constant strain, e.g. at γ = 1% and ω = 10rad/s, using a γ-value from the LVE range, i.e. **after a previously performed amplitude sweep** to determine the limit of the LVE range

a) Polymers: Measuring results and analysis
1) Overview: see Figure 8.42; and in detail referring to the spatial configuration of the macromolecules: (1) amorphous (Figures 8.36 and 8.37), (2) partially crystalline (Figures 8.38 and 8.39), (3a) with a high degree of cross-linking as with thermosets (Figures 8.40 and 8.41), and (3b) with a low degree of cross-linking as with elastomers
2) Important temperatures:
2a) The glass transition temperature T_g (optional determinations: either at the peak of the G"-curve or at the peak of the tanδ-curve; see Chapter 8.6.2.1a)
2b) The melting temperature T_m at the crossover point G" = G', showing afterwards G" > G' (Figure 8.39)
Note: Further analysis option for polymers: time/temperature shift (TTS) via frequency sweeps, and master curve according to Williams, Landel and Ferry (WLF, Figures 8.45 and 8.46, see also Chapter 12.3.2c2)

b) Crystallizing samples: Measuring results and analysis
Crystallization temperature T_k (when cooling) or melting temperature (when heating), respectively, both when reaching the crossover point G' = G" (Figure 8.43)

c) Gelation, hardening, curing: Measuring results analysis
1) Important temperatures (Figure 8.44):
1a) The melting temperature T_m when G' = G" (here as transition from G' > G" to G" > G')
1b) The temperature T_{CR} at the onset of a chemical reaction (minimum of the G'-curve)
1c) The sol-gel transition temperature T_{SG} when G' = G" (here as transition from G" > G' to G' > G"), also termed as the "gelation temperature" or "gel point"
2) The final values of G' and G" when the reaction is finished

12.4 Selection of the test type

For people working in an industrial laboratory it is important to have as much information as possible on the following aspects of a sample to be tested, [101, 251]:

• behavior at rest
• flow behavior
• time-dependent structure regeneration after applying a high shear load

12.4.1 Behavior at rest

Here, the sample should be tested under conditions coming close to the state of rest.

a) Determination of the yield stress value τ_y (or "yield point") and the flow stress value τ_f (or "flow point")
Note: Neither yield stress values nor flow stress values are material constants since they depend as well on the measuring conditions as well as on the analysis method.

Optional modes for testing and analysis (the scientific methods are M5a and b).
1) Rotational tests
M1) **Flow curves** with controlled shear rate or controlled shear stress; preset as in Figures 3.1 or 3.2, respectively. Flow curve presentation on a lin scale: The yield point is read off at the intersection point on the τ-axis (see Figure 3.21)
M2) **Flow curves** with controlled shear rate or controlled shear stress; preset as in Figures 3.1 or 3.2, respectively. Flow curve presentation on a lin scale: The yield point is calculated using a flow curve fitting model (e.g. Bingham, Figure 3.32; or Casson, or Herschel/Bulkley; see Chapter 3.3.6.4),
M3) **Flow curves** with controlled shear stress or controlled shear rate; preset similar to Figures 3.1 or 3.2, respectively. Flow curve presentation on a log scale: If a "plateau value" is occurring almost in parallel to the x-axis, the yield point is read off in the range of low shear rates (see Figure 3.22), for example, read-off at a previously defined low shear rate value (e.g. at $\dot{\gamma} = 0.01s^{-1}$)
M4) Test with controlled shear stress, preset similar to Figures 3.1 or 3.2, respectively. Presentation in a lg τ/lg γ diagram on a log scale, analysis using one or two fitting straight lines ("tangents"); alternatives:
M4a) Using a single tangent: The yield point as the upper limiting stress value of the linear elastic range (Figure 3.23)
M4b) The yield point at the "tangent crossover point" (Figure 3.24)

2) Oscillatory tests
M5) **Amplitude sweeps**; preset as in Figure 8.8; presentation versus τ (Figure 8.12), precondition: G' > G'' in the LVE range, and therefore, a "gel-like character" is occurring.
M5a) The **yield stress** τ_y is determined at the limit of the LVE range; for $\tau < \tau_y$ holds: This is the range of reversible-elastic deformation behaviour; here, the sample shows gel-like character.
M5b) The **flow stress** τ_f is determined at the crossover point G' = G''; for $\tau > \tau_f$ holds: This is the flow range, here, the sample shows the character of a liquid.

Note 1: For $\tau_y < \tau < \tau_f$ which is the range between yield stress and flow stress or the **"yield zone"**, respectively, holds: The sample still shows gel-like character, however, there may occur partially irreversible elastic deformation behavior now.

Note 2: The result may depend on the preset (angular) frequency which might even change the kind of the viscoelastic character. However, if in this case liquid-like behavior occurred **with G'' > G' in the LVE range**, there would be **indeed a yield point but** of course **no flow point** available.

b) If there is no flow point: the plateau value of the zero-shear viscosity η_0

Analysis of the function of the shear viscosity or complex viscosity in the low shear rate range, on a log scale; optional methods:

M1) **Rotational tests**: Preset as in Figures 3.1 or 3.2; analysis e.g. in the range of $\dot{\gamma}$ between 0.01 and $0.1s^{-1}$ (Figure 3.10)

M2) **Oscillatory tests, frequency sweeps**: Preset as in Figure 8.15; analysis e.g. in the range of ω between 0.01 and $0.1rad/s$ (Figure 8.17)

Note: The result may depend on the preset (angular) frequency which might even change the kind of the viscoelastic character. However, if in this case gel-like behavior occurred with G' > G'' in the range of low frequencies, there would be of course no longer an η_0-plateau value available.

c) Structural strength in terms of the G'-value

Performing amplitude sweeps; preset as in Figure 8.8; evaluation of the plateau value in the LVE range. If there is G' > G'', and therefore gel-like character (as in Figures 8.9, 8.11 and 8.12), the G'-value often is also called **"gel strength"**. Please note: The G'-value may depend on the frequency applied, and it is often increasing when presetting higher ω-values.

d) Long-term stability of dispersions and gels

Performing frequency sweeps within the LVE range; preset as in Figure 8.15; evaluation at a low frequency value, e.g. at $\omega = 0.01rad/s$ (Figures 8.21 and 8.22). A stable gel structure can be expected if G' > G'' and G' > 10Pa, which usually is enough to prevent sedimentation (see also Note 1 of Chapter 8.4.4a).

Note: Syneresis may occur if G' $\gg$ G'' since the gel body might be too rigid and inflexible then (see also Note 3 of Chapter 8.4.4a).

e) Material characterization: behavior at rest

Performing frequency sweeps within the LVE range; preset as in Figure 8.15; evaluation of the kind of viscoelastic character at rest, i.e., at a low frequency value (e.g. at $\omega = 0.01rad/s$). Question: Is there G' > G'' (gel-like structure, network of forces or chemical network, solid state; Figure 8.21: on the right-hand side), or G'' > G' (fluid character, liquid or sol state; Figure 8.21: on the left)?

12.4.2 Flow behavior

Rotational tests are performed mostly in the range of $\dot{\gamma}$ = 1 to $1000s^{-1}$; with controlled shear rate or controlled shear stress; preset as in Figures 3.1 or 3.2. Presentation: Flow and viscosity curves on a lin or on a log scale (Figures 3.26 to 3.31).

12.4.3 Structural decomposition and regeneration ("thixotropic behavior", e.g. of coatings)

Step test, consisting of three intervals at low/high/low shear conditions.

Optional methods (method M3 is recommended):

M1) Rotational test at a low/high/low shear rate $\dot{\gamma}$, preset as in Figure 3.42

Measuring result: Time-dependent viscosity function η (Figure 3.43)

M2) Oscillatory test at a low strain γ inside the LVE range/at a high strain γ outside the LVE range/again at a low strain γ inside the LVE range; and at a constant angular frequency ω in each interval; preset as in Figure 8.30

Measuring result: Time-dependent functions of G' and G'' (Figure 8.31)

M3) Oscillatory test at a low strain γ in the LVE range and ω = const/rotation/again at a low strain γ in the LVE range and ω = const; preset as in Figure 8.32

Measuring result: Time-dependent functions of G', G'' and η (Figure 8.33), or of η^* and η (Figure 8.34), respectively

13 Rheologists and the historical development of rheology

In order to understand how things have developed up to now it is useful to take a look on the history of rheology, and for those people who are interested in the development of engineering it is also an amusement. In general, never will be found a history book in which the complete evolution might be presented and therefore, the following is merely a list of selected dates, people and events which form part of the historical development of common science, and especially of rheology and rheometry. Some year specifications should be considered approximately. If no other references are given, many of the following dates are collected from the books of M. Reiner, R.I. Tanner and K. Walters, and from a lexicon on scientists. [303, 317, 352]

13.1 Development until the 19th century

600 BC: **Thales** of *Milet* (625? to 537 BC), philosopher. He assumes the existence of the laws of nature, trying to find **rational explanations for all phenomena**. [5]

400 BC: **Demokritos** of *Abdera* (460? to 375 BC), philosopher. He establishes a hypothesis on the **structure of matter** assuming **two principles: atoms as elementary particles and the emptiness**. All kind of matter and its changes are based on the composition and interactions of the atoms (Greek "atomos" means indivisible).

360 BC: **Aristoteles** (384 to 322 BC), philosopher. He creates the first systematic classification of the complete knowledge, based on rational explanations. In Europe, his work was re-discovered in the 12th century, above all thanks to Arabic translations and interpretations. [5]

300 BC: **Eukleides** of *Alexandria* (*Euclid*, 365? to 300? BC), mathematician. He collects the complete mathematic knowledge in 13 textbooks. Inspired by his studies on the geometry of various bodies, some rheologists later referred to "completely rigid bodies without any deformability" or "inelastic Euclidean bodies" (e.g. see *M. Reiner* 1931 and 1960). [125, 322]

300 BC: Foundation of **Bibliotheke** (library) **and Museion** (university) **in Alexandria** (Egypt), being the center of science for several centuries under Greek direction (until 565 AD). [75]

260 BC: **Archimedes** of *Syrakus* (285? to 212 BC), mathematician. He determines the **circle constant** as $3 + (10/71) < \pi < 3 + (10/70)$, thus $3.14085 < \pi < 3.14286$; in fact $\pi = 3.14159$. By the way, already around 2000 BC on a table of cuneiform characters from Babylon is mentioned the value $25/8 = 3.125$. He also studies the law of the lever (statics), hydrostatics (buoyancy, the "Archimedean principle"), and the use of water screws ("Archimedean screw"). [98, 353]

75 BC: *Titus* **Lucretius** *Carus* (98? to 55 BC), poet, philosopher. He explains in a didactic poem that **flow and gliding behavior of fluids** such as water, milk, wine, olive oil, honey, pitch (but also the speed of light and fire), and the **rigidity of solids** such as wood, stone, lead, iron, is related to the **size and shape of elementary particles** (in Latin **"corpora"**) and their interactions and entanglements, based on the hypotheses of *Demokritos* (see 400 BC) and **Epikuros** (341 to 271 BC). [362, 378]

1283 AD: The beginning of a "new time": Possibly the first **mechanical clock**, equipped with wheels and driven by a weight, is installed in the monastery of Dunstable. [9]

1445: *Johannes* **Gutenberg** (1400 to 1468), goldsmith, mirrormaker. He invents the modern **letterpress**, using single casted movable metal letters, and black printing ink consisting of a mixture of oils, resins,

Thomas G. Mezger: The Rheology Handbook
© Copyright 2011 by Vincentz Network, Hanover, Germany
ISBN 978-3-86630-864-0

pitch, waxes, soap and carbon black of pine wood. Using wooden screw presses, his team prints from 1452 to 1454 the first book, the Bible on 1282 pages, on parchment made of animal skins and on paper made of textile rags. In 1457, *Johannes* **Fust** and *Peter* **Schöffer** perform the first **multi-color print** using this technique. (Previous development: In about 600 AD first bookprinting in China, since about 740 using wooden and metallic blocks, in 1041/1045 *Pi Sheng* uses movable letters made of clay; in 1314 letters made of wood in Korea, and in 1392 made of copper; afterwards however, mostly wooden letters are in use.) Consequence: Since multiplication of information is possible now, **knowledge is available for many readers**. [145, 187, 355]

1490: **Leonardo** *da Vinci* (1452 to 1519), graphic artist, painter, sculptor, architect, engineer. Based on his own observations, he performs a lot of accurate **drawings of hydro- and aerodynamic phenomena** such as water waves, swirls, vortices, turbulences, e.g. of waterways and embankments. He studies also flow behavior of air under bird wings and flying apparatuses, and effects of air resistance. He investigates bending behavior of architectural structures such as wooden beams, columns, tapes, and wires, and biomechanical structures such as bones and muscles. Citation: "Visualization of invisible things". [160, 224]

1540: *Bernard* **Palissy** (1510 to 1589), potter. He uses the French terms **"visceux"** (viscous) in the sense of oily and **"pasteux"** (pasty) in the sense of dough-like, to characterize potter's clay. Besides, he develops a hypothesis on the genesis of fossils. [87, 378]

1575: *Guidobaldo* **del Monte** (1545 to 1607). Using models of machine parts, he develops **theorems in mechanics** which are **based on experiments**. [293]

1585: *Simon* **Stevin** (1548 to 1620), mathematician, engineer. He introduces **decimal fractions**, and develops the idea to decimalize all measures as well as the **parallelogram of forces**. [6]

1590: **Galileo** *Galilei* (1564 to 1642), mathematician, physicist, astronomer. He introduces the **systematic use of practical experiments on a scientific basis**, e.g. to characterize the behavior of liquids and solids; in 1590 he states a law of freely falling bodies showing a path in the form of a parabolic curve; in 1593, a hypothesis on the **elasticity of solids**, e.g. by studying bending behavior of wooden beams, in 1604, a **hypothesis of motion** trying to define velocity and acceleration. Besides, in 1592 he invents a simple thermometer, since 1609 he improves telescopes achieving up to 30 times magnification finally (telescopes are invented by *Hans* **Lippershey**, manufacturer of spectacles, in 1606). **He is the founder of classic experimental physics and of modern science which is based on quantitative mathematic analysis.** Citation: "The universe is written in the language of mathematics". [148, 141, 160, 331]

1637: *René* **Descartes** (1596 to 1650), philosopher, mathematician. He uses a right-angled three-dimensional **coordinate system** showing x-, y-, and z-axis (based on works of *François* **Viète** and *Pierre* **Fermat**), which later was called the "Cartesian coordinate system". Thus, he combines algebra and geometry, since now, **analytical functions can be presented in diagrams** in the form of data points P(x,y) or as function curves. Besides, he states that biological processes are based exclusively on mechanical ones, therefore, animals and humans being "living machines". [93, 230]

1640: *Evangelista* **Torricelli** (1608 to 1647), physicist, mathematician. He finds **"Torricelli's efflux formula"** for incompressible liquids, after applying Galileo's law of falling bodies to fluids and after own experiments using a container showing an outlet on bottom, with the flow velocity v, the gravitational constant g, the distance h between level line and outlet: $v = (2 \cdot g \cdot h)^{0.5}$. This is **one of the first basic studies in hydrodynamics**; however, considering only the flow behavior of "ideal fluids", drive force or pressure, respectively, and flow velocity, but not the flow resistance of the fluid. Besides, in 1643 using a mercury **thermometer**, he makes experiments on air **pressure** and vacuum. [234]

1649: *Blaise* **Pascal** (1623 to 1662), mathematician, physicist, philosopher, theologican. He reflects about "ideally inviscid fluids" without any flow resistance. Inspired by his ideas, some rheologists later referred to "ideally Pascalian liquids" (e.g. see *M. Reiner* 1931 and 1960). He studies the **effects of air pressure** and of pressure in liquids finding the **"law of communicating tubes"**, the dependence of air pressure on the altitude, and performs experiments with vacuum (based on the works of *Otto von* **Guericke**). Therefore later, the SI-unit of pressure and mechanical stress is termed **Pa**. Besides, in 1654 he publishes "Pascal's triangle" in mathematics to determine binomial coefficients; however, this concept is already known, for example, in China since at least 1303. Since 1642 he constructs a mechanical calculating machine for additions (as *Wilhelm **Schickard*** does in 1623, and *G.W. **Leibniz*** from 1671 to 1693 whose machine is designed also for multiplication and division, see 1684). [278, 12, 22, 26, 181]

1656: *Christiaan **Huygens*** (1629 to 1695), mathematician, physicist, astronomer. He states a **law of elastic collision**, and introduces the **moment of inertia**. Besides, using a spiral spring as a balance, he develops a **pendulum clock**, achieving in 1675 a constant oscillation frequency showing a deviation of only 1min per week, which is 100 times better than before; and in 1658 he designs a micrometer with a resolution of a few angular seconds. [9, 331]

1660: *Robert **Boyle*** (1627 to 1691), physicist, chemist. He makes experiments on "the spring of air" (meaning: elasticity), finding the relation between pressure and volume ("extension") of gases: p · V = const.

1676: *Robert **Hooke*** (1635 to 1703), physicist, architect. After performing own experiments on wires, springs and beams, he states **for solids proportionality between force and deformation**, the so-called **"Hookean elasticity law"** (in Latin: "ut tensio sic vis", meaning: as the extension, so is the force). This is **the basic law of solid-state physics** (see also *A. Cauchy* 1822). Besides, he studies thermodynamics, optics and biology, for example, coining the term "cellula", i.e. cell, when investigating microstructures of biological plants. He improves several scientific instruments, such as the **microscope** in 1667 (microscopes are invented by *Zacharias **Janssen*** in 1590, and improved by *Antoni **Leeuwenhoek*** in 1660, achieving a magnification of 275:1). [193, 160, 233, 265]

1684: *Gottfried W. **Leibniz*** (1646 to 1716), philosopher, mathematicien, diplomat. He finds a method of **differential calculus**. Besides, he constructs a mechanical calculating machine (see also *B. Pascal* 1649). [225, 353]

1687: *Isaac S. **Newton*** (1643 to 1727), nature philosopher. **For fluids**, he states **proportionality between flow resistance and flow velocity**, the so-called **"Newtonian viscosity law"**. He terms the proportionality factor as "defectus lubricitatus", meaning "lack of slipperiness" or "flow resistance", therefore assuming the existence of an "internal friction of fluids". This is **the basic law of fluid mechanics or aero- and hydrodynamics**, respectively (see also *G. Stokes* 1845). He develops a comprehesive system, the "Newtonian laws of mechanics". The 1st axiom: law of inertia, based on works of *Galileo Galilei* of 1638 and *R. Descartes* of 1637/1644. The 2nd axiom: **definition of force** or "law of actio", which is firstly formulated by *L. Euler* in 1736 in the form of F = m · a. The 3rd axiom: "actio = reactio". He discovers the gravitation force. He finds a method of **differential calculus**, probably in 1665/1671?, which he called "methodus fluxiorum" but he published it not before 1705 (besides *G.W. **Leibniz*** in 1676/1693; other ones preparing this field are *P. **Fermat***, published in 1679, *R. **Descartes***, *B. **Pascal*** in 1659, *B. **Cavalieri***, *C. **Huygens***, *J. **Gregory*** in 1667, *J. **Wallis***). Besides, he develops a hypothesis in optics on light and colors assuming "corpuscles" (light particles), and about dispersion of white light into a spectrum of colors; and in 1668 he designs a reflector telescope. [266, 156, 166, 233, 331, 337, 357]

1687: *Jacob **Bernoulli*** (1654 to 1705) and *Johann **Bernoulli*** (1667 to 1748), two brothers, both being professors in mathematics since 1687/1695. They improve the use of differential equations bringing the vaguely verbalized theorems of *R. Hooke* and *I. Newton* in an analytical form. *Jac. B.* states a **theory of elasticity**, and *Joh. B.* studies **hydrodynamic flow and efflux behavior** (1738/40), both are pathfinders for characterizing scientifically the behavior of fluids and solids since their work is **based on the fundamentals of mathematics**. This work is later continued by *D. Bernoulli* who is *Joh.'s* son (see 1738), and *L. Euler* who is a student of *Joh. B.* (see 1736). [35, 46, 133]

1714: *Daniel G. **Fahrenheit*** (1686 to 1736), physicist, instrument maker. He introduces the first calibrated **temperature scale** (in "Fahrenheit degrees"; with 0°F at -18°C approximately, which is the freezing point of a mixture of salt and ice; and 100°F at around the human body temperature (+37°C). [331]

1729: *Georg B. **Bülfinger*** (1693 to 1750). He states an **elasticity law for solids** which are not showing **ideally elastic behavior**: γ = a · τ^n. [58]

1736: *Leonhard **Euler*** (1707 to 1783), mathematician, physicist, astronomer. He **improves analytical methods of fluid mechanics and solid-state physics** finding new concepts on **deformation energy** and the **critical buckling load of long solid bars** using the **moment of inertia** (see *C. Huygens 1656*). He describes the flow behavior of fluids in a mathematical form, e.g. using "Euler's number of fluid mechanics" as the ratio of the forces occurring due to pressure p and inertia; with density ρ, and flow velocity v: Eu = $p/(\rho \cdot v^2)$. Hence, **"Euler's flow equation of inviscid liquids" of 1755** considers only the flow behavior of "ideal fluids", using drive force or pressure gradient, respectively, and flow velocity, but not the flow resistance of the fluid. He presents works on analytical geometry using functions f(x,y), and states **"Euler's number"** based on the natural logarithm (for n towards infinity or $n \to \infty$ holds): e = $[1 + (1/n)]^n$ = 2.7182818284..., and he finds "Euler's formula" combining exponential and trigonometric functions: e^{ix} = cos x + i sin x. [126, 46, 133, 160]

1738: *Daniel **Bernoulli*** (1700 to 1782), mathematician, physicist, doctor, botanist. He states a scientific concept of flow behavior based on mathematics, **"Bernoulli's equation** of **tube flow** of inviscid, incompressible fluids"** as a balance of the following pressures: static pressure p, pressure due to gravity, and dynamic stagnation pressure as p + $\rho \cdot g \cdot h$ + $(1/2) \rho \cdot v^2$ = const. This formula is useful to calculate the flow velocitiy via pressure measurements, however, still disregarding the flow resistance of the fluid. [34, 46]

1742: *Anders **Celsius*** (1701 to 1744), astronomer. He proposes a **temperature scale** in 100 "Celsius degrees", originally with 0°C at the boiling point of water, and 100°C at the melting point of ice. The scale was reversed later by *C. **Linné***.

1752: *Jean le Rond d'**Alembert*** (1717 to 1783), natural scientist. He studies the **motion of bodies** which are immersed **in a flowing liquid**, and the two-dimensional flow behavior of "inviscid fluids", therefore still disregarding the flow resistance of the fluid. Besides, he presents mathematical works about analytical geometry, differential equations, integral calculus, mechanics, acoustics, and optics. Together with *Denis **Diderot*** (1713 to 1784), philosopher, he publishes from 1751 to 1772 the **Encyclopédie** which is the first modern **lexicon** consisting of 17 text volumes and 11 illustrated picture volumes. [8, 48]

1767: *James **Watt*** installs the first low-pressure **steam machine** in an iron work. (Previous development: in 1690 atmospherical steam engines by *D. **Papin***, and in 1711 improved ones by *T. **Newcomens***.) This is the beginning of the "industrial revolution".

1784: *Charles A. de **Coulomb*** (1736 to 1806), physicist, engineer. He constructs a concentric cylinder apparatus (inner d = 42.9mm, L = 58.6mm, outer d = 128.6mm). The inner cylinder is hanging on a torsion wire, and after deflecting and releasing, it performs a free, oscillatory motion which is damped by the sample. He studies **torsional elasticity** of metal wires, however making no systematic tests on the "internal friction of liquids". He investigates the strength of materials e.g. in architecture. Besides, he studies friction between solids, finding in 1785 "Coulomb's friction laws", and he designs "Coulomb's torsion balance" to detect forces between electric charges or magnetic poles finding "Coulomb's law of electricity". [80, 46, 289, 331]

1794: In France is introduced the decimal graduation of time: per day 10 hours, per hour 100 minutes, and per minute 100 seconds. Thus, for a day: 1d = 10 · 100 · 100s = 100,000s; instead of 1d = 24 · 60 · 60s = 86,400s as we know it. A week has 10 days, and a month 3 weeks. This "revolution calendar" was repealed in 1806. [6]

1797: *Alois **Senefelder*** (1771 to 1834). He invents a **flat printing technology** using two stone plates, e.g. of chalk stone, performing since 1826 also multi-color prints on these kinds of stone presses. He calls this method **lithography** (i.e. "stone drawing"), using printing inks consisting of tallow fat from animals, natural waxes and resins, and carbon black. The prime time of lithograpy is around 1820 to 1850 when up to 15 different colors are printed step by step one upon another using the newly developed smooth "art paper" with casein as a binder of the paper coating. [355]

13.2 Development between 1800 and 1900

Since 1800: Use of **flow cups** for textile printing inks and oils. [378]

1801: Introduction of the decimal metric system for units in France which is developed by a commission of the Académie des Sciences de Paris since 1790 to reform all measures and weights; e.g. by *J.C. de **Borda***, *M.J. de **Condorcet***, *J.L. de **Lagrange***, *P.S. de **Laplace***, *A.L. de **Lavoisier***, and *G. **Monge***; establishing e.g. the kilogram, and defining the meter as "the circumference of the earth devided by 40,000,000". [92, 6, 31]

1801: *Carl F. Gauß* (or Gauss, 1777 to 1855), mathematician, physicist, astronomer, geodesist. He introduces the "Gaussian plane" to present complex numbers in the form of a diagram, e.g. the real part on the x-axis, and the imaginary part on the y-axis. He develops a concept for a system of measuring units. [150]

1803: *John Dalton* (1766 to 1844), chemist, physicist. He presents a **chemical atom theory** distinguishing the chemical elements by their **atom weights**; the **relative atom mass** is related to the hydrogen atom. [85]

1807: *Thomas Young* (1773 to 1829), doctor, physicist. He describes the **material-specific stiffness of solids**. Therefore later, the name "Young's modulus" is used as a synonym for the tensile elasticity modulus (see also *A. Cauchy* 1827). Besides, he develops a hypothesis of light and colors. [402, 160]

1811: *Sophie Germain* (1776 to 1831), mathematician. Her work on elastic surfaces is a basis for a new concept **on elasticity of metals**. [337]

1811: *Lorenzo R.A.C. Avogadro* (1776 to 1856), physicist, chemist. He studies the behavior of gases and specific heat, coining the term **"molecule"** (see also *J. Loschmidt* 1865: "Avogardro's number"). [13]

1811: *Friedrich König*. He constructs a **printing machine** consisting of a flat printing plate and an impression cylinder. This is the first **rapid printing press**, e.g. used for printing newspapers. [355]

1822: *Claude L.M.H. Navier* (1785 to 1836), mathematician, engineer, e.g. for bridge construction. Concerning fluid dynamics he enhances the **flow equation** of *L. Euler* (see 1736/1755) by a term **considering also** the internal friction of fluids, and therefore, also **the flow behavior of viscous liquids**, however, without further regards on the nature of the flow resistance; later called the **"Navier/Stokes equation"** (see also *G. Stokes* 1845). He develops a **theory of elasticity**, e.g. on bending behavior, after performing comprehensive mechanical experiments, on the strength of construction materials and others. He introduces **scientific fundamentals in engineering** while previously, material testing was performed usually by way of trial and error. [263, 65, 160]

1827: *Augustin L. Cauchy* (1789 to 1857), mathematician, physicist. He formulates the Hookean **elasticity law in a scientific three-dimensional form**, and therefore, he is considered **the founder of a modern concept on elasticity** (see also *R. Hooke* 1676). The **"Cauchy strain tensor"** presents a material **in the state of low deformation**. He describes the terms **shear stress** and **extension** (strain, deformation). He states the "Cauchy number of elastic bodies" as the ratio of the forces due to inertia and elasticity, with density ρ, angular frequency ω, and length L, using the **elasticity modulus** E: Ca $= \rho \cdot \omega \cdot L^2/E$, and the "Cauchy strain": $\varepsilon_c = (L - L_0)/L_0$ for tensile tests. In 1828, he develops a flow equation for viscous fluids. [71, 65, 160]

1827: *Robert Brown* (1773 to 1858), botanist. He observes trembling motion of colloid particles in dispersions, later called "Brownian motion". Firstly he saw it in biological cells of plants. (Later realized by *A. Einstein*, see 1905, as molecular collisions due to thermal motion). [29, 128]

1829: *Siméon D. Poisson* (1781 to 1840), mathematicien, physicist. He develops a **theory on elasticity**, describing also the relation between transversal and longitudinal extension of solids in tensile tests, as the "Poisson's ratio". In 1829, he develops a flow equation for viscous fluids. [292, 65, 160]

1835: *Wilhelm E. Weber* (1804 to 1891), physicist. He performs experiments using a **torsion pendulum** at a **free, damped oscillatory motion** and **creep tests** to study the behavior of suspensions and silk threads. He describes **"elastic after-effects"**, realizing that not all solids are showing ideally elastic, Hookean behavior, but time-dependent effects. Besides, he designs a first electromagnetic telegraph; he specifies several new units e.g. for voltage, electric current, and electric resistance; and in 1856 he determines – together with *R.H.A. Kohlrausch* – the light speed via electric measurements. [380, 18, 352]

1839: *Gotthilf H.L. Hagen* (1797 to 1884), hydraulic engineer. He designs a first **capillary viscometer**, enhancing J. Poiseuille's work about capillary flow. Since 1925, the relation of viscosity η, flow rate $\dot{V}$, pressure difference Δp, length L and radius R of a capillary is termed the **"Hagen/Poiseuille relation"**: $\eta = (\pi \cdot \Delta p \cdot R^4) / (8 \cdot L \cdot \dot{V})$. This relation was first published in 1856 by *G. Wiedemann*. [172, 233]

1840: *Jean L.M. Poiseuille* (1799 to 1869), doctor. He designs a **capillary viscometer** and performs experiments on the flow behavior of water, alcohol and blood. He also finds that η is proportional to R^4

(see *G. Hagen* 1939). Besides: "Poiseuille/Hartmann flow" of electrically conductive media in a magnet field (according to *Julius F. Hartmann*, 1881 to 1951). [291, 378]

1842: *Julius R. von Mayer* (1814 to 1878), doctor, natural scientist. He finds the **mechanical-caloric equivalent** by enhancing the work of *N.L.S. Carnot*. In 1845 he states the **law of energy conservation** as a law of universal validity, the so-called 1st law of thermodynamics (see also *J. Joule* 1845, and *H. Helmholtz* 1847). [331]

1843: *Claude Barré de Saint-Venant* (1797 to 1886). He publishes a **flow equation of viscous fluids with consideration of laminar and turbulent flow hehavior**. In 1847 he presents a work on the **elasticity limit of solids**, and in 1855 on **torsional behavior of solid bars** with non-circular cross-section. In 1868 he introduces the concept of **the "Saint-Venant model" for "plastic materials"** which later is illustrated in the form of a friction element or a "gliding shoe". [314, 65]

1845: *James P. Joule* (1818 to 1889), ale-brewer, physicist. Also he states the law of energy conservation (see also *J. Mayer* 1842, and *H. Helmholtz* 1847), and assumes after experiments on gases that **heat is not a substance** but is based on particle motion. [331]

1845: *Georges G. Stokes* (1819 to 1903), physicist, mathematician. Concerning hydrodynamics, he formulates **the flow equation in a scientific three-dimensional form**, and therefore, he is considered **the founder of a modern concept on flow dynamics**; later called the **"Navier/Stokes equation"** (see *I. Newton* 1687, *L. Euler* 1736/1755, *C. Navier* 1822). He clearly takes into consideration the "internal friction of flowing fluids". He performs scientific **falling-ball experiments**, finding in 1851 "**Stokes' law**" (practical use: e.g. sedimentation of particles): $F_R = 3 \cdot \pi \cdot d \cdot \eta \cdot v$ (with flow resistance force F_R, ball diameter d, viscosity η, velocity v of the falling ball). He studies **elasticity** describing the equilibrium state of solids under stress. Besides, he performs fundamental works on optics, spectroscopy, gravitation, and vector analysis (see also *G. S.* 1849). [348, 46, 107, 233, 286]

1847: *Hermann L. F. Helmholtz* (1821 to 1894), doctor, physicist. He **integrates the law of energy conservation into hydrodynamics**. In 1859 he performs experiments on "friction in liquids" (together with *G. von Piotrowski*), and on turbulent flow, developing in 1860 a formula on **tube flow** using a "gliding coefficient". In 1868 he discovers **similarities**. The same **differential equations** can be used as well for **hydrodynamic flow** as well as for the **flow of electric current or heat** in conducting materials. [179, 113, 331]

1848: *William Thomson* (*Kelvin*, 1824 to 1907), physicist. He states **the absolute temperature scale**, since based on the effectivity of the Carnot process, the **absolute zero point of temperature occurs at 0K** = -273.15°C. Besides, he studies thermodynamics (see *J. Mayer* 1842, *R. Clausius* 1850); searching for similarities of electric, magnetic and elastic effects (see also *W.T.K.* 1865). [359]

1848: *John Curtis* starts production of **chewing gums** on the basis of pinetree resins and beeswax, since 1870 also using liquorice. (Previous development: Already *Stone Age people* use to chew a mixture of birch resin, beeswax and honey. *The Mayas* of present South Mexico and *the Aztecs* chew chicle which is a thickened milky juice or latex from the sapodilla tree becoming rubber-like on air. Other ones chew also paraffin waxes. Later development: Since 1871, *Thomas Adams* produces chewing gum made of chicle. In 1892 *William Wrigley* begins to produce chewing gum, and by around 1910 his company is the market leader in the USA. As ingredients he uses a gum base consisting of natural mastic gum resins, rubber, milky juice from the gum arabicum tree, and gel made of cornstarch; sugar and corn syrup, aromas such as peppermint, and plasticizer; **later: synthetic resins** such as PVA, methyl ester (40 to 50%), **filling materials** such as chalk (20 to 30%), **waxes** (15 to 20%), **elastomers** such as butadiene copolymers (5 to 10%), **plasticizers** (2 to 5%), antioxidants, aromas, sugar or sweeteners. [74, 310]

1849/1860: *G. Stokes* and *E. Hagenbach* describe the **parabolic flow velocity distribution in capillaries**.

1850: *Rudolf J.E. Clausius* (1822 to 1888), physicist. Concerning aero-and hydromechanics, he finds the **kinetic gas theory with a new understanding of the nature of viscosity**, based on motion, collisions, **interaction of atoms and molecules**. Besides, he is a co-founder of **statistic mechanics** describing physical processes as a statistic function (see also *J. Maxwell* 1855). He introduces the term "kinetic energy" improving the so-called 2nd law of thermodynamics. He also performs experiments on steam engines. [286]

1853: *Charles N.* **Goodyear** (1800 to 1860), chemist, technician. He starts **production of hard rubbers** called "Ebonite". In 1839, he succeeds in the first **hot-vulcanization of unlinked rubber (caout-chouc)**, which results in a rubber which is stable and elastic even at low temperatures; US patent in 1844. (Previous development: In the 11. century, *the Mayas* use bouncing balls made of coagulated caoutchouc – in the Maya language called caa = wood, o-chu = to weep, to burst into tears, thus: "tears of wood" – for the latex collected from the trees; in 1495, also *Christopher* **Columbus** reports from Haiti about this. In 1765, *J.A.C.* **Charles** uses diluted rubber to seal textiles; and in 1783, brothers **Montgolfier** use rubber for their hot-air balloons. In 1819, *Thomas* **Hancock** succeeds in the **mastification of rubbers** and designs a kneading machine, receiving a patent in 1843 on the production of elastic tapes and tissues; *T.H.* see also 1992: trials on road pavement. In 1823, *Charles* **MackIntosh** gets a patent on rubber panels and produces raincoats, the so-called "mackintoshs". In 1845, *Robert W.* **Thomson** applies for a British patent on **air-filled rubber tires**; and in 1888, *John B.* **Dunlop**, veterinarian, invents this kind of tire again, Brit. patent 1888. See also *A. Peugeot* 1895). [216, 29, 30]

1854: *Thomas* **Graham** (1805 to 1869), physicist, chemist. He is the **founder of colloid chemistry** coining the term "colloid". In 1864, he **determines viscosity via membrane diffusion** using his so-called **"colloidoscope"**. [29, 128, 378]

1855: *James C.* **Maxwell** (1831 to 1879), physicist. Concerning aero- and hydrodynamics, he improves the kinetic gas theory in 1855/1867. In 1863/1865, he designs an **instrument to measure the "internal friction" or "thickdom" of gases**, in the form of a closed container filled with air or another gas, performing tests at different pressures and temperatures. He measures the degree of damping, using several parallel disks which are exited to an oscillatory torsional motion by a variable magnetic field. In 1866, he presents a paper about edge corrections of **disk measuring systems oscillating in a rotational apparatus**. He uses **statistic functions to describe transport phenomena such as viscosity**, diffusion and heat transport, **interpreting the nature of viscosity as an impulse transfer between layers of particles** (see also *R. Clausius* 1850). In 1867/1868, he introduces a **differential equation to characterize the behavior of viscoelastic liquids**, later called **"Maxwellian behavior"**. This is the first scientific concept to describe **"elastic reaction forces"** and relaxation behavior of viscous liquids, using a **"modulus of stress relaxation time"**. Materials showing very long relaxation times and highly viscous solids such as cold pitch and asphalt he terms as **"viscous solids"**, considering the **energy loss of flowing viscous fluids**. He improves vector and tensor analysis (the **"Maxwellian stress tensor"**). He tries to find **similarities** between mechanical, electric and magnetic processes: interpreting e.g. **viscosity** as internal friction of a liquid flowing through a porous membrane, compared to electric current flowing through an electric conductor; and **elasticity** as deformation of an impermeable membrane caused by a liquid whose motion is stopped in this way, compared to the behavior of a dielectric material or isolator which increases the capacitance of a capacitor when storing electric charge. Background: As well atoms and molecules as well as electrons are not discovered yet. However, material science is peripheral to Maxwell's major scientific interests since he focusses on finding a theory of electromagnetism; he is the founder of classic electrodynamics developing the corresponding "Maxwellian equations". [245, 233, 235, 286, 331]

1856: *Gustav H.* **Wiedemann** (1826 to 1899), physicist. He **publishes the "Hagen/Poiseuille relation"** (see also G. Hagen 1839), and discovers torsional deformation of current-carrying rod-like magnets. [386]

1857: *William H.* **Perkin** (1838 to 1907). In 1856, he develops the first synthetic dye, produced from coaltar which remains as a residue in coke-ovens and gas producing factories, e.g. for dyeing cotton, wool, linen and silk. He produces aniline violet colors, called "Mauvein" or "Perkin violet" (mauve is the color of the mallow flower). This is **the beginning** of the tar-color industry or **of industrial production of synthetic materials**, respectively. For the first time, besides the rich upper-class also many common people can afford colored clothes and household textiles. [33, 51, 74, 136]

1858: *William* **Bullock** constructs the first **rotary printing machine**, both the impression cylinder and the printing form are cylinders. [355]

1859: *A.V.* **Lourenço** observes **for an increasing molecule size** also an **increase in viscosity** (later, in the 1920s this is recognized as a polymerization process). [119]

1861: *Lipowitz* constructs a **penetrometer** or **"compressiometer"** to determine the rigidity of **gelatin gels**: A disk of d = 2 inches (= approx. 50mm), placed on the surface of a gel, is beginning to immerse into the gel when loading it more and more by small lead balls. The weight required is used as a measure of the gel strength. [378]

1863: "Treaty of Paris", leading to a global specification of weight measures for international postal use, based on the gram as the weight of a water cube of the volume of V = 1cm^3 at T = +4°C. This gives the decisive stimulus to the **worldwide introduction of the metric system** (see also 1960: SI system). [6]

1863: *Friedrich W. G. Kohlrausch* (1840 to 1910), physicist. He studies, as his father *R. Kohlrausch* did since 1847, the behavior of silk threads (cocoon), glass fibers and rubber threads using a **torsional creep apparatus**, testing the damping behavior of an oscillatory motion. [211, 233, 352]

1865: *Joseph Loschmidt* (1821 to 1895), physicist, chemist. He determines the so-called "Avogadro's number" which is the number of atoms or molecules per mol: $N_A = 6.022 \cdot 10^{23}$ (see *L. Avogadro* 1811).

1865: *W. Thomson (Kelvin)* performs experiments on the damping behavior of metals, describing their "elasticity and viscosity", i.e., the **behavior of viscoelastic solids**, later called "Kelvin/Voigt model", but he wrote down no formula (see also *O. Meyer* 1874, and *W. Voigt* 1892; and *W.T.K.* 1848). He studies also hydrodynamics, stating a concept on turbulent flow. [359, 155, 331]

1866: The beginning of the **"era of electricity"**. *Werner von Siemens* (1816 to 1892) invents an **effective generator** by intelligent wiring of a dynamo machine; this allows electricity to be produced economically and to be used for electric motors which are previously driven from batteries only. In 1881, he combines a steam engine directly with an electricity generator. (Previous development: In 1820 *Hans C. Oersted* discovers the correlation between flowing electric current and magnetism, and in 1831 *Michael Faraday* discovers the principle of electromagnetic induction and the conversion of motion into electricity and vice versa.) The consequences for measuring instruments: Electric driven devices improve the reproducibility of material testing, but in industrial laboratories they become widely spread not before the 1950s (see *M. Couette* 1888, *E. Hatschek* 1913, *F. MacMichael* 1915, and e.g. *Brookfield* 1945 and *Haake* 1954).

1870: Company **BASF** (Badische Anilin- und Soda-Fabrik), founded in 1865, starts **industrial production of colors based on coal-tar**, responsable is *Heinrich Caro* (1834 to 1910), a chemist. [67]

1872: *John W. Hyatt* (1837 to 1920) starts **industrial production of "Celluloid" ("nitro-cellulose")**, patent in 1869. Typical applications are combs, small containers, dolls, parts of false teeth, jewelry, photo and film materials, billard balls. (Previous development: In 1846 *Christian F. Schönbein*, chemist, isolates "highly viscous", i.e. polymeric, **cellulose nitrate** from cotton, the so-called "gun cotton" or **"Kollodium"**. Since 1862 production of "Parkesin" and "Parkesit" by *Alexander Parkes*, patent in 1865; see also 1903: viscose.) [29, 74, 108, 216]

1873: *Johannes D. van der Waals* (1837 to 1923), physicist. He states a kinetic theory of liquids, and develops the concept of **"van der Waals forces"** as **interactive forces in colloid science**. Besides, he finds the "van der Waals gas equation", and studies surface tension, capillarity, and thermodynamics. [371, 29, 128]

1874: *Oskar E. Meyer* (1834 to 1909), physicist. He firstly formulates the **differential equation** describing the **behavior of viscoelastic solids**: the so-called "Kelvin/Voigt model" (see also *W. Thomson-Kelvin* 1865, and *W. Voigt* 1892). He performs tests using oscillating disks. [250, 107, 155]

1874: *Ludwig Boltzmann* (1844 to 1906), physicist, mathematician. For viscoelastic materials, he develops the **"superposition principle"** which is the basis of the theory of **linear viscoelastic behavior**. He considers also an "elastic after-effect", thus, the influence of the pretreatment of a sample. [47, 114, 249]

around 1880: The beginning of the "era of oil". Vegetable oils, animal oils such as whale oil, natural fats and beeswax are increasingly being replaced in daily use by petrochemical products. Previously, besides "wood oils" from charcoal production, there are only little deposits of oils, e.g. from natural tarpits or heavy oil pools. However now, mineral oils and petroleum are used as illuminants, fuels, disinfectants and wood preservatives, and paraffins are used as cart-grease and mill-fat. Since the invention of **steam engines, lubricant consumption** has increased (see 1767 *J. Watt*). Although the invention of electrical light results in a decrease in petrol consumption, the importance of fuels and lubricants considerably increases with the number of automobiles, particularly since the 1920s (see 1882 *T. Edison*, and 1913 *H. Ford*). Consequence: More and more standardized products are required, and therefore, **quality control (QC) increasingly becomes important, e.g. to characterize flow behavior or firmness of samples in terms of "fluidity", "consistency", "plasticity" or "rigidity"**. Previous development: Since around 800 AD, oil springs are documented near Baku; and around 1000 AD there are **distillation plants** in China. Since 1834 near Baku, then in the Russian Empire and now in Azerbaijan, there are the first distillation plants for producing petroleum and paraffins; in 1846 **drilling for oil** by *V.N. Seymonov* near Baku; in 1858 oil drilling lead by *J. Miller* in "Oil Spring", Lambton County, ON, Canada; in 1858/59 hand-operated oil drilling lead by *G.C.K. Hunäus* in Wietze near Celle, Germany, however, his team is actually searching for brown coal; in 1859 oil drilling lead by *E.L. Drake* near Titusville, PA, USA, using a steam engine; since 1873 there is the first continuously operating distillation plant for crude oil near Baku, and this region is the world's largest exploration area before the "Russian Revolution" in 1917. [95, 131, 315]

1880: *T. Schwedoff* presents a **modified Maxwell model including a yield point** as a **flow curve model with rigidity at rest**. In 1890 he uses **concentric cylinders** (mean R = 35mm, gap 12.6mm) to measure the **"cohesion-of-liquids"**, i.e. viscosity, and the **"modulus-of-liquids"**, i.e. elastic effects, e.g. of gelatin solutions. He observes a **"relaxation-of-liquids"** of water after a preset strain, and **viscosity** which is **dependent on the rotational speed** e.g. of castor oils, glycerins and gelatin solutions; this is the first report on **shear-thinning**. He finds flow instability when using a **rotating inner cylinder**, thus, turbulence or **secondary flow effects**. [328, 107, 237]

1881: *M. Margules* proposes to use a concentric **cylinder apparatus** to measure viscosity, he develops the **calculation of the "coefficients of friction and gliding", i.e. viscosity, in the circular gap of a rotational measuring system**. [241, 107]

1882: *Otto Mohr* (1835 to 1918) develops **"Mohr's circle"** or the **"stress circle"**, as an illustrative presentation of the relation between the values of the normal stresses and shear stresses in a diagram on a linear scale. [253, 234]

1882: *B. Élie* constructs an apparatus using a **concentric sphere measuring system**. [107]

1882: *J. Perry* designs a concentric cylinder apparatus, in 1893 reporting about a **double-gap system** (mean R = 109 and 121mm, gap 10mm). He measures e.g. oil of sperm-whales. [285, 107, 289]

1882: *Thomas A. Edison* (1847 to 1931) opens the first public electric power plant in Brooklyn, near New York. In 1879 his team develops the first **light bulb** equipped with a carbon filament which is working "permanently", i.e. for 40 hours. For industry, this means that from now on factories can irrespectively produce day and night. Since 1898 *Carl Auer von Welsbach* produces the first useful light bulbs with a metallic filament made of osmium, however, the breakthrough is not before 1905 with the tungsten filament lamp. Firstly, Trinidad asphalt (bitumen) is used as an insulation material for underground electricity cables, later improved by adding linseed oil, paraffin and wax. [375, 405]

Public electricity supply, for example: After Stuttgart and London (both in 1882) the first block-unit power station starts up in Berlin in 1884. However, until 1915 merely by 6% of Berlin's inhabitants are supplied with electricity, and still in 1927, only half of the households in Germany are connected to public electricity supply. Correspondingly, this can also be assumed for the working conditions in laboratories. [164]

1883: *Osborne Reynolds* (1842 to 1912), physicist, engineer. He observes a **shear-thickening effect** of highly concentrated suspensions, showing "flow hardening" and **volume increase** when disturbing the closest sphere packing of the particles; these effects are also called "cubic dilatation", "transversal elasticity" or "plastic dilatancy". After experiments performed in a horizontal glass tube, observing the shape of a thread-like track of ink in flowing water, he states the **"Reynolds number"** as the relation between the forces of inertia and flow resistance **to characterize the transition between laminar and turbulent flow conditions**: $Re = (v \cdot L)/v$, with the velocity v, the "characteristic length" L (geometrical condition, e.g. the pipe's diameter), and the kinematic viscosity v (see also Chapters 11.3.1.3 and 10.2.2.4b). Besides, he studies static friction and corresponding lubrication conditions in bearings. [305, 276]

1884: *Carl O.V. **Engler*** (1842 to 1925), chemist. He designs a **flow cup viscometer** for testing mineral oils, e.g. for the German railroad institution; using the unit Engler degrees. 1°E is the flow time of a defined volume of water, e.g. of V = 200ml, at T = +20°C. [97, 378]

1884: *Svante A. **Arrhenius*** (1859 to 1927), chemist, astronomer. He develops the **"Arrhenius equation"** in the form of an exponential function, introducing an **"activation constant"** after studies of **kinetics of processes** in electrochemistry, finding the concept of dissociation of electrolytes into positive and negative ions. He is testing colloidal solutions and dispersions containing sulphur and proteins such as globulin and egg-albumin, in order to study the molecular mechanism of viscosity. [11]

1884: *Ottmar **Mergenthaler*** contructs the first hot-type printing press, called "Linotype", using letters made of lead which are casted automatically as a complete line. The use of this press is considerably **speeding-up the printing process**. [355]

1887: *Wilhelm **Ostwald*** (sen., 1853 to 1932), chemist, physicist, philosopher. He designs a **capillary viscometer**, the **"Ostwald viscometer"** which is widely spread. He builds the first **water bath thermostat with a feedback control system**. A gas-powered Bunsen burner heats a water bath from below and simultaneously causes an impeller, installed above the water bath, to rotate due to the rising hot air. On bottom of the shaft of the impeller is mounted a toy ship's propeller which is immersed into the water bath, improving the temperature distribution of the water by the occurring slow circulation. When the temperature increases, alcohol which is enclosed in a sealed glass tube is expanding, pushing liquid mercury on a certain path within a U-tube. As a consequence, the heating gas supply is reduced to a certain degree. Via this control process, the gas supply, and therefore the heating power, reduces or increases itselves. Background: Electrical current was not available everywhere at this time (see also 1882). *Ostwald* studies many fundamentals of scientific **colloid chemistry** and thermodynamics. Besides, he develops a concept of colors. [272, 106, 169, 325, 378]

1888: *Carl J. von **Bach*** rediscovers the "Bülfinger formula" (see 1729), as "Bach's **elasticity law**" for behavior being not ideally elastic. [15]

1888: *Maurice M. A. **Couette***, physicist. Using a self-made **rotational apparatus**, he measures viscosity, which he termed (in French) as "frottement des fluides" or "coefficient de fr. interne" meaning "friction of fluids" or "coefficient of internal friction". Testing water, he achieves only 10% error, other samples are air, rape-seed oil and mercury. His viscometer is driven by an **electric motor, with a speed-controlled outer cylinder showing a concentric narrow gap** (inner R = 72mm, gap 2.5mm). The torque is detected via the deflection of the inner cylinder which is attached on bottom of a **torsion wire**. A small mirror is mounted on bottom of the wire and in order to detect a good resolution of the deflection angle, a light beam is sent to the mirror and the reflected beam is detected. Since the flowing viscous sample deflects the inner cylinder, this type of instrument is also called a "drag-flow" viscometer. **This is the first useful rotational viscometer, with a direct drive for continuous operation at constant shear conditions.** *Couette* makes experiments using diverse materials and coatings for the cylinder walls such as glass, copper, tin, diverse lacquers, silver coatings, grease and paraffins to study the influence on the wall-slip and flow behavior. In 1890, he presents a **correction** procedure **for inlet flow effects in capillaries**, the so-called "méthode de Poiseuille". He **compares flow conditions** occurring **in both capillaries and cylinder systems**. [79, 107, 233, 289]

1888: *A. **Mallock***. Using concentric cylinders, he measures viscosity values of diverse liquids, obtaining with water about 5% error only. In order to reduce end effects, an air bubble is trapped in the bottom of the inner cylinder. He reports about an apparatus with a rotating inner cylinder (mean R = 48/82/88mm, gap 3.8/11/23mm), observing instable flow, i.e. turbulent flow (see also *T. Schwedoff* 1880). [238, 107]

1888: *E. **Bowen*** receives a patent for a **bitumen testing instrument**, which is improved in 1901 by *A.W. **Dow*** and later by *Forest* and ***Richardson***. To determine the viscosity of bitumen at T = +25°C, the **penetration** of a needle with a weight of m = 100 g is measured within t = 5s; specification of the "pen value" in 0.1mm (see also Chapter 11.2.9a/b). [405]

1888/1893: *J.J. **Thomson*** and *Emil **Wiechert*** describe **complex viscoelastic behavior** using a **combination of the models of Maxwell and Kelvin/Voigt**. [358, 385]

1892: *Woldemar **Voigt*** (1850 to 1919), physicist. He studies crystal physics and also the behavior of **viscoelastic solids**, generalizing *O. Meyer's* **differential equation** of 1874, which is later called the "Kelvin/Voigt model" (see also *W. Thomson/Kelvin* 1865). [376, 155]

1893: *C. **Barus*** reports about the **extrudate swell** or **die swell effect**. [108]

1894: *J. **Finger*** states the **"Finger strain tensor"** for samples under tensile strain and shear strain. [233]

1894: Market introduction of the so-called **"Poiseuille capillary viscometer"** (see also 1840).

1894: *T.E. **Thorpe*** and *J.W. **Rodger*** analyze "the relations between the viscosity (internal friction) of liquids and their chemical nature"; using a concentric cylinder device with a stationary inner cylinder. [360, 107]

1894: *Heinrich R. **Hertz*** (1857 to 1894), physicist. Working on the fundamentals of mechanics he develops a **theory of elasticity**. Besides, he studies the field of electrodynamics, generating electromagnetic waves by the "Hertz oscillator", showing the similar nature of these kinds of invisible waves and light waves, which is the basis for wireless information transmission. Please note: The old frequency unit **Hz** is not an SI-unit. [183]

1895: *Armand **Peugeot*** offers bicycles and **automobiles equipped with pneumatic tires made of rubber** (from *J.P. Dunlop*, see also 1853 *C.N. Goodyear*); ***Benz & Cie*** in Mannheim (Germany), in 1900 the world's largest automobile manufacturer with 603 automobiles sold per year, equips automobiles with "pneumatics" not before 1897. Before this time, iron tires or solid rubber tires are used. (Previous development: In 1867 *Nikolaus A. **Otto*** develops a modern combustion engine and in 1876 a four-stroke engine; in 1883 *Wilhelm **Maybach*** and *Gottlieb **Daimler*** construct a lightweight and fast-running gasoline-operated engine; in 1885 *G. Daimler* installs a gasoline motor on a bicycle as the first motor bike and also on a ship, and in 1886 *Carl **Benz*** designs the first automobile, a three-wheeler.) [30, 209, 405]

1896: ***Jago*** constructs a cylindric **extruder** equipped **with a piston** to measure the flow behavior of dough. [378]

around 1900: *A.W. **Dow*** develops a procedure for testing the **ductility of bitumen**. A specimen "in the shape of a biscuit (biscotti)" is immersed in a water bath at the measuring temperature, fixed at both ends and stretched at a constant speed until the occurring strand is breaking (see also Chapter 11.2.14 b2). [405]

around 1900: Rapid development of the chemical industry particularly in Germany; for example, the global share of synthetic **dyes** which are produced there increases from 50% in 1881 to now 80%. [51]

13.3 Development between 1900 and 1949

1901: *Willem H. **Keesom*** (1876 to 1956), physicist. He describes interionic electrostatic interactions between molecules and particles showing permanent dipoles, known as "Keesom forces" in **colloid science**. [246]

1901: **Forel** uses **the term "rheology"** (Greek "rhein" means to flow; see also 1929). [137]

1902: *J.H. **Poynting*** and *J.J. **Thomson*** firstly report about mechanical models using springs and dashpots (see also *J.J. Thomson* 1888). [294]

1902: *Carl H. **Meyer*** starts **production of synthetic resins**, thermoplastic phenols and formaldehyds; e.g. as "phenoplastic" **adhesives** (Previous development: In 1872, *Adolf **von Baeyer*** discovers the polycondensation of both materials.) [216]

1903: *Leo **Ubbelohde*** (1876 to 1964) develops a **capillary viscometer** "showing a hanging level", and uses a logarithmic division for **viscosity/temperature tables** η(T); determination of **the dropping point of lubricating greases**; conversion tables of viscosity values given in Engler degrees into the **absolute measuring system** (see 1884 *C. Engler*). He writes a four-volume "handbook on chemistry and technology of oils, greases, waxes and resins". [97, 369]

1903: *C.H. **Stearn*** and *C.F. **Topham*** start **industrial production of viscose**. Previous development: In 1880 *Joseph W. **Swan*** applies for a patent on **"synthetic silk"** based on cellulose; in 1890 *Louis-Marie H.B. Compte de **Chardonnet*** founds a factory for "Chardonnet silk" or "Reyon". In 1892, *C. **Cross**,*

*E. **Bevan*** and *C. **Beadle*** develop viscose, modified cellulose, cellulose acetate, and later "viscose silk"; see also 1872 Celluloid. [209, 216]

1904: *Frederick T. **Trouton*** (1863 to 1922). He makes experiments on pitch, mixtures of tar and pitch, waxes, and further highly viscous materials, and finds the **"Trouton relation" between extensional viscosity** η_E **and shear viscosity** η as $\eta_E(\dot{\varepsilon}) = 3 \cdot \eta(\dot{\gamma})$ which is true for ideally viscous behavior, and for non-Newtonian behavior is used the **"Trouton ratio"** TR = $\eta_E(\dot{\varepsilon})/\eta(\dot{\gamma})$; see also Chapter 10.8.4.1a. He names η_E the **"coefficient of viscous traction"**. [365, 233]

1904: *W. **Rubel*** invents the offset printing process. [209]

1905: *Albert **Einstein*** (1879 to 1955), theoretical physicist. Concerning rheology, he establishes the **viscosity equation of suspensions containing spherical particles**, or of solutions with molecules assumed to be spherical, **without physical-chemical interactions**, with viscosity η of the suspension, viscosity η_s of the suspension liquid and volume fraction Φ of the particles: $\eta = \eta_s (1 + 2.5 \cdot \Phi)$. He interprets the "Brownian motion" (see *R. Brown* 1827) as an effect of molecular motion occurring due to heat energy causing collisions of the suspended particles (colloids) and he uses this concept as an **argument for the existence of atoms and molecules**. This hypothesis can be seen as **the beginning of useful microstructural theories**. Besides, he develops the "hypothesis of energy quanta" for photons (light quanta), therefore enhancing the quantum theory. He states a hypothesis on the relation ("relativity") of time and space on the basis of the constancy of the light speed (of c = 299,792.458km/s). He finds also the "Einstein equation": E = m · c² describing the relation between energy E and mass m. [115, 32, 141, 331]

1907: *Ludwig **Prandtl*** (1875 to 1953), engineer, physicist. He studies fluid mechanics, aero- and hydrodynamics. In 1904, he develops the concept of boundary layers, and in 1910 the theory of **turbulent flow** using the "Prandtl number". He designs the **"Prandtl pipe"** to determine flow velocity in pipes via pressure values, and in 1907 to 1922 he investigates **subsonic** ("infrasonic") **and supersonic** ("ultrasonic") **flow**, he introduces model tests in **wind-tunnels**, he develops a concept on aircraft wings, and he studies also solid-state mechanics (*L.P.* see also 1924). [295, 46]

1909: *E.J. **Stormer*** constructs a **controlled torque rotational viscometer with a rotating inner cylinder** which is commercially available until the 1960s. Measuring systems are concentric cylinders, fork-shaped paddles, and flag-like vanes (see also Chapter 10.6.3: the "Krebs-Stormer viscometer" and "Krebs-spindles"). The stirrer shaft carries a pulley with a rolled up string which is moving downwards via a **weight-and-pulley drive** using weight pieces with a mass of e.g. 25 to 500g hanging on a string, moving downwards due to gravity. The velocity of the falling weight is slowed down by the viscosity of the sample. Using a stop watch, there are two options for the measurement when setting a certain weight: a) Time t is measured for a defined path of the falling weight downwards; or b) In a defined time t the path downwards is detected or the number of revolutions of the measuring cylinder is counted, respectively, and the rotational speed or angular velocity is calculated from this. Disadvantage of both options: The flow velocity obtained depends on the viscosity of the sample and vice versa, and therefore, viscosity values of non-Newtonian liquids cannot be compared in a scientific sense when using the same weight. [349, 18]

1909: *Jean B. **Perrin*** (1870 to 1942) makes experiments to determine Avogadro's number on the basis of *A. Einstein's* hypothesis on the existence of atoms (see 1905). [31]

1910: The beginning of the **"era of plastics"**. *Leo H. **Baekeland*** (1863 to 1944), chemist, industrialist, starts the **industrial production of "Bakelite"**; discovered in 1906, US patent in 1907/1909, German patent in 1908/1911: "heat-and-pressure patent" for phenol/formaldehyde resins, which is the first fully synthetic plastic material ("phenoplastics"). [16, 29, 102]

1912: *George F. C. **Searle*** (1864 to 1954). He designs a rotational viscometer with a concentrically rotating inner cylinder, driven by a weight-and-pulley system. [330, 18]

1913: *Emil **Hatschek*** (1869 to 1944). He **improves the** electrically driven **"Couette apparatus"** making this cylinder viscometer usable for practical studies. In diagrams is shown e.g. the angular velocity on the x-axis, and the "torsion" in form of the deflection angle which corresponds to the torque on the y-axis (see also *E.H.* 1928). [175, 378, 413]

1913: *Henry **Ford*** (1836 to 1947). Also he introduces **assembly-line work for large-scale industrial motorcar production**; in 1908, "model T" is launched of which 15 million are sold until 1927. Previous development: *Ransom **Olds*** takes an example from the slaughter houses of Chicago and introduces assembly lines for the mass production of automobiles, called "Oldsmobiles", his company produces 600 cars in 1901, and 2500 in 1902, and 5000 in 1904. In the beginning, drying time for several layers of base coat, first and second top coats takes up to four weeks per car. Therefore, company **DuPont** develops rapidly drying **nitrocellulose (NC) coatings** containing merely 30% of solid portion, which finally requires a drying time of "only" about 15h for all layers; these kinds of coatings are soluble in the cold state, but there is always a risk since they are very **inflammable. Since 1921, colored NC coatings** are available, up to then all cars are black (see also Chapter 13.5: 2000). **Since the 1920s, viscosity is tested using "Ford flow cups".** [74, 209]

1913: The first **synthetic varnishes appear on the industry market**, e.g. "acetylcellulose", "synthetic silk", urea resins; also as raw materials for color coatings. As a consequence, more and more **industrial companies are beginning to introduce scientific laboratory tests, e.g. for testing viscosity.** [102]

Note: The first company-owned laboratories are established already at the end of the 1890s, both for regular scientific analysis and also for routine tests in **quality control**, particularly to achieve a constant quality to **meet the requirements of national institutions**, such as railway, post or mail, and military. The first railway in England is operated from 1825 on by a private company, and most of the railway companies are owned by the state not before around 1900. Typical **lubricants** still consist of **natural products**; e.g. for a cart-grease "for heavy vehicles such as miller-cars" used until the 1950s, is documented 25 parts **ship's pitch**, 25 parts **fir resin**, 25 parts **wax** and 10 parts **pork fat** (the remaining 15 parts are not known). [400]

Craftsmen and painters continue to use very simple, manual tests to check flow behavior for daily use. Hot soluble paints, so-called **oil paints**, are produced by boiling oils such as linseed oil, fish oil and wood oil; natural resins such as shellac from the colonies, colophonium, balsam resin and wood resin; binders such as starch, cellulose, bone glue, bitumen and proteins; solvents such as turpentine; and pigments such as carbon black and white lead. All these components can be mixed and dispersed only in a molten state, i.e. at high temperature. **Adhesives** are often smelly mixtures of resins, rubber, turpentine, with cayenne pepper and lead oxide, also used as medical plasters. Usually, **viscosity is merely evaluated by the trial-and-error method, for example, by observing** the drip-off or flow-off behavior after a metal rod is dipped into the molten paint mass and then rapidly pulled out again. Typical **testing of mastic asphalt** on building sites around 1900 and still some decades later: The mixture in the mobile boiler vessel is approved for use if a previously immersed piece of wood can be removed without "too much" asphalt remaining on it. [74, 102, 405, 412]

1913: The first **symposium on "the viscosity of colloids"** takes place in London. Until then, the viscosity of solutions and "colloids" is usually considered a material constant to be calculated simply from measured raw data using the Hagen/Poiseuille relation. [378]

1915: *F.R. **MacMichael*** constructs an electrically driven, speed-controlled **rotational viscometer with a rotating cup**; speed n = 20min^{-1}, stationary inner disc of d = 60mm and 5mm thick, which is hanging on bottom of a **torsion wire to measure the torque**; viscosity specification in "MacMichael degrees". This type is sold until the 1940s, however, then with variable speed steps of n = 10 to 38min^{-1}. [232, 18]

1915: Start of **large-scale production of synthetic elastomers**, e.g. methyl-rubbers, by company **Bayer**. Previous development: In 1910 *Fritz C.A. **Hofmann***, chemist, receives a patent on **poly-isoprene** as the first useful "synthetic rubber". [29, 74, 216]

1915: *K.W. **Wagner*** introduces the loss factor tanδ also for mechanical oscillatory tests, similar to the loss occurring in electric oscillating circuits; he performs tests on rubbers at frequencies of f = 350 to 4800Hz and at T = -5 to +45°C. [352]

1916: *Eugen C. **Bingham*** (1878 to 1945) termes concentrated clay suspensions and oil paints as **"plastic materials"** showing certain rigidity at rest. For him, these kinds of materials are not liquids since they show a **"yield value"**. In 1919, *Bingham and H. **Green*** describe **"viscoplastic liquids"** using the **Bing-**

ham model, illustrated by a dashpot (according to *Newton*) and a friction element (acc. to *Saint Venant*) connected in parallel (see also *T. Schwedoff* 1880). These kinds of materials are also named "Bingham liquids", showing **"plastic-viscous behavior"** or "plastic-dynamic behavior" (see also Chapter 3.3.6.4). Sometimes, this model is presented with an additional spring in series (acc. to *Hooke*). [42, 276, 378]

1917: Foundation of the standardization committee of the German industry, firstly as NADI (Normen-Ausschuss der Deutschen Industrie), later as NDI (predecessor of DIN).

1920: *H. Green*, using his **"Microplastometer"**, observes **wall slipping effects of dispersions**. He calls this behavior **"plastic flow"** (*H.G.* see also 1916).

1920: *Hermann Staudinger* (1881 to 1965), chemist. **He describes polymers as** rigid, rod-shaped **large molecules**. Until then it was assumed for all liquids showing superstructures to consist of "associations" of small individual molecules as "colloids", comparable to the "micellar concept" of surfactants. For the first time he uses the (German) terms "makromolekulare assoziation" (in 1922) and "Makromolekül" (in 1924) for unlinked rubber (caoutchouc), respectively. **He is the founder of modern polymer science** introducing **capillary viscometry** as the method **to determine the molar mass** of polymers and the degree of polymerization. Citation: "Knowledge about macromolecular materials was essentially pushed by rheological measurements". [343, 119, 216, 378]

1920: *W.R. Hess* differentiates between starch solutions which are "soft, plastic bodies" showing a yield point, and those which are "elastic liquids". [184]

1921: *Arthur Eichengrün* develops the first modern **injection moulding machine for polymer melts**. [216]

1923: *Geoffrey I. Taylor* publishes calculations of the critical limiting value, where vortices occur in measuring systems **with rotating inner cylinder** due to **secondary flow effects** ("Taylor number" Ta, "Taylor vortices", and circular "Taylor cells"; see also Chapter 10.2.2.4a). [354, 233, 384]

1923: *Peter J.W. Debye* (1884 to 1966), physicochemist. He describes the interionic electrostatic **interactions** between permanent and induced dipoles of molecules and particles; "Debye forces" in **colloid science**. [29, 246]

1923: *Schalek* and *Szegvary* examine iron oxide dispersions, using the term "thixitropy"; literally: change due to contact or motion, here in the sense of phase change due to shearing (see also *Peterfi* 1927). [218, 377]

1923: *Farrow and Lowe* present a **model function of flow curves** for shear-thinning behavior, in the form of raw data.

1923/1925: *A. de Waele* and *Wolfgang Ostwald* (jun., 1883 to 1943) present a **model function of flow curves** for shear-thinning behavior without a yield point, also called the **"Power-Law", in the form of raw data** as $v = k \cdot p^m$, with the volume flow rate v and the pressure p when using capillary viscometers, or with the (angular) velocity v and the torque p, when using rotational viscometers. [96, 273, 29]

1924: The **Prandtl model** of "plastoelastic behavior", presented as a spring (acc. to *Hooke*) and a friction element (acc. to *Saint Venant*) connected in series. This is also termed a Prandtl/Reuss body, showing "elastic-plastic" or "plastic-static" behavior (*L.P.* see also 1907). [295, 215]

1924: *Heinrich Hencky* (1885 to1951) investigates "plastic deformation" and viscoelastic effects; definition of the **"Hencky strain"** $\varepsilon_H = \ln(L/L_0)$ as "natural extension", and the **"Hencky strain rate"** $\dot{\varepsilon}_H$ (see also Chapter 10.8.4.1b). [180, 276, 352]

1925: The **rotational "Turboviskosimeter"** developed by *Hans Wolff* and improved by *Hoepke* for testing the **consistency** of mineral oils, glycerol (85%), sugar solutions, paints such as nitro-cellulose (NC) solutions or consisting of oils such as linseed-oil, wood-oil, whale-oil, stand-oil (i.e., thickened oil by heating), benzine and pigments (e.g. zinc-white, lead-white, antimony-white, titan-dioxide, red iron-oxide, chrome-yellow, ocher). **A flag-like, S-shaped stirrer** is immersed into the liquid sample being **in a cylindric cup**. The stirrer is **driven by a weight-and-pulley system**, using weight pieces which are attached to a string. The velocity of the falling load is slowed down by the viscosity of the sample. There are two options: a) Using a certain weight G, time t is measured for a defined path s downwards (e.g between t = 5 and 35s for s = 1m); the disadvantage is here: Depending on the sample's viscosity, the tests are resulting in different rotational speeds or flow velocities; or b) By trial and error, the amount of weight is determined to reach the desired falling velocity v (i.e., path s per time t, for example, at v = 3 or 10 or 20cm/s); the advantage is here: Testing is performed always at a constant rotational speed, independent of the sample's viscosity. This is useful when comparing the flow behavior of different samples. Determination of the **"turbo-viscosity"** simply as V = k · G · t which is a relative value, with the device constant k depending on the stirrer and cup geometry used and also to the path s (statement: "V/10 shows the dimension of at least in Poise obtained with the Couette apparatus"). The following terms are used: **"plasticity"** for the degree of the viscosity dependence on the flow velocity, **"flow elasticity"** for not purely viscous flow behavior, and **"flow solidity"** for the yield point (in German: Fließfestigkeit). In diagrams is presented, for example, the flow velocity on the x-axis and the flow time t required for a defined path s on the y-axis. [413, 412]

1925: In Europe between 1920 and 1930, materials showing shear load-dependent viscosity are mostly regarded as structured liquids; *W. Ostwald jun.* uses the German term **"strukturviskos"** (literally "structure-viscous") to describe this kind of flow behavior. In the USA, these materials are seen as soft solids or **"pseudoplasts"**; this view is dominated by *E.C. Bingham's* opinion. [42, 276, 378]

1926: *Winslow H. Herschel* and *R. Bulkley* present a **model function of flow curves** showing a yield point. [182]

1926: Foundation of the International Federation of the National Standardizing Associates ISA (predecessor of ISO).

1927: *Hankoczy* constructs a **dough kneader** operating at a constant rotational speed, measuring the flow resistance in the form of torque values (see also *C.W. Brabender* 1930). [378]

1927: *S.B. Ellis* presents a **model function of flow curves showing zero-shear viscosity**. [120, 384]

1927: *A.E.H. Love* (1906 to 1927) publishes a textbook on elasticity.

1927: *Porter and Rao* present the **"Power-Law" using rheological parameters**. This leads to a better comparability of test results obtained when different instruments are used since this kind of analysis is not dependent on the device (see also *Ostwald/de Waele* 1923).

1927: *Peterfi* coins the term **"thixotropy"** (see also *Schalek* 1923). [352]

1927: **Furnace drying synthetic resins** are introduced requiring **research on temperature-dependent flow and curing behavior** (alkyd resins; developed by *Kienle*). [29, 102, 158]

1928: *Walter Bock, Eduard Tschunkur* and *Erich Konrad* develop an oil- and gasoline-resistant synthetic rubber, a **styrene rubber** ("Buna-S"), the first **automobile tires** with this rubber surface are launched in 1936. In the 1930s occurs **nitril rubber** ("Buna-N"), a **nitril butadiene rubber (NBR)** or acrylnitril-butadien rubber, since 1938 launched as "Perbunan". [29, 219, 398]

1928: *Faber* describes the **creep process** calling it "plastic flow".

1928: *E. Hatschek* performs tests using the enhanced **"Couette/Hatschek rotational viscometer"** with a torsion wire as the torque sensor, cylinder measuring systems of R_e = 55.2mm and R_i = 35 or 49.6mm, and a minimum speed of n = 1.2min^{-1} (*E.H.* see also 1913). [175]

1928: *Herbert M. F. Freundlich* (1880 to 1941) publishes a paper on **thixotropy**. [142, 366]

1928: *Otto Röhm* (1876 to 1939) succeeds in the polymerisation of methacrylates; start of **production of PMMA** (polymethyl-methacrylate) by company **Röhm & Haas**; since 1933 launched as "Plexiglas". [216]

1929: Foundation of the world's **first "Society of Rheology"** (SOR) in the USA by *E. Bingham*. He favors the **term "rheology"** (see also *Forel* 1901). [378]

The term **"rheology"** derives from the statement **"panta rhei"**, i.e., **"everything flows"**, which **is attributed to Herakleitos of Ephesos (Heraclitus**, 540? to 480 BC) by later authors. However, *Heraclitus* does not study the physical basis of flow processes, he is a **philosopher**. He choses the illustrative image of a river as a symbol of continuing motion to express that everything happening in the world is exposed to an eternal circulation of changes, undergoing alternation and is becoming and fading, moving on and progressing, nothing remaining permanently the same, since stopping or standing still is equivalent to death. Similar ideas can be found in many natural-philosophical systems all over the world, e.g. of the Indians in Northern America, or in East Asia. For example, the citation: **"The fact that all is changing is the**

only thing that does not change" from one of the five classic Chinese books called **Yijing** (or I-Ging), i.e., the "book of the changes", of which great parts came into existence between the 11th and 5th century BC. A corresponding statement is given in the Book Zhuangzi, i.e. "Master Zhuang", from the assumed author **Zhuang Zhou** who lived around 350 BC; this book was composed between 400 and 100 BC. Also later philosophers published similar ideas, e.g. Arthur **Schopenhauer** (1788 to 1860), inspired by philosophy from India, putting it in words like "the only constant thing is change", and everywhere is "eternal progress like an endless flow". [23, 312, 374]

1929: Karl **Weissenberg** (1893 to 1976) and B. **Rabinowitsch** present a **correction** to calculate the **"true shear gradient" of non-Newtonian liquids in capillary viscometers**. Besides, he works in the field of X-ray crystallography designing the "Weissenberg X-ray goniometer" (K.W. see also 1947 and 1951, and Chapter 10.8.2.5). [383]

1929: Using **capillary viscometry**, Elmer O. **Krämer** coins the term **limiting viscosity value** of "solvated colloids" (meaning: of **diluted polymer solutions**), if the concentration value approximates the value zero (c → 0). Current terms used are "intrinsic viscosity" or "Staudinger Index". [212]

1929: R.V. **Williamson** presents a **model function of flow curves**. [276]

1929: H. **Jeffreys** terms the behavior of e.g. bitumen as **"elastoviscous"**, meaning viscoelastic liquid; and he regards materials showing Kelvin/Voigt behavior (see 1890) as **"firmoviscous"**, meaning viscoelastic solid. [202]

1929: E.W. **Dean** and G.H.B. **Davis** present their method for determining the **viscosity index VI** in order to charaterize **temperature-dependent behavior of petrochemicals** such as mineral oils, by measuring the kinematic viscosity at the two temperatures T = +40 and +100°C (today as ISO 2909; see also Chapter 11.4.1.3). [88, 95]

1930: Fritz W. **London** (1900 to 1954), physicist. He describes **interactions** between non-polar molecules and particles due to fluctuating dipoles; "London forces" in **colloid science**, which are of quantum-mechanical nature (F.W.L. see also 1938). [29, 128, 246]

1930: Werner **Kuhn** (1899 to 1963) finds the **coil structure of macromolecules**, instead of the rod-shaped structure as previously assumed. He studies particularly unlinked rubbers, polystyrene and gelatin (see H. Staudinger 1920 and H. Mark 1932). [214, 119]

1930: C.W. **Brabender** contructs **dough kneaders**, termed "Amylograph", "Farinograph" and "Extensograph" (based on Hankoczy's prototypes, see 1927). [50, 378]

1930: Start of **production of polystyrene** (PS) by company **IG Farben**. [327]

1931/1950: A. **Nadai** publishes the textbooks "Plasticity", presenting a **theory on "plastic behavior"**, and "Theory of Flow and Fracture of Solids".

1931: Markus **Reiner** (1886 to 1976) presents a **classification of all possible rheological properties**, devided in six groups (M.R. see also 1945, 1960 and 1968). [303, 378]

a) rigid or **Euclidean body**: completely inelastic, absolutely not deformable

b) deformable body: b1) completely elastic or **Hookean body**, b2) not completely elastic body

c) viscous liquid: c1) non-Newtonian liquid, c2) classic viscous or **Newtonian liquid**

d) ideal or **Pascalian liquid**: inviscid, without viscosity, completely friction-less

1930s: Discussions by E. **Bingham**, M. **Reiner** and **Straub** about the behavior of reinforced concrete, marble and limestone: Do these materials show viscous flow according to Newton or plastic flow according to Saint Venant?

1931: Melvin **Mooney** (1893 to 1968) presents the **"shearing disk viscometer"** equipped with a **parallel-disk measuring system** (M.M. see also 1934 and 1940). [255, 233]

1931: **Paints based on synthetic alkyd resins** are launched, showing increased resistance. [74]

1931/1935: Start of **large-scale industrial production of PVC** (polyvinyl-chloride) by company **IG Farben**; since Georg **Wick** develops a process to produce PVC **at temperatures above 150°C**. Previous development: In 1912 Fritz **Klatte** suceeds in the synthesis of vinylchloride. [216, 327]

1932: *Hermann* **Mark** investigates e.g. cellulose, silk, unlinked rubber, polystyrene; he finds that **polymer macromolecules are flexible**, in contrast to the "macromolecular theory" (see also *H. Staudinger* 1920 and *W. Kuhn* 1930). The type of intramolecular bonds is no longer up for discussion, now the emphasis is on the intermolecular bonding forces. He assumes that the specific viscosity is proportional to the molar mass and develops of the **"Mark/Houwink relation"** (see also Chapter 11.4.1.2). In 1930 his book is published about "the structure of highly polymeric organic natural substances". [244, 44, 132]

1933: *Fritz* **Höppler**, chemist, working for company **Haake**, producing starch. He designs a **falling-ball viscometer**, which is available on the market since 1934. Later he constructs the "consistometer" in the form of a penetrometer. [194, 324]

1933: *R.* **Eisenschitz** and *Wladimir* **Philippoff** perform **undamped, continuous oscillatory tests** to study the behavior of liquids such as glycerol, gum arabicum, honey, cellulite in dioxin, and of "colloidal systems" as well as of "plastic materials showing flow rigidity". They use an **"electromagnetic viscometer"** with a mechanical **vibrator system**: Immersing an oscillating needle into a liquid sample and presetting a force via electromagnets, they determine both the deflection amplitude using a micrometer screw and the frequency values occurring between f = 30 and 630Hz via the acoustic analysis of the tone in a telephon receiver; calibration by use of standard fluids. They analyze only the real part η' of the complex viscosity as a function of the frequency (but they do not detect the elastic part or the phase shift δ). In 1936, *W.P.* presents a **model function of flow curves** (*W.P.* see also 1968). [116, 288, 237, 352]

1934: *M.* **Mooney** and *R.H.* **Ewart** use a **cone-and-plate measuring system**, as a "conicylindrical viscosimeter" (*M.M.* see also 1931 and 1940). [255, 233]

1934: *Robert* **Murjahn** develops **water-based (WB) acrylic emulsion paints**, as polymer dispersions containing water and color pigments. In contrast to conventional solvent-borne paints, WB polymer dispersions do not show ideally viscous flow behavior. However, useful WB paints are available not before the 1950s or even the 1980s (see also 1975 and 1987). [102]

1935: *J.M.* **Burgers** presents a **combination of Maxwell model and Kelvin/Voigt model for viscoelastic materials**, the so-called "Burgers model" (see also *J. Thomson* 1888). [60]

1935: First machines are introduced for **blowing plastic hollow-ware**. [216]

1936: *H.* **Eyring** presents a **model function of flow curves**, which is later called **Eyring/Prandtl/ Ree** EPR or Ree/Eyring model. [130, 300, 276]

1936: Development of **unsaturized polyester** (UP) **resin systems**. [102]

1937: *R.* **Houwink** studies elasticity and "plasticity" of typical technical materials of his time: resins, varnishes, paints, guttapercha, balata (a rubber-like natural resin), rubber, cellulose, starch, dough, asphalt, clay, sulphur and glass. [195]

1938: *George W.* **Scott Blair** (1902 to 1987) publishes **the first textbook with the term "rheology" in the title**. Nevertheless, commonly still terms such as plasticity, fluidity and consistency are used (*G.S.B.* see also 1942). [329, 378]

1938: *H.* **Roelig** constructs an **apparatus to mechanically generate harmonically oscillating forces**, using a combination of a pre-stressed spring and an eccentrically rotating mass. He **determines the phase shift between the preset force and the resulting deformation by an optical system**. [308, 384]

1938: *Pjotr L.* **Kapiza** (1894 to 1984), discovers **suprafluidity** or **superfluidity** of Helium showing an **inviscid phase** (without viscosity) at T < 2.13K (based on works by *Satyendra N.* **Bose** of 1924, and a concept of *F.* **London** of 1938, see also 1930, *Heinz* **London**, and later in 1941 also *Lew D.* **Landau**. [331]

1938/39: *Rayleigh* performs a **long-term loading experiment** over 20 months on a glass sample at room temperature. He finds that glass shows no yield point since it is creeping. His conclusion: Even materials with very high viscosity show **viscoelastic** and not "plastic" **behavior** (to the discussion on the nature of glass: see also Note 2 in Chapter 8.6.2.2a).

1939: *C.F.* **Goodeve** discusses possible mechanisms of **thixotropic behavior**, either as orientation or as reduction of the size of particles when shearing. [44]

1939: *A.P. **Aleksandrov*** and *Y.S. **Lazurkin*** construct an **apparatus to mechanically generate harmonically oscillating forces**, using a cam compressing a plate spring, in the frequency range of f = 0.017 to 33 Hz. They **determine the resulting deformation amplitude by an optical system** (but without detection of the phase shift). [7, 384]

1939: *Eric W. **Fawcett*** and *Reginald **Gibson*** succeed in the **large-scale synthesis of poly-ethylene** (PE) via high-pressure polymerization. Previous development: In 1933 synthetis of PE; in 1937 company **I.C.I. starts production of PE** in a pilot plant; see also *K. Ziegler* 1952: low-pressure PE. [216]

1939: Synthesis of **polyethylene (PE) waxes**. Previous development: Originally, **natural waxes** such as beeswax, shellac wax, wool wax, carnauba wax, candelilla wax and rice oil wax are used. Since 1830, paraffins are produced from beech wood and coal tar, and since 1860 paraffins from crude oil, brown coal, peat, and oil shale. Typically, waxes are used as lubricants, corrosion protection and for treating flooring; and in the past also for drive tapes of engines made of leather and for horse hoofs. [74]

1939: Start of **large-scale production of synthetic polyamides** PA 66, called "Nylon". In 1929, there are first trials by *Wallace H. **Carothers***, US patent 1937/1938 for spinnable PA fibers. Since 1940 nylons (stockings) are launched by company **DuPont**. In 1938/1940, *Paul Schlack* gets a German patent for spinnable PA, called "Perlon", produced by company **IG Farben**. **These developments give a strong impetus to the polymer industry, and therefore, also to polymer rheology.** [68, 108, 216, 327]

1930s/1940s: Due to the Second World War and due to a large number of immigrants who have to leave Europe for political reasons, **polymer research** in the USA is booming, **aiming to replace natural materials by synthetic polymers**. For example, there is a risk of supply with shellac resins as the base substance for isolation material. Therefore are founded, for example, the "Synthetic Rubber Project", the "Polymer Research Institute", the "Journal of Polymer Science" and the "Journal of Applied Polymer Science". **Practice-oriented rheology and rheometry** begins to spread **in industry**, e.g. in the form of **viscosity measurements**. Industrial production of synthetic polymers and rubbers causes an increase in **studies in the field of viscoelasticity.** [132, 237]

1940: *M. **Mooney*** presents a **theory of rubber elasticity** (*M.M.* see also 1931 and 1934). [255]

1941: *Konrad E.O. **Zuse*** (1910 to 1995), engineer. He presents the first **program-controlled electro-mechanical digital computer** "Z3", equipped with 2000 relays, operating with binary floating point numbers. Previous development: In 1936 *Zuse's* mechanical computer "Z1", and in 1939 the electro-mechanical "Z2"; in 1944 *Howard **Aiken*** invents independently a computer; in 1946 *J.W. **Mauchly*** and *J.P. **Eckert*** present the "ENIAC", an "electronic numerical integrator and computer" with 18,000 electron tubes and in a calculation rate of 5000 additions per second. [408, 22, 267]

1941: *F. **Patat*** and *G. **Seydel*** construct a **viscometer, driven by a weight-and-pulley system**, using weights between m = 0.5 and 40g, to achieve "tangential pressures" of 15 to 12,500dyn/cm^2 (i.e. shear stresses of 1.5 to 1250Pa), and concentric cylinder measuring systems of d_i = 10/25/40 or 60mm and d_e = 15/30/44 or 100mm, or cylindric-conical bobs of d = 14/25 or 50mm; later sold as "Kämpf-Viskosimeter". It is equipped with an **electromagnetic friction-free bearing** of the spindle, and a temperature controlled jacket for T = -60 to +150°C using a liquid bath. By a light barrier combined with an electrically switching stopwatch of an accuracy of 0.1 or 0.01s, the time is determined of a weight to fall 100cm downwards. Options: **"Struktur-Viskosimeter"** or **"Thixotrometer** device" to measure time-dependent viscosity, **"thixotropy", "dilatancy"** or **"elastic after-effects"**. Using a rotary plate of d = 170mm on which the measuring cup is attached, and driven by an electric synchronic drive at a constant speed of n = 70 or 140min^{-1}, a two-step procedure is performed: at first a pre-shear step as "stirring" at a constant speed of the rotary plate, i.e. here, the cup is rotating while the bob is stationary (without measuring a torque since there is no sensor available); then a second step using the rotating bob which is now driven by the falling weight on a string, while the bottom plate with the cup stands still. [280, 341]

1941: *Rex **Whinfield*** and *James **Dicksen*** develop **PET** (polyethylene-terephthalate), a polymer which is impermeable for gases. [74]

1942: Start of **industrial production of silicones** (SI) being flexible also in the cold state. Previous development: In 1900 *Frederic S. **Kipping*** discovers SI, and in 1940 *Eugene **Rochow*** finds an industrial synthesis process for SI; see also 1960. [216]

1942: *Harry **Kloepfer*** presents **pyrogenic silica**, since 1944 large-scale production by company **Degussa**, launched as "Aerosil". Originally developed as a reinforcing filler for car tires, it is also used as a rheological additive, e.g. as a viscosifier or gellant showing thixotropic properties. [91, 335]

1942: *George W. Scott Blair* presents a **classification of all possible rheological properties**, devided in nine groups (*G.S.B.* see also 1938), [329, 378]:

a) **idealelastic**: a1) Hookean, a2) non-Hookean

b) non-idealelastic: completely recoverable

c) **plastic flow**: c1) **plastoelastic**, c2) plastoinelastic, c3) Bingham

d) viscous, non-Newtonian: d1) **viscoelastic**, d2) viscoinelastic

e) **viscous flow**: Newtonian

1943: Industrial application of **polyurethanes** (PU); in 1937 patent by *Otto **Bayer*** (1902 to 1982); use as elastic synthetic fibers, called "Elastan". [25, 173, 247]

1945/1948: *M. **Reiner*** and *R.S. **Rivlin*** **propose to use the theory of viscous liquids also for polymer melts** (M.R. see also 1931, 1960 and 1968). [303]

1945: Company **Brookfield** starts serial production and sale of **rotational viscometers** based on developments since 1934. Using an **electric drive**, controlled speed with two speed steps, a **spiral spring as a torque sensor**, and spindles in the form of disks and pins, these kinds of devices are widely spread (see also Chapter 10.6.2 and Figure 10.11).

1946: *M.S. **Green*** and *A.V. **Tobolsky*** present the **"transient network model" for unlinked polymer molecules**, as a network in a time-dependent, transitional state. [220]

1946: *R.J. **Russell*** measures **normal stresses**, using **cone-and-plate** and **parallel-disk measuring systems**. [233]

1946: Start of **production of PTFE** (poly-tetrafluor-ethylene) by company **DuPont**, called "Teflon" (development of PTFE in 1938 by *Roy **Blanket***). [74, 327]

1946: Foundation of the **International Organization for Standardization ISO** (see also 1926 ISA).

1947: *K. **Weissenberg*** observes "transversal elasticity in elastic liquids" and the "rod-climbing effect" on a rotational axis, later called the **"Weissenberg effect"**. He firstly describes scientifically the effects of **normal stresses in shear flow** (*K.W.* see also 1929 and 1951). [233]

1947: The first **epoxy resins** (EP) occur on the market; since 1938 they are technically applicable. [54, 158]

1947: *John **Bardeen*** (1908 to 1991), *Walter H. **Brattain*** (1902 to 1987), and *William B. **Shockley*** (1910 to 1989) discover the **transistor effect**. Previos development: In 1939 *Walter **Schottky*** states theoretical basics of the semi-conductor technology. Transistors on silicium basis are replacing more and more the classic electron tube, opening the door for the **miniaturization in electronics**. [74]

1948: The first "International Congress on Rheology" takes place in Scheveningen.

1948: *R.S. **Rivlin*** presents his **neo-Hookean basic equation**.

1948: The **"Hercules Hi-Shear Viscometer"** (HHSV), with a concentric cylinder system at a shear gap of 0.1 to 5mm, and a rotating inner cylinder at a controlled speed of n = 5 to 6500min[-1] (specification: shear rate range up to 110,000s[-1]). The viscous drag of the sample causes a deflection of the cup producing a torque which is detected. [363, 384]

1948: *J. **Rabinow*** publishes the first paper on the **magneto-rheological effect**. [297, 237]

1949: *L'Hermite* coins the term **"anti-thixotropy"**, meaning rheopexy.

1949: *W.M. **Winslow*** publishes the first paper on the **electro-rheological effect**. [391, 237]

1949: *E. **Hodgkinson*** presents a **model function for non-idealelastic behavior**: $\tau = a \cdot \gamma + b \cdot \gamma^2$ (see also *G. Bülfinger* 1729, and *C. Bach* 1888).

In the 1940s: It becomes widely **accepted** amongst scientifically working rheologists **that the behavior of both liquids and solids can be described as viscoelastic behavior**.

13.4 Development between 1950 and 1979

1951: *Karl* **Weissenberg** presents the **"Rheogoniometer"** (R1) **for rotational and oscillatory tests**, with controlled strain or speed mode, torsion wire or torsion bar, **normal force sensor**, cone-and-plate measuring systems with rotating bottom plate, as the first rheometer with an **air bearing**. **This instrument can be regarded as the first scientific rheometer** (*K.W.* see also 1929 and 1947). [383, 174, 233, 237, 307]

1952: *Karl* **Ziegler** develops **low-pressure polyethylene** (PE) at company **Hoechst**. Due to the reduced costs of this technology, PE becomes the first mass-produced polymer (see also 1939 high-pressure PE). [327]

1953: *Alfred G.* **Epprecht** presents a **rotational viscometer** with controlled speed (see also 1960). [122]

1953/1956: *P.E.* **Rouse** and *B.H.* **Zimm** present the **"bead-and-spring model" for unlinked polymer molecules** (B.Z. see also 1962). [311, 404]

1953: *Hermann* **Schnell** develops **polycarbonate** (PC), in 1958 launched by company **Bayer**, as "Makrolon". [74]

1954: The **"Ferranti-Shirley viscometer"** with cone-and-plate system, a single constant speed, a spiral spring as the torque sensor, and an electric sensor with a light signal to check the contact between cone and bottom plate (e.g. in ASTM D3245). [233, 384]

1954: Company **Haake** presents its first **rotational viscometer** (type "Rotovisco RV"), with controlled speed (see also *F. Höppler* 1933).

1954: *Giulio* **Natta** succeeds in the synthesis of **polypropylene** (PP) at company **Montecatini**. [216, 327]

1954: *Arthur S.* **Lodge** (1942 to 2007) presents an **extended transient network model** for unlinked polymer molecules (see also *M.S. Green* 1946). [229]

1955: **WLF relation** and **time/temperature shift** (TTS method) of unlinked polymers by *M.L.* **Williams**, *R.F.* **Landel** and *John D.* **Ferry** (*J.D.F.* see also 1970). [389]

1956: *F.* **Moore** and *L.J.* **Davies** use a "double-viscometer" which is a **double-gap cylinder system**, with a point bearing for the rotating hollow cylinder on top of a stationary inner cylinder which is mounted in the center of the cup. [256, 384]

1957: *E.B.* **Bagley** presents an equation for the **correction of inlet flow disturbances** occurring **in capillary viscometers**. [17]

1957: Due to improvements, **polyurethane resins** are increasingly used in industry (see also 1943, PU). [173, 247]

1958: *W.P.* **Cox** and *E. H.* **Merz** present the **Cox/Merz relation**: Unlinked polmers show similar behavior in the low-shear range as can be illustrated by the same shape of the curves of complex viscosity from frequency sweeps/oscillatory tests and shear viscosity from flow curves/rotational tests (see also Chapter 8.8 and Figure 8.47). [82]

1958: The **"Weissenberg Rheogoniometer"** is now available on the market (as types "R7", "R8", etc.).

1958: *A.W.* **Sisko** presents a **model function of flow curves showing infinite-shear viscosity**. [338, 155, 366]

1959: *N.* **Casson** presents a **model function of flow curves showing a yield point**, using a square root function; originally it was developed for printing inks. [70, 384]

1959: *I.M.* **Krieger** and *T.J.* **Dougherty** present a **model function of viscosity curves showing zero-shear viscosity and infinite-shear viscosity**. [213]

1959: In automotive industry, **electrostatic spray coatings** are introduced, **requiring exactly adjusted viscosity values for the automatic coating equipment**. [102]

1960: Company **Contraves** presents its first **rotational viscometer** (type "Rheomat RM"), speed-controlled, which is an enhanced "Epprecht Viscometer" (see also 1953).

1960: Introduction of the **SI system of units** (French: Système International d'unités). This m/k/s system is based on the following basic units:

m (**meter:** length); kg (**kilogram:** mass); s (**second:** time); K (**Kelvin:** temperature); **mol** (amount of a material, 1mol equals 6.022 · 10²³ atoms or molecules, which is Avogadro's number); A (**Ampere:** electric current); cd (**candela:** light intensity).

For example in Germany by legislation, SI units have to be used since 1978. Nowadays the SI system is established all over the world, besides of Liberia, Birma/Myanmar, and the USA. [59, 6, 397]

1960: *M.* **Reiner** presents his actualized **classification of all possible rheological properties**, devided in five groups (*M.R.* see also 1931, 1945 and 1968), [303]:

a) rigid body or **Euclidean body**, showing an "infinitely high" shear modulus

b) elastic body or **Hookean body**

c) **plastic body according to Saint Venant**

c) viscous liquid or **Newtonian fluid**

d) ideal fluid or **Pascalian fluid**, showing an "infinitely low" viscosity

1960: **Silicone resins** (SI) are introduced to the market (see also 1942, silicones). [102]

1960: For the first time, the global production of **synthetic rubber** surpasses the production of natural rubber. [29]

1961: Company **Brookfield** presents a **process viscometer with a pneumatic drive** using compressed air (type "Viscosel").

1961: *P.* **Rehbinder** presents a **creep apparatus** for presetting low-stresses, specified is a detection range of shear rates of lower than $10^{-6}s^{-1}$. [18]

1961: *K.* **Steiger**-*Trippi* and *A.* **Ory** present a **model function of flow curves**. [346]

1962: *B.H.* **Zimm** and *D.M.* **Crothers** construct a **rotational viscometer** with concentric cylinders made of glass, the outer one is stationary and the inner one, containing small steel pellets, is driven from outside by a rapidly rotating magnet. The preset is a constant, very low torque in the range of M_{min} = 10^{-9}Nm = 1nNm, for testing dilute aqueous solutions (*B.Z.* see also 1953; see also 1969 *A. Williams*). [404, 215, 237]

1962: **Polyamid resins** are introduced to the market by Company **DuPont** (see also 1939). [216]

1964: Development of electrostatic **powder coatings** as an alternative to liquid coatings. However, useful powder coatings are available not before the 1990s (see also 1997). [74, 102]

1965: *M.M.* **Cross** presents a **model function for viscosity curves showing zero-shear viscosity and infinite-shear viscosity**. [83, 384]

1965: *Hanswalter* **Giesekus** presents an **equation for behavior in the non-linear range** (*H.G.* see also 1985). [155]

1965: *W.W.* **Graessley** presents an **entanglement theory for unlinked polymer molecules**. [163]

1967: *Sam* **Edwards** presents an entanglement theory for unlinked polymer molecules.

1967: *Friedrich R.* **Schwarzl** presents a concept of **linear viscoelastic transformation**, using the **data conversion** method. [327]

1968: *M.* **Reiner** and *W.* **Philippoff** present a **model function of flow curves** (*M.R.* see also 1931, 1945 and 1960, and *W.P.* see also 1933).

1968: *Jack J.* **Deer** presents a **CS rheometer** for controlled stress tests, based on a re-designed Weissenberg Rheogoniometer (see 1951 and 1958); it is equipped with an air bearing, and a pneumatic **air turbine drive** to preset torque values which are therefore depending on the flow-rate of the air (min/max. torque in 1970: 0.1/10mNm, and in 1978: 0.01/10mNm). *Peter* **Finlay** adds the first **electronics**, later improvements are based on microprocessors. Since 1972 is developed an electric **asynchronous induction drive** which is on the market since 1982. Here, the rotating component of the motor is a copper cup, later replaced by an aluminum cup, which is dragged due to inductive effects generated by a rapidly rotating electromagnetic field, and therefore, this kind of drive is also called a **"drag-cup motor"**. Asynchronous means: The motion of the rotor shows a certain delay compared to the exciting field. **Detection of the deflection angle or the rotational speed**, respectively, of the measuring bob, **at first by use of a capacitive distance detection** via an inclined plane, a snail-cam **and an inductive transducer,** and **later by an optical encoder via a sectored disk** which is **"cutting" a light beam into countable impulses when rotating** (see also company Carrimed 1983). [89, 18]

1968: **Van der Wal** constructs a **"torsional creep apparatus"** which is mechanically driven by weights, with an **electromagnetic angle detection**, the resolution is specified with $\Delta\varphi = 0.01° = 0.16$mrad. [327]

1968: *A.* **Képès** designs the **"Balance Rheometer"** with an eccentrically rotating hemispherical measuring system, generating oscillations mechanically in a frequency range of f = $3 \cdot 10^{-5}$ to 20Hz, [205, 152, 215, 233]. The measuring curves are analyzed manually via **"Lissajous figures"** in the form of ellipses when measuring in the LVE range (according to *Jules A.* **Lissajous**, 1822 to 1880; see also Chapter 8.3.6). [227]

1968: *P.J.* **Carreau** presents a **model function of viscosity curves showing zero-shear viscosity and infinite-shear viscosity**. [69]

1968: *G.C.* **Berry** and *T.G.* **Fox** present a relation between **zero-shear viscosity** η_0 **and molar mass M** for concentrated polymers; η_0 is proportional to M when M < M_{crit} and η_0 is proportional to $M^{3.4}$ when M > M_{crit}, with the critical molar mass M_{crit} (see also Chapter 3.3.2.1a). [37, 220]

1968: Technical use of the **liquid crystals**, e.g. for electro-optical displays (LCDs). [74]

1968: Production of **UV-curing and electron beam (EB)-curing adhesives and coatings** which therefore are immediately dry (see also 1982 and 1995). [74]

1969: *J.* **Meissner** constructs an **extensional rheometer** in the form of a **uniaxial stretching apparatus** to measure extensional viscosity η_E. [233]

1969: *A.R.* **Williams** improves the **"Zimm/Crothers Viscometer"** (see *B. Zimm* 1962), using the stator of an alternating current (AC) induction motor instead of a rotating magnet; the advantage is the possibility to preset variable shear stress values now. [18]

1969: Market introduction of **epoxy resin systems as powder coatings** (see also EP 1947, and powder coats 1964). [102]

1970: *Philip* **Sherman** presents a device, with a cup driven by a weight-and-pulley system, using **ribbed cylinders to avoid slip**, with the inner cylinder hanging on a torsion wire. At first, the angular deflection is detected visually via a light beam and a mirror, and later is used an electronic transducer to detect the rotational angle, e.g. for creep tests; specification of a minimal shear stress value of 0.5Pa. [18]

1970: Company **Rheometrics** presents its first **rheometer** (type "RMS"); with an air bearing and a **synchronous drive** for controlling deflection angle and rotational speed, respectively.

1970: *J.D.* **Ferry** describes linear viscoelastic transformation using **data conversion** (*J.D.F.* see also 1955). [134]

1971: *Pierre G.* **de Gennes** (1933 to 2007) presents the **"reptation model"** to explain the **motion of unlinked polymer molecules**, as a slow motion comparable to creeping snakes. [90]

1975: Increasing use of **solvent-reduced**, odorless and water-dilutable synthetic **coatings and adhesives as water-based (WB) emulsions**, e.g. urethane-modified alkyd resins (PU, see also 1943 and 1957; and WB coats, see 1934 and 1987). Until the 1970s, the majority of coatings and adhesives are still produced using natural raw materials such as resins and unlinked rubber, dissolved in organic solvents such as toluene, xylene or acetone. In contrast to solvent-borne coatings, **WB coatings do not show ideally viscous behavior**. The demand for rotational viscometers increases, since testing with flow cups no longer meets the requirements. [74, 200]

1976: Standards are published regarding the **geometry of measuring systems** such as concentric cylinders, cone-and-plate, and parallel-plate (DIN 53018 and DIN 53788).

1978: *M.* **Baumgärtel** and *H. H.* **Winter** present a **software program for data conversion**, using **fitting functions for relaxation spectra and retardation spectra**. [24, 392]

In the 1970s: Analog controlled programmers for viscometers are occurring in industrial labs, and together with **analog xy online recorders**, users can now perform complete flow curves tests automatically without interruption, plotting the result simultaneously; these kinds of diagrams are often called "rheograms". Until then, the single rotational speeds had to be switched stepwise, manually. There are e.g. external programmers available as separate units of company Haake since 1971, recorders and programmers of Contraves since 1974/1976, recorders of Brookfield since 1976, and the Ferranti viscometer has a built-in programmer. As a consequence of this development in industry labs, **single-point tests are** more and more **replaced by measurement and analysis of flow curves**, and this applies increasingly also to QC tests.

13.5 Development since 1980

In the 1980s: Using computers and software, testing and analysis options are extended enormously by digital control of instruments, storage, analysis and presentation of measuring data. There are e.g. software programs available on cassettes using **magnetic recording tapes** to analyze and plot stored data, of company Haake since 1978 and of Rheometrics since 1982. Control of viscometers is possible via external computers using a D/A-A/D data converter of Haake since 1986; A/D means analog-to-digital and D/A vice versa. There are viscometers with microprocessor-controlled keyboards of Brookfield since 1984. At first the instruments are connected to the computers by special **interface controllers** as a separate modular unit, e.g. of Haake the "Rheocontroller RC" for the HP 85 computer, of Contraves the "Rheoconverter" for the HP 9815 and the "Rheoanalyzer RA" which is "IBM compatible", or of Physica the "System Interface SI".

1980: Publication of a **DIN standard** on the geometry of cylinder systems showing a narrow gap, the so-called **"DIN measuring systems"** (DIN 53019).

1980: *Wolfgang Gleissle* presents the 1st and 2nd **mirror relation**, illustrating the similar shapes of the following curve functions; the mirror axis is right-angled to the x-axis on which is presented the shear rate and the time, respectively, [157, 222, 233, 276]:

a) The steady-state shear viscosity $\eta(\dot\gamma)$ and the transient shear viscosity $\eta^+(\dot\gamma, t)$,

b) The steady-state 1st normal stress coefficient $\Psi_1(\dot\gamma)$ and the transient 1st normal stress coefficient $\Psi_1^+(\dot\gamma, t)$

1980: Companies **Apple** and **Commodore** present the first **"home computers"** which are successful on the market, since 1983 with the operating element "mouse". (Previous development: In 1959 by *Jack Kilby* and independently in 1961 by *Robert Noyce* the first integrated circuits IC; in 1967 a **hand-held calculator** of **Texas Instruments**; in 1971 a single-chip **microprocessor** by *M. Hoff, F. Faggin, S. Mazor, M. Shima*, all of company **Intel**; in 1974 a **programmable pocket calculator** of **Hewlett-Packard**; in 1975 *Steve Wozniak* develops a **computer with display and keyboard** which is the predecessor of the first marketable computers which he sells together with *Steve Jobs* as a product of **Apple**; later VLSI, very large-scale integration, of millions of transistors per circuit and in 1986 the "megabit RAM", random access memory; in 1981 company **IBM** launches the **"personal computer"** using the MS-DOS disk-organization system from **Microsoft** of *Bill Gates*, and control circuits, so-called "chips", of **Intel**; this "IBM standard" succeeds more and more in measuring technics, and **"IBM compatible computers"** and corresponding software programs are controlling increasingly also the rheometers in the industrial labs.)

1981: Company **Contraves** presents a speed-controlled **viscometer** which is the first one **with an integrated microprocessor to measure torque values as a function of the power consumption** of the electric drive (type "Rheomat 108"). [18]

1981: Company **Rheometrics** presents the **"force re-balance transducer"** to compensate the deflection of the torsion element which is a mechanical torque sensor. It is aimed to perform tests with "zero deflection". [233]

1982: Industrial use of **UV-curing adhesives** (see also 1968 and 1995).

1982: Publication of a **DIN standard** on the geometry of **double-gap cylinder measuring systems** (DIN 54453).

1983: Company **Carrimed** presents its first **rheometer** (type "CS 50"), a "controlled stress" device with air bearing and asynchronous drive, **controlled by an Apple microcomputer**; it is an enhanced "Deer Rheometer" (see 1968). [18, 89]

1983: Company **Bohlin** presents its first **rheometer** (type "VOR"), with air bearing and DC motor for controlled angle or speed tests; based on developments by *Leif Bohlin* since 1974. [233]

1985: *Howard A. Barnes* and *Ken Walters* claim that yield points are not actually existing, since they are detected only because the instruments used are showing limited angle resolution; hence they term it **"apparent yield stress"**. [18, 233]

1985: *H. Giesekus* and *R.B. Bird* present **models of non-linear deformation behavior** (*H.G.* see also 1965). [155]

1985: Company **Physica** presents its first **rotational viscometer** (type "LC"), with a DC motor for controlled speed tests.

1986: *M. Doi* and *S.F. Edwards* present an **enhanced reptation model** to describe the motion of unlinked polymer molecules (see *P.G. de Gennes* 1971). [103]

1986: *Hans-Martin Laun* presents the following relation: $2 \cdot G'(\omega) = N_1(\dot{\gamma})$ for $\dot{\gamma} = \omega \rightarrow 0$; thus, for the same low values of ω (from oscillation/frequency tests) and $\dot{\gamma}$ (from rotation/flow curves), twice the value of G' corresponds to the value of the 1st normal stress difference N_1. [223]

1986: Company **Bohlin** presents a controlled stress **rheometer** (type "CS 10"), with air bearing and asynchronous drive.

1986: Company **IBM** launches the first **laptop**; with 5.4kg it weighs 8kg less than the previous model.

1987: Company **Herberts** develops **water-based** (WB) **coatings** with clearly reduced solvent amount to be used as base coats for **OEM** (original equipment manufacturer) **automotive serial coating**. (Development: In 1981 market introduction of **water-based fillers**, and in 1992 of **water-based clear coats**; see also 1934 and 1975; and repair coats "until 2000".) [74, 102]

1988: *J. Des Cloizeaux* presents a **double-reptation model** to describe the motion of unlinked polymer molecules. [94]

1989: *N.W. Tschoegl* presents a calculation program for **data conversion**. [367]

1989: *J. Honerkamp* and *J. Weese* present a software program for **data conversion**. [192]

1991: Publication of the **international standard ISO 3219** on the geometries of concentric cylinder and cone-and-plate measuring systems.

1992: After a request from the Federal Highway Administration (FHA) in the USA, the American Association of State Highway and Transportation Officials **(AASHTO)** starts the **Strategic Highway Research Program (SHRP)**. The aim is to increase the working life of road surfaces without affecting the performanace. This results in specifications for **evaluating the viscoelastic behavior of binding agents for asphalt used for road construction**, e.g. of **polymer-modified bitumen** (PmB) used since the 1980s. Recommended are **oscillatory tests using a "Dynamic Shear Rheometer" (DSR)**, see also the example in Chapter 8.5.2: SHRP test and ASTM D7175. (Previous development: Since around 2000 BC **the Sumerians** in Mesopotamia, today's Iraq, use **natural bitumen as adhesives, sealants and building materials, e.g. in the form of asphalt mortars**. Since around 800 AD, oil wells are documeted in Baku. Since 1000, there are reports about tarpits and "mountain tar", i.e. **heavy oil ponds**, e.g. in Europa. Bitumen is used as a **sealant in ship construction**, and in 1498 *C. Columbus* orders his ships to be sealed with material from the **asphalt lake on the Caribean island of Trinidad**. In 1746 *Haskin* receives a patent for the distillation of raw tar from coal, terming the residue **coal tar pitch** as "caput mortuum", in Latin, meaning "dead head" or "dead man". Around 1800, first trials in Europe bringing **asphalt mastic and asphalt/tar mixtures on road surfaces** as a measure against dirt, e.g. from horses. However, the mixtures remain sticky for a too long time and the carriages with their iron-covered wheels are leaving deep tracks. In 1823, *T. Hancock* receives a patent for a mixture of natural rubber latex and stone as road pavement, however, the practical realization is too expensive; *T.H.* see also 1853: rubber kneading machine. In 1838/39, on a test road in London wooden blocks still prove to be better than paving stones and asphalt plates. Since 1850 **mastic asphalt** as **"stamped asphalt"** is used, which is compacted using iron handstampers and rollers heated by coke fire. However, because the surface of these kinds of roads is too smooth, there is a risk of horses to slip.

Since 1876, in America **"rolled asphalt"** or **"sand asphalt"** is used, firstly made of heavy **tar and pitch** which is available as a by-product **from gas power stations and coking plants**, and later made of **bitumen from crude-oil refineries**. At the beginning, this asphalt is **compacted using heavy steam rollers**. In 1902, on the initiative of the medical doctor *E. Guglielminetti,* **coal tar** is sprayed on a 20km long part of the road between Nice and Monte Carlo, but there is no long-term success. Since the 1920s however, in many towns there can increasingly be found asphalted areas. Since 1967, systematic tests are performed on roads paved with **bitumen-oil-rubber mixtures**, as **"rubber-modified asphalt" RmA**, e.g. containing grinded used-up tires. Since the 1980s, **mixtures** are used **containing polyethylene PE, polybutadiene PB,** or **polystyrene PS as "elastomer-modified bitumen" EmB**, sometimes described as a "mycelium-structure".) [1, 28, 351, 405]

1992: Company **Haake** presents a controlled stress **rheometer** (type "RheoStress RS 100"), **with automatic lift system**, air bearing and asynchronous drive.

1992: Company **Reologica** presents a controlled stress **rheometer** (type "StressTech"), with air bearing and asynchronous drive.

1993: Start of the **"Internet"** or www (world-wide web).

1993: Company **TA Instruments** acquires company Carrimed (see 1983). Since 1996, TAI is a subsidiary of the company Waters Corporation.

Since 1995: Industrial use of **UV-curing water-based emulsions**, e.g. as wood coatings (UV-systems see also 1968 and 1982). [200]

1995: Company **Physica** presents the first **completely digitally controlled rheometer** (type "UDS 200"), with air bearing, **electronically commutated (EC) synchronous drive** with **digital control** of the torque, and very fast digital control of deflection angles and speeds. [222]

1996: Company **Thermo Electron Corporation** acquires company Haake (see 1933, 1954, 1992), later changes of names to Thermo Haake, Thermo, and in 2007 to Thermo Scientific as a part of Thermo Fisher; still selling the Haake rheometers.

1997: Company **Herberts** develops **powder coatings as clear coats for OEM automotive serial coating**. (Previous development: In 1995 powder coats are launched as fillers, and in 1996 as clear coats in the form of **powder slurries** by company **BASF Coatings**; powder coats see also 1964.) [102]

until 2000, **development of automotive coatings and their application techniques**, e.g. of **repair coats** (in Western Europe), [247]:

– since 1920: using nitrocellulose (NC) coats for both base coats and top coats, application with brushes (see also 1913 *H. Ford*)

– since 1940: spray application (see also 1959, electrostatic spray coats)

– since 1950: NC fillers, short-oil alkyd resin base coats, and medium-oil alkyd or thermoplastic (TPA) top coats (see also 1931, alkyd paints)

– since 1970: UPE fillers (unsaturated polyesters), wash primers (WPs), alkyd or two-component polyurethane (2c PU) base coats, and 2c PU top coats (PU see also 1943 and 1957)

– since 1980: UPE fillers, WPs, 2c PU or epoxy base/fillers, CAB base coats (cellulose-acetyl-butyrate), and 2c PU clear coats (EP see also 1947)

– since 2000: UPE fillers, WPs, 2c PU or epoxy base/fillers, water-based (WB) base coats, and 2c PU clear coats (as a high-solids coats containing only a limited amount of solvent; see also 1987: WB wet coats for OEM automotive coating)

2000: Company **BASF** uses **roboters** in corporation with external rheometers for an automatic performance of both sampling and cleaning of the measuring systems used; in 2006 company **Bosch Lab Systems** launches a complete **robot-controlled system**, i.e. both rheometer and robot are controlled by the same system; and in 2007 company **Anton Paar** presents the "high-throughput rheometer HTR" as a completely integrated system.

2002: Company **TA Instruments** acquires company Rheometrics (see 1970 and 1981).

2003: **Organically modified silanes** are available as functional coatings, being hybrid materials consisting of an inorganic sol/gel component and an organic polymer, and **deliberately also nanoparticles** are used; e.g. for scratch-proof automotive **clear coatings**. [54, 236, 335]

2003: Company **Malvern Instruments** acquires company Bohlin (see 1983 and 1986), Malvern is part of company Spectris, still selling the Bohlin rheometers.

2004: Change of company name Physica to **Anton Paar Germany** (see 1985 and 1995), APG is part of company Anton Paar since 1996, still selling the Physica rheometers.

2004: Company **Anton Paar** presents the "TruGap" option, as an **automatic control or adjustment of the gap dimension** of parallel-plate and cone-and-plate measuring systems.

2010: With DIN 51810-2 on determining the **yield point and flow point of lubricating greases**, and DIN 54458 on testing the behavior of **viscoelastic adhesives** particularly regarding **stability and stringiness**, standards are published describing modern, application-related rheological test methods by use of oscillatory rheometers (see also the examples of Chapters 8.3.4.3 and 8.5.2.2d).

14   Appendix

14.1 Symbols, signs and abbreviations used

a) Latin characters (small letters)

a	[m]	cone tip truncation (CP measuring systems)
a	[m]	specimen thickness (e.g. rectangular solid bars)
a_T	[1]	horizontal temperature shift factor
b	[m]	specimen width (e.g. rectangular solid bars)
b_T	[1]	vertical temperature shift factor
c		constant, coefficient
c	[kg/m³]	concentration
cmc	[mol/l]	critical micelle concentration
c_{crit}	[kg/m³]	critical concentration
c_L	–	end-effect correction factor (cylinder measuring systems)
c_p	[J/g · K]	specific heat capacity
d, d_{max}	[m]	diameter, maximum diameter
d_p	[m]	particle diameter
e	–	Euler's number (e = 2.718 281...)
f	[Hz]	frequency
f_{DE}	[Hz]	frequency, di-electric analysis DEA
f_i	–	individual frequency factors (multiwave tests)
f_T	[1]	friction coefficient (tribology)
f(x)		function, e.g. as a time-dependent function f(t)
g	[m/s²]	gravitation constant (g = 9.81m/s²)
h	[m]	gap dimension, layer thickness, fluid level
h_{max}	[m]	maximum gap width (e.g. with a CP measuring system)
i	–	(individual) counting number (e.g. with i = 1 to k)
k	–	maximum counting number (e.g. for i = 1 to k)
lg		logarithm (to the basis 10)
ln		natural logarithm (to the basis e, Euler's number)
m	[kg]	mass
m_{nom}	[kg]	nominal load as a mass (MFR testers)
n	[min⁻¹]	rotational speed
n_{act}	[min⁻¹]	actual (rotational) speed
n_c	[min⁻¹]	critical (rotational) speed
n_{min}, n_{max}	[min⁻¹]	minimum, maximum (rotational) speed
p	–	exponent, e.g. of model functions
p	[MPa]	pressure (0.1MPa = 1bar)
p_F	[MPa]	flow pressure (e.g. plasticity testers)
pH		pH-value (hydrogen ion exponent, chemical reaction as acidic or alkaline)
p_{ref}	[MPa]	reference pressure
Δp	[MPa]	pressure difference

Thomas G. Mezger: The Rheology Handbook
© Copyright 2011 by Vincentz Network, Hanover, Germany
ISBN 978-3-86630-864-0

r	[m]	radius
r_m	[m]	mean radius (e.g. of PP measuring systems)
s	–	curve slope
s	[m]	deflection, distance, path
s_A	[m]	deflection amplitude (e.g. linear shear curemeters)
$\dot{s}$	[m/s]	time derivative of the deflection (ds/dt)
t	[s]	time
t_{CR}	[s]	start time of a chemical reaction
t_i	[s]	incubation time (e.g. curemeters)
t_s	[s]	scorch time (Mooney viscometers)
t_{SG}	[s]	time point of the sol/gel transition ("gel point")
t_x	[s]	vulcanisation or cure time to reach a turnover (curemeters)
t_5 , t_{35}	[s]	scorch time (Mooney viscometers)
v	[m/s]	velocity
vGP		van Gurp/Palmen plot (frequency tests)
v_m	[m/s]	mean velocity
v_s	[m/s]	sliding speed (tribology)
w	[%]	relative amount, percentage, weight fraction
x, y, z		coordinates (e.g. in a Cartesian coordinate system)
x	[%]	turnover (vulcanisation reaction; e.g. curemeter)

b) Latin characters (capitals)

A	[m²]	area
A/D		analog/digital converter (for electric signals)
AFRP		aramid-fibre reinforced plastics
AGC		automatic gap control
AGS		automatic gap setting
API		American Petroleum Institute
ASTM		American Society for Testing and Materials (standards, USA)
B	[T]	magnetic flux density (in Tesla; electro-rheology)
BBR		bending beam rheometer
BBU		Bingham build-up (structure regeneration, "thixotropy")
BMS		ball measuring systems
BPT		borderline pumping temperature (when testing mineral oils)
BS		British standards
BU		Brookfield units (relative viscosity values)
C		constant, coefficient
CC		concentric (or coaxial) cylinders (measuring systems)
CCS		cold cranking simulator (low-temperature tests on oils)
CFRP		carbon-fibre reinforced plastics
CMT		combined motor transducer (rheometer setup)
CP		cone-and-plate (measuring systems)
CRI	[1/s]	cure rate index (curemeters)
CSD (or CD)		controlled shear deformation
CSR (or CR)		controlled shear rate
CSS (or CS)		controlled shear stress
CV		capillary viscometer(s)
C_{H}	[N/m]	spring constant
C_{KF}	[m²/s²]	instrument constant of falling ball viscometers
C_{N}	[Ns/m]	dashpot constant

C_{sr}	[min/s]	MS constant (conversion between rot. speed and shear rate)
C_{ss}	[Pa/Nm]	MS constant (conversion between torque and shear stress)
D	[s⁻¹]	(previously used for) shear rate
D/A		digital/analog converter (for electric signals)
DDI	[%]	Daniel dilatancy index
DEA		di-electric analysis
DG		double-gap (of a cylinder measuring system)
DGT	[°C]	dynamic gel temperature
DIN		Deutsches Institut für Normung (German standards organization)
DMA		dynamic-mechanical analysis
DMTA		dynamic-mechanical thermoanalysis
DSC		differential scanning calorimetry
DSO		direct strain oscillation
DSP		digital signal processors
DSR		dynamic shear rheometer
DTMA		dynamic thermomechanical analysis
DTT		direct tension tester (for testing bitumens)
D_2 and D_3		dashpots in a Burgers model
E	[Pa]	tensile modulus (or Young's modulus, or elasticity m. in tension)
E	[V/m]	electric field strength (electro-rheology)
EC		electronically commutated (rheometer motor)
EN		European standards
ERF		electro-rheological fluid
E*, E', E"	[Pa]	complex elasticity modulus, storage modulus, loss modulus (in tension, compression, bending)
E_A	[kJ/mol]	flow activation energy (e.g. in Arrhenius relation)
E_{ad}	[J/m²]	energy of separation (tack-test), area-related
F	[N]	force
FFT		fast Fourier transform
FIC		flow-induced crystallization
FP	[cm³/g]	(Daniel) flow point
FRR	–	flow rate ratio (melt flow tester, e.g. MVR)
FTS		frequency/temperature shift
F_A	[N]	axial force (in the direction of the axis)
F_{ext}	[N]	external force
F_f	[N]	friction force (tribology)
F_G	[N]	gravitational force, (force of the) weight
F_{int}	[N]	internal force, cohesion force, structural strength
F_L	[N]	load (e.g. compressive force in tribology)
F_N	[N]	normal force (acting vertical on the reference plane)
F_P	[N]	pressure force (e.g. in capillaries)
F_R	[N]	flow resistance (force)
G	[Pa]	shear modulus
GFRP		glass-fibre reinforced plastics
GH		Gardner/Holdt (GH viscosity, bubble viscometers)
GI, GIT		gelation index, GI temperature (ASTM D7110)
GPC		gel permeation chromatography (molar mass determination)
G*, G', G"	[Pa]	complex shear modulus, storage modulus, loss modulus
G_e	[Pa]	equilibrium shear modulus (relaxation tests)
G_i	[Pa]	shear moduli of individual Maxwell or Kelvin/Voigt elements
G_i', G_i''	[Pa · m]	interfacial storage modulus, interfacial loss modulus

G_{max}	[Pa]	maximum shear modulus
G_P	[Pa]	plateau value of the shear modulus (frequency tests)
G_{RP}	[Pa]	value of the shear modulus on the rubber-elastic plateau
G_0	[Pa]	instantaneous shear modulus (creep and relaxation tests)
G_1, G_2	[Pa]	shear moduli of the springs in a Burgers model (creep tests)
H	[m]	gap dimension, height
H	[A/m]	magnetic field strength (magneto-rheology)
H	[J]	enthalpy
H		intensity, amount (e.g. as $H(\lambda)$, continuous relaxation time spectrum)
HS, HSV		high shear, high-shear viscosity
HTHS		high temperature, high shear
I	[A]	electric current, amperage
ICA		International Confectionery Association (chocolate tests)
IOCCC		International Office of Cocoa, Chocolate and Sugar Confectionery
IP		Institute of Petroleum (standards)
IR		interfacial rheology
ISO		International Standards Organization
IUPAC		International Union of Pure and Applied Chemistry
IV	[cm^2/g]	intrinsic viscosity (capillary viscometry)
I_p	[m^4]	polar geometrical moment of inertia
I_t	[m^4]	torsional geometrical moment of inertia
J	[Pa^{-1}]	(shear) compliance (creep tests)
J^*, J', J''	[Pa^{-1}]	complex compliance, and its real and imaginary part
J_e	[Pa^{-1}]	equilibrium (shear) compliance (creep tests)
J_g	[m^3/kg]	Staudinger index (capillary viscometry)
J_i	[Pa^{-1}]	creep compliance of individual Maxwell or Kelvin/Voigt elements
J_{max}	[Pa^{-1}]	maximum compliance
J_r	[Pa^{-1}]	recovery (shear) compliance (creep tests)
J_v	[m^3/kg]	Staudinger function (capillary viscometry)
J_0	[Pa^{-1}]	instantaneous (shear) compliance (creep tests)
KU		Krebs units (relative viscosity values)
L	[m]	length
LAOS		large-amplitude oscillatory shear
LC		liquid crystal(s)
LCB		long-chain branching (polymers)
LS, LSV		low shear, low-shear viscosity
LTHS, LTLS		low temperature high shear, low temperature low shear
LVE		linear viscoelastic
LVN	[cm^3/g]	limiting viscosity number (capillary viscometry)
L' and L''	[m]	dimensions of a cylinder measuring system
L_0	[m]	initial length
L_p	[nm]	persistence length (surfactants)
M	[mNm]	torque (1Nm = 1000mNm)
M	[g/mol]	(average) molar mass
MDR		moving-die rheometer (rubber industry)
ME		Mooney/Ewarts (e.g. as ME measuring systems)
MFI		melt flow index
MFR		melt mass-flow rate (e.g. according to ISO 1133)
MMD		molar mass distribution
MRF		magneto-rheological fluid
MRV		mini-rotary viscometer (for testing mineral oils)

MS		measuring system
MU		Mooney units (shearing disk viscometer)
MVR		melt volume-flow rate (e.g. according to ISO 1133)
M_A	[mNm]	amplitude of the torque (oscillatory tests)
M_{crit}	[g/mol]	critical molar mass
M_f	[mNm]	torque caused by friction effects
M_i	[mNm]	torque caused by inertia effects
M_{min}, M_{max}	[mNm]	minimum, maximum torque
M_O	[mNm]	operation torque
M_{rel}	[%]	relative torque
M_S	[mNm]	start torque
M_t	[mNm]	torque at time point t (e.g. with torsional curemeters)
M_{tot}	[mNm]	total torque
NC		nitro-cellulose (e.g. coatings)
NDT		non-destructive testing
NF	[N]	normal force
NFC		normal force control
NLGI		National Lubricating Grease Institute (USA)
N_1 and N_2	[Pa]	1st and 2nd normal stress difference
$N_{1, LM}$	[Pa]	1st normal stress difference, using Lodge/Meissner relation
ODC		oscillating disk curemeter (rubber industry)
OEM		original equipment manufacturer (e.g. cars)
OICC		Office International du Cacao et du Chocolat
OIT		oxidative induction time
ORO		oscillation/rotation/oscillation
PCI		Powder Coatings Institute (Northern America)
PCM		phase-change material
Pen		penetration test, or pen. number
PI		plasticity index
PIV (PTV)		particle image (tracking) velocimetry (for flow profiles)
PP		parallel-plate (measuring systems)
PPI		pseudoplastic index
PSSI		permanent shear stability index (acc. to ASTM D6022)
PYV		pseudo yield value (Laray falling rod viscometer)
Q	[J]	quantity of heat, thermal energy
QA or QC		quality assurance, quality control
$\dot{Q}$	[J/s] = [W]	time-dependent heat flow rate (e.g. with DSC tests; dQ/dt)
R	[m]	radius
Re, Re_c	[1]	Reynolds number, critical Reynolds number
RBU		rate of build-up (structure regeneration)
RIS		Redwood-no.1-seconds (flow cup)
R_e	[m]	external radius (e.g. of the cup of a cylinder MS)
R_G	[kJ/mol · K]	gas constant ($R_G = 8.314 \cdot 10^{-3}$ kJ/mol · K)
R_i	[m]	internal radius (e.g. of the bob of a cylinder MS)
R_s	[m]	radius of the shaft of a cylinder MS
R & B		ring-and-ball (tests; e.g. for testing bitumen)
R & D		research and development
S	[N/μm]	stiffness (and compliance: 1/S [μm/N])
SAE		Society of Automotive Engineers (viscosity classification of oils)
SALS		small-angle light scattering (rheo-optics)
SAM		self-assembling monolayer (surfactants)

SANS		small-angle neutron scattering (rheo-optics)
SAOS		small-amplitude oscillatory shear (i.e. in the LVE-range)
SAXS		small-angle X-ray scattering (rheo-optics)
SF		shortness factor (Laray falling rod viscometer)
SHRP		Strategic Highway Research Program (asphalt testing)
SI		système international d'unitées (internat. system of units)
SIS		shear-induced structure
SMT		separated motor transducer (rheometer setup)
SP	[°C]	softening point
SR	[1]	swell ratio (die swell, extrusion)
SRI	[-]	structure recovery index
STBF		solid torsion bar fixture
SUS		Saybolt universal seconds (flow cup)
SVM		Stabinger viscometer
S_1 and S_2		springs in the Burgers model (creep tests)
T	[°C], [K]	temperature (conversion: T [K] = T [°C] + 273.15K)
Ta	[1]	Taylor number
TA		thermal analysis or thermoanalysis; texture analyzer
TAPPI		Technical Association of the Pulp and Paper Industry (Northern America)
TBS		tapered bearing simulator (for testing mineral oils)
TEB	[MJ/m³]	tensile energy to break, volume-related
TEMPs		tissue engineered medical products
TG		thermogravimetry
TI		thixotropy index
TMA		thermomechanical analysis
TPE		thermoplastic elastomers
TPV		tapered plug viscometer (for testing mineral oils)
TR		Trouton's Ratio (of extensional and shear viscosity)
TTS		time/temperature shift
T_C	[°C]	counter temperature (e.g. for Peltier elements)
T_{CR}	[°C]	temperature at the onset of a chemical reaction
T_d	[°C]	dew point
T_g	[°C]	glass transition temperature
T_i	[°C]	individual temperature (e.g. using the TTS method)
T_k	[°C]	crystallization temperature
T_m	[°C]	melting temperature
T_M	[°C]	measuring temperature
T_{Mi}	[°C]	individual measuring temperature (e.g. for TTS)
T_{ref}	[°C]	reference temperature
T_{SG}	[°C]	sol/gel transition temperature
U	[V]	voltage
UV		ultra-violet (light)
V	[m³]	volume
VE, VES		viscoelastic, e.g. viscoelastic surfactants
VN	[cm³/g]	viscosity number (capillary viscometry)
VR	[1]	viscosity ratio (e.g. with Laray falling rod viscometers)
V_p	[m³]	particle volume
$\dot{V}$	[m³/s]	volume flow rate (dV/dt), throughput
WAXS		wide-angle X-ray scattering (rheo-optics)
WLF		Williams/Landel/Ferry (TTS method, mastercurve)

WP	[cm³/g]	(Daniel) wet point
W_p	[m³]	polar section modulus
W_t	[m³]	torsional section modulus
YP, YS	[Pa]	yield point, yield stress

c) Greek characters

α	[°] or [rad]	cone angle of CP measuring systems
α	[°]	angle of the bob apex (cylinder measuring systems)
α_p	[MPa⁻¹]	viscosity/pressure coefficient ($1\,MPa^{-1} = 0.1\,bar^{-1}$)
α_{TE}	[10⁻⁶/K]	coefficient of thermal expansion
β	[°]	slope angle of the creep curve (range of steady-state flow)
γ	[%], [1]	deformation or strain (with 100% = 1)
γ_A	[%]	strain (or deformation) amplitude (oscillatory tests)
γ_e	[%]	elastic reformation (creep tests)
γ_i	[%]	deformation of an individual Kelvin/Voigt element (creep tests)
γ_L	[%]	limiting strain value of the LVE range (amplitude sweeps)
$\gamma_{min}, \gamma_{max}$	[%]	minimum, maximum deformation
γ_r	[%]	reformation (recoverable deformation; creep tests)
γ_v	[%]	finally remaining viscous deformation (creep tests)
γ_0	[%]	preset strain (relaxation tests), or instantaneous deformation (creep tests)
γ_1	[%]	pre-strain before the effective step strain (relaxation tests)
$\dot{\gamma}$	[s⁻¹]	shear rate (strain rate; $d\gamma/dt$)
$\dot{\gamma}_A$	[s⁻¹]	shear rate amplitude (oscillatory tests)
$\dot{\gamma}_c$	[s⁻¹]	critical shear rate
$\dot{\gamma}_{cc}, \dot{\gamma}_{cp}$	[s⁻¹]	shear rate in the cylindric and conical gap of a ME system
$\dot{\gamma}_e$	[s⁻¹]	elastic part of the shear rate (e.g. using a Maxwell model)
$\dot{\gamma}_i$	[s⁻¹]	shear rate of an individual Maxwell element (relaxation tests)
$\dot{\gamma}_m$	[s⁻¹]	mean shear rate (PP measuring systems)
$\dot{\gamma}_{min}, \dot{\gamma}_{max}$	[s⁻¹]	minimum, maximum shear rate
$\dot{\gamma}_{rep}, \dot{\gamma}_i, \dot{\gamma}_e$	[s⁻¹]	representative, internal, external shear rate (cylinder MS)
$\dot{\gamma}_v$	[s⁻¹]	viscous part of the shear rate (e.g. using a Maxwell model)
$\dot{\gamma}_w$	[s⁻¹]	shear rate at the wall (e.g. in a pipe or capillary)
$\dot{\gamma}_0$	[s⁻¹]	shear rate in the steady-state flow range (creep tests)
$\ddot{\gamma}$	[s⁻²]	acceleration of the strain (2nd derivative with respect to time)
δ	[°]	phase shift angle, loss angle
δ_{cc}	[1]	ratio of the radii (concentric cylinder measuring systems)
δ_{DE}	[°]	phase shift angle (dielectric analysis, DEA)
$\tan\delta$	[1]	loss factor, damping factor
$\tan\delta_{DE}$	[1]	loss factor, damping factor (dielectric analysis, DEA)
Δ		difference (e.g. in the form of ΔT or $\Delta\eta$)
ε	[1], [%]	elongation (tensile tests; with 100% = 1)
$\varepsilon^*, \varepsilon', \varepsilon''$	[F/m]	permittivity, its real part and imaginary part (dielectric analysis)
ε_B	[%]	elongation at break (tensile tests)
ε_H	[%]	Hencky strain (tensile tests)
$\dot{\varepsilon}$	[s⁻¹]	extensional strain rate (tensile tests)
$\dot{\varepsilon}_H$	[s⁻¹]	Hencky strain rate (tensile tests)
η	[Pas]	(shear) viscosity
η^*, η', η''	[Pas]	complex viscosity, and its real and imaginary part
η^+	[Pas]	transient viscosity

η_E	[Pas]	extensional viscosity
η_i	[Pas]	viscosity of individual Maxwell or Kelvin/Voigt elements
η_i	[Pas · m]	interfacial viscosity
η_{in}	[m³/kg]	logarithmic viscosity number (or inherent viscosity; capillary viscometry)
η_r	[1]	relative viscosity (or viscosity ratio; capillary viscometry)
η_{red}	[m³/kg]	reduced viscosity (or viscosity number; capillary viscometry)
η_{rep}	[Pas]	representative viscosity (ISO cylinder measuring systems)
η_s	[Pas]	viscosity of a solvent, or of a suspension liquid
η_{sp}	[1]	specific viscosity (or viscosity relative increment; capillary viscometry)
η_0	[Pas]	zero-shear viscosity
η_∞	[Pas]	infinite-shear viscosity
$[\eta]$	[m³/kg]	intrinsic viscosity (or limiting viscosity number; capillary viscometry)
λ	[s]	relaxation time
λ_i	[s]	relaxation time of individual Maxwell elements
λ_w	[nm]	wave length (rheo-optics)
Λ	[s]	retardation time
Λ_i	[s]	retardation time of individual Kelvin/Voigt elements
μ	[1]	Poisson's ratio
ν	[m²/s]	kinematic viscosity (1mm²/s = 10^{-6}m²/s)
π	–	circle constant (π = 3.141 592)
Π	[J/As]	Peltier coefficient (temperature control)
ρ	[kg/m³]	density (1g/cm³ =1000kg/m³)
ρ_{fl}	[kg/m³]	density of a fluid (e.g. using falling ball viscometers)
ρ_k	[kg/m³]	density of the ball (using falling ball viscometers)
σ	[Pa]	tensile or compressive stress
Σ		summation sign ("sigma")
τ	[Pa]	shear stress
τ_A	[Pa]	shear stress amplitude (oscillatory tests)
τ_{cc}, τ_{cp}	[Pa]	shear stress in the cylindric and conical gap of a ME system
τ_e	[Pa]	equilibrium shear stress (relaxation tests), or elastic part of the shear stress (e.g. using a Maxwell model)
τ_f	[Pa]	flow stress, flow point
τ_i	[Pa]	shear stress of an individual Maxwell or Kelvin/Voigt element
τ_m	[Pa]	mean shear stress (PP measuring systems)
$\tau_{rep}, \tau_i, \tau_e$	[Pa]	representative, internal, external shear stress (cylinder MS)
τ_v	[Pa]	viscous part of the shear stress (e.g. using a Maxwell model)
τ_w	[Pa]	shear stress at the wall (e.g. in a pipe, capillary)
$\tau_{xx}, \tau_{yy}, \tau_{zz}$	[Pa]	normal stresses
τ_y	[Pa]	yield stress, yield point, yield value
τ_0	[Pa]	preset shear stress (creep tests)
τ_1	[Pa]	shear stress value before the strain step (relaxation tests)
$\dot{\tau}$	[Pa/s]	change in shear stress with time (1st derivative with respect to time; e.g. using a Maxwell model)
$\ddot{\tau}$	[Pa/s²]	acceleration of the shear stress (2nd derivative with respect to time)
φ	[Pas⁻¹]	fluidity (inverse value of viscosity)
φ	[rad], [°]	deflection or displacement angle (1rad = 1000mrad; 1rad corresponds to 57.3° approx.)
φ_A	[mrad]	amplitude of the deflection angle (oscillatory tests)
$\dot{\varphi}$	[rad/s]	time derivative of the deflection angle

Φ	[1]	volume concentration, volume fraction
Ψ_1, Ψ_2	[Pa $\cdot$ s^2]	1st and 2nd normal stress coefficient
$\Psi_{1,0}, \Psi_{2,0}$	[Pa $\cdot$ s^2]	1st and 2nd zero-normal stress coefficient
ω	[rad/s], [s^{-1}]	angular velocity (rotation), angular frequency (oscillation)
ω_c	[rad/s], [s^{-1}]	critical angular frequency (e.g. for Re and Ta numbers)
ω_{co}	[rad/s], [s^{-1}]	angular frequency at the cross-over point (e.g. G' = G")
ω_{DE}	[rad/s], [s^{-1}]	angular frequency (dielectric analysis DEA)
ω_i	[rad/s], [s^{-1}]	individual angular frequency (e.g. with multiwave tests)
ω_{min}	[rad/s], [s^{-1}]	minimum angular frequency
ω_r	[rad/s], [s^{-1}]	reduced angular frequency (WLF mastercurve)

14.2 The Greek alphabet

small letter	capital letter	Greek name	English pronunciation	equivalent to English
α	A	Alpha	ulfa	a
β	B	Beta	beeta (or báta)	b
γ	Γ	Gamma	gamma	g
δ	Δ	Delta	delta	d
ϵ	E	Epsilon	epseelon (like in end)	e
ζ	Z	Zeta	zeeta (or zata)	z
η	H	Eta	eeta (or ata)	ee
ϑ	θ	Theta	theeta (or thata)	th
ι	I	Jota	i-ota	i
κ	K	Kappa	kappa	k
λ	Λ	Lambda	lambda	l
μ	M	My	mu (or mew)	m
ν	N	Ny	nu (or new)	n
ξ	Ξ	Xi	ksee (or zi)	x
o	O	Omikron	o-mecron (or omicron)	(short) o
π	Π	Pi	pi (or pee)	p
ρ	P	Rho	ro	r
σ	Σ	Sigma	seegma	s
τ	T	Tau	tou	t
υ	Y	Ypsilon	why	y
φ	Φ	Phi	fi (or fee)	f
χ	X	Chi	ki (or kee)	ch
ψ	Ψ	Psi	psee (or si)	ps
ω	Ω	Omega	o-mega (or omeega)	(long) o

14.3 Conversion table for units

The mentioned values are rounded to three significant figures (more information, also about "antique" parameters, see e.g. in [6, 98, 306]).

a) Length L (ISO unit: m)
1m = 10dm (decimeters) = 100cm = 1000mm = $10^6\mu$m = 10^9nm (nanometers)
1in (inch) = 25.4mm = 2.54cm
1mil = (1/1000)in = 25.4μm
1ft (foot) = 12in = 0.305m

1yd (yard) = 36in = 3 ft = 0.914m
1mile = 1.61km
1Å (Angström) = 10^{-10}m = 0.1nm

b) Area A (ISO unit: m²)

$1m^2$ (square meters) = 10^4cm^2 = 10^6mm^2
$1in^2$ = $6.45cm^2$
$1ft^2$ = $144in^2$ = $929cm^2$ = $0.0929m^2$

c) Volume V (ISO unit: m³)

1 m^3(cubic meters) = 10^6cm^3 = 10^9mm^3
$1m^3$ = 1000 l
$1cm^3$ = 1ml
1 l = $1dm^3$ = $0.001m^3$ = $10^{-3}m^3$
$1in^3$ = $16.4cm^3$
$1ft^3$ = 1cf (cubic foot) = 28.3 l = $0.0283m^3$
1pt imp (imperial pint) = (1/8)gal imp = 0.568 l
1qt imp (imperial quart) = 2pt imp = 1.14 l
1gal imp (imperial gallon) = 4.55 l
1fl oz (fluid ounce) imp = 28.4ml
1 liq pt US (liquid pint) = (1/8)gal US = 0.473 l
1 liq qt US (liquid quart) = 2 liq pt US = 0.946 l
1gal US (US gallon) = 3.79 l
1fl oz (fluid ounce) US = 29.6ml
1barrel = 159 l (for petroleum products)

d) Volume flow rate $\dot{V}$ (ISO unit: m³/s)

$1m^3$/h = 1000 l/h = 16.7 l/min = 0.278 l/s = 2.78 · $10^{-4}m^3$/s
1 l/s = $10^{-3}m^3$/s = 60 l/min = 3600 l/h = $3.6m^3$/h
1cfm (cubic feet per minute) = $1ft^3$/min = $1.70m^3$/h

e) Mass m (ISO unit: kg)

1kg = 1000g = 10^6mg
1t = 1000kg = 1Mg (megagrams) = 10^6g
1oz (ounce) = 28.4g
1 lb (pound) = 16oz = 454g

f) Density ρ (ISO unit: kg/m³)

$1kg/m^3$ = 1g/l
$1000kg/m^3$ = $1g/cm^3$ = 1g/ml = 1kg/l
1 lb/ft^3 = $16.0kg/m^3$
1 lb/in^3 = $27.7g/cm^3$

g) Force F (ISO unit: N)

1N (newton) = 1kg · m/s^2
1kp = 1000p = 1kgf (kilogramforce) = 9.81N
1lbf (poundforce) = 4.45N
1dyne = 10^{-5}N

h) Torque M (ISO unit: Nm)

1Nm (newtonmeter) = 1000mNm = 10^6μNm = 10^9nNm
1Ncm = 0.01Nm = 10mNm
1kpm = 1kgf · m = 9.81Nm
1dyne · cm = 10^{-7}Nm = 0.1μNm

1 lbf · in = 0.113Nm
1 lbf · ft = 1.36Nm

i) Mechanical stress τ, and pressure p (ISO unit: Pa or N/m²)

1Pa (pascal) = $1N/m^2$ = $1kg/m \cdot s^2$
1hPa (hektopascal) = 100Pa
1kPa (kilopascal) = 1000Pa
1MPa (megapascal) = 1000kPa = 10^6Pa
1GPa (gigapascal) = 1000MPa = 10^9Pa
$1N/mm^2$ = 1MPa = 10^6Pa
$1N/cm^2$ = 10kPa = 10^4Pa
$1kp/mm^2$ = $1kgf/mm^2$ = $9.81N/mm^2$ = 9.81MPa
$1kp/cm^2$ = $1kgf/cm^2$ = 98.1kPa
$1dyne/cm^2$ = 1μbar = 0.1Pa
1bar = 1000mbar = $10^6dyne/cm^2$ = $1.02kp/cm^2$ = $10N/cm^2$
1bar = 10^5Pa = 100kPa = 0.1MPa
1mbar = 1hPa = 100Pa
1at = $1kp/cm^2$ = 98.1kPa
1atm = 101kPa
1Torr = 1mm Hg = 133Pa
1psi (pounds per square inch) = $1 lbf/in^2$ = 6890Pa
1000psi = 68.9bar = 6.89MPa
1psf (pounds per square foot) = $1 lbf/ft^2$ = 47.9Pa
1kipsi = $1kp/in^2$ = 15.2kPa

k) Angle φ (recommended: rad)

1° = 60' (minutes) = 3600" (seconds) = 17.5mrad
1' = 0.0167° = 0.291mrad
1" = $2.78 \cdot 10^{-4°}$ = 4.85μrad
360° = 2πrad = 6.28rad
1rad = 1000mrad = 10^6μrad = 57.3°

l) Time t (ISO unit: s)

1s = 1000ms = 10^6μs = 10^9ns (nanoseconds) = 10^{12}ps (picoseconds)
1h = 60min = 3600s
1d (day) = 24h = 1440min = 86,400s
1y (year) = 365 (or 366)d = 8760 (or 8784)h
1y (astro.) = approx. 365.25d

m) Velocity v (ISO unit: m/s)

1m/s = 60m/min = 3.6km/h
1km/h = 0.621miles/h = 0.278m/s
1mph (miles per hour) = 1.61km/h = 0.447m/s
1ft/min = 0.305m/min = $5.08 \cdot 10^{-3}$m/s

n) Rotational speed n or angular velocity ω (ISO unit: rad/s or s⁻¹)

For rotation at a circular motion:
1rps (revolution per second) = 1/s = $1s^{-1}$ = 60/min
1rpm (revolution per minute) = 1/min = $1min^{-1}$ = 1/(60 s) = $0.0167s^{-1}$
n = $1min^{-1}$ corresponds to ω = $(2\pi \cdot n)/60$ = 0.105rad/s (or s^{-1})
Instead of rad/s can be used s^{-1}, if the values are related to the "unity circle" with the radius
R = 1, as it is common use in mathematics and physics.
ω = 1rad/s corresponds to n = $(60/2\pi) \cdot \omega$ = $9.55min^{-1}$

o) Frequency f or angular frequency ω (ISO unit: rad/s or s⁻¹)

1) For oscillation at a linear motion:

1vps (vibration per second) = 1cps (cycle per s) = $1/s$ = $1s^{-1}$ = 1Hz = 60/min

1vpm (vibration per minute) = 1cpm (cycle per min) = $1/min$ = $1min^{-1}$ = $1/(60\,s)$ = $0.0167s^{-1}$ = 0.0167Hz

2) For oscillation at a circular motion:

f = 1Hz corresponds to $\omega = (2\pi \cdot f) = 6.28rad/s$ (or s^{-1})

Instead of rad/s can be used s^{-1}, if the values are related to the "unity circle" with the radius R = 1, as it is common use in mathematics and physics.

ω = 1rad/s corresponds to $f = (\omega/2\pi) = 0.159min^{-1}$

p) Temperature T (ISO unit: K, but °C is more useful for the daily work)

K (in *Kelvin*); °C (in *degrees Celsius*); °F (in *degrees Fahrenheit*)

For a temperature interval ΔT applies: 1K = 1°C = 1.8°F

0K = -273°C = -460°F

0°C = +273K = +32°F

0°F = -17.8°C = +255K

T [°C] = T [K] - 273K

T [°C] = (5/9) · (T [°F] - 32)

T [°F] = (9/5) · T [°C] + 32

q) Energy E, and work W (ISO unit: J)

$1J = 1Ws = 1Nm = 1(kg \cdot m^2)/s^2$

1kWh = 3600kJ

1kpm = 9.81Nm

1kcal = 4.19kJ

$1erg = 10^{-7}J$

r) Power P (ISO unit: W)

$1W = 1J/s = 1Nm/s = 1(kg \cdot m^2)/s^3$

1kW = 1kJ/s

1hp (horse power) = 0.736kW

s) Viscosity

1) Kinematic viscosity ν (ISO unit: m²/s)

$1m^2/s = 10^6mm^2/s = 10^4cm^2/s$

1St (stokes) = 100cSt (centistokes) = $1cm^2/s$

$1cSt = 1mm^2/s$

2) (Shear) Viscosity η (ISO unit: Pas)

1Pas (pascalsecond) = 1000mPas (millipascal second)

$1Pas = 1(N \cdot s)/m^2 = 1kg/(m \cdot s)$

1P (poise) = 100cP (centipoise) = 0.1Pas

1cP = 1mPas

15 References

15.1 Publications and books

[1]   AASHTO design. T315-04 (Am. Assoc. of State Highway and Transportation Officials), SHRP product 1007 (Strategic Highway Research Program), Standard method for determining the rheological properties of asphalt binders using a dynamic shear rheometer DSR, 2004 (almost identical to ASTM D7175-2005)

[2]   ACA Systems Oy, Sotkuma, product information, www, 2005

[3]   A&D Co., Tokyo, product information, www, 2005

[4]   *Aguirre-Mandujano, E., Lobato-Calleros, C., Beristain, C.I., Garcia, G.S., Vernon-Carter, E.J.*, Microstructure and viscoelastic properties of low-fat yoghurt structured by monoglyceride gels, J. LWT/Food Sci. & Techn. 42, 2009

[5]   *Albaum, E.A.*, Die Vordenker, J. Geo Epoche "Das alte Griechenland", 2004

[6]   *Alder, K.*, The measure of all things, Free Press, New York, 2002

[7]   *Aleksandrov, A.P., Lazurkin, Y.S.*, in: J. Tech. Phys. USSR, 1939

[8]   *Alembert, J.L. d'*, Essai d'une nouvelle théorie de la résistance des fluides, David, Paris, 1752; Opuscules mathématiques, 1768

[9]   *Andrewes, W.J.H.*, Eine kurze Geschichte der Zeitmessung, J. Spektr. d. Wiss. Spezial "Zeit", 2003

[10]   Anton Paar GmbH, Graz, product information (e.g. automatic micro-falling ball viscometer AMV, Physica rheometers, Stabinger viscometer SVM, TruGap function), www, 2010; e-learning CDs, Basics of rheology, p.1: Rotation, 2009, p.2: Oscillation, 2010

[11]   *Arrhenius, S.A.*, Recherches sur la conductibilitée galvanique des électrolytes, 1884; Über die innere Reibung verdünnter wässriger Lösungen, Z. Physik. Chem.1, 1887; Lärobok i teoretisk elektrokemi, 1900; Viskosität und Hydratation kolloidaler Lösungen, Medd. K. Vet. Akad. Nobelinst., 1915; Kemien och det moderna livet, 1919

[12]   *Attali, J.*, "Blaise Pascal" (biography), Klett, Stuttgart, 2006 (in German; orginal: Blaise Pascal ou le genie français, Fayard, 2000)

[13]   *Avogadro, L.R.A.C.*, Essai d'une manière de déterminer les masses relatives des molécules des corps, 1811; Fisica dei corpi ponderabili ossia Trattaio della constituzione generale dei corpi, 1837/1841

[14]   *Aydin, O., Dragon, A.*, Acrylatdispersionen für Haftklebstoffe, J. W. d. Farben, 09/2002

[15]   *Bach, C.*, Maschinenelemente, 1881; Elastizität und Festigkeit, 1889

[16]   *Baekeland, L.H.*, US patent 1907/1909; DR patent 1908/1911; Beschreibung des "Druck und Hitze"-Patents, Chemiker Z., 1909; also in: Ind. Eng. Chem., 1909

[17]   *Bagley, E.B.*, End corrections in the capillary flow of polyethylene, J. Appl. Phys., 1957; The separation of elastic and viscous effects in polymer flow, Trans. Soc. Rheol., 1961

[18]   *Barnes, H.A., Walters, K.*, The yield stress myth?, Rheol. Acta, 1985; *Barnes, H.A., Hutton, J.F., Walters, K.*, An introduction to rheology, Elsevier: Amsterdam, 1989; *Barnes, H.A., Schimansky, H., Bell, D.*, 30 years of progress in viscometers and rheometers, J. Appl. Rheol., 1999; *Barnes, H.A.*, A handbook of elementary rheology, Univ. of Wales Inst. Non-Newtonian Fluid Mechanics: Aberystwyth, 2000; *Barnes, H.A., Bell, D.*, Controlled-stress rotational rheometry – a historical review, Korea-Australia Rheol. J, 2003; *Barnes, H.A.*, The "yield stress myth" paper – 21 years on, J. Appl. Rheol. 17, 2007

[19]   *Barrett, D.M., Garcia, E., Wayne, J.E.*, Textural modification of processing tomatoes, CRC Press, 1998

[20]   Bartz, W.J., Viskosität und Fließverhalten, Handbuch der Tribologie und Schmierungstechnik, Bd. 7, Expert, Renningen, 1994

[21]   BASF AG, Technische Information, www, 2005; Wachs in der Wand sorgt für Wohlfühlklima, J. W. d. Farben 3, 2010

[22]   *Bauer, F.L.*, Kurze Geschichte der Informatik, Wilhelm Fink, München, 2007

[23]   *Bauer, W.*, Geschichte der chinesischen Philosophie, Beck, München, 2001

[24]   *Baumgärtel, M., Winter, H.H.*, Determination of discrete relaxation and retardation time spectra from dynamic mechanical data, Rheol. Acta, 1989; Interrelation between continuous and discrete relaxation time spectra, J. Non-Newt. Fluid Mech., 1992

[25]   *Bayer, O.*, J. Angew. Chem. 59, 1947

[26]   *Béguin, A.*, "Pascal" (biography), Rowohlt, Hamburg, 1992 (original in French)

Thomas G. Mezger: The Rheology Handbook
© Copyright 2011 by Vincentz Network, Hanover, Germany
ISBN 978-3-86630-864-0

[27]   *Belitz, H.D.*, Lehrbuch der Lebensmittelchemie, Springer, Berlin, 1992 (4th ed.)

[28]   *Bellin, P.*, Die Ergebnisse der Bitumen- und Asphaltforschung des SHRP T3: Westtrack, Z. Bitumen 1, 2001

[29]   *Beneke, K.*, Die Entwicklung der Kolloidwissenschaften, Mitteilg. Kolloid-Ges., 1996; Biographische Daten bedeutender Kolloidwissenschaftler, in: *G. Lagaly* et al, Dispersionen und Emulsionen, Steinkopff, Darmstadt, 1997

[30]   Benz & Cie, Die Benzwagen, Mannheim, 1913 (reprint: Wellhöfer, Mannheim, 2008)

[31]   *Beretta, M.*, "Lavoisier", J. Spektr. d. Wiss. Biografie, ed. E. Bellone, 1999 (original in Italian: Le Science, Milano, 1998)

[32]   *Bergia, S.*, "Einstein", J. Spektr. d. Wiss. Biografie, ed. E. Bellone, 2005 (original in Italian: Le Science, Milano, 1998)

[33]   *Berhorst, R.*, Die Giftmacher - künstlich hergestellte Farbstoffe, in: Geo Epoche "Die industrielle Revolution", Gruner, Hamburg, 2008

[34]   *Bernoulli, D.*, Hydrodynamica sive de viribus et motibus fluidorum commentarii ("Hydrodynamica"), Dulsecker, Straßburg, 1738

[35]   *Bernoulli, Joh.*, Hydraulica, 1740

[36]   *Bernzen, M.*, Determination of the mechanical and thermal stability of skin cream with oscillation and freeze-thaw cycle tests, appl. rep, APG, 2009

[37]   *Berry, G.C., Fox*, T.G., The viscosity of polymers and their concentrated solutions, Adv. Polym. Sci., 1968

[38]   *Berry, S.*, Was treibt das Leben an?, Rowohlt, Reinbek, 2007

[39]   *Bertholdt, U.*, Rheology of heat-set web offset ink emulsions, ECJ, 05/2007

[40]   *Beyer, H., Walter, W.*, Lehrbuch der organischen Chemie, Hirzel, Stuttgart, 1991/2004 (22nd/23th ed.)

[41]   *Biederbick, K.*, Kunststoffe, Kamprath-Reihe "kurz u. bündig", Vogel, Würzburg, 1974

[42]   *Bingham, E.C.*, An investigation of the laws of plastic flow, Bull. U.S. Bur. of Standards, 1916; *Bingham, E.C., Green, H.*, Paint - a plastic material and not a viscous liquid, Proc. ASTM, 1919; Fluidity and plasticity, McGraw-Hill, New York, 1922; The history of the Society of Rheology (from 1924 to 1944), 1944

[43]   Biode, Westbrook, product information, www, 2005

[44]   *Björkmann, U.*, The nonlinear history of fibre flow research, J. Appl. Rheol. 18, 2008

[45]   *Boger, D.V., Walters, K.*, Rheological phenomena in focus, Elsevier, Amsterdam, 1993

[46]   *Böhme, G.*, Strömungsmechanik nicht-newtonscher Fluide, Teubner, Stuttgart, 1981

[47]   *Boltzmann, L.*, Zur Theorie der elastischen Nachwirkung, Sitzungsber. Kgl. Akad. Wissensch., Math.-Naturwiss. Classe, Wien, 1874; Über einige Probleme der Theorie der elastischen Nachwirkung und über eine neue Methode Schwingungen mittels Spiegelablesung zu beobachten ohne den schwingenden Körper mit einem Spiegel von erheblicher Masse zu belasten, ibid., 1877; Zur Theorie der Gasreibung, T. 1 bis 3, ibid., 1880/81

[48]   *Borek, J.*, "Denis Diderot" (biography), Rowohlt, Reinbek, 2000

[49]   *Borup, B., Standke, B., Waßmer, C.*, Sol-gel - VOC-free, ECJ, 11/2007

[50]   *Brabender, C.W.*, Studies with the faringograph for predicting the most suitable types of American export wheats and flours for mixing with European soft wheats and flours, Cereal Chem., 1932; Six years of farinography, Cereal Chem., 1936; *Brabender, C.W., Muller, G., Köster, A.*, Der Amylograph, ein Apparat zur Messung der Backfähigkeit von Roggen, Zeitschr. d. Ges. f. Getreide-, Mühlen- und Bäckereiwesen, 1937

[51]   *Bram, G., Nguyen, T.A.*, Der Siegeszug der Farbstoffindustrie, J. Spektr. d. Wiss. Spezial "Farben", 2000

[52]   *Brandau, T.*, Mikrokapseln - eine runde Sache (Lebensmitteltechnik), J. Nachr. Chemie, 05/2009

[53]   *Brandrup, J., Immergut, E.H., Grulke, E.A.* (all eds.), Polymer handbook, Wiley, New York, 2003 (4th ed.)

[54]   *Brock, T., Groteklaes, M., Mischke, P.*, Lehrbuch der Lacktechnologie, Vincentz, Hannover, 2009 (3rd ed.); *Brock, T.*, Wie wird in Zukunft lackiert - Trends und Perspektiven, in: Aktuelles z. Chemie d. Farben u. Lacke, Reihe High Chem hautnah, Bd 3, GDCh, Frankfurt/M, 2008

[55]   Brookfield Inc., Middleboro, product information, www, 2005

[56]   *Brummer, R., Griebenow, M., Hetzel, F., Schlesiger, V., Uhlmann, R.*, Rheological swing test to predict the temperature stability of cosmetic emulsions, Proc. 11th IFSCC Internat. Congr. Berlin, Verl. Chem. Ind., 2000

[57]   *Bühne, S., Woocker, A., Linzmaier, A.*, Orientierung von Effektpigmenten, J. Farbe & Lack 11, 2008; Getting the best effect pigment orientation, J. ECS 09, 2009

[58]   *Bülfinger, G.B.*, De solidorum resistentia specimen, Commentarii Academiae Scientiarum Imperialis Petropolitanae, St. Petersburg, 1729

[59]   Bureau International des Poids et Mesures, Le système international d'unités, 6[c] éd., Sèvres, 1991

[60]   *Burgers, J.M.*, Mechanical considerations, model systems, phenomenological theories of relaxation and of viscosity, 1st report on viscosity and plasticity, Nordemann, Amsterdam, 1935

[61]   BWV (Bodenseewasserversorgung), Die Wasseraufbereitung in Sipplingen, VDI-Z., Düsseldorf, 2003

[62]   Byk Gardner, Geretsried, product information, www, 2005

[63]   Cambridge Viscosity Systems, Medford, product information, www, 2006

[64]   *Cannon, M.R., Manning, R.E., Bell, J.D.*, The kinetic energy correction and a new viscometer, Analytical Chemistry, 1960

[65]   *Cannone, M., Friedlander, S.*, "Navier" - blow-up and collapse, Notices AMS, 2003

[66]   Carl Zeiss AG, Magazin Innovation No. 14 to 21, Oberkochen, 2004 to 2008

[67]   *Caro, H.*, Über die Entwicklung der Theerfarbenindustrie, 1893

[68]   *Carothers, W.H., Berchet, G.J.*, Amides from ε-amino caproic acid, J. Am. Chem. Soc., 1930; *Carothers, W.H.*, Synthetic fibers, US-Patent (DuPont), 1937/38

[69]   *Carreau, P.J.*, Thesis, Univ. Wisconsin, 1968; *Carreau, P.J.*, Rheological equations from molecular network theories, Transactions Soc. Rheol., 1972

[70]   *Casson, N.*, A flow equation for pigment-oil suspensions of the printing ink type, in: Rheology of disperse systems, ed. C.C. Mill, Pergamon Press, New York, 1959

[71]   *Cauchy, A.L.*, Cours d'analyse, 1821; De la pression ou tension dans un corps solide, 1827; Sur les équations differentielles ou de mouvement pour les points matériels, 1829; Exercise d'analyse et de physique mathématique, 1840/1847; Leçons de calcul différentiel et de calcul intégral, 1840/1861

[72]   CCFRA (Campden & Chorleywood Food Research Association), workshop "Rheology and Texture Testing", Chipping Campden, 2003

[73]   Ceast S.p.A., Pianezza, product information, www, 2005

[74]   Clariant, documentation: 150 Jahre Forschung und Entwicklung in der Chemie, 2006

[75]   *Clauss, M.*, Alexandria - eine antike Weltstadt, Klett-Cotta, Stuttgart, 2003

[76]   Coatings Handbook, Pressure-sensitive adhesives, 2001

[77]   Contraves AG, Zürich, product information, 1979

[78]   *Couaraze, G., Grossiord, J.L.*, Initiation à la rheologie, Lavoisier, Paris, 1991

[79]   *Couette, M.M.*, Oscillations tournantes d'un solide de révolution en contact avec un fluide visqueux, Compt. Rend. Acad. Sci. Paris, 1887; Sur un nouvel appareil pour l'étude du frottement des fluides, ibid., 1888; La viscosité des liquides, Bull. Sci. Phys., 1888; Corrections relatives aux extrémités des tubes dans la méthode de Poiseuille, J. de Phys., 1890; Etudes sur le frottement des liquides, Ann. de Chimie et Physique, 1890

[80]   *Coulomb, C.A.*, Recherches théoretiques et éxperimentales sur la force de torsion et sur l'élasticité des fils du métal, 1787; in: Mémoires de l'Institut (Savants Etrangers), 1801

[81]   *Coussot, P.*, Rhéologie des laves torrentielles, La Houille Blanche, 1994; Lois d'écoulement des laves torrentielles boueuses, ibid., 1994

[82]   *Cox, W.P., Merz, E.H.*, Correlation of dynamic and steady flow viscosities, J. Polym. Sci., 1958

[83]   *Cross, M.M.*, Rheology of non-Newtonian fluids: A new flow equation for pseudoplastic systems, J. Colloid Interface Sci., 1965

[84]   *Dahlquist, C.A.*, in: Adhesion fundamentals and practice, McLaren, London, 1996

[85]   *Dalton, J.*, A new system of chemical philosophy, 1803

[86]   *Daniel, F.K.*, Determinations of mill base compositions for high speed dispersers, JPT 1966

[87]   *Daston, L.*, Kultur der Neugier, J. Spektr. d. Wiss. Spezial "Renaissance", 2004

[88]   *Dean, E.W., Davis, G.H.B.*, Chem. & Met. Eng. 36, 1929

[89]   *Deer, J.J.*, Controlled stress rheometry - early days, Bull. Brit. Soc. Rheol., 1994

[90]   *De Gennes, P.G.*, Reptation of a polymer chain in the presence of fixed obstacles, J. Chem. Phys., 1971

[91]   Degussa, Schriftenreihe: Pigmente (Nr. 11); Grundlagen von Aerosil (Nr. 16); Hydrophobe Aerosil-Typen und ihr Einsatz in der Lackindustrie (Nr. 18), Fine Particles (Nr. 18); Untersuchungsmethoden für synthetische Kieselsäuren und Silikate; Hanau, 2005

[92]   *Delambre, J.B.J.*, Base du système métrique, Paris, 1806/1810 (3 vol.)

[93]   *Descartes, R.*, Discours de la méthode pour bien conduire sa raison et chercher la verité dans les sciences, 1637 (publ. 1644)

[94]   *Des Cloizeaux, J.*, Europhys. Lett., 1988

[95]   Deutsche BP AG, Das große Buch vom Erdöl, Reuter & Klöckner, Hamburg, 1989

[96]   *De Waele, A.*, Viscometry and plastometry, Oil and Color Chem. Assoc. J., 1923

[97]   DGMK (Dt. Wissensch. Ges. f. Erdöl, Erdgas u. Kohle e.V.), calendar, Hamburg, 2007

[98]   *Dilke, O.A.W.*, Mathematics and measurement, Brit. Museum Publ., London, 1987

[99]   *Dinger, D.R.*, Ceramic processing, E-zine, www, 2005

[100]   DIN e.V., Rheologie-Normen, DIN-Taschenbuch 398, Berlin, 2007

[101]   DIN technical report 143 of working group NPF/NAB-AK 21.1 "Rheology", Modern rheological test methods (fundamentals and round-robin test), part 1: Yield point, Beuth, Berlin, 2005; part 2: Thixotropy, 2010 (draft)

[102]   *Dohnke, K.*, Die Lack Story - 100 Jahre, Verb. d. Lackindustrie (ed.), Dölling, Hamburg, 2000

[103]   *Doi, M., Edwards, S.F.*, Theory of polymer dynamics, Clarendon, Oxford, 1986; Doi, M., Introduction to polymer physics, Clarendon, Oxford, 1996

[104]   *Domke, W.*, Werkstoffkunde und Werkstoffprüfung, Girardet, Essen, 1975

[105] Domnick, J., Steger, R., On-line Viskositätsmessung, Papers 9th DFO Europ. Autom. Coat. Conf., Luxemburg, 2002

[106] Domschke, J.P., Lewandrowski, P., "Wilhelm Ostwald" - Chemiker, Wissenschaftstheoretiker, Organisator (biography), Urania, Leipzig, 1982

[107] Dontula, P., Macosko, C.W., Scriven, L.E., Origins of concentric cylinders viscometry, J. Rheol., 2005

[108] Doraiswamy, D., The origins of rheology, a short historical excursion, Bull. Brit. Soc. Rheol., 2002

[109] Dörfler, H.D., Grenzflächen und kolloid-disperse Systeme, Springer, Berlin, 2002

[110] Dornhöfer, G., Moderne Verfahren zur Messung der Kältefließfähigkeit von Schmierfetten - Prüfmethoden im direkten Vergleich, J. Tribol. & Schmierungstech. 03/2002

[111] Dreiss, C.A., Wormlike micelles - where do we stand, recent developments, linear rheology and scattering techniques; Soft Matter 3, 2007

[112] Dubbel, Taschenbuch für den Maschinenbau, eds. K.H. Grote, J. Feldhusen, Springer, Berlin, 2007 (22nd ed.)

[113] Eckart, W.U., Gradmann, C., "Hermann Helmholtz", J. Spektr. d. Wiss. Dossier "Große Physiker", 2004

[114] Ehrenstein, G.W., Riedel, G., Trawiel, P., Praxis der Thermischen Analyse, Hanser, München, 1998; Ehrenstein, G.W., Polymer-Werkstoffe, Hanser, München, 1978/1999

[115] Einstein, A., Eine neue Bestimmung der Moleküldimensionen, Diss., Bern, 1905/Ann. Physik, 1906; Über die von der molekulartheoretischen Theorie der Wärme geforderte Bewegung von in ruhenden Flüssigkeiten suspendierten Teilchen, ibid., 1906; Zur Theorie der Brownschen Bewegung, ibid., 1906; Berichtigung zu meiner Arbeit: "E. n. Best. d. Moleküldim.", ibid., 1911

[116] Eisenschitz, R., Rabinowitsch, B., Weissenberg, K., in: Mitt. Dtsch. Mat. Prüf. Anst., 1929; Eisenschitz, R., Philippoff, W., Eine neue Methode zur Bestimmung mechanischer Materialkonstanten von Kolloiden, Die Naturwissenschaften, 1933

[117] Elcometer Instruments Ltd., Manchester, product information, www, 2005

[118] Elementis Inc., Hightstown, product information, 2006

[119] Elias, H.G., Große Moleküle, Springer, Berlin, 1985

[120] Ellis, S.B., Thesis, Lafayette College, 1927

[121] Enidine, Orchard Park, product information

[122] Epprecht, A.G., Die Viskosität plastischer Flüssigkeiten, Chem. Rundschau, 1956; Die Bestimmung der Fliessgrenze an plastischen Substanzen, Revue Internat. de la Chocolaterie, 1957; Die Thixotropiemessung mit dem Rheomat 15, Chem. Rundsch., 1957

[123] Eriksson, I., Bolmstedt, U., Axelsson, A., Evaluation of a helical ribbon impeller as a viscosity measuring device for fluid foods with particles, J. Appl. Rheol. 12, 2002

[124] Erni, P., Fischer, P., Windhab, E.J., Stress- and strain-controlled measurements of interfacial shear viscosity and viscoelasticity at liquid/liquid and gas/liquid interfaces, Rev Sci Instr, vol 74, Am Inst Physics, 11/2003; Erni, P., Fischer, P., Heyer, P., Windhab, E.J., Kusnezow, V, Läuger, J., Rheology of gas/liquid and liquid/liquid interfaces with acqueous and biopolymer subphases, Prog Colloid Polym Sci 129, 2004

[125] Eukleides (Euclid), Ta stoicheia (elements of geometry), 300 BC (first print in Latin: E. Ratdolt, Venezia, 1482)

[126] Euler, L., Mechanica, 1736; Rechenkunst, 1738/1740; Introductio in analysin infinitorum, 1748; Scientia navalis, 1749; Institutiones calculi differentialis, 1755; Principes géneraux du mouvement des fluides, Mém. Acad. Sci., Berlin, 1755; Theoria motus corporum ("the second Mechanica"), 1765; Vollständige Anleitung zur Algebra, 1770; Théorie complette de la construction et de la manoevre des vaisseaux, 1773

[127] Eurocommit (Europ. Committee f. Ink Testing Methods), working group rheology, Frankfurt/M, 2004

[128] Everett, D.H., Basic principles of colloid science, Royal Soc. Chemistry, London, 1988

[129] Ewoldt, R.H., Hosoi, A.E., McKinley, G.H., New measures for characterizing nonlinear viscoelasticity in large amplitude oscillatory shear, J. Rheol., 11/2008

[130] Eyring, H., Viscosity, plasticity and diffusion as examples of absolute reaction rates, J. Chem. Phys., 1936

[131] Fardon, J., Eyewitness guides: Oil, Dorling Kinderley, London, 2007

[132] Feichtinger, J., Die Wiener Schule der Hochpolymerforschung in England und Amerika - Emigration, Wissenschaftswandel und Innovation; Univ. Graz (www), 2001

[133] Fellmann, E.A., "Leonhard Euler" (biography), Rowohlt, Reinbek, 1995

[134] Ferry, J.D., Viscoelastic properties of polymers, Wiley, New York, 1961/70/80 (3rd ed.)

[135] Fikentscher, H., J. Cellulosechemie, 13, 1932

[136] Finlay, V., Colour - travels through the paintbox, Hodder, London, 2002

[137] Fischer, E.K., Note on the origin of the term "rheology", J. Coll Sci 3, 1948

[138] Fischer, P., Windhab, E.J., Rheologie der Lebensmittel - Grundlagen und Applikationen, ETH Zürich, Inst. Food Techn., Zürich, 2003

[139] Fitzgerald, J.V., Matusik, F.J., Applications of a vibratory viscometer, J. Am. Lab., 06/1976

[140] Flucon Fluid Control, Clausthal-Zellerfeld, product information, www, 2009

[141] Fölsing, A., "Galileo Galilei" - Prozess ohne Ende (biography), Piper, München, 1983; "Albert Einstein" (biography), Suhrkamp, Frankfurt/M, 1993

[142] Freundlich, H., Über Thixotropie, Kolloid-Zeitschrift, 1928; Thixotropie, Hermann, Paris, 1935

[143] *Friedrich, C., Hofmann, B.*, Nichtkorrekte Aufgaben in der Rheometrie, Rheol. Acta, 1983
[144] *Fussbroich, P.*, Moderne Drug Delivery Systeme, Spezialpraktikum Liposomen u. Emulsionen, FH Sigmaringen, 2007
[145] *Füssel, S.*, "Johannes Gutenberg" (biography), Rowohlt, Reinbek, 2003
[146] *Gaede, P.M.* (ed.), Die 100 wichtigsten Erfindungen, J. Geo kompakt Nr. 18, 2009
[147] *Gahleitner, M., Sobczak, R.*, Bedeutung der Nullviskositätsbestimmung für das Modellieren von Fließkurven, J. Kunststoffe, 1989; Modifiziertes Carreau-Modell, ibid., 1991
[148] *Galilei, G.*, Dialogo dei massimi sistemi del mondo Tolemaico e Copernicano, 1624/1630; Discorsi e dimonstrazioni matematiche intorno a due nuove science attenenti alla mecanica e i movimenti locali, 1638
[149] *Gallenkamp, A. et al.*, Viscometer, J. Sci. Instrum., 1934; product information (Gallenkamp universal torsion viscometer), 1988
[150] *Gauß (Gauss), C.F.*, Disquisitiones arithmeticae, 1801
[151] *Gerngross, O., Goebel, E.*, Chemie und Technologie der Leim- und Gelatinefabrikation, Steinkopff, Dresden, 1933
[152] *Gerth, C.*, Rheometrie, in: Ullmann's Encyklopädie der technischen Chemie, Bd.5: Analysen und Meßverfahren, Verl. Chemie, Weinheim, 1980
[153] *Gerthsen, C., Kneser, H.O., Vogel, H.*, Physik – ein Lehrbuch, Springer, Berlin, 1989 (16th ed.)
[154] *Giacomin, A.J., Dealy, J.M.*, Using large-amplitude oscillatory shear, in: A.A. Collyer, D.W. Clegg, Rheological Measurement, Chapman, London, 1998 (2nd ed.)
[155] *Giesekus, H.*, Die Elastizität von Flüssigkeiten, Rheol. Acta, 1966; Phänomenologische Rheologie, Springer, Berlin, 1994
[156] *Gleick, J.*, "Isaac Newton" (biography), Harper, London, 2004
[157] *Gleissle, W.*, Two simple time-shear rate relations combining viscosity and first normal stress coefficient in the linear and non-linear flow range, ed. G. Astarita, G. Marucci, L. Nicolais, Proc. 8th Internat. Congr. Rheol. in Naples, Plenum Press, 1980
[158] *Goldschmidt, A., Streitberger, H.J.*, BASF Handbuch Lackiertechnik, Vincentz, Hannover, 2002
[159] Goodyear, product information, 2000
[160] *Gordon, J.E.*, The science of structures and materials, Sci. Am. Library, New York, 1988
[161] Göttfert Werkstoff-Prüfmaschinen GmbH, Buchen, product information, www, 2005
[162] *Gradzielski, M.*, Rheologie von viskoelastischen Tensidlösungen, Sinterface/Anton-Paar seminar, Potsdam, 2008
[163] *Graessley, W.W.*, Molecular entanglement theory of flow behaviour in amorphous polymers, J. Chem. Phys., 1965; The entanglement concept in polymer rheology, Adv. Polym. Sci., 1974
[164] *Grasmück, G.*, Der Wunsch jeder Hausfrau – ein Elektroherd, in: Die elektrische Gesellschaft, Badisches Landesmuseum, Karlsruhe, 1996
[165] *Guggisberg, D., Eberhard, P.*, Konsistenz von stichfestem Joghurt, J. dmz, 2004; *Guggisberg, D., Zehntner, U.*, Welche rheologischen Methoden eignen sich für Joghurt, ALP Science Nr. 514, Bern, 2007
[166] *Guicciardini, N.*, "Newton", J. Spektr. d. Wiss., ed. E. Bellone, 1999 (original in Italian: Le Science, Milano, 1998)
[167] *Gupta, R.K., Sridhar, T.*, Elongational rheometers, in: A.A. Collyer, D.W. Clegg, Rheological Measurement, Chapman, London, 1998 (2nd ed.)
[168] *Gurp, M.v., Palmen, J.*, Time-temperature superposition for polymeric blends, Rheol. Bull., 1998
[169] *Guth, P.*, Eine gelebte Idee – Wilhelm Ostwald und sein Haus "Energie" in Großbothen, Hypo & Dt. Werkbund Sachsen, 1999
[170] *Gysau, D.*, Fillers for paints, Vincentz, Hannover, 2006
[171] Haake Messtechnik, Karlsruhe, product information (e.g. on consistometers), 1962
[172] *Hagen, G.H.L.*, Über die Bewegung des Wassers in engen zylindrischen Rohren, Pogg. Ann. Physik, 1839; Handbuch der Wasserbaukunst, 1841/1861
[173] *Halpaap, R., Meier-Westhues, U., Richter, F.*, Polyurethane – 50 Jahre Chemie der Lackpolyisocyanate, in: Aktuelles z. Chemie d. Farben u. Lacke, Reihe High Chem hautnah, vol. 3, GDCh, Frankfurt/M, 2008
[174] *Harris, J.*, "Karl Weissenberg", Bull. Brit. Soc. Rheol., 1993
[175] *Hatschek, E.*, The general theory of viscosity of two-phase systems, Trans. Faraday Soc., 1913; The viscosity of colloidal solutions, Proc. Phys. Soc. London, 1916; The viscosity of liquids, London, 1928; Die Viskosität der Flüssigkeiten, Steinkopff, Leipzig, 1929
[176] *Heilen, W., et al.*, Additive für wässrige Lacksysteme, Vincentz, Hannover, 2009
[177] *Hell, S.W.*, Nanoskopie mit fokussiertem Licht, Physik J. 6, Nr.12, 2007
[178] *Hellenthal, W.*, Physik für Mediziner und Biologen, WVG, Stuttgart, 2007 (8th ed.)
[179] *Helmholtz, H.*, Über discontinuierliche Flüssigkeits-Bewegungen, 1868; *Helmholtz, H., Piotrowski, G.v.*, Über die Reibung in Flüssigkeiten die Tropfen bilden, Akad. Sitzungsber., Wien, 1860
[180] *Hencky, H.*, Zur Theorie plastischer Deformation und der hierdurch im Material hervorgerufenen Nebenspannungen, Z. f. angew. Math. u. Mechanik (ZAMM), 1924; Die Bewegungsgleichungen beim

nichtstationären Fließen plastischer Massen, ZAMM, 1925; Das Superpositionsgesetz eines endlich deformierten relaxationsfähigen elastischen Kontinuums und seine Bedeutung für die exakte Ableitung von Gleichungen für die zähe Flüssigkeit, Ann. Physik, 1929; The law of elasticity for isotropic and quasi-isotropic substances by finite deformation, J. Rheology, 1931

[181]   *Heppe, I.*, Was einst neu und gut war kam aus China, J. PM "China", 1994

[182]   *Herschel, W.H., Bulkley, R.*, Measurement of consistency as applied to rubber benzene solutions, Proc. ASTM 82, 1925; Konsistenzmessungen von Gummi-Benzol-Lösungen, Kolloid- Z., 1926

[183]   *Hertz, H.R.*, Prinzipien der Mechanik, 1894

[184]   *Hess, W.R.*, Die innere Reibung gelatierender Lösungen, Kolloid-Z., 1920

[185]   *Heyer, P., Läuger, J.*, Correlation between friction and flow of lubrication greases in a new tribometer device, Lubrication Sci 21, Wiley, 2009; *Heyer, P., Pop, L., Läuger, J.*, Overview of test methods in electro-rheology using a standard rotational rheometer, Proc. 11th Int. Conf. on Electrorheol. Fluids & Magnetorheol. Susp., eds. S. Odenbach, D. Borin, Dresden, 2008

[186]   *Hirsch, C., Struck, S., Cavaleiro, P.*, Flüssiger Pigment-Manager, J. Farbe & Lack, 06/2006

[187]   *Hobhouse, H.*, Seeds of change: Five plants that transformed mankind, Sidgwick & Jackson, London, 1985

[188]   *Hoffmann, H., Rehage, H., Platz, G., Schorr, G., Thurn, H., Ulbricht, W.*, Investigations on a detergent system with rodlike micelles, Colloid & Polymer Sci., 1982; *Hoffmann, H., Thunig, C., Schmiedel, P., Munkert, U., Ulbricht, W.*, The rheological behaviour of different viscoelastic surfactant solutions, Tenside Surf. Det., Hanser, München, 1994; *Hoffmann, H., Ulbricht, W.*, Mikrostrukturen und Fließverhalten viskoelastischer Tensidlösungen, Nachr. Chem. Tech. Lab. 1, 1995

[189]   *Hoffmann, M., Krömer, H., Kuhn, R.*, Polymeranalytik I, Thieme, Stuttgart, 1977

[190]   *Hohenadel, R., Rehm, T., Mieden, O.*, Polyvinylchlorid PVC, J. Kunststoffe, 10/2005

[191]   *Holzhausen, U., Millow, S.*, Thermische Analyse, in: Kittel, Lehrbuch der Lacke und Beschichtungen, vol. 10, ed. H.D. Otto, Hirzel, Stuttgart, 2006, 2nd ed.

[192]   *Honerkamp, J., Weese, J.*, A nonlinear regularization method for the calculation of relaxation spectra, Rheol. Acta, 1993

[193]   *Hooke, R.*, Micrographia, 1665; De potentia restitutiva, John Marty, London, 1678

[194]   *Höppler, F.*, Das Höppler-Konsistometer, Forschungsbericht Prüfgerätewerk Medingen (FPM), 1940; Rheologische und elastometrische Messungen an Kautschukprodukten, FPM, 1941; Viskosität, Plastizität, Elastizität und Kolloidik der Bitumina, FPM, 1942

[195]   *Houwink, R.*, Elasticity, plasticity and the structure of matter, Cambridge Univ. Press, 1937; Elastizität, Plastizität und Struktur der Materie, Steinkopff, Dresden, 1938

[196]   *Hu, Y.T., Lips, A.*, Kinetics and mechanism of shear banding in an entangled micellar solution, J. Rheol. 49(5), 09/2005; *Hu, Y.T., Wilen, L.*, Is the constitutive relation for entangled polymers monotonic?, J. Rheol. 51 (2), 03/2007

[197]   ICA (International Confectionery Association), Publication list of methods, no. 46: Viscosity of cocoa and chocolate products (2001), Bruxelles, (www) 2009

[198]   *Ilschner, B., Singer, R.F.*, Werkstoffwissenschaften und Fertigungstechnik, Springer, Berlin, 2005 (4th ed.)

[199]   IOCCC (International Office of Cocoa, Chocolate and Sugar Confectionery), Viscosity of cocoa and chocolate products, 2001

[200]   *Irle, C.*, Polyurethanlacke - Domäne in der Holzbeschichtung, in: Aktuelles z. Chemie d. Farben u. Lacke, Reihe High Chem hautnah, vol. 3, GDCh, Frankfurt/M, 2008

[201]   IUPAC (International Union of Pure and Applied Chemistry), information, www, 2005

[202]   *Jeffreys, H.*, The earth, Cambridge Univ. Press, 1929

[203]   *Joseph, D.D.*, Fluid dynamics of viscoelastic liquids, Springer, New York, 1990

[204]   Kema (Keramikmaschinen Görlitz), product information, 1984

[205]   *Képès, A.*, Unpubl. paper presented at 5th Internat. Congr. Rheol., Kyoto, 1968

[206]   *Khademhosseini, A., Vacanti, J.P., Langer, R.*, Progress in tissue engineering, J. Sci. Am., 05/2009

[207]   Kinematica AG, Littau-Luzern, product information, www, 2009

[208]   Kittel, Lehrbuch der Lacke und Beschichtungen, vol. 4: Lösemittel, Weichmacher und Additive, ed. M. Ortelt, Hirzel, Stuttgart, 2007

[209]   *Klemm, F.*, Geschichte der Technik, Rowohlt, Hamburg, 1983

[210]   *Kohl, P.*, Ruhe im Betrieb - Maßnahmen des Schallschutzes, J. Masch.-Markt, 3/2006

[211]   *Kohlrausch, R. (sen.)*, Nachtrag über die elastische Nachwirkung beim Cocon und Glasfaden, Pogg. Ann. Physik, 1847; *Kohlrausch, F.W. (jun.)*, Über die elastische Nachwirkung bei der Torsion, ibid., 1863; Beiträge zur Kenntnis der elastischen Nachwirkung, ibid., 1866; Experimental-Untersuchung über die elastische Nachwirkung bei der Torsion Ausdehnung und Biegung, ibid., 1876

[212]   *Krämer, E.O., Williamson, R.V.*, Internal friction and the structure of "solvated" colloids, J. Rheology, 1929; *Krämer, E.O.*, Molecular weights of celluloses and cellulose derivatives, J. Industr. Eng. Chem., 1938

[213]   *Krieger, I.M., Dougherty, T.J.*, A mechanism for non-Newtonian flow in suspensions of rigid spheres, Trans. Soc. Rheol., 1959

[214]   *Kuhn, W.*, Über qualitative Deutung der Viskosität und Strömungsdoppelbrechung von Suspensionen, Kolloid-Z. 62, 1933

[215]   *Kulicke, W.M.*, Fließverhalten von Stoffen und Stoffgemischen, Hüthig & Wepf, Basel, 1986; *Kulicke, W.M., Clasen, C.*, Viscosimetry of polymers and electrolytes, Springer, Berlin, 2004

[216]   Kunststoff-Museums-Verein e.V. (ed.), Faszination Kunststoff, Düsseldorf, 1989

[217]   *Laba, D.* (ed.), Rheological properties of cosmetics and toiletries, Dekker, N.York, 1993

[218]   *Lagaly, G., Schulz, O., Zimehl, R.*, Dispersionen und Emulsionen, Steinkopff, Darmstadt, 1997

[219]   Lanxess, Leverkusen, product information, 2009

[220]   *Larson, R.G.*, The structure and rheology of complex fluids, Oxford Univ. Press, N.York, 1999

[221]   Lauda GmbH & Co KG, Lauda, product information, www, 2005

[222]   *Läuger, J., Huck, S.*, Real controlled stress and controlled strain experiments with the same rheometer, Proc. 8th Int. Congr. Rheol., Cambridge, 2000; *Läuger, J., Bernzen, M.*, Getting the zero-shear viscosity of polymer melts with different rheological tests, Ann Trans Nordic Rheol Soc 8/9, 2000/2001; *Läuger, J., Wollny, K., Huck, S.*, Direct strain oscillation – a new method enabling measurements at very small shear stresses and strains, Rheol. Acta, 2002; *Läuger, J., Wollny, K., Stettin, H., Huck, S.*, A new device for the full rheological characterization of magneto-rheological fluids, Int. J. Mod. Phys. B 19, 2004; *Läuger, J., Heyer, P.*, Validation of empirical rules for a standard polymer solution by different rheological tests, Ann Trans Nordic Rheol Soc 15, 2007; *Läuger, J., Heyer, P.*, Interfacial shear rheology of coffee, Proc 5th Int. Symp. Food Rheology & Structure ISFRS, Zürich, 2009; *Läuger, J., Stettin, H.*, Differences between stress and strain control in the non-linear behavior of complex fluids (LAOS), Rheol. Acta, 2010

[223]   *Laun, H.M.*, Prediction of elastic strains of polymer melts in shear and elongation, J. Rheol., 1986; *Laun, H.M., Schmidt, G., Gabriel, C., Kieburg, C.*, Reliable plate-plate magnetorheometry based on validated radial magnetic flux density profile simulations, Rheol Acta, Springer, 2008

[224]   *Laurenza, D.*, "Leonardo da Vinci" – Künstler, Forscher, Ingenieur, J. Spektr. d. Wiss. Biographie, ed. E. Bellone, 2000 (original in Italian: Le Scienze, Milano, 1999)

[225]   *Leibniz, G.W.*, Nova methodus pro maximis et minimis, 1684; De geometria infinitorum, 1686

[226]   *Li, X., Cao, H.L., Pan, F.Y., Weng, L.Q., Song, S.H., Huang, Y.D.*, Preparation of body armour material of Kevlar fabric treated with colloidal silica nanocomposite, J. Plastics, Rubber & Composites, 2008

[227]   *Lissajous, J.A.*, Mémoire sur l'étude optique des mouvements vibratoires, Ann. Chim. Phys., 1857

[228]   *Litters, T., Koch, B.*, Untersuchungen zum Kältefließverhalten von Schmierfetten mit modernen Rotations-viskosimetern - Einfluss von Verdicker und Basisöl auf die viskoelastischen Eigenschaften bei -40°C, J. Schmierstoffe & Schmierungstechnik, 2006

[229]   *Lodge, A.S.*, A network theory of flow birefringence and stress in concentrated polymer solutions, Trans. Faraday Soc., 1956; L., Stress relaxation after a sudden shear strain, Rheol. Acta, 1975; *Lodge, A.S. Meissner, J.*, J. Rheol. Acta, 1976; L., On-line measurement of elasticity and viscosity in flowing polymeric liquids, J. Rheol. Acta, 1996; L., Normal stress differences from hole pressure measurements, in: *A.A. Collyer, D.W. Clegg*, Rheological Measurement, Chapman, London, 1998 (2nd ed.); L., A rheo-lodgical half century, J. Non-Newt. Fluid Mech., 2007

[230]   *Lojacono, E.*, "René Descartes", J. Spektr. d. Wiss. Biographie, ed. E. Bellone, 2001 (original in Italian: Le Scienze, Milano, 2000)

[231]   *Lootens, D., Jousset, P., Dagallier, C., Hébraud, P., Flatt, R.J.*, The "Dog Tail Test" – a quick and dirty measure of yield stress - application to polyurethane adhesives, J. Appl. Rheol. 19, 2009

[232]   *MacMichael, F.R.*, in: J. Ind. Eng. Chem., 1915; US patent no. 1281042, 1918

[233]   *Macosko, C.W.*, Rheology – principles, measurements and applications, Wiley-VCH, New York, 1994

[234]   *Magnus, K., Müller, H.H.*, Grundlagen der Technischen Mechanik, Teubner, Stuttgart, 1974

[235]   *Mahon, B.*, The man who changed everything – the life of James Clerk Maxwell, Wiley, Chichester, 2004

[236]   *Maleika, R., Nennemann, A., Pyrlik, O.*, Sol-Gel-Beschichtungen und Hybridsysteme, in: Aktuelles z. Chemie d. Farben u. Lacke, Reihe High Chem hautnah, vol. 3, GDCh, Frankfurt/M, 2005

[237]   *Malkin, A., Askadsky, A., Chalykh, A., Kovriga, V.*, Experimental methods of polymer physics, Mir, Moskow, 1983; *Malkin, A., Kuznetsov, V.V.*, Linearization as a method for determining parameters of relaxation spectra, Rheol. Acta, 2000; *Malkin, A., Isayev, A.I.*, Rheology – concepts, methods, and applications, Chem Tec, Toronto, 2006

[238]   *Mallock, A.*, Determination of the viscosity of water, Proc. R. Soc. London, 1888; Experiments on fluid viscosity, Philos. Trans. R. Soc., 1896

[239]   Malvern Instruments Ltd., Malvern, product information (e.g. Rosand capillary rheometers), www, 2005

[240]   *Manneville, S.*, Review: Recent experimental probes of shear banding, Rheol Acta 10, 2007

[241]   *Margules, M.*, Über die Bestimmung des Reibungs- und Gleitungskoeffizienten aus ebenen Bewegungen einer Flüssigkeit; Über die Bewegung zäher Flüssigkeiten und über Bewegungsfiguren, Wien, 1881

[242]   *Marin, G.*, Oscillatory rheometry, in: A.A. Collyer, D.W. Clegg, Rheological Measurement, Chapman, London, 1998 (2nd ed.)

[243]   *Marin-Santibanez, B.M., Perez-Gonzalez, J., Vargas, L.de, Rodriguez-Gonzalez, F., Huelsz, G.*, Rheometry
        - PIV of shear-thickening wormlike micelles, Am. Chem. Soc., 03/2006

[244]   *Mark, H.F.*, Physik und Chemie der Cellulose, Springer, Berlin, 1932

[245]   *Maxwell, J.C.*, On the viscosity or internal friction of air and other gases, 1866; On the dynamical theory of
        gases, Phil. Trans. Roy. Soc., 1867; Constitution of solids, Encyclopaedia Britannica, 1878

[246]   *Meichsner, G., Mezger, T., Schröder, J.*, Lackeigenschaften messen und steuern, Vincentz, Hannover, 2003

[247]   *Meier-Westhues, U.*, Polyurethane - Lacke, Kleb- und Dichtstoffe; Vincentz, Hannover, 2007

[248]   *Mendez-Sanchez, A.F., Lopez-Gonzalez, M.R., Rolon-Garrido, V.H., Perez-Gonzalez, J., de Vargas, L.*, Instabili-
        ties of micellar systems under homogeneous and non-homogeneous flow conditions, Rheol Acta 42, 2003

[249]   *Menges, G.*, Werkstoffkunde der Kunststoffe, Hanser, München, 1979/1998 (4th ed.)

[250]   *Meyer, O.E.*, Über die Reibung in Flüssigkeiten, J. de Crelle, 1861/1863; Theorie der elastischen Nachwir-
        kung, Pogg. Ann. Physik, 1874; *Neumann, F., Meyer, O.E.*, Vorlesungen über die Theorie der Elasticität der
        festen Körper und des Lichtäthers, Teubner, Leipzig, 1885

[251]   *Mezger, T.*, Rheological characterization of foodstuffs, CCFRA workshop "Rheology and texture testing",
        Chipping Campden, 2003; *Mezger, T.*, Rheologie von Lacken und Farben, Elementis/APG-Seminar, Köln,
        2003/2010; *Mezger, T., van Peji, D.*, Modern rheology measurements in today's coatings industry, ECC
        Nürnberg, Vincentz, Hannover, 2005; *Mezger, T.*, Rheologische Prüfungen, in: Kittel, Lehrbuch der Lacke
        und Beschichtungen, vol. 10, ed. H.D. Otto, Hirzel, Stuttgart, 2006, 2nd ed.; *Mezger, T.*, Bestimmung der
        Fließgrenze mit diversen Messmethoden, DGK User Meeting (Dt. Ges. f. Kosmetik), Hamburg, 2006;
        Mezger, T., Nano- und Mikrostrukturen von Tensidsystemen und deren rheologisches Verhalten,
        Background Info APG, 2009

[252]   *Miller, R., Liggieri, I.*, Interfacial rheology, in: Progress in colloid and interface science, vol. 1, Brill, 2009

[253]   *Mohr, O.*, Über die Darstellung des Spannungszustandes und des Deformationszustandes eines Körpere-
        lementes, J. Zivilingenieur, 1882

[254]   *Monk, C.H.J.*, A rotary viscometer for thinning paint samples, JOCCA, 1958; Routine measurement of the
        viscosity of paint samples, ibid., 1966

[255]   *Mooney, M.*, Explicit formulas for slip and fluidity, J. Rheol., 1931; *Mooney, M., Ewart, R.H.*, The conicylin-
        drical viscosimeter, Physics, 1934; *Mooney, M.*, A theory of large elastic deformation, J. Appl. Phys., 1940;
        Secondary stresses in viscoelastic flow, J. Colloid Sci., 1951; in: Rheology - theory and applications, ed.
        F.R. Eirich, Academic Press, 1958

[256]   *Moore, F., Davies, L.J.*, The consistency of ceramic slips, Trans. Brit. Ceram. Soc., 1956

[257]   *Morrison, F.A.*, Understanding rheology, Oxford University Press, New York, 2001

[258]   *Müller, B.*, Additive kompakt, Vincentz, Hannover, 2009

[259]   *Muller, F., Schurtenberger, P.*, SAXSess - Nanoparticles influence the liquid-crystalline structure of soap,
        AP Appl. Note, 2006

[260]   *Müller, M., Tyrach, J., Brunn, P.O.*, Rheological characterization of machine-applied plasters, ZKG
        International, 1999

[261]   *Nakken, T., Tande, M., Elgsaeter, A.*, Measurements of polymer induced drag reduction and polymer
        scission in Taylor flow using standard double-gap sample holders with axial symmetry, J. Non-Newt. Fluid
        Mech., 2000

[262]   Nametre, Metuchen, product information (e.g. Viscoliner), 1994

[263]   *Navier, C.L.M.H.*, Mémoire sur les lois du mouvement des fluides, Acad. Sci., Paris, 1822; Mémoire sur les
        lois de l'équilibre et du mouvement des corps solides élastiques, ibid., 1829

[264]   *Neukom, J.*, Studien über das Geliervermögen von Pektinstoffen und anderen hochmolekularen Polyoxy-
        verbindungen, Promotion ETH, Zürich, 1949

[265]   *Newth, E.*, Die Jagd nach der Wahrheit, dtv, München, 2000 (original in Norwegian: Jakten påsannheten,
        Tiden Norsk, Oslo, 1996)

[266]   *Newton, I.*, Philosophiae naturalis principia mathematica ("Principia"), London, 1687

[267]   Nixdorf Museumsforum (ed.), Museumsführer, Paderborn, 2000

[268]   *Oberbach, K.*, Kunststoff-Taschenbuch, Hanser, München, 2001 (28th ed.)

[269]   *O'Connor, A.E., Willenbacher, N.*, The effect of molecular weight and temperature on tack properties of
        model polyisobutylenes, Int. J. of Adhesion & Adhesives 24, 2004

[270]   *Oehler, H., Lellinger, D., Alig, I.*, Filmbildung von Dispersionen mit Ultraschall verfolgen, J. Farbe & Lack,
        07/2005

[271]   OICCC (Office International du Cacao et du Chocolat), Viscosity of chocolate - determination of Casson
        yield value and Casson plastic viscosity, J. Rev. Int. Choc., 1973

[272]   *Ostwald, Wi. (sen.)*, Lehrbuch der allgemeinen Chemie, Steinkopff, Leipzig, 1891; Grundriss der allgemei-
        nen Chemie, Engelmann, Leipzig, 1899

[273]   *Ostwald, Wo. (jun.)*, Zur Systematik der Kolloide, Koll. Z., 1907; Über die Geschwindigkeitsfunktion der
        Viskosität disperser Systeme, ibid., 1925; *Ostwald, W., Auerbach, R.*, Über die Viskosität kolloider
        Lösungen im Struktur-, Laminar- und Turbulenzgebiet, ibid., 1936

[274] *Ottersbach, J.*, Bedruckstoff und Farbe., Verl. Beruf u. Schule, Itzehoe, 1995

[275] Outlast Technologies Inc., product information "Thermocules", www, 2009

[276] *Pahl, M., Gleissle, W., Laun, H.-M.*, Praktische Rheologie der Kunststoffe und Elastomere, VDI, Düsseldorf, 1995

[277] *Parker, A., Vigouroux, F.*, Texture profiling with the vane for characterizing the rheology of shear-sensitive soft foods, 3. Internat. Sympos. Food Rheol. & Structure, Zürich, 2003

[278] *Pascal, B.*, Essai pour les coniques, 1640; Récit de la grande éxperience de l'équilibre des liqueurs, 1649; Traités de l'équilibre des liqueurs et de la pesenteur de la masse de l'air, Paris, 1663

[279] *Pashias, N., Boger, D.V., Summers, J., Glenister, D.J.*, A fifty cent rheometer for yield stress measurements, J. of Rheol., 1996

[280] *Patat, F. Seydel, G.*, Über ein Rotationsviskosimeter zur Untersuchung strukturviskoser Stoffe, Z. Die chem. Fabrik, 1941

[281] *Patton, T.C.*, Paint flow and pigment dispersion, Wiley, New York, 1978 (2nd ed.)

[282] PCI (Powder Coating Institute), North America, information, www, 2005

[283] *Pechold, W., Blasenberg, S.*, Koll. Z. u. Z. f. Polym., 1970

[284] *Perona, P.*, Bostwick degree and rheological properties – an up-to-date viewpoint, J. Appl. Rheol., 2005

[285] *Perry, J.*, Practical Mechanics, Cassell, London, 1883; Liquid friction, Philos. Mag., 1893

[286] *Peruzzi, G.*, "Maxwell" – der Begründer der Elektrodynamik, J. Spektr. d. Wiss. Biographie, ed. E. Bellone, 2000 (original in Italian: Le Science, Milano, 1998)

[287] *Pethrick, R.A.*, Rheological studies using a vibrating probe, in: Collyer, A.A., Clegg, D.W., Rheological Measurement, Chapman, London, 1998 (2nd ed.)

[288] *Philippoff, W.*, Dynamische Untersuchungen an kolloiden Systemen, 1934; Ein elektromagnetisches Viskosimeter mit Schwingungsbeanspruchung, Hirzel, Leipzig, 1934; Über die Strömung von Stoffen mit Fließfestigkeit, Koll. Z., 1936; Viskosität der Kolloide, Steinkopff, Leipzig, 1942

[289] *Piau, J.M., Bremond, M., Couette, J.M., Piau, M.*, "Maurice Couette" un des fondateurs de la rhéologie, Cahiers de Rhéologie, 1994; "Maurice Couette" one of the founders of rheology, Rheol. Acta, 1994; *Piau, J.M., Piau, M.*, The relevance of viscosity and slip at the wall of the early days in rheology and rheometry and stability in concentric cylinder viscometry, J. Rheol., 2005; *Piau, J.M., Piau, M.*, Le Prix "Maurice Couette" du GFR, Rhéologie, vol. 8, 2005

[290] *Pochettino, A.*, in: Nuovo Cimento, 1914

[291] *Poiseuille, J.L.M.*, Recherches expérimentales sur le mouvement des liquides dans les tubes de très petits diamètres, Acad. Sci., Paris, (1840) publ. 1846

[292] *Poisson, S.D.*, Mémoire sur les equations generales de l'équilibre et du mouvement des corps solides élastiques et des fluides, J. École Polytechnique, Paris, 1829

[293] *Popplow, M.*, Die Emanzipation der Technik, J. Spektr. d. Wiss. Spezial "Renaissance", 2004

[294] *Poynting, J.H., Thomson, J.J.*, Properties of matter, Charles Griffin, London, 1902

[295] *Prandtl, L.*, Über Flüssigkeitsbewegung bei sehr kleiner Reibung, 3. Int. Mathematiker-Kongress, Heidelberg, 1904; Bericht über Untersuchungen zur ausgebildeten Turbulenz, ZAMM, 1925; Neuere Ergebnisse zur Turbulenzforschung, Z. VdI, 1933; Strömungslehre, Vieweg, Braunschweig, 1960

[296] *Prasad, V, Semwogerere, D., Weeks, E.R.*, Confocal microscopy of colloids, J. Phys. - Condens. Matter 19, 2007

[297] *Rabinow, J.*, The magnetic fluid clutch, AIEE Transactions, 1948

[298] *Rabinowitsch, B.*, Über die Viskosität und Elastizität von Solen, Z. Phys. Chem., 1929

[299] *Rao, M.A.*, Rheology of Fluid and semisolid foods, Springer, New York, 2007 (2nd ed.)

[300] *Ree, T., Eyring, H.*, Theory of non-Newtonian flow: 1. Solid plastic system, J. Appl. Phys., 1955

[301] *Rehage, H.*, Rheologische Untersuchungen an viskoelastischen Tensidlösungen, Diss., Bayreuth, 1982

[302] *Reimann, V. Joos-Müller, B., Dirnberger, K., Eisenbach, C.D.*, Assoziativverdickern auf der Spur, J. Farbe & Lack, 05/2007

[303] *Reiner, M., Riwlin, R.*, Über die Strömung einer elastischen Flüssigkeit im Couette-Apparat, Kolloid-Z., 1927; *Reiner, M., Weissenberg, K.*, A thermodynamical theory of the strength of material, Rheology Leaflet no. 10, 1939; *Reiner, M.*, Twelve lectures on theoretical rheology, North Holland, Amsterdam, 1949; Deformation, strain and flow, Lewis, London, 1960 (in German: Rheologie in elementarer Darstellung, Hanser, München, 1968/1969, 2nd ed.); Selected papers on rheology, Amsterdam, 1975

[304] *Reitberger, H.*, Rheologie UV-härtender Materialien, J. Welt der Farben, 11/2006

[305] *Reynolds, O.*, An experimental investigation of the circumstances which determine whether a motion of water shall be direct or sinuous and the law of resistance in parallel channels, Philos. Transactions, 1883; On the dynamical theory of incompressible viscous fluids and the determination of the criterion, ibid., 1895; The structure of the universe, Cambridge Univ. Press, 1903

[306] Robert Bosch GmbH, Kraftfahrtechnisches Taschenbuch, Stuttgart, 2006 (25th ed.)

[307] *Roberts, J.E.*, The early development of the rheogoniometer, Bull. Brit. Soc. Rheol., 1993

[308] *Roelig, H.*, in: Proc. Rubber Technol. Conf., Cambridge, 1938

[309] *Römpp* Lexikon "Lacke und Druckfarben", ed. U. Zorll, Thieme, Stuttgart, 1998

[310] *Röse, B.*, Kontinuierliches Aufbereiten von Gum Base, ZfL 7/8, 1995

[311] *Rouse, P.E.*, A theory of linear viscoelastic properties of dilute solutions of coiling polymers, J. Chem. Phys., 1953

[312] *Russell, B.*, A history of western philosophy, Simon & Schuster, New York, 1945

[313] *Sack, O., Mezger, T.*, Wirkung rheologischer Additive in wässrigen Beschichtungssystemen - ein Überblick, J. Welt d. Farben, 03/2010

[314] *Saint Venant, A.J.B. de*, Mémoire sur l'équilibre des corps solides, les limits de leur élasticité, et sur les conditions de leur résistence quand les désplacements éprouvé par leurs points ne sont pas très pétit, Acad. Sci., Paris, 1847

[315] *Salesch, M.*, Kurz gefasste Geschichte der Erdölförderung in Baku, Ölpost 007, ed. Dt. Erdölmuseum eV, Wietze, 2007

[316] *Saucy, D., Bobsein, B.*, Tolerating surfactants, Europ. Coat. J., 11/2007

[317] *Sauermost, R.* (ed.), Lexikon der Naturwissenschaftler, Spektrum, Heidelberg, 2000

[318] *Schäffler, H., Bruy, E., Schelling, G.*, Baustoffkunde, Vogel, Würzburg, 1975

[319] *Schatzmann, M., Bezzola, G.R., Minor, H.E., Windhab, E.J., Fischer, P.*, Rheometry for large-particulated fluids - analysis of the ball measuring system and comparison to debris flow rheometry, Rheol. Acta, 2009

[320] *Schnablegger, H., Singh, Y.*, A practical guide to SAXS, Anton Paar, Graz, 2006

[321] *Schoff, C.K., Kamarchik, P.*, Rheological measurements, in: Encyclopedia of Polymer Science and Technology, vol. 11, ed. H.F. Mark, Wiley, 2004 (3rd ed.)

[322] *Schönbeck, J.*, "Euklid", Serie vita mathematica, ed. E.A. Fellmann, Birkhäuser, Basel, 2003

[323] Schott-Geräte, Mainz, product information, www, 2005; Theorie u. Praxis der Kapillarviskosimetrie, 1999

[324] *Schramm, G.*, Einführung in Rheologie und Rheometrie, Haake, Karlsruhe, 1995

[325] *Schreier, W.*, Einige Sehenswürdigkeiten in der Wilh.-Ostwald-Gedenkstätte in Großbothen, in: Physica et Historia, ed. S. Splinter et al, Acta Hist. Leop. 45, Halle/S., 2005

[326] *Schulze, B., Brauns, J., Schwalm, I.*, Neuartiges Baustellen-Messgerät zur Bestimmung der Fliessgrenze von Suspensionen, J. Geotechnik, 1991

[327] *Schwarzl, F.R.*, Polymermechanik, Springer, Berlin, 1990

[328] *Schwedoff, T.*, in: J. de Phys., 1880; Recherches experimental sur la cohesion des liquides, part 1: Module des liquides, J. ibid, 1889, part 2: Viscosité des liquides, ibid., 1890

[329] *Scott Blair, G. W.*, Introduction to Industrial Rheology, Churchill, London, 1938; A survey of general and applied rheology, Pitman, London, 1944/1949; Foodstuffs – their plasticity, fluidity and consistency, 1953; Elementary rheology, Acad. Press, London, 1969

[330] *Searle, G.F.C.*, A simple viscometer for very viscous liquids, Proc. Cambridge Philos. Soc., 1912

[331] *Segrè, E.*, From falling bodies to radio waves - classical physicists and their discoveries, Freeman, New York, 1984

[332] *Selby, T.W.*, The use of the scanning Brookfield technique to study the critical degree of gelation of lubrificants at low temperatures, SAE paper 9100746, Soc. of Automotive Eng., 1991

[333] *Sentmanat, M.*, Measuring the transient extensional rheology of polyethylene melts using the SER universal testing platform, J. Rheol. 49(3), 2005

[334] *Seo, K.S., Posey-Dowty, J.D.*, Device to measure the solidification properties of a liquid film and method therefor, US pat. appl., publ. no. US 2005/0132781 A1, 2005

[335] *Sepeur, S., Laryea, N., Goedicke, S., Groß, F.*, Nanotechnologie - Grundlagen und Anwendungen, Vincentz, Hannover, 2008

[336] Sheen Instruments Ltd., Kingston-upon-Thames, product information, www, 2005

[337] *Singh, S.*, Fermat's last theorem, Fourth Estate, London, 1997

[338] *Sisko, A.W.*, The flow of lubricating greases, Industrial and Engineering Chemistry, 1958

[339] *Smith, D.N.*, Polyurethane thickeners for waterborne coatings, Coatings Technology Handbook, RC Press, Bocaraton, 2006

[340] Sofraser, Villemandeur, product information, 1989

[341] Spindler & Hoyer KG, product information (e.g. viscometer acc. to Dr. Kämpf), Göttingen, 1953/1959

[342] Stable Micro Systems, Godalming, product information, www, 2005

[343] *Staudinger, H.*, Über Polymerisation, Ber. dt. chem. Ges., 1920; Die hochmolekularen organischen Verbindungen von Kautschuk und Cellulose, Springer, Berlin, 1932; Organische Kolloidchemie, Vieweg, Braunschweig, 1950

[344] *Steffe, F.S.*, Rheological methods in food process engineering, Freemann, East Lansing, 1996 (2nd ed.)

[345] *Stehr, W.*, Friction phenomena at lowest sliding speeds - tribological fingerprint of test specimen, 14th Int. Coll. of Tribology, Ostfildern, 2004; *Stehr, W., Dobler, K.*, Of fried sausages and bearing damage - on tribology, Tillwich, Horb, 2008

[346] *Steiger-Trippi, K., Ory, A.*, Zur Interpretierung von Fliesskurven, J. Pharmac. Acta Helevetiae, 1961

[347] *Steiner, E.H.*, A new rheological relationship to express the flow properties of melted chocolate, Rev. Int. Choc., 1958

[348] *Stokes, G.G.*, Report on the recent researches in hydrodynamics, Report Brit. Assoc., 1846; On some cases of fluid motion, Trans. Cambridge Philos. Soc., 1849 (1843); On the theories of the internal friction of fluids in motion and of the equilibrium and motion of elastic solids, ibid., 1849 (1845); On the effect of the internal friction of fluids on the motion of pendulums, ibid., 1851 (1850)

[349] *Stormer, E.J.*, in: Trans. Amer. Ceramic. Soc., 1909

[350] *Svejda, P.*, Prozesse und Applikationsverfahren, Serie: Moderne Lackiertechnik, Vincentz, Hannover, 2003

[351] *Tabb., J.R., Kulash, D.*, Strategic Highway Research Program, Washington, 1992

[352] *Tanner, R.I.*, Engineering rheology, Oxford Univ. Press, New York, 1985 /2000; *Tanner, R.I., Walters, K.*, Rheology – an historical perspective, Elsevier, Amsterdam, 1998; *Tanner, R.I.*, Note on the beginnings of sinusoidal testing methods, Korea-Australia Rheol. J., 2002; *Tanner, R.I., Tanner, E.*, "Heinrich Hencky" – a rheological pioneer, Rheol. Acta 42, 2003

[353] *Taschner, R.*, Das Unendliche, Springer, Berlin, 2006 (2nd ed.)

[354] *Taylor, G.I.*, Stability of a viscous liquid contained between two rotating cylinders, Philos. Trans. Roy. Soc. London, 1923; Fluid friction between rotating cylinders, Proc. Roy. Soc., London, 1936

[355] *Teschner, H.*, Offsetdrucktechnik, Fachschriftenverlag, Fellbach, 1997 (10th ed.)

[356] *Thermo Electron GmbH*, Karlsruhe, product information (e.g. melt flow indexer), www, 2005

[357] *Thiele, R.*, Frühe Meister des unendlich Kleinen, J. Spektr. d. Wiss. Spezial "Renaissance", 2004

[358] *Thomson, J.J.*, Applications of dynamics to physics and chemistry, MacMillan, London, 1888

[359] *Thomson, W. (Kelvin)*, On the elasticity and viscosity of metals, Proc. Roy. Soc., 1865; Papers, London, 1890

[360] *Thorpe, T.E., Rodger, J.W.*, On the relation between the viscosity (internal friction) of liquids and their chemical nature, Philos. Trans. Roy. Soc., London, 1894

[361] *Tillner, S., Mock, U.*, Mikrokapseln kitten Risse, J. Farbe & Lack, 10/2007

[362] *Titus Lucretius Carus*, De rerum natura/Welt aus Atomen, Reclam, Stuttgart, 2000

[363] *Triantafillopoulos, N.*, Measurement of fluid rheology and interpretation of rheograms, Paper Trade J., 1948; and: Kaltec Scientific, Novi, product information, 1993/www 2005

[364] *Trinkle, S., Friedrich, C.*, Van Gurp-Palmen-plot: a way to characterize polydispersity of linear polymers, Rheol. Acta, 2001; *Trinkle, S., Walter, P., Friedrich, C.*, Van Gurp-Palmen-plot II: classification of long chain branched polymers by their topology, ibid., 2002

[365] *Trouton, F.T., Andrews, E.S.*, in: Phil. Mag., 1904; T., On the viscosity of pitch-like substances, Proc. Roy. Soc., London, 1905; T., On the coefficient of viscous traction and its relation to that of viscosity, ibid., 1906

[366] *Tscheuschner, H.D.*, Rheologische Eigenschaften von Lebensmittelsystemen, in: D. Weipert et al., Rheologie der Lebensmittel, Behr's, Hamburg, 1993

[367] *Tschoegl, N.W.*, The phenomenological theory of linear viscoelastic behavior, Springer, New York, 1989

[368] *Tyrach, J.*, Rheologische Charakterisierung von zementären Baustoffsystemen, Diss., Techn. Fak. Univ. Erlangen-Nürnberg, 2001

[369] *Ubbelohde, L.*, Tabellen zum Englerschen Viskosimeter, Hirzel, Leipzig, 1907/1918/1930; Das einfachste und genaueste Viskosimeter und andere Geräte mit dem hängenden Niveau, J. Öl und Kohle, 1936; Zur Viskosimetrie, Hirzel, Stuttgart, 1965

[370] *Uhlherr, P.H.T., et al.*, Yield stress fluid behavior on an inclined plane, Advances in rheology, ed. B. Mena et al., Mexico City, 1984

[371] *Van der Waals, J.D.*, Over de continuiteit van den gas- en vloeistoftoestand, PhD work, 1873

[372] *Vilgis, T.*, Die Molekül-Küche, Hirzel, Stuttgart, 2007 (6th ed.)

[373] *Vinogradow, G.V., Malkin, A.Y.*, Rheology of polymers, Prentice-Hall, Englewood Cliffs, 1983 (original: Mir, Moskwa, 1979)

[374] *Vogt, M.*, Dumonts Handbuch Philosophie, Dumont-Monte, Köln, 2003

[375] *Vögtle, F.*, "Thomas Alva Edison" (biography), Rowohlt, Reinbek, 2004 (5th ed.)

[376] *Voigt, W.*, Über innere Reibung fester Körper insbesondere der Metalle, Ann. Phys., 1892

[377] *Wahrig, G., Hermann, U., et al.*, Deutsches Wörterbuch, Mosaik/Bertelsmann, 1982

[378] *Wassermann, L.*, From Heraklit to W. Scott Blair, J. Rheology, 04/1991; Historische Aspekte der Lebensmittelrheologie, in: D. Weipert et al., Rheologie der Lebensmittel, Behr's, Hamburg, 1993

[379] *Wassner, E., Laun, H. M.*, Grundlagen der Rheologie von Polymerschmelzen, Tutorial BASF, Ludwigshafen, 2002

[380] *Weber, W.E.*, Über die Elasticität von Seidenfäden, Ann. Phys. Chem., 1835; Über die Elasticität fester Körper, ibid., 1841

[381] *Weidisch, R.*, Entwicklung von super-elastischen Kunststoffen, J. Laborpraxis, 06/2007

[382] *Weipert, D., Tscheuschner, H.-D., Windhab, E.*, Rheologie der Lebensmittel, Behr's, Hamburg, 1993

[383] *Weissenberg, K.* (citation by *B. Rabinowitsch*), Über die Viskosität und Elastizität von Solen, Z. Phys. Chem., 1929; *Weissenberg, K.*, Die Mechanik deformierbarer Körper, Abh. Preuß. Akad. Wiss., Phys.-math. Klasse No 2, de Gruyter, Berlin, 1931; A continuum theory of rheological phenomena, Nature, 1947; Rheology of hydrocarbon gels, Proc. Roy. Soc., 1950; The testing of materials by means of the Rheogoniometer, Sangamo Controls Ltd, 1964

[384] *Whorlow, R.W.*, Rheological techniques, Ellis Horwood, New York/Simon & Schuster, Hemel Hempstead, 1980/1992 (2nd ed.)

[385] *Wiechert, E.*, Gesetz der elastischen Nachwirkung für constante Temperatur, Ann. Phys. Chem., 1893

[386] *Wiedemann, G.*, Über die Bewegung der Flüssigkeiten im Kreise der geschlossenen galvanischen Säule und ihre Beziehungen zur Elektrolyse, Pogg. Ann., 1856

[387] *Wildmoser, H., Scheiwiller, J., Windhab, E.J.*, Impact of disperse microstructure on rheology and quality aspects of ice cream, Proc. 3. Int. Sympos. Food Rheol. & Structure, ed. P. Fischer et al., Zürich, 2003

[388] *Willenbacher, N., Hanciogullari, H., Radle, M.*, New laboratory test to characterize immobilization and dewatering of paper coating colors, Tappi J., 1999

[389] *Williams, M.L., Landel, R.F., Ferry, J.D.*, The temperature dependence of relaxation mechanisms in amorphous polymers and other glass-forming liquids, J. Am. Chem. Soc., 1955

[390] *Windhab, E.*, Ausgewählte Beispiele der ingenieurstechnischen Anwendung der Rheologie, in: D. Weipert et al., Rheologie der Lebensmittel, Behr's, Hamburg, 1993; *Windhab, E.J.*, Rheology and microstructure – keys for quality processing of food systems, Proc. 3. Int. Sympos. Food Rheol. & Structure, ed. P. Fischer et al., Zürich, 2003; *Windhab, E.J.*, What makes for smooth creamy chocolate?, J. Physics Today, 06/2006

[391] *Winslow, W.M.*, Induced fibration of suspensions, J. Appl. Phys., 1949

[392] *Winter, H.H.*, Analysis of dynamic mechanical data: Inversion into a relaxation time spectrum and consistency check. J. Non-Newt. Fluid Mech., 1997; *Winter, H.H., Mours, M., Baumgärtel, M., Soskey, P.R.*, Computer-aided methods in rheometry, in: A.A. Collyer, D.W. Clegg, Rheological Measurement, Chapman, London, 1998 (2nd ed.)

[393] *Winterstein, G.*, Das steifplastische Strangpressen von Keramik, J. Sprechsaal, 1990

[394] *Wissbrun, K.F.*, Transient rheometry, in: A.A. Collyer, D.W. Clegg, Rheological Measurement, Chapman, London, 1998 (2nd ed.)

[395] *Witt, M.*, Novel plate rheometer configuration allows monitoring real-time wood adhesive curing behavior, J. Adhesion Sci. Technol., vol. 18, no. 8, 2004

[396] *Wollny, K.*, New rheological test method to determine the dewatering kinetics of suspensions, J. Appl. Rheol., 2001; *Wollny, K., Läuger, J., Huck, S.*, Magneto sweep – a new method for characterizing the viscoelastic properties of magneto-rheological fluids, ibid., 2002

[397] Wörterbuch der Chemie, comp. by B. Frunder, E. Hillen, U. Rohlf, dtv, München, 1995

[398] *Wrana, C.*, Introduction to polymer physics, Lanxess, Leverkusen, 2009

[399] *Wunderlich, W.*, Ein erfolgreiches Jahr 2006, J. GIT Kunststoff Trends, 09/2007

[400] Württembergisches Landesmuseum, Museum für Kutschen, Chaisen, Karren in Heidenheim a.d. Brenz; Stuttgart, 1989

[401] *Yasuda, K., Armstrong, R.C., Cohen, R.E.*, Shear-flow properties of concentrated solutions of linear and star branched polystyrenes, Rheol. Acta, 1981

[402] *Young, T.*, A course of lectures on natural philosophy and the mechanical arts, 1807

[403] *Zilles, J.*, in: Kittel, Lehrbuch der Lacke und Beschichtungen, vol. 10: Analysen und Prüfungen, ed. H.-D. Otto, Hirzel, Stuttgart, 2006

[404] *Zimm, B.H.*, J. Chem. Phys., 1956; *Zimm, B.H., Crothers, D.M.*, Simplified rotating cylinder viscometer for DNA, Proc. Nat. Acad. Sci., 1962

[405] *Zirkler, E.*, Asphalt – ein Werkstoff durch Jahrtausende, Giesel, Isernhagen, 2001; Straßengeschichte – Mobilität durch Jahrtausende, ibid., 2003

[406] *Zorll, U., et al.*, Fließverhalten und Viskosimetrie, in: Lehrbuch der Lacke und Beschichtungen, vol. 8, p.1, ed. H. Kittel, Colomb, Stuttgart, 1980

[407] *Zosel, A.*, Lack- und Polymerfilme, Vincentz, Hannover, 1996; Built to last, J. Adhesive Age, 08/2000

[408] *Zuse, K.E.O.*, Der Computer – mein Leben (autobiography), 1970

[409] Zwick GmbH & Co. KG, Ulm, product information (e.g. extrusion plastometers), www, 2005

[410] *Schmidt, M., Knorz, M., Wilmes, B.*, A novel method for monitoring real-time curing behaviour, J. Wood Sci Technol, 07/2010; *Schmidt, M., Knorz, M.*, Gluing of European beech and Douglas fir for load bearing timber structures, World Conf. on Timber Eng., 2010

[411] *Chu, Z., Feng, Y., Su, X., Han, Y.*, Wormlike micelles and solution properties of a C22-tailed amidosulfobetaine surfactant, J. Langmuir, 26 (11), 7783-7791, 2010

[412] *Stolzenbach, H.G.*, Von der Arbeit im Sudhaus zur Lackchemie bei Herberts, Born, Wuppertal, 1996

[413] *Wolff, H.*, Über viskoses und elastisches Fließen von Anstrichstoffen, Kolloid-Z., H. 1, 1931; *Wolff, H., Zeidler, G.*, Über die Viskosität von Ölfarben als Ausdruck ihrer Struktur, Kolloid-Z., H.1, 1936

15.2 ISO standards

ISO means **International Standards Organisation**.

ISO 35: Natural rubber latex concentrate, prevulcanized latex: Mechanical stability. 2004
ISO 37: Rubber, vulcanized and thermoplastics: Tensile stress-strain properties. 2005
(ISO 174) Plastics (resins of vinyl chloride): Viscosity number in dilute solution (e.g. Fikentscher K-value). 1974 (withdrawn 1995)
ISO 178: Plastics: Flexural properties. 2001
ISO 289: Rubber, unvulcanized: Using a shearing-disc viscometer (p. 1 to 4; Mooney viscosity; pre-vulcanization characteristics). 1999/2005
ISO 294: Plastics: Injection moulding of test specimens of thermoplastic materials (p. 1: general principles, and moulding of multipurpose and bar test specimens; p. 2: small tensile bars; p. 3: small plates; p. 4: moulding shrinkage; p. 5: preparation of standard specimens for investigating anisotropy). 1996/2002
ISO 307: Plastics (dilute solutions of polyamides): Viscosity number. 2007
ISO 472: Plastics: Vocabulary. 1999
ISO 527: Plastics: Tensile properties (parts 1 to 5: general principles; moulding and extrusion plastics, films and sheets, fibre-reinforced composites). 1993/2009
(ISO 537) Plastics: Torsion pendulum. 1980 (withdrawn 1989)
ISO 812: Rubber: Low-temperature brittleness (by impact). 2006
(ISO 815) Rubber: Compression (elastic properties). 1991 (withdrawn 2008)
ISO 1133: Plastics: Melt mass-flow rate (MFR) and melt volume-flow rate (MVR) of thermoplastics. 2005
(ISO 1157) Plastics (cellulose acetate in dilute solution): Viscosity number & v. ratio. 1990 (withdrawn 2001)
ISO 1432: Rubber: Low temperature stiffening (Gehmann test; for T = RT to -150°C). 1988
(ISO 1517) Paints and varnishes: Surface drying test (ballotini method, using glass pearls). 1973 (withdrawn 2010, see ISO 9117-3)
ISO 1628: Plastics: Viscosity of polymers in dilute solution using capillary viscometers, 1990/2009; p. 1 to 6: general principles; viscosity, v. number, limiting v. number LVN; etc.)
ISO 1652: Rubber latex: Apparent viscosity by the Brookfield test method. 2004
ISO 2006: Rubber latex, synthetic: Mechanical stability, 2009; p. 1: High-speed method (stirring, using a disk with d = 36mm and 1,6mm thick; at a rotational speed of 14,000min⁻¹); p. 2: Moderate-speed method under (relatively high shear) load.
(ISO 2123) Sodium and potassium silicates for industrial use: Dynamic viscosity. 1972 (withdrawn 2002)
ISO 2137: Petroleum products (lubricating grease, petrolatum): Cone penetration. 2007
ISO 2176: Petroleum products (lubricating grease): Dropping point (with grease cup, for T = 288 to 400°C). 1995
ISO 2207: Petroleum waxes: Congealing point (formation of a gel structure when cooling). 1980
ISO 2285: Rubber: Tension under constant elongation, and creep under constant tensile load. 2007
ISO 2431: Paints and varnishes: Flow time using flow cups. 1999
ISO 2555: Plastics (resins in the liquid state or as emulsions or dispersions): Apparent viscosity by the Brookfield test method. 1989
ISO 2884: Paints and varnishes: Viscosity using rotary viscometers; p. 1: Cone-and-plate viscometer operated at a high rate of shear, 1999; p. 2: Disc or ball viscometer operated at a specified speed, 2003
ISO 2909: Petroleum products: Calculation of the viscosity index VI from kinematic viscosity. 2002
ISO 2921: Rubber: Low-temperature characteristics, temperature-retraction procedure (TR test). 2005
ISO 2930: Rubber, raw natural: Plasticity retention index (PRI). 2009
ISO 3013: Petroleum products: Freezing point of aviation fuels (optical method). 1997
ISO 3015: Petroleum products: Cloud point (optical method). 1992
ISO 3016: Petroleum products: Pour point (sagging method). 1994
ISO 3104: Petroleum products (transparent and opaque liquids): Kinematic viscosity, calculation of dynamic viscosity. 1997
ISO 3105: Glass capillary kinematic viscometers: Specifications and operating instructions. 1994
ISO 3146: Plastics: Melting behaviour (melting temperature or melting range) of semi-crystalline polymers by capillary tube and polarizing-microscope methods. 2002
ISO 3219: Plastics (polymers/resins in the liquid state or as emulsions or dispersions): Viscosity using a **rotational viscometer with a defined shear rate** (definition of the **concentric cylinder** and **cone-and plate measuring geometries with a narrow shear gap**). 1993
ISO 3384: Rubber, vulcanized or thermoplastic: Stress relaxation in compression at ambient and at elevated temperatures. 2005
ISO 3386: Polymeric materials (cellular flexible): Stress-strain characteristics in compression, p. 1: Low-density materials, 1986; p. 2: High-density materials, 1997
ISO 3417: Rubber: Vulcanization characteristics with the oscillating disc curemeter. 2008

ISO 3448: Industrial liquid lubricants (mineral oils, hydraulic fluids): ISO classification (using capillary viscometers). 1993

ISO/TR 3666: Viscosity of water. 1998

ISO 4575: Plastics (PVC pastes): Apparent viscosity using the Severs rheometer. 2007

ISO 4625: Binders for paints and varnishes, softening point; p. 1: Ring-and-ball method; p. 2: Cup-and-ball method. 2004

(ISO 4663) Rubber: Dynamic behaviour of vulcanizates at low frequencies; torsion pendulum method. 1986 (withdrawn 2006)

ISO 4664: Rubber, vulcanized and thermoplastic: Dynamic properties, 2005/2006; p. 1: General guide (free and forced vibration methods); p. 2: Torsion pendulum methods at low frequencies.

ISO 6388: Surface active agents: Flow properties using a rotational viscometer. 1989

ISO 6502: Rubber: Guide to the use of curemeters. 1999

ISO 6721: Plastics: Dynamic mechanical properties. 1995/2008

part 1: General principles (shear, tensile, compression moduli, Poisson's ratio, test modes),

part 2: Torsion-pendulum method (damped oscillation),

part 3: Flexural vibration – resonance-curve method,

part 4: Tensile vibration – non-resonance method,

part 5: Flexural vibration – non-resonance method,

part 6: Shear vibration – non-resonance method,

part 7: Torsional vibration – non-resonance method,

part 8: Longitudinal and shear vibration – wave-propagation method,

part 9: Tensile vibration – sonic-pulse propagation method,

part 10: Complex shear viscosity using a **parallel-plate oscillatory rheometer (polymer melts**, G^*, G', G''). 1999

ISO 6743: Lubricants, industrial oils, classification, 1981/2006; parts 1 to 15 & 99

ISO 6873: Dental gypsum products (e.g. needle penetration). 1998

ISO 6892: Metallic materials, tensile testing, 2009/2010; p.1: At room temperature; p.2 (DIS): At elevated temperatures

ISO 6914: Rubber: Ageing characteristics by measurement of stress relaxation in tension (using a stress relaxometer, preset of elongation). 2008

ISO 7884: Glass: Viscosity and viscometric fixed points. 1987; p. 1: Principles; p. 2: Viscosity by rotational viscometer; p. 3: Viscosity by fibre elongation viscometer; p. 4: Viscosity by beam bending; p. 5: Working point by sinking bar viscometer; p. 6: Softening point; p. 7: Annealing point and strain point by beam bending; p. 8: Dilatometric transformation temperature

ISO 8130: Powder coatings: p. 5: Flow properties of a powder/air mixture, 1992; p. 6: Gel time of hot curing powder coatings, 1998; p. 11: Inclined-plate flow test, 1997

ISO 8619: Plastics (phenolic powder): Flow distance on a heated glass plate. 2003

ISO 9000: Quality management systems, fundamentals and vocabulary. 2005

ISO 9117: Paint, varnishes: Drying tests, 2009/2010; p. 1: Through-dry state and through-dry time; p. 2: Pressure test for stackability; p. 3: Surface drying test using ballotini

(ISO 9371) Plastics: Phenolic resins in the liquid state or in solution, viscosity. 1990 (withdrawn 1998)

ISO 9514: Paints and varnishes: Pot-life of multicomponent coating systems. 2005

ISO 10414: Petroleum, natural gas: Field testing of drilling fluids (viscosity, gel strength, with viscometers, etc.), 2002/2008; p. 1: Water-based fluids; p. 2: Oil-based fluids

ISO 10426: Petroleum, natural gas: Cements, materials for well cementing, 2003/2009; p. 1: Specification; p. 2: Testing well cements; p. 3: Deepwater well cement formulations; p. 4: Foamed cement slurries at atmospheric pressure; p. 5: Shrinkage and expansion at atm. p. 6: Static gel strength (under simulated well conditions)

ISO 11357: Plastics: Differential scanning calorimetry (DSC), 1999/2009; p. 1 to 5: general principles; glass transition temperature; temp. and enthalpy of melting and crystallization; characteristic reaction-curve temp. and times, enthalpy of reaction and degree of conversion

ISO 11358: Plastics: Thermogravimetry (TG) of polymers ; p. 2: Activation energy. 1997

ISO 11359: Plastics: Thermomechanical analysis (TMA), 1999; p. 1: Penetration temperature; and p. 2: General principles; coefficient of linear thermal expansion and glass transition temperature.

ISO 11409: Plastics: Phenolic resins - heats and temperatures of reaction by differential scanning calorimetry (DSC). 1993

ISO 11443: Plastics: Fluidity of plastics using capillary and slit-die rheometers (extrusion rheometry). 2005

ISO 12058: Plastics: Viscosity using a falling-ball viscometer; p. 1: Inclined-tube method, 1997

ISO 12115: Fibre-reinforced plastics (thermosetting moulding compounds and prepregs): Flowability, maturation and shelf life. 1998

ISO 12644: Graphic technology: Rheological properties of paste inks and vehicles using the falling rod viscometer. 1996

ISO 13503: Petroleum and natural gas industries (completion fluids and materials); 5 parts; p. 1: Viscous properties of completion fluids (for drilling fluids and well cements; by rotational viscometer with concentric cylinders; viscosity at the shear rates 25/50/75/100s^{-1}; analysis using the "power law"). 2005

ISO 13737: Petroleum products and lubricants: Low-temperature cone penetration of lubricating greases. 2004

ISO 14129: Fibre-reinforced plastic composites: In-plane shear stress/shear strain response, including the in-plane shear modulus and strength, using the plus or minus 45 degree tension test method. 1997

ISO 14130: Fibre-reinforced plastic composites: Apparent interlaminar shear strength using the short-beam method. 2003

ISO 14446: Binders for paints and varnishes: Viscosity of cellulose nitrate solutions and classifications of such solutions. 1999

ISO 14678: Adhesives: Resistance to flow (sagging, after application and during cure). 2005

ISO 16862: Paints and varnishes: Evaluation of sag resistance. 2003

ISO 20965: Plastics: Transient extensional viscosity of polymer melts (specimens stretched uniaxially under constant strain rate and constant temperature, strain rate range of 0.01 to 1s^{-1}, at Hencky strains up to 4, temperatures up to 250°C, extensional viscosity range 10kPas to 10MPas). 2005

15.3 ASTM standards

ASTM means **American Society for Testing and Materials**, and ASTM terms its standards meanwhile as **"ASTM International Standards"**.

ASTM C109, C187, C230, C266, C270, C349, C496, C780, C1437: Hydraulic cement pastes, mortars, mortar masonry, concrete (flow table, flow & flow rate, cone penetration, compressive strength, splitting tensile strength, consistency by Vicat needles, time of setting by Gillmore needles). 2004/2009

ASTM C965: Viscosity of glass above the softening point (rotational viscometer). 2007

(ASTM C1276) Viscosity of mold powders above their melting point using a rotational viscometer. 2005 (withdrawn 2009)

ASTM D5: Penetration of bituminous materials. 2006

ASTM D36: Softening point of bitumen (ring-and-ball apparatus). 2009

ASTM D87: Melting point of petroleum wax (cooling curve). 2009

ASTM D88: Saybolt viscosity. 2007

ASTM D97: Pour point of petroleum products. 2009

ASTM D113: Ductility of bituminous materials (tensile test). 2007

ASTM D115: Solvent containing varnishes used for electrical insulation (e.g. viscosity using a rotational visometer). 2007

ASTM D127: Drop melting point of petroleum wax (including petrolatum). 2008

ASTM D217: Cone penetration of lubricating grease. 2007

ASTM D244: Emulsified asphalts (e.g. Saybolt Furol viscosity). 2009

ASTM D333: Clear and pigmented lacquers (e.g. by Ford flow cup). 2007

ASTM D341: Viscosity-temperature charts for liquid petroleum products (by capillary viscometer). 2009

ASTM D365: Soluble nitrcellulose base solutions (e.g. by Ford flow cup). 2005

ASTM D395: Rubber property, compression set (A: Constant force; B: Constant deflection). 2003 (2008)

ASTM D412: Vulcanized rubber, thermoplastic rubbers, thermoplastic elastomers – tension (tensile test, elongation). 2006

ASTM D445: Kinematic viscosity of transparent and opaque liquids (and calculation of dynamic viscosity; by glass capillary viscometer). 2009

ASTM D446: Specifications and operating instructions for glass capillary kinematic viscometers. 2007

ASTM D562: Consistency of paints measuring Krebs unit (KU) viscosity using a Stormer-type viscometer. 2001 (2005)

ASTM D566: Dropping point of lubricating grease. 2002 (2009)

ASTM D618: Conditioning plastics for testing. 2008

ASTM D624: Tear strength of conventional vulcanized rubber and thermoplastic elastomers (TPE). 2000 (2007)

ASTM D638: Tensile properties of plastics (with ≥ 1.0mm in thickness). 2008

ASTM D746: Brittleness temperature of plastics and elastomers by impact. 2007

ASTM D747: Apparent bending modulus of plastics by means of a cantilever beam. 2008

ASTM D789: Solution viscosities of polyamide PA (e.g. by Pipet viscometer or Brookfield v.). 2007

ASTM D790: Flexural properties of unreinforced and reinforced plastics (etc.). 2007

ASTM D803: Tall oil (part 7: Bubble time method). 2002 (2008)

ASTM D816: Rubber cements (natural and reclaimed rubber, synthetic elastomers, e.g. by Zahn flow cup). 2006

(ASTM D856) Testing pine tars and pine-tar oils. 1987 (withdrawn 1991)

ASTM D882: Tensile properties of thin plastic sheeting (with < 1.0mm, and films with ≤ 0.25mm in thickness). 2009

ASTM D937: Cone penetration of petrolatum. 2007

ASTM D938: Congealing point of petroleum waxes including petrolatum (when cooling). 2005

ASTM D946: Penetration-graded asphalt cement for use in pavement construction. 2009

ASTM D1004: Tear resistance (Graves tear) of plastic film and sheeting (with ≤ 0.25mm in thickness). 2009

ASTM D1043: Stiffness properties of plastics as a function of temperature by means of a torsion test. 2009

ASTM D1053: Rubber property, stiffening at low temperatures: flexible polymers and coated fabrics (by torsion). 1992 (2007)

ASTM D1074: Compressive strength of bituminous mixtures. 2009

ASTM D1076: Rubber-concentrated, ammonia preserved, creamed, and centrifuged natural latex (e.g. viscosity by rotational viscometer). 2006

ASTM D1084: Viscosity of adhesives (by consistency cup, rotational viscometer, Stormer v., Zahn cup). 2008

ASTM D1092: Apparent viscosity of lubricating greases (by capillary method, extruder). 2005

(ASTM D1131) Rosin oils (bubble time method). 1997 (withdrawn 2003)

ASTM D1200: Viscosity by Ford viscosity cup (for paints, varnishes, lacquers). 2005

ASTM D1238: Flow rates of thermoplastics by extrusion plastometer (MFR and MVR). 2004

ASTM D1243: Dilute solution viscosity of vinyl chloride polymers (inherent viscosity, by capillary viscometers). 1995 (2008)

ASTM D1321: Needle penetration of petroleum waxes. 2004

ASTM D1349: Rubber-standard temperatures for testing. 2009

ASTM D1403: Cone penetration of lubricating grease using one-quarter and one-half scale cone equipment. 2002 (2007)

ASTM D1417: Rubber (synthetic) latices (e.g. viscosity by rotational viscometer, and Mooney viscosity). 2003

ASTM D1439: Sodium carboxymethylcellulose (e.g. viscosity by rotational viscometers). 2003 (2008)

ASTM D1478: Low-temperature torque of ball bearing grease (rotational test by a ball bearing standard device). 2007

ASTM D1545: Viscosity of transparent liquids by bubble time method (acc. to Gardner-Holdt, e.g. oils and resins). 2007

(ASTM D1559) Resistance of plastic flow of bituminous mixtures using Marshall apparatus. 1989 (withdrawn 1998)

ASTM D1601: Dilute solution viscosity of ethylene polymers (by capillary viscometer, determining inherent, intrinsic, reduced, relative, specific viscosity, viscosity ratio). 1999 (2004)

ASTM D1640: Drying, curing, film formation of organic coatings at room temperature. 2003 (2009)

ASTM D1646: Rubber-viscosity, stress relaxation, and pre-vulcanization characteristics (by Mooney viscometer). 2007

ASTM D1665: Engler specific viscosity of tar products. 1998 (2009)

ASTM D1725: Viscosity of resin solutions (bubble time method). 2004

ASTM D1754: Effect of heat and air on asphaltic materials (thin-film oven test, TFOT). 2009

ASTM D1795: Intrinsic viscosity of cellulose. 1996 (2007)

ASTM D1823: Apparent viscosity of plastisols and organisols at high shear rates by extrusion viscometer. 1995 (2009)

ASTM D1824: Apparent viscosity of plastiosols and organosols at low shear rates (by Brookfield viscometer). 1995 (2002)

ASTM D1831: Roll stability of lubricating grease (consistency changes by cone penetration). 2000 (2006)

ASTM D1922: Propagation tear resistance of plastic thin sheeting (with ≤ 0.25mm in thickness) by pendulum method (Elmendorf apparatus). 2009

ASTM D1986: Apparent viscosity of polyethylene wax (by Brookfield thermosel system). 1991 (2007)

ASTM D2084: Rubber property: vulcanization by oscillating disk cure meter. 2007

ASTM D2137: Rubber property: brittleness point of flexible polymers and coated fabrics (by impact). 2005

ASTM D2140: Calculating carbon-type composition of insulating oils of petroleum origin (electrical insulating oils; using viscosity and v.-gravity constant VGC). 2008

ASTM D2161: Conversion of kinematic viscosity to Saybolt Universal viscosity or to Saybolt Furol viscosity. 2005

ASTM D2162: Basic calibration of master (capillary) viscometers, viscosity oil standards. 2006

ASTM D2166: Unconfined compressive strength of cohesive soil. 2006

ASTM D2170: Kinematic viscosity of asphalts (bitumens). 2007

ASTM D2171: Viscosity of asphalts by vacuum capillary viscometer. 2007

ASTM D2196: Rheological properties of non-Newtonian materials by rotational viscometer. 2005

(ASTM D2236) Torsion pendulum test: plastics. 1981 (withdrawn 1984, replaced by ASTM D4065)

ASTM D2240: Rubber property: Durometer hardness (indentation hardness). 2005

ASTM D2243: Freeze-thaw resistance of water-borne coatings (to low-temperature cycling; each cycle for 17h at T = -18°C and for 7h at RT; up to 5 cycles; afterwards visual and viscosity test by Stormer- or Brookfield-type viscometer). 1995 (2008)

ASTM D2265: Dropping point of lubricating grease over a wide temperature range. 2006

ASTM D2270: Calculating viscosity index VI from kinematic viscosity at 40 and 100°C (for petroleum products). 2004

ASTM D2344: Short-beam strength of polymer matrix composite materials and their laminates (by three-point bending test). 2006

ASTM D2364: Hydroethyl-cellulose (e.g. viscosity by rotational viscometer). 2001 (2007)

ASTM D2370: Tensile properties of organic coatings (elongation, tensile strength, stiffness). 1998 (2002)

ASTM D2386: Freezing point of aviation fuels (optical method). 2006

ASTM D2396: Powder-mix time of PVC resins using a torque rheometer. 1994 (2004)

ASTM D2422: Standard classification of industrial fluid lubricants by viscosity system (by capillary viscometer). 1997 (2007)

ASTM D2493: Viscosity-temperature charts for asphalts. 2009

ASTM D2500: Cloud point of petroleum products (and biodiesel fuels). 2009

ASTM D2501: Calculation of viscosity-gravity constant (VGC) of petroleum oils. 1991 (2005)

ASTM D2532: Viscosity and viscosity change after standing at low temperature of aircraft turbine lubricants (by capillary viscometer). 2003

ASTM D2538: Fusion of PVC compounds using a torque rheometer (by a "high-shear mixer" with "roller-style blades", e.g. for "shear stability tests"). 2002

ASTM D2556: Apparent viscosity of adhesives having shear rate-dependent flow properties (e.g. by rotational viscometer). 1993 (2005)

ASTM D2573: Field vane shear test in cohesive soil (shear strength of geomaterials like fine-grained clays, silts, sands, gravels, mine tailings, organic muck, e.g. in drilled boreholes). 2008

(ASTM D2602) Viscosity measurement by cold cranking simulator. 1986 (withdrawn in 1993, replaced by ASTM D5293)

ASTM D2603: Sonic shear stability of polymer-containing oils (by sonic oscillator, for hydraulic fluids). 2001 (2007)

ASTM D2669: Apparent viscosity of petroleum waxes compounded with additives (hot melts; by Brookfield-type viscometer). 2006

ASTM D2850: Unconsolidated-undrained triaxial compression test on cohesive soils. 2003 (2007)

ASTM D2857: Dilute solution viscosity of polymers. 1995 (2007)

ASTM D2872: Effect of heat and air on a rolling film of asphalt (rolling thin-film oven test RTFOT). 2004

(ASTM D2882) Petroleum and non-petroleum hydraulic fluids in constant volume vane pumps. 2000 (withdrawn 2003)

ASTM D2884: Yield stress of heterogeneous propellants by cone penetration method. 1993 (2007)

ASTM D2979: Pressure-sensitive tack of additives using an inverted probe machine. 2000

ASTM D2983: Low-temperature (LT) viscosity of lubricants using a Brookfield-type viscometer (at low-shear rate). 2009

ASTM D3039: Tensile properties of polymer matrix composite materials. 2008

ASTM D3117: Wax appearance point of distillate fuels. 2003

ASTM D3121: Tack of pressure-sensitive adhesives by rolling ball (for low-tack adhesives). 2006

(ASTM D3205) Viscosity of asphalt with cone and plate viscometer 1986 (withdrawn 2000)

(ASTM D3232) Consistency of lubricating greases at high temperatures (using rotational viscometer with trident system). 1988 (withdrawn 1994)

ASTM D3236: Apparent viscosity of hot melt adhesives and coating materials (by rotational viscometer). 1988 (2009)

ASTM D3245: Pumpability of industrial fuel oils (e.g. by Ferranti viscometer). 2003

ASTM D3346: Rubber property – processability of SBR with the Mooney viscometer. 2007

ASTM D3364: Flow rates for PVC with molecular structural implications (by extrusion). 1999 (2004)

ASTM D3381: Specification for viscosity-graded asphalt cement for use in pavement construction (viscosity, penetration). 2009

ASTM D3418: Transition temperatures of polymers by differential scanning calorimetry (DSC). 2008

ASTM D3461: Softening point of asphalt and pitch (Mettler cup-and-ball method). 1997 (2007)

ASTM D3468: Liquid-applied Neoprene and CS-PE used in roofing and waterproofing (e.g viscosity by rotational viscometer). 1999 (2006)

ASTM D3591: Logarithmic viscosity number of PVC in formulated compounds. 1997 (2003)

ASTM D3716: Emulsion polymers for use in floor polishes (e.g viscosity by rotational viscometer). 1999 (2008)

ASTM D3791: Effects of heat on asphalts (e.g viscosity by rotational viscometer). 2004

(ASTM D3794) Coil coatings (e.g. by Zahn viscometers). 2000 (withdrawn 2009)

ASTM D3829: Borderline pumping temperature of engine oil (BPT, by mini-rotary viscometer). 2002 (2007)

ASTM D3835: Polymeric materials by a capillary rheometer (melt viscosity via extrusion). 2008

(ASTM D3945) Shear stability of polymer-containing fluids using a Diesel injector nozzle. 1993 (withdrawn 1998)

ASTM D3954: Dropping point of waxes (also for paraffins and microcrystalline polyethylenes). 1994 (2010)

ASTM D4016: Viscosity of chemical grouts by Brookfield viscometer (laboratory method). 2008

ASTM D4040: Viscosity of paste printing and vehicles by falling-rod viscometer. 2005

ASTM D4065: Dynamic mechanical properties of plastics (viscoelastic characteristics, oscillatory tests for frequencies from 0.01 to 1000Hz, for elastomers & solid plastics with elastic modulus from 0.5MPa to 100GPa under flexural, tensile, compressive, torsional and linear-shear load). 2006

ASTM D4092: Terminology for plastics: dynamic mechanical properties (for solutions, melts and solids of polymeric materials). 2007

ASTM D4142: Testing epoxy resins. 1989 (2009)

ASTM D4212: Viscosity by dip-type viscosity cups. 1999 (2005)

ASTM D4287: High-shear viscosity using a cone/plate viscometer (e.g. "ICI-type"; for paints, varnishes and related products). 2000 (2010)

ASTM D4300: Ability of adhesive films to support or resist the growth of fungi (e.g. evaluation of biodegradation and viscosity by rotational viscometer). 2001 (2008)

ASTM D4318: Liquid limit, plastic limit, plasticity index of soils (strain-controlled loading, strength, stress-strain relation; earth works, exvacations, foundation construction, underground works). 2005

ASTM D4359: Test method for determining whether a material is a liquid or a solid. 1990 (2006)

ASTM D4402: Viscosity of asphalt at elevated temperatures by rotational viscometer (for T = +38 to 260°C). 2006

ASTM D4440: Plastics, dynamic mechanical properties: **melt rheology (oscillatory tests** in torsional shear mode). 2008

ASTM D4473: Plastics, dynamic mechanical properties: **Cure behavior (of thermosetting resins; oscillatory tests** in torsional shear or in compression mode). 2008

ASTM D4483: Precision for test method standards in the rubber and carbon black industries (by Mooney viscosity). 2005

ASTM D4539: Filterability of Diesel fuels by low-temperature flow test (LTFT; cold filter plugging point). 2009

ASTM D4565: Physical and environmental performance properties of insulations and jackets for telecommunications wire and cable (melt and compound flow, tensile and elongation, cable torsion, cold bend, heat distorsion). 1999 (2004)

ASTM D4603: Inherent viscosity of poly-ethylene terephthalate (PET) by glass capillary viscometer. 2003

(ASTM D4624) Apparent viscosity by capillary viscometer at high temperature and high shear rates (for petroleum products, fluid lubricants, engine oils). 1998 (withdrawn 2004)

ASTM D4648: Laboratory miniature vane shear test for saturated fine-grained clayey soil (shear strength, shear value). 2005

ASTM D4683: Viscosity at high shear rate and high temperature by tapered bearing simulator (for engine oils). 2009

ASTM D4684: Yield stress and apparent viscosity of engine oils at low temperature (by mini-rotary viscometer). 2008

ASTM D4693: Low-temperature torque of grease-lubricated wheel bearings. 2007

ASTM D4741: Viscosity of oils at high temperature and high shear rate by tapered-plug viscometer (for engine oils). 2006

ASTM D4878: PU raw materials: viscosity of polyols (e.g. by Brookfield or glass capillary viscometers). 2008

ASTM D4889: PU raw materials: viscosity of crude or modified isocyanates (e.g. by rotational viscometer). 2004

ASTM D4957: Apparent viscosity of asphalt emulsion residues and non-Newtonian bitumens by vacuum capillary viscometer. 2008

ASTM D4989: Apparent viscosity (flow) of roofing bitumens by parallel plate plastometer. 1990 (2008)

ASTM D5001: Lubricity of aviation fuels by the ball-on-cylinder lubricity evaluator BOCLE (friction, wear, boundary lubrication properties on rubbing steel surfaces). 2008

ASTM D5018: Shear viscosity of coal-tar and petroleum pitches (by rotational viscometer). 1989 (2009)

ASTM D5023: Plastics: Dynamic mechanical properties, in flexure (three point bending; for solid rectangular bars of resins and composites). 2007

ASTM D5024: Plastics: Dynamic mechanical properties, in compression (for solid cylindrical specimens of resins and composites). 2007

ASTM D5026: Plastics: Dynamic mechanical properties, in tension (for solid rectangular specimens of resins and composites). 2006

ASTM D5099: Rubber (compounded, unvulcanized; and thermoplastic elastomers): measurement of processing properties using a capillary rheometer (CR; flow viscosity by extrusion; A: Piston type CR; B: Screw extrusion type CR). 2008

(ASTM D5119) Automotive engine oils in the CRC L-38 spark-ignition engine (viscosity stability and viscosity loss of multiviscosity-graded oils). 2002 (withdrawn 2003)

ASTM D5125: Viscosity of paints and related materials by ISO flow cups. 2005

ASTM D5133: Low temperature, low shear rate, viscosity/temperature dependence of lubricating oils by temperature-scanning technique (by rotational viscometer; gelation index GI, LTLS viscosity). 2005

ASTM D5225: Solution viscosity of polymers by differential (capillary) viscometer. 2009

ASTM D5275: Fuel injector shear stability test (FISST) for polymer containing fluids. 2003

ASTM D5279: Plastics: Dynamic mechanical properties, in torsion (of solid bars). 2008

ASTM D5289: Rubber property, vulcanization by rotorless cure meters. 2007

ASTM D5293: Apparent viscosity of engine oils between -5 and -35°C by the cold-cranking simulator (CCS; and enhanced range: -1 to -40°C). 2010

ASTM D5329: Sealants and fillers, hot-applied, for joints and cracks in asphaltic and Portland cement concrete pavements (by cone penetration). 2009

ASTM D5418: Plastics: Dynamic mechanical properties, in flexure (dual cantilever beam). 2007

ASTM D5478: Viscosity of materials by a falling needle viscometer (varnishes, paints). 2009

ASTM D5481: Apparent viscosity at high temperature and high shear rate (HTHS) by multicell capillary viscometer (for petroleum products). 2004

ASTM D5621: Sonic shear stability of hydraulic fluids. 2007

ASTM D5771: Cloud point of petroleum products (and biodiesel fuels; stepped cooling method; optical detection). 2005

ASTM D5772: Cloud point of petroleum products (and biodiesel fuels; linear cooling rate method, optical detection). 2005

ASTM D5773: Cloud point of petroleum products (and biodiesel fuels; constant cooling rate method, optical detection). 2007

ASTM D5853: Pour point of crude oils (cold flow). 1995 (2006)

ASTM D5895: Drying or curing during film formation of organic coatings using mechanical recorders. 2003 (2008)

(ASTM D5934) Modulus of elasticity for rigid and semi-rigid plastic specimens by controlled rate of loading using three-point bending. 2002 (withdrawn 2009)

ASTM D5950: Pour point of petroleum products (automatic tilt method; for T = -57 to +51°C). 2002 (2007)

ASTM D6022: Calculation of permanent shear stability index PSSI (using viscosities before and after a shearing procedure). 2006

ASTM D6049: Rubber property: Viscous and elastic behavior of unvulcanized raw rubbers and rubber compounds by compression between parallel plates. 2003 (2008)

ASTM D6080: Viscosity characteristics of hydraulic fluids (e.g. by a Brookfield-type viscometer). 1997 (2007)

ASTM D6084: Elastic recovery of bituminous materials (and waxes) by ductilometer (ductility). 2006

ASTM D6128: Bulk solids using the Jenike shear cell (powder, translational shear tester, cohesive strength, yield strength, wall friction, internal friction angle). 2006

ASTM D6204: Rubber, unvulcanized rheological properties using rotorless shear rheometers. 2007

ASTM D6267: Apparent viscosity of hydrocarbon resins at elevated temperatures (by a Brookfield-type viscometer). 2008

ASTM D6278: Shear stability of polymer containing fluids by European Diesel injector apparatus (percent viscosity loss at 100°C). 2007

ASTM D6371: Cold filter plugging point of Diesel and heating fuels. 2005

ASTM D6373: Specification for performance graded asphalt binder. 2007

ASTM D6382: Dynamic mechanical analysis (DMA) and thermogravimetry (TG) of roofing and waterproofing membrane material. 1999 (2005)

ASTM D6521: Accelerated aging of asphalt binder using a pressurized aging vessel (PAV). 2008

ASTM D6601: Rubber, cure and after-cure dynamic properties by rotorless shear rheometer. 2002 (2008)

ASTM D6648: Flexural creep stiffness (or compliance) of asphalt binder by bending beam rheometer (BBR; temperature range of T = -36 to 0°C; low-temperature stress-strain-time response within the linear viscoelastic range). 2008

ASTM D6723: Fracture properties of asphalt binder in direct tension (DT; temperature range of T = +6 to -36°C; failure strain, fracture, thermal cracking, critical cracking temperature). 2002

ASTM D6746: Green strength of raw rubber or unvulcanized compounds (extensional test; "green" for unvulcanized, uncured raw rubber). 2008

ASTM D6773: Bulk solids by Schulze ring shear tester (powder, soil, translational shear tester, yield strength, wall friction, internal friction angle, flowability, caking). 2008

ASTM D6821: Low temperature viscosity of drive line lubricants in a constant shear stress viscometer. 2002 (2007)

ASTM D6895: Rotational viscosity of heavy duty Diesel drain oils at 100°C (shear rate range of 10 to 300s^{-1}, and viscosity range of 12 to 35mPas; comparison at at s.r. of 100s^{-1}). 2006

ASTM D6896: Yield stress and apparent viscosity of used engine oils at low temperature (by a mini-rotary viscometer). 2003 (2007)

ASTM D7042: Dynamic viscosity and density of liquids by Stabinger viscometer (and calculation of kinematic viscosity). 2004

ASTM D7110: Viscosity-temperature relationship of used soot-containing engine oils at low temperatures (by rotational viscometer, LTLS viscosity, gelation index GI, GI temperature). 2005

ASTM D7175: Rheological properties of **asphalt binder by** dynamic shear rheometer (DSR, i.e. an **oscillatory rheometer**, parallel-plate geometry; SHRP method). 2008

ASTM D7271: Viscoelastic properties of paste ink vehicle using an oscillatory rheometer (measuring G', G'', and tan(delta) via frequency sweeps of printing inks). 2006

ASTM D7346: No-flow point of petroleum products (no-flow point temperature; in the range of T = -95 to 45°C). 2007

ASTM D7395: Cone/plate viscosity at a 500s⁻¹ shear rate (for paints). 2007

ASTM D7397: Cloud point of petroleum products (miniaturized optical method). 2008

ASTM D7467: Specification for Diesel fuel oil, biodiesel blend (e.g. cloud point). 2009

ASTM D7483: Dynamic and kinematic viscosity of liquids by oscillating piston viscometer. 2008

ASTM E28: Softening point test of resins derived from naval stores by ring-and-ball apparatus. 1999 (2009)

ASTM E102: Saybolt Furol viscosity of bituminous materials at high temperatures. 1993 (2009)

(ASTM E380) Use of the international system of units (SI): The modern metric system (in 2002 replaced by ASTM-SI-10)

ASTM E473: Standard terminology relating to thermal analysis (TA) and rheology. 2009

ASTM E793: Enthalpies of fusion and crystallization by differential scanning calorimetry (DSC). 2006

ASTM E794: Melting and crystallization temperatures by thermal analysis (DSC for T = +120 to 600°C, and DTA for T = +25 to 1500°C). 2006

ASTM E831: Linear thermal expansion of solid materials by thermomechanical analysis (TMA for T = +120 to 900°C). 2006

ASTM E1142: Standard terminology relating to thermophysical properties. 2007

ASTM E1356: Assignment of the glass transition temperatures by differential scanning calorimetry (DSC). 2008

ASTM E1640: Glass transition temperature (T_g) by dynamic mechanical analysis (DMA). 2009

ASTM E1867: Temperature calibration of dynamic mechanical analyzers (range T = -150 to +500°C). 2006

ASTM E2254: Storage modulus calibration of dynamic mechanical analyzers (for T = -100 to +300°C; using reference materials with G' = 1 to 200GPa). 2009

ASTM E2425: Loss modulus conformance of dynamic mechanical analyzers (using PMMA as a reference material at T = +21°C). 2005

ASTM E2510: Torque calibration or conformance of rheometers. 2007

15.4 DIN, DIN EN, DIN EN ISO and EN standards

DIN, DIN EN, DIN EN ISO and EN means **Deutsche Industrie Norm, European standards** (European norms) and **International Standards Organisation. DIN EN ISO** and **EN ISO**: see ISO standards in Chapter 15.2. DIN-standards: see also [100].

DIN 1342: Viscosity (**rotational tests**); p. 1: **Rheological terms**, p. 2: Newtonian fluids, p. 3: non-Newtonian fluids. 2003

(DIN) **EN 1426**: Bitumen and bituminous binders: Needle penetration. 2007

(DIN) **EN 1427**: Bitumen and bituminous binders: Softening point, ring and ball method. 2007

(DIN) **EN 1871**: Road marking materials: Marking colors, cold plastic masses and hot plastic masses, physical properties. 2008

DIN 1995: Bitumen and bituminous binders: requirements; p. 4: Cold bitumen. 2005

DIN 1996: Bituminous materials for road building, etc. 1971/1990; p. 11: Marshall-stability and M.-flow value; p. 13: penetration test using a plane stamp (besides: p. 1 to 6, and p. 9 to 14)

DIN 4094-4: Subsoil, field testing, p. 4: Field vane test. 2002

DIN 10331: Butter hardness (e.g. penetration tests). 1996

(DIN) **EN 12092**: Adhesives: Determination of viscosity. 2002

(DIN **EN 12591**) Bitumen, binders: Specifications for paving grade bitumen. 1999 (withdrawn)

(DIN) **EN 12595**: Bitumen and bituminous binders: Kinematic viscosity. 2007

(DIN) **EN 12596**: Bitumen and bituminous binders: Viscosity using vacuum-capillaries. 2007

(DIN) **EN 12607**: Bitumen and bituminous binders, 2007: Resistance to hardening under the influence of heat and air; p. 1: RTFOT method; p. 2: TFOT method; p. 3: RFT method

(DIN) **EN 12846**: Bitumen and bituminous binders, efflux time by an efflux viscometer; 2008; p. 1: Bituminous emulsions; p. 2: Cutted and fluxed bituminous binders

(DIN) **EN 13179**: Filler aggregate used in bituminous mixtures; 2000; p. 1: Delta ring and ball test; p. 2: Bitumen number

(DIN) **EN 13302**: Bitumen, binders: viscosity of bitumen using a rotational viscometer (spindle/cylinder; for T = +40 to 230°C; speed preset depending on resulting torque). 2010

DIN 13316: The mechanics of idealelastic bodies (terms, parameters, symbols). 1980

(DIN 13342) Non-Newtonian liquids. 1976 (withdrawn, replaced by DIN 1342-3)

DIN 13343: Linear **viscoelastic materials** (**terms**, constitutive equations, basic functions; **creep tests, relaxation tests, oscillatory tests**). 1994

(DIN) **EN 13357**: Bitumen and bituminous binders: Efflux time of petroleum cut-back and fluxed bitumens. 2003

(DIN) **EN 13398**: Bitumen and bituminous binders: Elastic recovery of modified bitumen. 2009

(DIN) **EN 13880**: Hot applied joint sealants; 2003/2004. p. 1: Density at 25°C; p. 2: Cone penetration at 25°C; p. 3: Ball penetration and elastic recovery (resilience); p. 4: Heat resistance, change in cone penetration value; p. 5: Flow resistance (flow distance); p. 10: Adhesion and cohesion following continuous extension and compression

(DIN) **EN 14023**: Bitumen and bituminous binders: Specification of polymer modified bitumens. 2010

(DIN) **EN 14214**: Automotive fuels, fatty acid methyl esters (FAME) for Diesel engines: Requirements and test methods (e.g. density at T = +15°C; reqired kinematic viscosity using a glass capillary viscosimeter at +40°C: ν = 3.5 to 5.0mm^2/s). 2009

(DIN) **EN 14770**: **Bitumen** and bituminous binders: Complex shear modulus and phase angle, using a dynamic shear rheometer (DSR; and MS-PP, d = 8 to 25mm, H = 0.5 to 2mm; pretest as an amplitude sweep at γ = 0.5 to 10% to determine the LVE range; testing: **frequency sweep** at f = 0.1 to 10Hz, or **temperature sweep** between T = +5 and +85°C; specification of G* and δ). 2006

(DIN) **EN 15324**: Bitumen and bituminous binders: Equiviscous temperature (EVT) based on low-shear viscosity (LSV) using a dynamic shear rheometer (DSR) in low frequency oscillation mode. 2005

(DIN) **CEN/TS 15325**: Bitumen and bituminous binders: Zero-shear viscosity (ZSV) using a shear stress rheometer in creep mode. 2008

DIN 16945: Reaction resins, reactants and reaction moulding materials: test methods (e.g. falling-rod). 1989

DIN 18122: Soil: Consistency limits; p. 1: Liquid limit and plastic limit, 1997; p. 2: Shrinkage limit, 2000

DIN 18137: Soil, shear strength: p. 1: Concepts, test conditions, 2010; p. 2: Triaxial test, 1990; p. 3: Direct shear test, 2002

(DIN) **EN 23015**: Petroleum products: Cloud point (corresponds to ISO 3015). 1992/1994

DIN 29971: Aerospace: Unidirectional carbon fibre epoxy sheet and tape prepreg (DMA tests). 1991

DIN 32513: Soldering pastes (e.g. viscosity test using a rotational viscometer). 2005

DIN 50100: Continuous vibration test (definitions, procedure, symbols, evaluation). 1978

DIN 50125: Metallic materials: Tensile test pieces. 2009

DIN 51005: Thermal analysis (TA): Terms. 2005

DIN 51006: Thermal analysis (TA): Thermogravimetry (TG). 2005

DIN 51007: Thermal analysis (TA): Differential thermoanalysis (DTA). 1994

DIN 51045: Thermal expansion of solids, 2009: p. 1: Basic rules; p. 2/3: Fired and non-fired fine ceramic materials (dilatometer method); p. 4: Fired heavy ceramic materials.

DIN 51366: Mineral oil hydrocarbons: Kinematic viscosity by Cannon-Fenske viscometer for opaque liquids. 1977

(DIN 51372) Mineral oils: Kinematic viscosity using the BS/IP-U-tube viscometer for opaque liquids. (withdrawn)

DIN 51377: Lubricants: Apparent viscosity of motor oils at low temperature between -5 and -35°C using the cold-cranking simulator. 2003

DIN 51398: Lubricants: Apparent viscosity at low temperature using a Brookfield viscometer (liquid bath method). 1983

DIN 51502: Lubricants: Designation, marking. 1990

(DIN 51511) SAE viscosity grades of engine lubricating oils. 1985 (withdrawn)

(DIN 51512) SAE viscosity grades of lubricating oils for automotive gears. 1988 (withdrawn)

(DIN 51519) ISO viscosity grades of liquid lubricants used in industry. 1998 (withdrawn)

DIN 51524: Pressure fluids (hydraulic oils); 2006. p. 1 to 4: Minimum requirements oils HL, HLP, HVLP (e.g. temperature-dependent viscosity)

(DIN 51550) Determination of viscosity, general principles. 1978 (withdrawn)

(DIN 51560) Measurement using an Engler viscometer. (withdrawn)

(DIN 51561) Measurement using a Vogel-Ossag viscometer (capillary v.). (withdrawn)

DIN 51562: Kinematic viscosity using an Ubbelohde viscometer (capillary v); p. 1: Specification, measuring procedure, 1999; p. 2: Micro-Ubb. v., 1988; p. 3: Viscosity relative increment at short flow times, 1985; p. 4: Viscometer calibration. 1999

DIN 51563: Mineral oils: Viscosity-temperature relation. 1976

(DIN 51564) Calculation of the viscosity index from the kinematic viscosity. (withdrawn)

(DIN 51569) Measurement with a Vogel-Ossag viscometer (capillary v). (withdrawn)

DIN 51579: Testing of paraffins: Needle penetration. 2010

DIN 51580: Testing of paraffins: Cone penetration. 2008

DIN 51603: Liquid fuels (fuel oils); 2003/2008. p. 1 to 6: Minimum requirements fuel oils EL, S, SA, EL-A (e.g. viscosity)

DIN 51805: Flow pressure of lubricating greases using the Kesternich method. 1974

DIN 51810: Lubricating greases; p. 1: Shear viscosity by rotational viscometer with a cone-and-plate measuring system, 2007; p. 2: Flow point by oscillatory rheometer with a parallel-plate m. system. 2010

DIN 51818: Consistency classification of lubricating greases (NLGI grades). 1981/1988

DIN 52007: Bituminous binders: Viscosity, 2006; p. 1: General principles and evaluation; part 2: Ball-draw (drawn-sphere) viscometer

DIN 52013: Bitumen and bituminous binders: Ductility. 2007

(DIN 52023) Bitumen and coal tar pitch, 1989; p. 1: Flow time of binders by standard tar viscometer; p. 2: Equal temperature flow time (EVT). (withdrawn)

DIN 52377: Plywood: Modulus of elasticity in tension and tensile strength. 1978

DIN 53012: Capillary viscometry of Newtonian liquids. 2003

DIN 53014: Capillary viscometers with circular and rectangular slit cross-section for determining flow curves, 1994; p. 1: Principles; p. 2: Systematic deviations, corrections

DIN 53015: Viscosity using a rolling-ball viscometer according to Höppler. 2001

DIN 53017: Temperature coefficient of viscosity of liquids. 1993

(DIN 53018) Dynamic viscosity of Newtonian fluids using rotational viscometers, 1976 (since 2001 replaced by DIN 53019); p. 1: Principles (cylinders), p. 2: Sources of error and corrections

DIN 53019: Viscometry – viscosities and flow curves using rotational viscometers; p. 1: Principles and **measuring geometries**, 2008; p. 2: Viscometer calibration and determination of the uncertainty of measurement, 2001; p. 3: Errors of measurement and corrections (inclusive Mooney/Ewart measuring system, Taylor- and Reynolds-vortices), 2008. (DIN 53019 includes the old DIN 53018 of 1976 and DIN 53019 of 1980: Concentric **cylinders** in general and the "standard cylinder geometry", **cone-and-plate**, and **parallel-plate**.)

DIN 53177: Binders for paints and varnishes: Dynamic viscosity of liquid resins, resin solutions and oils using an Ubbelohde capillary viscometer. 2002

DIN 53184: Binders for paints and varnishes: Unsaturated polyester resins (e.g. gelation and curing time). 1999

(DIN 53211) Paints, varnishes, coatings: Flow time using the DIN cup. 1987 (withdrawn in 1996, and replaced by DIN EN ISO 2431)

(DIN 53214) Paints, varnishes, coatings: Flow curves (rheograms) and viscosities using rotational viscometers. 1982 (withdrawn in 2003)

DIN 53222: Viscosity by falling-rod viscometer. 1990

DIN 53229: Paints, varnishes, coatings: Viscosity at high shear rates using rotational viscometers (concentric cylinders, cone-and-plate). 2004

(DIN 53440) Oscillatory bending test: Dynamic-elastic properties of specimens showing the shape of rods and tapes (withdrawn)

(DIN 53445) Polymer materials: Torsion pendulum test (free, damped oscillation). 1986 (withdrawn in 2004)

(DIN 53457) Elasticity modulus using tensile, compression, and bending tests (withdrawn)

(DIN 53493) Flow value of ultra-high molecular polyethylen (UHMW-PE) form masses. 1980 (withdrawn)

DIN 53504: Rubber: Tensile strength at break, tensile stress at yield, elongation at break, stress values in tensile tests. 2009

DIN 53512: Rubber: Rebound resilience (Schob pendulum). 2000

DIN 53513: Elastomers: Viscoelastic properties under forced vibration at non-resonant frequencies (using tensile and linear-shear tests). 1990

(DIN 53514) Rubber: Plasticity according to Baader by testing under compression and heat, "defo-tester". (withdrawn; however still existing as UNE 53514: Vulcanized rubber: Temperature rise and resistance to fatigue using a compression flexometer. 2005)

DIN 53523: Rubber, elastomers, 1976/1991: Shearing disk viscometer according to Mooney. p. 1: Sample preparation; p. 2: Requirements; p. 3: Mooney viscosity; p. 4: Scorching behavior (onset of vulcanization)

DIN 53529: Rubber, elastomers: curemetry, 1983/1991; p. 1: General basics; p. 2: Vulcanisation characteristics and reaction kinetics by cure isotherms; p. 3: Rotorless curemeters; p. 4: Oscillating disk curemeters

DIN 53535: Rubber, elastomers: General requirements for dynamic testing (using tensile, compression, torsional and linear-shear testers). 1982

DIN 53545: Elastomers: Low temperature behavior (using DMA testers). 1990

(DIN 53726) Plastics: Viscosity number and K-value of vinyl chloride VC polymers (using Ubbelohde capillary viscometers; Fikentscher K-value). 1983 (withdrawn)

(DIN 53727) Plastics: Viscosity number of thermoplastics in dilute solution, polyamides PA (using Ubbelohde capillary viscometers). 1980 (withdrawn)

DIN 53728: Plastics: Viscosity of diluted solutions (using Ubbelohde capillary viscometers); p. 3: V. number of PETP and PBTP, 1985 (p. 1: V. number of cellulose acetate, 1985; p. 4: Polyethylene PE and polypropylene PP, v. number and Staudinger index "which was up to then called limiting viscosity number", 1975; p. 1 and p. 4 are withdrawn)

(DIN 53735) Plastics: Melt flow index of thermoplastics. 1988 (2005 withdrawn, replaced by DIN EN ISO 1133)

DIN 53752: Plastics: Coefficient of linear thermal expansion. 1980

DIN 53765: Plastics, elastomers, thermal analysis: DSC method (dyn. scan. calorimetry). 1994

(DIN 54453) Dynamic viscosity of anaerobic adhesives using a rotational viscometer (and a **double-gap measuring system**). 1982 (2004 withdrawn)

DIN 54458: Structural adhesives: Flowability and application behaviour of viscoelastic adhesives using oscillatory rheometry. 2010

DIN 54801: Plastics: Apparent viscosity at high shear rates of PVC-pastes by Severs capillary viscometer. 1979

DIN 54811: Plastics: Flow behavior of polymer melts using capillary rheometers (pressure by piston, screw, or gas). 1984

(DIN 55990) Coatings: Powder coatings (e.g. gelation time). 1979 (withdrawn)

DIN 65583: Aerospace, fibre-reinforced materials: Glass transition of fibre composites under dynamic load. 1999

15.5 Important standards for users of rotational rheometers

For users of rotational and oscillatory **rheometers, important** are above all the following standards.

a) Geometries of measuring systems
ISO 3219: Concentric **cylinders** CC and **cone-and-plate** CP (for measuring absolute viscosity values of fluids when performing rotational tests)
ISO 6721-10: **Parallel-plate** PP
DIN 53019: Cylinders, cone-and-plate, and parallel-plate (for cylinders and CP almost identical to ISO 3219)

b) Polymer testing
ASTM D4440: **Polymer melts** via oscillatory tests
ASTM D4473: **Curing of resins** via oscillatory tests
ASTM D5279: **Solid bars** via torsional oscillatory tests

c) Testing of other kinds of measuring samples
DIN EN 14770: **Bitumen** via oscillatory tests (more comprehensive tests as in ASTM D7175)
ASTM D7175: **Asphalt binders** using a very simple oscillatory test ("SHRP")
DIN 51810: **Lubricating grease**, p. 1: Rotational test; p. 2: Oscillatory test
DIN 54458: Structural **adhesives** via combined oscillatory tests

d) Rheological terms
DIN 1342: Flow behavior and rotational tests
DIN 13343: Viscoelastic behavior, creep tests, relaxation tests and oscillation tests

Author

Thomas G. Mezger, born 1954, successfully completed a degree (Diplom-Ingenieur) in Chemical Engineering at Stuttgart University, Germany. He gained a wealth of experience in the field of practical rheology in the Sales and Product Management departments of the rheometer manufacturers, from 1984 to 1988 at Contraves and afterwards at Physica Messtechnik, today Anton Paar Germany. For over twenty years, he has held seminars for students and employees of companies and institutes from a wide range of technical branches. These seminars cover the basics of rheology and rheometry as well as the useful transfer of this knowledge into industrial practice. As a result of this dialog with customers "The Rheology Handbook" came into existence.

Thomas G. Mezger: The Rheology Handbook
© Copyright 2011 by Vincentz Network, Hanover, Germany
ISBN 978-3-86630-864-0

Index

A

abbreviations 383
absolute measuring systems 255
acetone 28
acoustic damping 104, 208
acoustic wave viscometer 305
activation energy 84
active hood 247, 253, 347
Adams consistometer 293
adhesive dispersions 23f, 64
adhesive failure/fracture energy (tack) 285
adhesives 91, 101, 109, 141, 155, 172, 184, 187, 220,
    252, 269, 282, 301, 329, 369
adhesives (tack) 284
aero-gels 187
aero-sols 23, 24
AFNOR cups 314
agglomerates 44, 169
aggregates 214, 219, 231
agrochemicals 226
air bearings (rheometer) 343
air bubbles 19, 240
air viscosity 28
aluminum 92, 179, 266f
amorphous polymers 190
amorphous solids (glass) 198
amorphous structure 273
amphoteric surfactants 213
amplitude sweeps 146, 284
amplitude sweeps (frequency-dependent) 149, 154
amplitude sweeps (guideline) 351
amplitude sweeps (G'' peak) 149
amylopectine 195
amylose 193
anchor stirrer 262
angle units 342, 393
Angström (unit) 214, 392
angular frequency 136, 159, 394
angular velocity 235, 342, 393
anisotropic properties 218, 229, 266, 283
anisotropic properties (rheo-optics) 271f , 275
anti-thixotropy 71, 375
apparent viscosity 38, 45
apparent yield point 50
application rate (AR, blade coating) 24
approximation functions (flow curves) 59
approximation functions (viscosity/temperature) 83,
    204, 207

aqueous dispersions 64ff, 223, 231
Arrhenius relation 84, 366
asphalt 23, 177, 298f, 307, 309, 369, 380
association colloids 214
associative polymers 228
associative thickeners 67, 223ff
ASTM standards (list) 409
ASTM viscosity classification (lubricants) 317
atom nuclei, position (SANS) 278
automatic gap control (AGC) 244, 251, 263
automatic gap setting (AGS) 124, 244, 251
automotive adhesive 141, 155, 184
automotive bumpers 105
automotive coatings 25, 381
Avogadro's number 364, 377
axial force (tribology) 287

B

ball bearings (viscometers) 343
ball bearing test (tribology) 288
ball measuring systems 261
ball-shaped spindles 258
bamboo 106
bars (specimens) 262
bearings (rheometers) 343
behavior at rest (guideline) 354
bending beam rheometer (BBR) 309
bending testers 308
bentonite (gellant) 65
bi-cone (interfacial rheology) 278
bi-layers (surfactants) 217
Bingham build-up (BBU) 78, 301
Bingham model function 38, 54, 61, 369
biological tissues 149, 170, 218, 225, 274, 283
bio-materials 105, 158
bio-membranes 149, 218f
bio-tech products 149, 225
birefringence (rheo-optics) 271, 275
bitumen 28f, 91, 147, 177, 197, 298f, 307ff, 366f, 380
bitumen number (pen.) 299
blade coating 23f, 332
blade stirrer 262
blood 23, 28, 105
blood plasma 28f
blood vessels 105, 283
Bloom gelometer 300
bonds 54
bones 92, 105

Thomas G. Mezger: The Rheology Handbook
© Copyright 2011 by Vincentz Network, Hanover, Germany
ISBN 978-3-86630-864-0

R

CPSIA information can be obtained at www.ICGtesting.com
Printed in the USA
LVOW11*1821020913

350628LV00038B/269/P

9 783866 308640

UNIVERSITY OF BRADFORD

MAKING KNOWLEDGE WORK™

Spine damage noted 27/03/17 HLM

LIBRARY

This book should be returned not later than the last date stamped below.
The loan may be extended on request provided there is no waiting list.

FINES ARE CHARGED ON OVERDUE BOOKS

ONE WEEK ONLY		

L45

350EE